Handbook of Snow, Ice and Glaciers

Handbook of Snow, Ice and Glaciers

Edited by **Joe Carry**

SYRAWOOD
PUBLISHING HOUSE
New York

Published by Syrawood Publishing House,
750 Third Avenue, 9th Floor,
New York, NY 10017, USA
www.syrawoodpublishinghouse.com

Handbook of Snow, Ice and Glaciers
Edited by Joe Carry

International Standard Book Number: 978-1-68286-256-8 (Hardback)

Printed in China.

Contents

Preface

This book has been a concerted effort by a group of academicians, researchers and scientists, who have contributed their research works for the realization of the book. This book has materialized in the wake of emerging advancements and innovations in this field. Therefore, the need of the hour was to compile all the required researches and disseminate the knowledge to a broad spectrum of people comprising of students, researchers and specialists of the field.

The cryosphere has a significant impact on the climate of our planet. This book presents a detailed study of the cryosphere with an emphasis on snow and glaciers. It covers a diverse range of case studies contributed by internationally renowned scientists and experts of the field. The extensive range of topics presented it this text include surface elevation of glaciers, cloud physics, mass changes in ice caps, influence of precipitation patterns, etc. This book will help students in better understanding the theories and concepts revolving around snow, ice and glaciers. It will serve as a reference guide for academicians and professionals alike.

At the end of the preface, I would like to thank the authors for their brilliant chapters and the publisher for guiding us all-through the making of the book till its final stage. Also, I would like to thank my family for providing the support and encouragement throughout my academic career and research projects.

Editor

1

Regional melt-pond fraction and albedo of thin Arctic first-year drift ice in late summer

D. V. Divine[1,2], M. A. Granskog[1], S. R. Hudson[1], C. A. Pedersen[1], T. I. Karlsen[1], S. A. Divina[1], A. H. H. Renner[1,3], and S. Gerland[1]

[1]Norwegian Polar Institute, Fram Centre, 9296 Tromsø, Norway
[2]Department of Mathematics and Statistics, Faculty of Science and Technology, University of Tromsø, 9037 Tromsø, Norway
[3]Institute of Marine Research, Sykehusveien 23, 9019 Tromsø, Norway

Correspondence to: D. V. Divine (dmitry.divine@npolar.no)

Abstract. The paper presents a case study of the regional (≈ 150 km) morphological and optical properties of a relatively thin, 70–90 cm modal thickness, first-year Arctic sea ice pack in an advanced stage of melt. The study combines in situ broadband albedo measurements representative of the four main surface types (bare ice, dark melt ponds, bright melt ponds and open water) and images acquired by a helicopter-borne camera system during ice-survey flights. The data were collected during the 8-day ICE12 drift experiment carried out by the Norwegian Polar Institute in the Arctic, north of Svalbard at 82.3° N, from 26 July to 3 August 2012. A set of > 10 000 classified images covering about 28 km^2 revealed a homogeneous melt across the study area with melt-pond coverage of ≈ 0.29 and open-water fraction of ≈ 0.11. A decrease in pond fractions observed in the 30 km marginal ice zone (MIZ) occurred in parallel with an increase in open-water coverage. The moving block bootstrap technique applied to sequences of classified sea-ice images and albedo of the four surface types yielded a regional albedo estimate of 0.37 (0.35; 0.40) and regional sea-ice albedo of 0.44 (0.42; 0.46). Random sampling from the set of classified images allowed assessment of the aggregate scale of at least 0.7 km^2 for the study area. For the current setup configuration it implies a minimum set of 300 images to process in order to gain adequate statistics on the state of the ice cover. Variance analysis also emphasized the importance of longer series of in situ albedo measurements conducted for each surface type when performing regional upscaling. The uncertainty in the mean estimates of surface type albedo from in situ measurements contributed up to 95 % of the variance of the estimated regional albedo, with the remaining variance resulting from the spatial inhomogeneity of sea-ice cover.

1 Introduction

A new thin-ice Arctic system requires reconsideration of the set of parameterizations of mass and energy exchange within the atmosphere–sea ice–ocean system used in modern coupled general circulation models (CGCMs) including Earth system models. Such a reassessment would require a comprehensive collection of measurements made specifically on first-year pack ice with a focus on the summer melt season, when the difference from typical conditions for the earlier multiyear Arctic sea-ice cover becomes most pronounced (Perovich et al., 2002a; Grenfell and Perovich, 2004; Perovich and Polashenski, 2012).

Surface albedo is one of the major physical quantities controlling the intensity of the energy exchange at the atmosphere–sea ice–ocean interface and the heat balance of sea ice (e.g., Doronin and Kheisin, 1977; Maykut, 1982; Curry et al., 1995). Knowledge of the surface albedo for different types of sea ice, as well as its spatial and seasonal variability, is crucial for obtaining adequate representations of the sea-ice cycle in the CGCMs (e.g., Holland et al., 2012; Björk et al., 2013; Karlsson and Svensson, 2013).

During summer, the net positive heat balance of sea ice causes substantial transformation in the state of the ice cover. Water runoff from melting snow and upper ice layers tends to form puddles in depressions in the sea-ice surface (e.g.,

Zubov, 1945; Untersteiner, 1961; Nazintsev, 1964; Fetterer and Untersteiner, 1998). These melt ponds spread rapidly and, on level first-year ice (FYI), can cover up to 75 % of the surface during the initial stage of surface melt (Hanesiak et al., 2001; Grenfell and Perovich, 2004; Polashenski et al., 2012). As the albedo of a melt pond is markedly lower than that of the bare or snow-covered sea ice (e.g., Doronin and Kheisin, 1977; Grenfell and Maykut, 1977; Fetterer and Untersteiner, 1998; Perovich et al., 2002b; Grenfell and Perovich, 2004), the spatial distribution of melt ponds and leads has clear implications for the spatial aggregate albedo (Perovich, 2005) and accelerated summer decay of sea ice.

Field observations suggest a pronounced difference in the seasonal evolution of first-year sea-ice albedo compared with that of multiyear ice. The surface of multiyear sea ice typically features more rough topography and thicker snow cover, leading to a limited potential melt-pond coverage (e.g., Eicken et al., 2002, 2004; Perovich and Polashenski, 2012). Thicker ice underneath the melt-pond bottom leads to generally higher spatial albedo, lower transmission and lower energy absorption on melting multiyear ice (Eicken et al., 2002; Perovich, 2005; Hudson et al., 2013; Nicolaus et al., 2012). As a result, the summer albedo of multiyear ice cover is systematically higher than that of younger ice throughout the entire melt season, inducing an additional ice age–albedo feedback (Perovich and Polashenski, 2012).

The relatively small spatial scale of a typical pond system, typically few tens to thousands of m^2 (e.g., Tschudi et al., 2001; Perovich et al., 2002b; Hohenegger et al., 2012), large intersite variability in melt-pond coverage and the overcast conditions prevailing in the summer Arctic promote the use of low-altitude airborne methods for studying the morphological and optical properties of the sea-ice cover. Although remote sensing of summer sea ice utilizing various satellite-based sensors has made considerable progress throughout the last decades (e.g., Markus et al., 2003; Rösel et al., 2012; Tschudi et al., 2008; Kim et al., 2013), these aerial surveys can provide valuable high-resolution validation data for the emerging algorithms. Combining the spatial data on surface types with in situ measurements of incident/reflected solar radiation (albedo) and turbulent heat fluxes for different types of surfaces may in turn provide estimates of the regional-scale surface energy balance of sea ice. A number of such studies have been conducted in the past with a focus on spatial and temporal evolution of fractional melt-pond coverage, pond-size probability density (e.g., see Perovich et al., 2002b, for a review), and their relationship with the pre-melt surface topography (Derksen et al., 1997; Eicken et al., 2004; Petrich et al., 2012) and surface albedo. Depending on the instrumentation setup used, the spatial ranges covered varied from tens of meters to hundreds of kilometers, on the order of the typical scale of a GCM grid cell.

Safety and logistical challenges associated with these types of studies result in the relevant surface-based field data preferentially representing thicker first-year sea ice at the initial stages of melt and/or sea ice from coastal areas, where the sediment load may modify the spectral albedo and melt pattern. Limited data exist for thinner, less than 1 m thick, Arctic first-year ice that is expected to occupy a substantial part of the Arctic basin in the future if (and when) the projected transition to a nearly seasonal ice cover has occurred.

A comprehensive set of observations of the energy balance of melting Arctic first-year sea ice was conducted during an 8-day ice station in July–August 2012. Hudson et al. (2013) presented results from in situ measurements obtained during the drift experiment. This paper shows the analysis of the regional morphological properties of the sea-ice surface, inferred from aerial surveys. The in situ measurements of broadband albedo and the derived regional spatial distribution of surface types are used to obtain an estimate of the regional albedo of Arctic first-year ice in the advanced stage of melt. The upscaling scheme applied in the study treats all major observed quantities as random variables. Corroborated with the respective areal data on sea-ice thickness, the analysis provides the probability density functions on the regional albedo together with the albedo of thin (70–90 cm) first-year ice with a well-developed melt-pond cover.

The paper is organized as follows. Section 2 presents the geographical setting, instrument setup, image-processing techniques, details on the upscaling technique applied and uncertainties in the key variables we used for estimating the regional albedo. Section 3.1 shows the spatial variability of melt-pond and open-water fractions inferred from six helicopter ice-survey flights. The along-track albedo variability and the regional and sea-ice albedo estimates are then presented in Sect. 3.2. Finally the results of the work are discussed and summarized in Sect. 4.

2 Data and methods

2.1 ICE12 drift experiment

The energy balance of melting thin first-year Arctic sea ice was a focus of the 8-day ICE12 drifting ice floe experiment on R/V *Lance*, conducted from 26 July to 3 August 2012, north of Svalbard in the southwestern Nansen Basin (82.3° N, 21.5° E). Figure 1 shows the *Lance* drift track that was in an area of very close (≥ 90 %) drift ice. The corresponding operational ice chart produced by the Norwegian Ice Service of the Norwegian Meteorological Institute (NMI, www.met.no) from 1 August is shown superimposed onto the map. The ice floe (ICE12 floe hereafter) that *Lance* was moored to during the drift had a size of approximately ∅ 600 m and a modal ice thickness of 0.8 m, deducted from drillings and measurements using a Geonics EM-31 electromagnetic induction device (Hudson et al., 2013). The floe was mainly represented by level ice, with ridging over less than 10 % of the area. Based on airborne surveys of ice thickness using another electromagnetic induction device, the EM

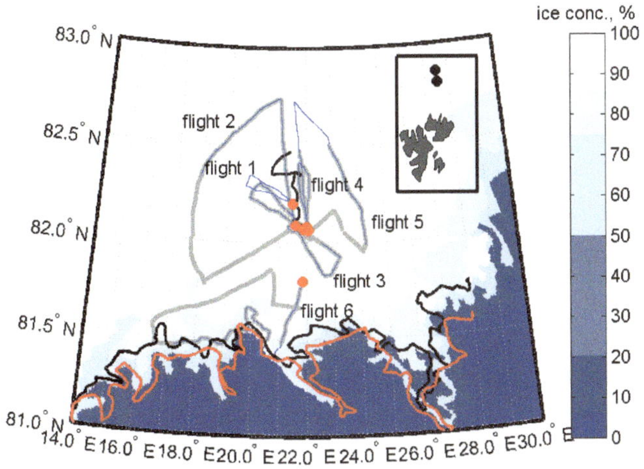

Figure 1. Map of the study area showing the track of the ICE12 floe during the drift north of Svalbard from 26 July to 3 August 2012 (solid black line); an inset map in the upper right corner also shows the start and end drift coordinates relative to the Svalbard archipelago. The black and red curves outline the ice edge on 2 days, 31 July and 2 August, defined as 40 % ice concentration based on ice charts from the Norwegian Meteorological Institute (NMI). The NMI ice chart from 1 August is shown as the reference. The grey and blue lines show the segments of six helicopter ice-reconnaissance flight tracks with EM bird and ICE camera data, respectively (see Table 1). Red dots mark the starting points for the flights.

bird (Haas et al., 2009), and analysis of aerial photography, the floe was found to be representative for the area. The sea ice was in the latter stage of melt, covered by melt ponds some 15–30 cm deep with steep margins. The majority of ponds were connected to complex networks, often with an outlet to the ocean. Some of the ponds had actually melted through the ice slab, corresponding to stage III of surface-melt and melt-pond evolution according Eicken et al. (2002).

2.1.1 In situ broadband albedo measurements

The broadband albedo of the sea-ice surface was measured in situ during the ICE12 drift experiment using a mobile instrument platform for measuring the radiation budget on sea ice (Hudson et al., 2012, 2013). Observations of the surface radiative fluxes were done on seven representative transects with a 5 m interval over a total of 490 m. Hudson et al. (2013), using the surface type classification technique from Renner et al. (2013), discriminated between four major types of sea-ice surface in the ICE12 floe area: open water and bare ice and dark and bright ponds. The latter refers to light blue ponds with thicker, more reflective ice underneath. The measurements were grouped according to the surface types to yield the mean albedos for the dark ponds $\overline{\alpha_{\mathrm{dp}}} = 0.15$ and light ponds $\overline{\alpha_{\mathrm{bp}}} = 0.34$, respectively, and of bare white ice $\overline{\alpha_{\mathrm{bi}}} = 0.55$ (see Table 1 in Hudson et al., 2013, for more details and Table S1 in the Supplement presented here). The albedo of open water/leads was set to the commonly used $\overline{\alpha_{\mathrm{ow}}} = 0.066$ (Pegau and Paulson, 2001). We note that cloudy conditions prevailed during the drift experiment, ensuring relative homogeneity in illumination in the study area.

2.1.2 Low-altitude imagery of sea ice during ICE12 experiment

The imaging of the sea-ice surface during the cruise was done using a recently designed ICE camera system mounted on a Eurocopter AS-350 helicopter. The hardware component of the system includes two downward-facing Canon EOS 5D Mark II digital photo cameras equipped with Canon 20 mm f/2.8 USM lenses, a combined SPAN-CPT GPS/INS unit by Novatel and LDM301 by Jenoptik, a laser distance measurement device used as an altimeter in the setup. These components were housed in a single aerodynamic enclosure and mounted outside the helicopter. The single-point horizontal positioning accuracy for the system was within 1.5 m, and the uncertainty in the altitude over the sea ice was estimated to be < 0.3 m, which corresponds to a typical scale of sea-ice draft variability.

Since the ICE camera was designed as a component of a photogrammetric setup, the image shooting rate was set to one frame per second per camera yielding two captured images per second. This was sufficient to ensure about 50–70 % overlap between successive images for flights at an altitude of 35–40 m and with a velocity of 30–40 m s^{-1} – parameters typical for EM bird flights. We fixed the camera lenses' focal lengths to infinity. For every captured image, the position, attitude and altitude of the event were logged in the system. The cameras' own 128 GB compact flash cards stored the captured images; the card size was sufficient for the system to shoot continuously for about 1 h, taking about 4500 images per camera in raw Canon format. A subset of some 10 300 images with minimal (< 10 %) or no overlap captured during six longer survey flights was selected for further processing and used in the presented study. To form this subset, every second image from one of the cameras was used. Figure 1 shows the selected flight tracks. Results of the data analysis from these flights together with in situ observations are reported below and also summarized in Tables 1 and 2.

2.1.3 Image and navigation data processing

For a typical flight altitude of about 35 m over the sea ice, the camera lenses used in the setup provide a footprint of about 60 by 40 m. With the image sensor geometry at its native resolution this corresponds to a pixel size on the ground of about 1 cm. For typical helicopter roll (pitch) angles of about −2° (1°), the distortion of the image plane from an ideal rectangular one and the associated uncertainty in the image area of less than 1 % was considered insignificant; therefore no correction for pitch and roll was applied to the images.

Table 1. Summary statistics on the state of sea-ice cover along the six processed helicopter flight tracks from the ICE12 cruise. The open-water coverage f^{s}_{ow} and melt-pond fraction f^{s}_{mp} (relative to sea-ice area) are the whole swath-based estimates rather than averages of the respective values from individual images presented in the corresponding figures. The values of f^{s}_{ow} and f^{s}_{mp} for flight 2, shown in parentheses, are the respective estimates based on the images processed using the method of Renner et al. (2013). The bottom entry shows the regional aggregate values derived from flights 1–5.

Flight number	Date	GMT start–end times	N images	Transect length (area), km (km^2)	f^{s}_{ow} %	f^{s}_{mp} %
1	31.07.12	7:36–8:10	1031	67 (2.4)	7	26
2	01.08.12	7:22–8:34	1902	139 (5.0)	10 (9)	24 (27)
3	01.08.12	16:45–18:03	2237	154 (5.7)	14	25
4	02.08.12	11:21–12:00	993	78 (2.5)	14	24
5	02.08.12	13:21–14:45	2121	170 (5.2)	12	26
6	03.08.12	14:43–16:04	1979	165 (7.4)	30	20
Regional aggregate	–	–	8284	608 (20.8)	12	25

Table 2. Summary statistics on the aggregate surface albedo α^{s} and sea-ice albedo α^{s}_{si} along the six processed helicopter flight tracks from the ICE12 cruise and the respective regional estimates α^{r} and α^{r}_{si}. The total regional albedo is calculated with and without flight 6 data taken into account. The numbers in parentheses in the albedo column denote the respective block bootstrap 95 % confidence interval on the estimates.

Flight number	Aggregate albedo (α^{s})	Aggregate albedo sea ice (α^{s}_{si})
1	0.41 (0.39; 0.43)	0.44 (0.42; 0.46)
2	0.40 (0.38; 0.43)	0.45 (0.42; 0.47)
3	0.38 (0.36; 0.41)	0.44 (0.41; 0.46)
4	0.39 (0.36; 0.41	0.44 (0.42; 0.46)
5	0.39 (0.37; 0.41)	0.44 (0.41; 0.46)
6	0.32 (0.29; 0.35)	0.44 (0.42; 0.47))
α^{r} (1–5)	0.39 (0.37; 0.41)	–
α^{r} (1–6)	0.37 (0.35; 0.40)	–
α^{r}_{si} (1–6)	–	0.44 (0.42; 0.46)

Image correction for camera lens distortion is necessary prior to any further analysis of the acquired images. We used generic lens correction and vignetting correction procedures with a polynomial lens distortion model implemented in Adobe Lightroom® software.

The large array of data to be analyzed promoted the use of a simplified image-processing technique. In order to discriminate between open water, bare ice and melt ponds, we applied a three-step object identification and classification procedure. This involved:

a. image segmentation/binarization using Otsu's method, which chooses the threshold to minimize the intra-class variance of the black and white pixels (Otsu, 1979);

b. boundary tracing on the binarized images by the Moore–Neighbor tracing algorithm modified by Jacob's stopping criteria (Gonzalez etal., 2010);

c. object classification (open water, bare ice or melt pond) using thresholding in the red channel intensity.

Due to the relatively high contrasts between the different surface types during summer melt, this relatively simplistic approach appeared to work well with a minimum of supervision required during the processing of the sequences of images captured by the camera system. All procedures were implemented in Matlab using the "image processing" toolbox (MATLAB, 2012).

For each flight track of length N images, the method yielded the series of $\left\{f^{i}_{mp}, f^{i}_{ow}, S^{i}\right\}, i = 1, \ldots, N$, with f^{i}_{mp} denoting the image fractional melt-pond coverage with respect to the sea-ice area, f^{i}_{ow} the open-water fraction and S^{i} standing for the respective area of image i. Figure 2 demonstrates an example of the object classification procedure for an image captured during flight 1 (Table 1). The edges of the melt-pond objects are accurately identified. Note that we left out the darker objects with an area less than 0.5 m^2 because the contribution of these objects to the total melt-pond coverage was found to be negligible. The identified set of objects of three types is then used for calculating along the track summary statistics on melt-pond coverage and open-water fraction. The parts of the image not classified as melt ponds or open water were considered as bare sea ice. For the case in Fig. 2, f_{ow} was calculated to be 8 % and f_{mp} was 16 % with respect to the total sea-ice area.

Figure 2. Example of the image segmentation procedure showing an image captured during flight 1 from an altitude of 35 m. The dimensions of the scene are 60.5 by 40.5 m. Black contours highlight the edges of melt ponds; the green contour outlines the open-water area; blue is for the smaller patches of sea ice within melt-pond/open-water objects. For this particular scene the melt-pond fraction f_{mp} (relative to sea-ice area) and open-water fraction f_{ow} are 16 and 8 %, respectively.

2.2 Accounting for uncertainties in the variables used

2.2.1 Error models for melt-pond and open-water fractional coverage

Error models on f_{mp} and f_{ow} are built on the additional analysis of 1622 images from flight 2 using the classification method of Renner et al. (2013). The technique involves a semi-automated surface type classification and manual supervision of the processed images, allowing more reliable results at the cost of increased labor intensity. Processing of the images used in this verification procedure yielded the image-based fractional coverage of the four surface classes: dark ponds, bright ponds, open water and bare ice. This data set was used as a reference to estimate the uncertainty in the corresponding quantities derived from the larger image set and to assess the probability density of the ratio of the areas of dark to bright ponds at the regional scale.

Imagewise intercomparison of f_{mp} and f_{ow} values demonstrated an average bias of f^{b}=0.03, with $\sigma_{f^{b}}$=0.04 in the fraction of melt ponds between the images processed using the technique of Renner et al. (2013) and the simplified approach applied in this study. Inspection of images revealed that the algorithm presented in Sect. 2.1.3 sometimes underestimates the melt-pond coverage by identifying some bright ponds as bare white ice. Likewise, some of the darkest melt ponds were sometimes misidentified as open water/leads. The error model for f^{i}_{mp} and f^{i}_{ow} of an image i is therefore defined as

$$\left\{p(f^{i}_{mp}),\ p(f^{i}_{ow})\right\} = \tag{1}$$
$$\begin{cases} p(f^{i}_{mp}) \sim p(f^{i}_{mp} + \mathcal{N}(f^{b},\sigma^{2}_{f^{b}})\,|\,\mathcal{N}(f^{b},\sigma^{2}_{f^{b}}) \geq 0) \\ p(f^{i}_{ow}) \sim p(f^{i}_{ow} - (1-f^{i}_{ow})\mathcal{N}(f^{b},\sigma^{2}_{f^{b}})\,|\,\mathcal{N}(f^{b},\sigma^{2}_{f^{b}}) < 0) \end{cases},$$

where parameters of the Gaussian distribution were estimated from the data.

The areal ratio of dark to bright ponds, r, was estimated using a bootstrap technique (Efron and Tibshirani, 1993) involving sampling with replacement from the same complementary data set of classified images, followed by a re-estimation of the sought r for each bootstrap replicate. The proportion of the drawn to the replaced data points (i.e., classified images) within each replicate was set to 2/1 with all the images being equally weighted. The resulting distribution of the mean areal r derived from 10 000 replicates was approximated by a Gaussian probability density function (pdf) with $p(r) \sim \mathcal{N}(2.8, 0.15^2)$.

2.2.2 In situ broadband albedo as a random variable

Uncertainties in the average in situ albedo $\overline{\alpha_j}$ are estimated empirically from available data for each surface type j. During the ICE12 experiment we obtained 50 individual albedo measurements over bare white ice, 12 over dark melt ponds and 1 over a bright pond. This yields sample standard deviations (SDs), σ^{sp}_{α}, on single point measurements of 0.05 and 0.04 for bare white ice and dark ponds, respectively (see Table S1 for details). Using a simplistic error model assuming independent measurements with random Gaussian errors, we calculate the uncertainty of the measurement-based average albedo of surface type j as

$$\sigma_{\overline{\alpha_j}} = \frac{\sigma^{sp}_{\alpha_j}}{\sqrt{m_j}} + \frac{\sigma^{ins}_{\alpha_j}}{\sqrt{m_j}}, \tag{2}$$

where m_j refers to the number of available albedo measurements in the surface type under consideration. The single measurement instrumental error, $\sigma^{ins}_{\alpha_j}$, was set to $0.1\overline{\alpha_j}$, where the coefficient 0.1 stems from a declared 5 % measurement uncertainty yielding a total uncertainty of 10 % for the ratio of reflected-to-incoming radiation (i.e., albedo), again assuming the errors are independent. For the "bright pond" category, where only one albedo measurement was available with no significant influence from other surface types, we assigned an uncertainty of $0.1\overline{\alpha_{bp}}$ although we acknowledge that this value can be a biased estimate. For the open-water albedo uncertainty, a value of 0.0066, derived from 24 measurements, was adopted from Pegau and Paulson (2001). Table S1 shows the resulting values of $\sigma_{\overline{\alpha_j}}$ for the four surface classes. The mean albedo of every surface type j can now be considered as a t-distributed random variable with m_j degrees of freedom, distributed as $p(\overline{\alpha_j}) \sim \overline{\alpha_j} + t_{m_j}\sigma_{\overline{\alpha_j}}$.

The use of t-distribution accounts for a larger spread in the estimate of the true mean when dealing with the relatively small sample sizes. For bright ponds, the Gaussian approximation was used instead to prevent the occasional generation of albedo values outside the admissible range of [0, 1] due to heavy tails of the t-distribution with one degree of freedom.

This approach should be considered a simplification, as it reduces the whole variety of surface types with different optical characteristic to only four major surface types. However we expect that the imposed range of random variability in a particular surface-type albedo covers the natural variation of this parameter, thereby accounting indirectly for the effects of numerous additional factors like the thickness of ice, surface-state and small-scale morphology, pond depth and ice thickness beneath the pond as well as changing light conditions.

2.3 Bootstrap aggregate albedo

The aggregate albedo of a spatial mosaic of surface types is generally defined as (Perovich, 2005)

$$\alpha = g(\alpha_j, f_j) := \sum_j \alpha_j f_j;\ \{\alpha_j, f_j\} \in [0, 1], \tag{3}$$

where summation is over all surface types used, here $j = \{\mathrm{ow}, \mathrm{bi}, \mathrm{bp}, \mathrm{dp}\}$, with the corresponding fractional coverage f_j. Note that for convenience we use the fractional total melt-pond coverage, f_{mp}, relative to the sea-ice area. Coefficients f_{bp} and f_{dp} are defined as fractions of bright and dark melt ponds with regard to the relative melt-pond coverage, i.e., $f_{\mathrm{bp}} = (1/(1+r))f_{\mathrm{mp}}$ and $f_{\mathrm{dp}} = (r/(1+r))f_{\mathrm{mp}}$. This transforms Eq. 3 for α to

$$\begin{aligned} \alpha &= \alpha_{\mathrm{ow}} f_{\mathrm{ow}} + \alpha_{\mathrm{bi}}(1 - f_{\mathrm{mp}})(1 - f_{\mathrm{ow}}) + \\ &+ \alpha_{\mathrm{bp}} f_{\mathrm{bp}}(1 - f_{\mathrm{ow}}) + \alpha_{\mathrm{dp}} f_{\mathrm{dp}}(1 - f_{\mathrm{ow}}). \end{aligned} \tag{4}$$

For any arbitrary set $\left\{f^i_{\mathrm{mp}}, f^i_{\mathrm{ow}}, S^i\right\}, i = 1, \ldots, N$, the set-based aggregate albedo α^{s} is therefore calculated in the same way as the local estimate using Eq. 4, with the values of $f^{\mathrm{s}}_{\mathrm{ow}}$ and $f^{\mathrm{s}}_{\mathrm{mp}}$ derived as

$$f^{\mathrm{s}}_{\mathrm{ow}} = \sum_i S^i f^i_{\mathrm{ow}} / \sum_i S^i$$

$$f^{\mathrm{s}}_{\mathrm{mp}} = \sum_i S^i (1 - f^i_{\mathrm{ow}}) f^i_{\mathrm{mp}} / \sum_i S^i$$

and referring to the set-based estimates of open-water and melt-pond fractions.

Deriving particular values of interest from the analysis of individual sea-ice images is analogous to sampling from a random data field with an a priori unknown theoretical distribution and a covariance structure. Any empirical statistic calculated from a set of analyzed images is therefore a derivative of the available data sample and should be considered an estimate accurate to within some unknown probability density.

Since the probability distribution of the local, image-based albedo α^i is non-Gaussian, the large number of available samples makes the bootstrapping (i.e., sampling with replacement) technique (Efron and Tibshirani, 1993) an optimal choice to assess the probability density and the accuracy of the estimated image-set albedo. In our setting, the sets are formed of the swaths of images prone to the presence of autocorrelation in the variables used. It suggests the use of the moving block bootstrap approach (Kunsch, 1989).

For each flight the application of this method to the sequence of $\left\{f^i_{\mathrm{ow}}, f^i_{\mathrm{mp}}, S^i\right\}$ involves the following steps:

1. The series of $\left\{f^i_{\mathrm{ow}}, f^i_{\mathrm{mp}}, S^i\right\}$ of length N is split into $N - K + 1$ overlapping blocks of length K; the block length is determined empirically from the data using the procedure described in the next subsection.
2. N/K blocks are drawn at random, with replacement, from the constructed set of $N - K + 1$ blocks, and their sequence numbers are registered.
3. M bootstrap samples are drawn from the subset of N/K blocks; albedo for the four different surface types and the values for f^i_{ow}, f^i_{mp} and r can at this step be drawn at random from the respective probability distributions defined in Sects. 2.2.1 and 2.2.2; the set- or swath-based albedo α^{s} is then calculated for each sample using Eq. 4.

Steps 2–3 are repeated L times to generate $L \times M$ estimates of the swath-based aggregate albedo α^{s}. The assigned values of $\{L, M\} = 200$ yield a total of 40 000 samples of α^{s} combined to generate the bootstrap pdf of the swath-based α^{s}. The 95 % confidence interval ($\mathrm{CI}_{0.95}$) on the estimate is then calculated as $\{2.5, 97.5\}$ % of the empirical bootstrap pdf of α^{s}.

2.3.1 Estimating the image block length *K* using the Markov chain

Accounting for the autocovariance in the analyzed data is implemented following the Nychka et al. (2000) modification of the Mitchell et al. (1966) formula

$$N_{\mathrm{eff}} = N \frac{1 - \phi - 0.68/\sqrt{N}}{1 + \phi + 0.68/\sqrt{N}}, \tag{5}$$

where N_{eff} stands for the effective number of degrees of freedom ("effective sample size"); in general, $N_{\mathrm{eff}} < N$ due to the presence of autocorrelation in a series. This approach implicitly assumes that the analyzed sequence can be adequately described as a realization of the discrete first-order autoregressive process with the autoregressive parameter ϕ.

For each classified image i treated as an individual data sample, further categorization into "ice" or "open water" was

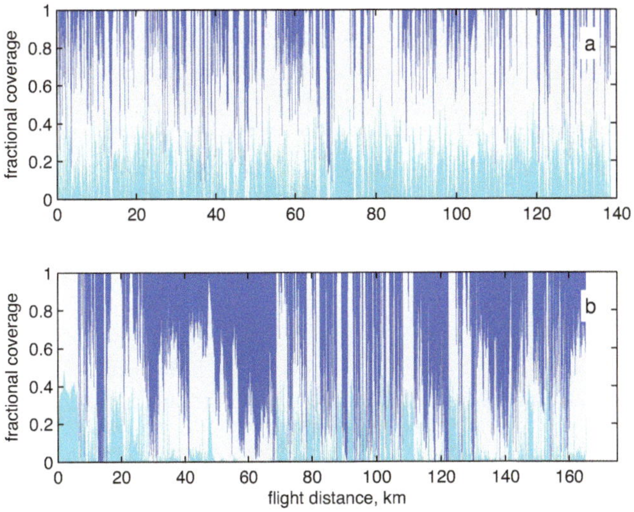

Figure 3. Along-track distribution of fractional melt-pond coverage f^i_{mp} (light blue), bare ice f^i_{bi} (light grey) and open-water fraction f^i_{ow} (blue), all relative to the image area, for flights 2 (very close drift ice, **a**) and 6 (marginal ice zone, **b**). With a swath width of 35–40 m, the covered area corresponds to roughly 0.35–0.40 km^2 per 10 km flight track.

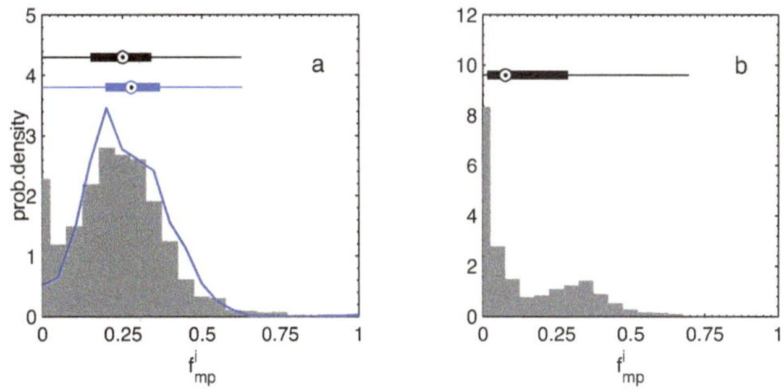

Figure 4. Empirical probability density of f^i_{mp} along the flight tracks 2 (**a**) and 6 (**b**) relative to the sea-ice area. For flight 2 image-based mean f_{mp} of 25 % and the quartiles $Q_{1,2,3}$ of 15, 25 and 34 %, respectively, as shown by the box plot, image-averaged $f_{\text{ow}} =$ 9 %. The blue line and blue box plot in (**a**) show the estimates of the same quantities of $f_{\text{mp}} = 28$ % and $Q_{1,2,3} = \{19, 28, 37\}$ from flight 2 images processed using the method of Renner et al. (2013). For flight 6 image-based mean $f_{\text{mp}} = 15$ % and $Q_{1,2,3} = \{1, 7, 28\}$, image-averaged $f_{\text{ow}} = 37$ %. The whiskers on box plot highlight the 1.5 times interquartile range to cover some 99 % of the observations in total.

applied. Such binarization into the two major surface classes is related to their dominant contribution to the swath-based albedo variance. The images within one flight track that have both open water and sea ice are categorized using a threshold in local open-water fraction. The value for the threshold f^{t}_{ow} was set to 5 %, which for the typical flight altitude would correspond to an opening in sea-ice cover at least a few meters wide, i.e., a very small fracture according to WMO sea-ice nomenclature (World Meteorological Organization, 1970).

Fitting the Markov chain of first-order to the derived binary sequence of surface states comprising one complete flight yields the transition matrix **T**. Its largest entry, which in our case characterizes the likelihood of retaining the "ice" state between two successive images, is used as the sought parameter ϕ – a simplistic metric of spatial autocorrelation in the surface state for the analyzed flight track.

3 Results and discussion

3.1 Regional melt-pond coverage and open-water distribution during ICE12 drift

This section presents the results of the analysis of sea-ice imagery along the six selected flight tracks that took place during the ICE12 cruise (Table 1). All but one flight (flight 1, on 31 July) were combined EM bird/ICE camera flights, which fixed the helicopter flight altitude to approximately 35 m above the sea-ice surface except for some shorter periods of climbing to 150–200 m for EM bird calibration.

Figures 3 and 4 show the summary statistics of melt-pond and sea-ice/open-water fractions along the tracks of flights 2 and 6, derived using the technique presented in Sect. 2.1.3. The data for the other four flights are presented in the Supplement in Figs. S1, S3, S5 and S7. Note that for flights 1–5, carried out from 31 July to 2 August, the results are similar, with a typical f_{mp} of about 26 % relative to the sea-ice area and a similarity in the shapes of the respective pdf. In 50 % of these images, the observed f_{mp} was between 15 and 36 %. We found that in some occasions the melt ponds could cover as much as 66 % of the ice surface within the image frame, yet for some 10 % of images with sea ice in the field of view, the sea-ice surface exhibited no or very little melt-pond coverage ($f^i_{\text{mp}} < 4$ %). The average open-water fraction of $f^i_{\text{ow}} = 11$ % was characteristic of very close drift ice and varied for the analyzed images between 0 and 8 % in 50 % of cases, with fewer than 1 % of images showing 100 % open water. This variability lies within the uncertainty of the estimates and corresponds well to the respective operational ice charts for the area (see Fig. 1).

Flight 6, on 3 August, was conducted while moving southwards out of the close drift ice. The flight track traversed the marginal ice zone (MIZ) with extensive areas/strips of open water. Thus the estimates of f_{ow} (30 %) and f_{mp} (20 %) for flight 6 are substantially different from those inferred from survey flights conducted the previous days in the close pack ice (see Fig. 4).

The EM bird surveys conducted during flights 2–6 further corroborate the inference of regional-scale homogeneity in the properties of the sea-ice cover. The probability density functions on sea-ice thickness presented in Fig. 5 suggest fairly similar shapes of the distributions, with the modal ice

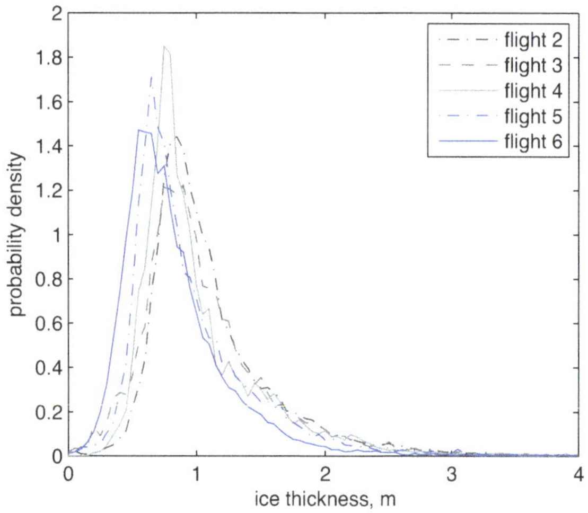

Figure 5. Probability density on sea-ice thickness for flights 2–6 derived from EM bird measurements.

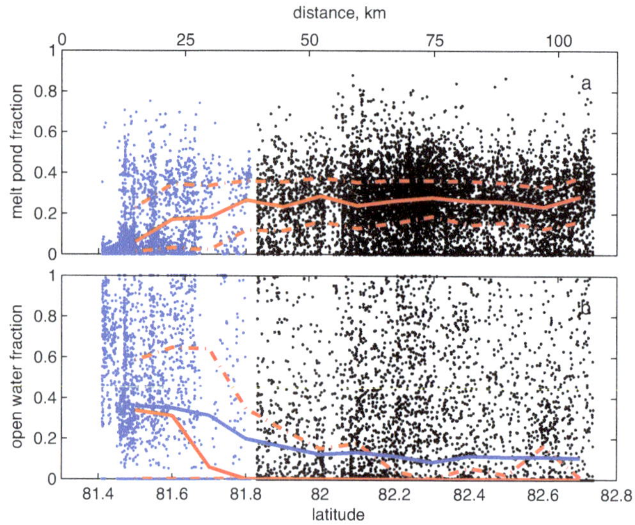

Figure 6. Latitudinal distribution in f^i_{mp} **(a)** and f^i_{ow} **(b)** summarized from the six flight tracks. Black dots highlight the f^i_{mp} and f^i_{ow} inferred from analysis of imagery from flights 1–5; blue dots are for the corresponding values from flight 6. Red solid and dashed lines show the moving median and the quartiles $Q_{1,3}$, respectively, estimated in the window of 0.1° latitude width. For comparison the blue line in **(b)** also shows the moving average to highlight the skewness of the respective image-based probability density.

thickness ranging within 0.7–0.9 m for flights 2–5. The pdf for flight 6 reveals a tendency towards generally thinner ice, with a modal ice thickness of about 0.6 m. We note, however, that there can be a negative bias associated with a much higher open-water coverage observed during this flight.

Figure 6 summarizes the latitudinal distribution in melt-pond fraction and open-water coverage in the study area. Due to the nearly east–west orientation of the MIZ within the study area, Fig. 6 reflects the variability in these parameters towards the sea-ice edge. We note that in the time between the first and fifth flights the ice drifted southwards some 20 km, somewhat smearing the actual distribution in this direction. Flight 6 in turn provided a snapshot across the marginal ice zone. The figure reveals a fairly stable melt-pond coverage across a range of latitudes associated with very close drift ice during the experiment. In the ≈ 30 km wide MIZ the melt-pond coverage shows a gradual decline to values below 10 % close to the edge of the ice pack, in parallel with an increase in the open-water fraction. The transition occurs when the mean open-water fraction exceeds a threshold of approximately 20 % and is most likely associated with a generally more intense melt and a decrease in the typical ice floe size in the area. As the ice floes tend to break up preferentially along the existing melt ponds and melt channels, subsequent transformation of ponds into open water leads to a decreased f_{mp} in the MIZ.

3.2 Bootstrap swath-based and regional albedo estimates

The bootstrap technique described in Sect. 2.3 is applied to the flight-track data of surface type variability $\left\{f^i_{\mathrm{mp}}, f^i_{\mathrm{ow}}, S^i\right\}$, $i = 1, \ldots, N$ and in situ albedo measurements from the ICE12 drift experiment to yield the upscaled estimates of swath-based α^{s} and a regional albedo of the study area α^{r}. In addition we use the same technique to calculate the albedo of the ponded sea ice alone (α^{si}). Figures 7a and 8a show local (i.e., based on individual images) aggregate albedo estimates, α^i, made from the helicopter imagery along the two selected flights with contrasting surface conditions presented in Sect. 3.1. The results for other tracks are presented in the Supplement and further summarized in Table 2. Note that in this case the image-based albedo variability is estimated from the data treated "as is" without taking the uncertainties into account.

Figure 7b and corresponding figures in the Supplement demonstrate fairly similar pdfs of local aggregate surface albedo for the flight tracks 1–5, suggesting a homogeneous state of sea-ice cover in the area within approximately 80 km of the ICE12 floe. We note that the empirical probability density functions of local albedo are skewed substantially towards zero due to the contribution of open-water areas. This suggests that an estimate of the regional-scale albedo of melting sea ice pack made by simple averaging of the respective quantities from a sequence of local scenes can be negatively biased. This may have implications for areal estimates of the surface energy budget both in observational and modeling studies.

Panels c in Figs. 7, 8, S2, S4, S6 and S8 display the generated bootstrap probability density of the swath-based α^{s} for the six flights. Table 2 shows the calculated values of the average swath-based albedos and their respective bootstrap $\mathrm{CI}_{0.95}$. The respective values of ϕ from the transition matrix

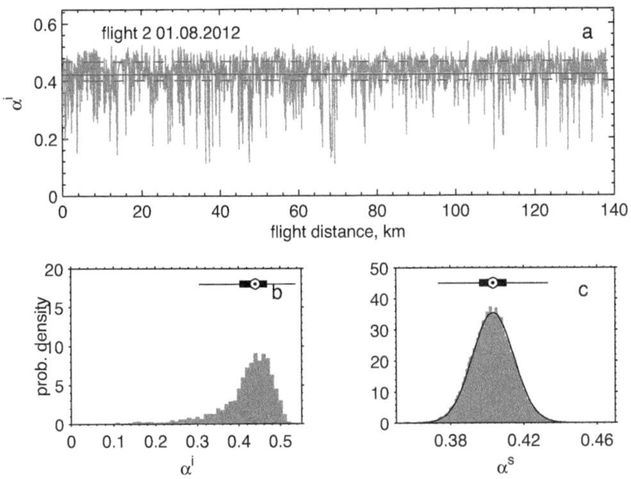

Figure 7. (a) Image-based aggregate surface albedo (α^i) along flight track 2 shown in Fig. 3. Solid blue line is for the image-based track average albedo of 0.42, and dashed lines show the quartiles $Q_{1,3}$ of (0.40,0.47) of the respective α^i probability density shown in **(b)**. Note skewness of the distribution towards lower albedo values and asymmetric position of the mean with respect to the 25 and 75 %; **(c)** bootstrap swath-based aggregate albedo α^s probability density, and the solid line shows the fitted normal pdf $\mathcal{N}(0.40, 0.01^2)$. The box plots on **(b)** and **(c)** use the same conventions as in Fig. 4.

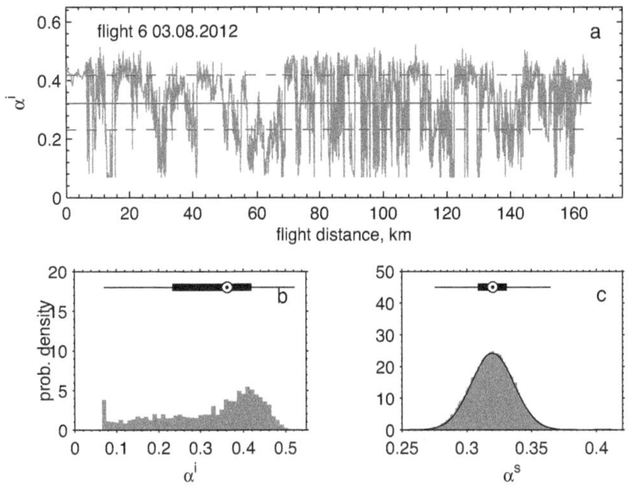

Figure 8. Same as in Fig. 7 but for flight 6 shown in Fig. 3. Solid blue line is for the image-based track average albedo of 0.32, and dashed lines show the 25 and 75 % (0.23,0.42) of the respective α^i probability density shown in **(b)**; **(c)** bootstrap swath-based aggregate albedo α^s probability density, and solid line shows the fitted normal pdf $\mathcal{N}(0.32, 0.02^2)$.

varied in the range of 0.78–0.88, whereas the probability of retaining the "open-water" state was lower: 0.51–0.57. These results are summarized in Table 3. The block length K was calculated as a ratio of N/N_{eff}, yielding a block size of 9–12 images for four of the six transects, which corresponded to approximately 500–700 m of the flight track. For the tracks with the lowest (flight 1) and highest (flight 6) open-water fractions the derived block lengths were 18 and 7 images, respectively.

For all tracks the α^s probability density is approximately Gaussian, with 95 % confidence according to the Lilliefors goodness-of-fit test of composite normality (Conover, 1999). The respective fits are shown together with the bootstrap pdfs in Figs. 7c, 8c, S2, S4, S6 and S8. The calculated standard deviations of the fitted Gaussian distributions are $\sigma_{\alpha^s} = 0.01$ for flights 1–5 and $\sigma_{\alpha^s_6} = 0.02$ for flight 6.

Flight tracks 1–5 demonstrate similar values of the swath-based aggregate albedo α^s of about 0.39, all lying within the estimated confidence intervals (see Table 2). This suggests that the data from these five flights can be combined to provide the regional-scale albedo estimate for the ice pack outside the MIZ. This is implemented using the same technique applied to the concatenated sequence of $\left\{f^i_{\mathrm{mp}}, f^i_{\mathrm{ow}}, S^i\right\}$ for all flight tracks but flight 6. When flight 6, representing mainly the marginal ice zone, is included in calculations, it decreases α^r to a value of 0.37. The latter is related to the presence of extensive open-water areas in the some 30 km wide MIZ. The results of calculations are presented in Fig. 9a, and Table 2 further summarizes the results of the analysis.

Table 3. Auxiliary data for the processed flight tracks used in the calculation of the flight-track albedo. $\mathbf{T}_{11}$ and $\mathbf{T}_{21}$ denote elements of the transition matrix of the fitted first-order Markov model and the respective estimated image block lengths.

Flight number	$\mathbf{T}_{11}$ $\phi_{\mathrm{ice}\to\mathrm{ice}}$	$\mathbf{T}_{21}$ $\phi_{\mathrm{ow}\to\mathrm{ow}}$	block length
1	0.88	0.57	18
2	0.83	0.53	12
3	0.78	0.48	8
4	0.80	0.49	9
5	0.82	0.52	10
Regional aggregate	0.82	0.51	10
6	0.76	0.25	7

The effect of open-water areas on the spatial albedo is demonstrated in Fig. 9b showing the bootstrap pdfs of sea-ice albedo α^s_{si} for tracks 1–6. We note that the spread in the inferred α^s_{si} pdfs between the individual tracks is much less pronounced compared to the respective α^s. The regional bootstrap α^r_{si} of about 0.44 thereby provides a good estimate of the albedo for melting sea ice about 0.7–0.9 m thick for the entire study area.

The data on f_{ow} and f_{mp}, merged from all six flights, were further binned in 0.1° wide latitudinal bins in a way similar to what was presented in Sect. 3.1. We calculated the bootstrap areal and sea-ice albedo for each latitudinal subset to yield the latitudinal distribution of these quantities. Figure 10

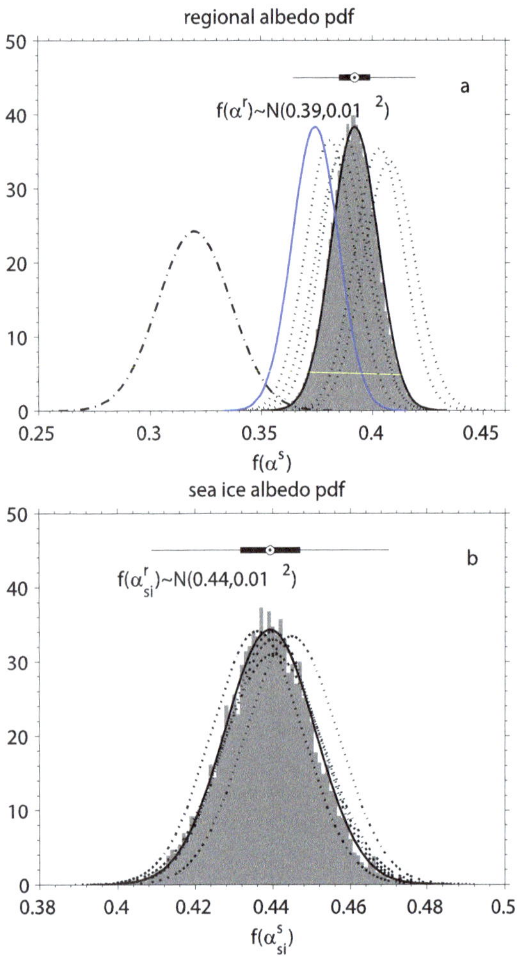

Figure 9. Regional (**a**) and sea-ice (**b**) bootstrap albedo pdfs obtained from merging the data from flights 1–5 (**a**) and 1–6 (**b**). Solid black lines highlight the fitted Gaussian pdf with the parameters indicated in the panel. Dotted black lines show for the reference the fitted Gaussian pdfs for α^s from flights 1–5 (**a**) and 1–6 (**b**). Black dash-dotted and solid blue lines in (**a**) also show the bootstrap albedo pdfs for flight 6 and the regional albedo derived from merging the data from all 6 flights together, respectively. The box plots on the top of the panels use the same conventions as in Fig. 4.

presents the results, demonstrating fairly stable values of α^s and α^s_{si} for the area outside the MIZ, in accordance with the corresponding results on f_{ow} and f_{mp} from Fig. 6. Within the MIZ increasing (decreasing) values of f_{ow} (f_{mp}) towards the ice edge drive opposite trends in the bootstrap albedos α^s and α^s_{si}. This suggests that a decrease in melt-pond fraction towards the ice edge may have a weak compensating effect on the areal albedo, slowing down the sea-ice surface melt in the MIZ. For solar radiation conditions observed during the drift experiment (Hudson et al., 2013), the net effect of increasing the sea-ice albedo to about 0.5 would be to remove roughly 5 W m^{-2} of solar energy input, enough to reduce melt by about 1.5 cm of pure ice over the period of

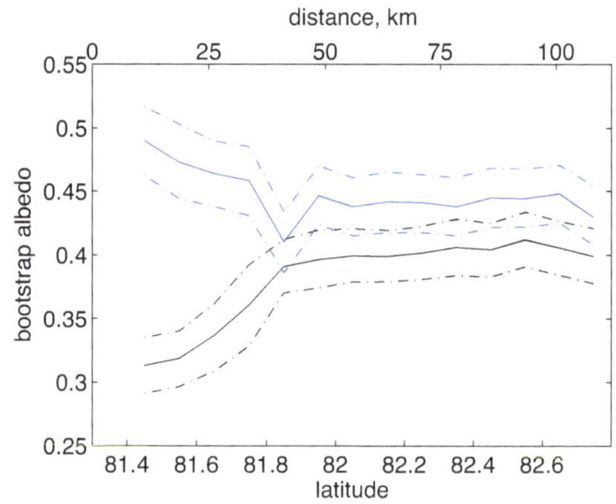

Figure 10. Latitudinal distribution in areal (black) and sea-ice (blue) bootstrap albedo derived from the six flight tracks. Dash-dotted lines show the respective 95 % confidence intervals on the estimate.

the experiment. One should note, however, that the upscaling results in this area with a more intense bottom and lateral sea-ice melt should be interpreted with caution. Potential for bias in the EM sea-ice thickness measurements due to effects of open water in the footprint of the EM instrument and a large dependence of sea-ice albedo on thickness for the thinner ice makes the application of the in situ albedo measurements made outside the MIZ less certain.

In order to infer the relative contribution of the spatial variability in melt-pond/open-water coverage and the uncertainty of in situ albedo measurements to the overall variance of the swath-based and regional albedo estimates, we also repeated the numerical experiments with the albedo of surface types treated as constants. The result demonstrated a substantial reduction in the standard deviations σ_{α^s} and σ_{α^r} down to 0.003 and 0.002, respectively. This indicates that in the defined framework, about 90 % of the estimated variance of α^s and 95 % in α^r is due to variability and uncertainties in the in situ albedo measurements. Only a minor part of the variance is due to all other errors and variability accounted for in the model.

3.3 Assessing the aggregate scale for ICE camera imagery

The notion of aggregate scale for an environmental variable refers to the minimal spatial scale at which the contribution of local sampling variability to its total variance is diminished (Moritz et al., 1993). The concept is directly related to the weak law of large numbers, provided that the samples are drawn from a stationary distribution. Knowledge of this scale is crucial for an accurate upscaling of local measurements and subsequently linking them to larger-scale climate models. We note that in a hierarchy of spatial scales, the present study focuses specifically on the range of meters to hundreds of kilometers, which encompasses the scales typical for in situ measurements up to regional models and CGCM.

The aggregate scale for the regional albedo was estimated using sets of (pseudo-)independent samples of different size drawn from the whole collection of classified images. The sample size varied from 10 to 1000 images, and for each sample size 10 000 subsets were drawn at random, without replacement, to gain the necessary statistics on the aggregate albedo distribution as a function of sample size and total sample area. As the image areas within each sample were not identical due to variations in the flight altitude, the average total area for each sample size was used. Images with an area over 6000 m^2, corresponding to a flight altitude above 55 m, were not included in the analysis.

Figure 11 (black lines) shows the fraction of sample-based aggregate albedo estimates, falling within the interval of ± 1 and ± 2 standard deviations of the regional aggregate albedo (Table 2), as a function of sample area. The results demonstrate a rapid growth in the proportion of accurate estimates of the regional albedo with an increase in the number of images drawn for analysis. The curves level out when the total sample area exceeds the threshold of about 0.7 km^2, when some 95 % of the subset-based estimates lie within the interval of 2 SD of the regional bootstrap albedo. One should emphasize that these estimates are specific to this study's setup, time period and region. For the range of flight altitudes typically sustained during the operation of the EM bird, the 0.7 km^2 aggregate scale corresponds to a set of at least 300 independent images spatially representative of the study region.

In order to simulate higher flight altitudes and examine the effect of smaller sample sets and/or sub-kilometer scale spatial autocorrelation in the state of sea-ice cover on the estimate of the aggregate scale, the numerical experiment was repeated with successive images combined into blocks of different length. The validity of this experiment relies on the assumption of smaller-scale anisotropy in statistical properties of the sea-ice surface. The red and grey lines in Fig. 11 show the fraction of the accurate estimates of the regional albedo for image blocks of length 10 and 25 images, respectively. Results suggest an increase in the aggregate scale to values above 2 km^2 which would correspond to sets of at least 80 (30) area-representative images captured from an altitude of about 100 (170) m. Notably the estimated thresholds (aggregate scales) have an order of magnitude similar to the respective estimate of > 1 km^2 obtained by Perovich et al. (2002b) during the SHEBA experiment in a different region of the Arctic.

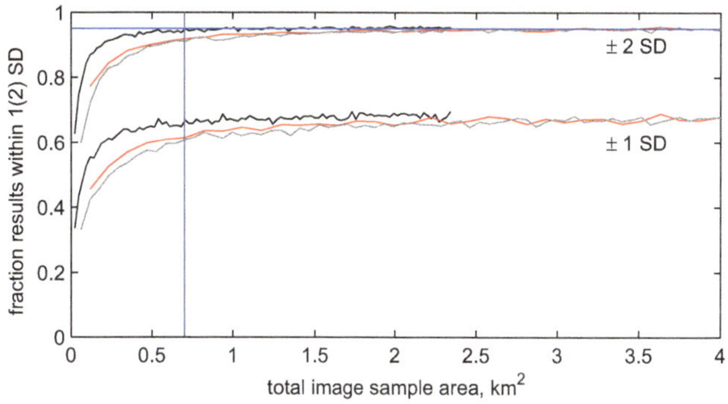

Figure 11. Fraction of image subset-based aggregate albedo values within the interval of ± 1 and ± 2 SD of the bootstrap estimated regional albedo as a function of total image (sample) area. The subsets are formed of image blocks of length 1 (black), 10 (red) and 25 (grey) images. The solid blue lines highlight the 0.95 fraction and 0.7 km^2 aggregate scale for subsets formed of single image blocks.

4 Conclusions

The formation of melt ponds on summer sea ice alters its optical properties over a broad range of wavelengths. This has implications for the surface energy balance and summer sea-ice decay as well as for practical issues of the remote sensing of sea ice. The study of sea-ice topography and the associated processes at these smaller scales was therefore identified to be of crucial importance for a better understanding of the seasonal evolution of the ice pack at a pan-Arctic scale and improvement of sea-ice parameterizations in GCMs (Eicken et al., 2004). Yet the considerable regional and intraseasonal variability of summer first-year ice albedo stipulates the need for further regional-scale studies of this parameter and its relation to other key physical factors characterizing the current state of sea-ice cover. Moreover, the recent progress made in the area of field data assimilation suggests even the regional-scale studies similar to the one presented here can potentially be valuable for improving the skills of GCMs in making seasonal sea-ice forecasts (Schröder et al., 2014; Castro-Morales et al., 2014).

Analysis of imagery and EM bird ice-thickness data from six low-altitude ice-survey flights conducted during the ICE12 drift experiment north of Svalbard at 82.3° N in late July/early August 2012 revealed a regional-scale homogeneity in the state of ice cover in the area of the drift track outside the MIZ. Within this area, with an extent of ≈ 150 km, the observed melt-pond fraction varied from 15 to 36 % in 50 % of cases, around the median of f^i_{mp}=26 %, relative to the sea-ice area. Accounting for the inferred bias of the image-processing technique, a value of $f^i_{mp} = 29$ % should be considered a realistic regional estimate for the 70–90 cm thick ice observed during the campaign. We note that in some occasions the melt ponds could cover as much as 66 % of the ice surface. For some 10 % of images with sea ice in the field of view, the sea-ice surface exhibited no or very little melt-pond coverage ($f^i_{mp} < 4$ %), possibly associated with the snow-free sea ice formed in the leads late in the winter season (Eicken et al., 2004). Within the 30 km wide MIZ, f^i_{mp} showed a decline towards the ice edge to an average value below 10 %, which we linked to more intense melt leading to a transformation of melt ponds into open water and to a decrease in the typical floe size.

The regional spatial albedo and albedo of pack ice have been obtained from the observational data on the distribution of surface types and the respective broadband albedos using the block bootstrap technique. The method implicitly accounts for uncertainties due to sampling in the spatial domain with a priori unknown variability, surface type classification errors and in situ albedo measurements. The set of more than 10 000 classified images representing a total of 28 km^2, combined with a series of in situ broadband albedo measurements conducted on sea ice, was used to produce the regional aggregate albedo estimate of 0.37 (0.35; 0.40). Elimination of the MIZ with its higher open-water fraction from the computations would increase α^r to a value of 0.39 (0.37; 0.41), still within the estimated confidence bounds. The respective value of α^r_{si} of 0.44 (0.42; 0.46) for the observed first-year pack ice shows little dependence on the data subset used. The inferred homogeneous latitudinal distribution of both α^r and α^r_{si} reflects the homogeneity of the melt-pond and open-water fractions in the study area. The tendency towards decreasing α^r and increasing α^r_{si} is observed only within the MIZ, as a result of corresponding changes in f_{mp} and f_{ow}.

The regional melt-pond fraction observed in this campaign is well within the range of variability of this parameter that was reported in the previous studies on the topic both for the multiyear and first-year ice, including landfast ice, in a similar stage of melt (e.g., Fetterer and Untersteiner, 1998; Tschudi et al., 2001; Perovich et al., 2002b, 2009; Eicken et al., 2004, see also a summary on previous observations in Polashenski et al., 2012). We also observe a consistency with the decadal (2000–2011) average of the remote sensing based retrievals of this parameter for the corresponding latitude and period of the year (Rösel et al., 2012); yet the termination of the data set in 2011 prevented us from making a direct comparison for the study area.

Analysis of the relevant literature indicates that our aggregate albedo estimates are systematically lower than the values for melting FYI reported in a number of other ship-based and aerial studies from matching latitudes and this time of year. The bare level ice albedo of 0.55 we used is lower than the estimates of 0.6–0.65 typically used for bare first-year ice (e.g., Grenfell and Maykut, 1977; Grenfell and Perovich, 2004), which is most likely to be attributed to the thinner, 70–90 cm thick, ice we observed. The melt-pond albedo (specifically prevalent dark ponds) measured during the campaign was already at the lower edge of previously observed values of 0.1–0.4 (e.g., Tschudi et al., 2001; Perovich et al., 2002b, a; Lu et al., 2010) as well as analytical approximations (Makshtas and Podgorny, 1996). Since the pond formation during melt is considered the main mechanism driving an overall decrease of the aggregate sea-ice albedo, we attribute a generally lower value of $\alpha^r_{si} = 0.44$ to a late melt stage and the associated darker ponds on the surface. The lower aggregate albedo of melting first-year ice of 0.37 reported by Nicolaus et al. (2012) based on the results of the trans-Arctic cruise ARK-XXVI/3 in 2011 and measured albedos from Perovich (1996) is related to a substantially higher first-year ice melt-pond fraction (0.43) that we did not observe in our study. This discrepancy nevertheless highlights a substantial regional and intraseasonal/interannual variability in the parameters used in upscaling to a regional aggregate estimate. We note also that the derived relatively low values for a regional melting first-year ice albedo highlights the need for a reassessment/improvement of many existing albedo parameterizations used in the sea-ice modules of GCMs. Although it has been identified as one of the research priorities more than a decade ago (e.g., Curry et al., 2001; Eicken et al., 2004), a number of models still rely on far too high albedos for melting first-year ice (see e.g., Johnson et al., 2012), with implications for the modeled seasonal sea-ice cycle.

The use of a large collection of classified images from the area allowed an assessment of the aggregate scale for the regional albedo of about 0.7 km^2, which corresponds to at least 300 representative images of the study area captured by the ICE camera setup from an altitude of 35–40 m. Higher flight altitudes would require fewer classified images, though the area covered must be larger. We emphasize that these estimates are linked with the setup configuration used as well as the state of sea-ice cover during the ICE12 experiment. This result suggests that gaining adequate regional statistics on f^s_{ow}, f^s_{mp} and hence α^r, provided a spatial homogeneity of sea ice, would require a relatively limited number of processed images, with implication for the labor intensity of the procedure.

The results indicate that about 95 % of the uncertainty in our regional albedo estimate is due to variability in the in situ albedo measurements. This variability is related to both the natural local variability of this parameter due to, e.g., underlying ice thickness or pond depth, as well as to the uncertainty stemming from the measurement technique itself. This indicates the need for a series of local measurements carried out for each surface category as a necessary prerequisite for a high-quality regional upscaling. A particular focus should be on melt-pond albedo evolution at the latter stages of ice decay, when the ice beneath the ponds gets thin, the ponds begin to melt through, and their albedo approaches that of open water.

Processing and analysis of the data from 2012 is an ongoing effort. The plans for further work include a detailed analysis of the spatial melt-pond distribution and a joint analysis of EM bird ice thickness data, optical melt pond characteristics and ridging of sea ice. As the setup was designed to enable the capability of producing 3-D reconstructions of the sea-ice surface topography, some scenes were selected for a detailed analysis of the surface morphology. Gaining statistics on small-scale sea-ice topography is considered necessary (Eicken et al., 2004) for better understanding and modeling the evolution of first-year ice during melt.

Acknowledgements. We thank the crew of R/V *Lance* and Airlift as well as other scientists and engineers on board for their assistance in carrying out the measurements. Funding was provided by the Centre for Ice, Climate and Ecosystems (ICE) at the Norwegian Polar Institute via the ICE-Fluxes project. This work was also supported by ACCESS, a European project within the Ocean of Tomorrow call of the European Commission Seventh Framework Programme, grant no. 265863, and the Research Council of Norway through the EarthClim (207711/E10) project. M. Nicolaus (AWI, Bremerhaven) and an anonymous reviewer are acknowledged for their constructive comments on the first version of the manuscript.

Edited by: M. van den Broeke

References

Björk, G., Stranne, C., and Borenäs, K.: The Sensitivity of the Arctic Ocean Sea Ice Thickness and Its Dependence on the Surface Albedo Parameterization, J. Climate, 26, 1355–1370, doi:10.1175/JCLI-D-12-00085.1, 2013.

Castro-Morales, K., Kauker, F., Losch, M., Hendricks, S., Riemann-Campe, K., and Gerdes, R.: Sensitivity of simulated Arctic sea ice to realistic ice thickness distributions and snow parameterizations, J. Geophys. Res.-Oceans, 119, 559–571, doi:10.1002/2013JC009342, 2014.

Conover, W.: Practical nonparametric statistics, Wiley series in probability and statistics, Wiley, New York, USA, 3rd edn., 584 pp., 1999.

Curry, J. A., Schramm, J. L., and Ebert, E. E.: Sea Ice-Albedo Climate Feedback Mechanism., J. Climate, 8, 240–247, doi:10.1175/1520-0442(1995)008<0240:SIACFM>2.0.CO;2, 1995.

Curry, J. A., Schramm, J. L., Perovich, D. K., and Pinto, J. O.: Applications of SHEBA/FIRE data to evaluation of snow/ice albedo parameterizations, J. Geophys. Res.-Atmos., 106, 15345–15355, doi:10.1029/2000JD900311, 2001.

Derksen, C., Piwowar, J., and LeDrew, E.: Sea-Ice Melt-Pond Fraction as Determined from Low Level Aerial Photographs, Arctic Alpine Res., 29, 345–351, 1997.

Doronin, Y. and Kheisin, D.: Sea Ice, Amerind Publishing Company, Office of Polar Programs and the National Science Foundation, Washington D.C., USA, 323 pp., 1977.

Efron, B. and Tibshirani, R.: An introduction to the bootstrap, CRC press, New York, USA, 57, 456 pp., 1993.

Eicken, H., Krouse, H. R., Kadko, D., and Perovich, D. K.: Tracer studies of pathways and rates of meltwater transport through Arctic summer sea ice, J. Geophys. Res.-Oceans, 107, 8046, doi:10.1029/2000JC000583, 2002.

Eicken, H., Grenfell, T. C., Perovich, D. K., Richter-Menge, J. A., and Frey, K.: Hydraulic controls of summer Arctic pack ice albedo, J. Geophys. Res.-Oceans, 109, C08007, doi:10.1029/2003JC001989, 2004.

Fetterer, F. and Untersteiner, N.: Observations of melt ponds on Arctic sea ice, J. Geophys. Res.-Oceans, 103, 24821–24835, doi:10.1029/98JC02034, 1998.

Gonzalez, R. C., Woods, R. E., and Eddins, S. L.: Digital Image Processing using MATLAB, Gatesmark Publishing, USA, 2nd edition, 827 pp., 2009.

Grenfell, T. C. and Maykut, G. A.: The optical properties of ice and snow in the Arctic Basin, J. Glaciol., 18, 445–463, 1977.

Grenfell, T. C. and Perovich, D. K.: Seasonal and spatial evolution of albedo in a snow-ice-land-ocean environment, J. Geophys. Res.-Oceans, 109, C01001, doi:10.1029/2003JC001866, 2004.

Haas, C., Lobach, J., Hendricks, S., Rabenstein, L., and Pfaffling, A.: Helicopter-borne measurements of sea ice thickness, using a small and lightweight, digital EM system, J. Appl. Geophys., 67, 234–241, doi:10.1016/j.jappgeo.2008.05.005, 2009.

Hanesiak, J. M., Barber, D. G., de Abreu, R. A., and Yackel, J. J.: Local and regional albedo observations of arctic first-year sea ice during melt ponding, J. Geophys. Res.-Oceans, 106, 1005–1016, doi:10.1029/1999JC000068, 2001.

Hohenegger, C., Alali, B., Steffen, K. R., Perovich, D. K., and Golden, K. M.: Transition in the fractal geometry of Arctic melt ponds, The Cryosphere, 6, 1157–1162, doi:10.5194/tc-6-1157-2012, 2012.

Holland, M. M., Bailey, D. A., Briegleb, B. P., Light, B., and Hunke, E.: Improved Sea Ice Shortwave Radiation Physics in CCSM4: The Impact of Melt Ponds and Aerosols on Arctic Sea Ice, J. Climate, 25, 1413–1430, doi:10.1175/JCLI-D-11-00078.1, 2012.

Hudson, S. R., Granskog, M. A., Karlsen, T. I., and Fossan, K.: An integrated platform for observing the radiation budget of sea ice at different spatial scales, Cold Reg. Sci. Technol., 82, 14–20, doi:10.1016/j.coldregions.2012.05.002, 2012.

Hudson, S. R., Granskog, M. A., Sundfjord, A., Randelhoff, A., Renner, A. H. H., and Divine, D. V.: Energy budget of first-year Arctic sea ice in advanced stages of melt, Geophys. Res. Lett., 40, 2679–2683, doi:10.1002/grl.50517, 2013.

Johnson, M., Proshutinsky, A., Aksenov, Y., Nguyen, A. T., Lindsay, R., Haas, C., Zhang, J., Diansky, N., Kwok, R., Maslowski, W., Häkkinen, S., Ashik, I., and de Cuevas, B.: Evaluation of Arctic sea ice thickness simulated by Arctic Ocean Model Intercomparison Project models, J. Geophys. Res.-Oceans, 117, C00D13, doi:10.1029/2011JC007257, 2012.

Karlsson, J. and Svensson, G.: Consequences of poor representation of Arctic sea-ice albedo and cloud-radiation interactions in the CMIP5 model ensemble, Geophys. Res. Lett., 40, 4374–4379, doi:10.1002/grl.50768, 2013.

Kim, D.-J., Hwang, P., Chung, K., Lee, S., Jung, H.-S., and Moon, W.: Melt pond mapping with high-resolution SAR: the first view, Proc. IEEE, 101, 748–758, doi:10.1109/JPROC.2012.2226411, 2013.

Kunsch, H. R.: The Jackknife and the Bootstrap for General Stationary Observations, Ann. Stat., 17, 1217–1241, doi:10.1214/aos/1176347265, 1989.

Lu, P., Li, Z., Cheng, B., Lei, R., and Zhang, R.: Sea ice surface features in Arctic summer 2008: Aerial observations, Remote Sens. Environ., 114, 693–699, doi:10.1016/j.rse.2009.11.009, 2010.

Makshtas, A. and Podgorny, I.: Calculation of melt pond albedos on arctic sea ice, Polar Res., 15, 43–52, 1996.

Markus, T., Cavalieri, D. J., Tschudi, M. A., and Ivanoff, A.: Comparison of aerial video and Landsat 7 data over ponded sea ice, Remote Sens. Environ., 86, 458–469, doi:10.1016/S0034-4257(03)00124-X, 2003.

MATLAB: version 8.0.0.783 (R2012b), The MathWorks Inc., Natick, Massachusetts, USA, 2012.

Maykut, G.: Large-Scale Heat Exchange and Ice Production in the Central Arctic, J. Geophys. Res., 87, 7971–7984, 1982.

Mitchell, J., Dzerdzeevskii, B., and Flohn, H.: Climatic change: report of a working group of the Commission for Climatology, Technical Note 79, World Meteorological Organisation, 1966.

Moritz, R. E., Curry, J., Thorndike, A., and Untersteiner, N.: SHEBA: a Research Program on the Surface Heat Budget of the Arctic Ocean, Report 3, Arctic Syst. Sci.: Ocean-Atmos.-Ice Interact., 34 pp., 1993.

Nazintsev, Y.: The heat balance of the surface of the multiyear ice cover in the central Arctic, Trudy AANII, 267, 110–126, 1964 (in Russian).

Nicolaus, M., Katlein, C., Maslanik, J., and Hendricks, S.: Changes in Arctic sea ice result in increasing light transmittance and absorption, Geophys. Res. Lett., 39, L24501, doi:10.1029/2012GL053738, 2012.

Nychka, D., Buchberger, R., Wigley, T., Santer, B. D., Taylor, K. E., and Jones, R. H.: Confidence Intervals for Trend Estimates With Autocorrelated Observations (unpublished manuscript), available at: http://citeseerx.ist.psu.edu/viewdoc/download?doi=10.1.1.33.6828&rep=rep1&type=pdf (last access: 4 February 2015), 2000.

Otsu, N.: A Threshold Selection Method from Gray-Level Histograms, Systems, Man and Cybernetics, IEEE T. Syst. Man. Cyb., 9, 62–66, doi:10.1109/TSMC.1979.4310076, 1979.

Pegau, S. W. and Paulson, C. A.: The albedo of Arctic leads in summer, Ann. Glaciol., 33, 221–224, doi:10.3189/172756401781818833, 2001.

Perovich, D. K.: The optical properties of sea ice, US Army Corps of Engineers, Cold Regions Research & Engineering Laboratory, 33 pp., available at: www.dtic.mil/cgi-bin/GetTRDoc?AD=ADA310586, 1996.

Perovich, D. K.: On the aggregate-scale partitioning of solar radiation in Arctic sea ice during the Surface Heat Budget of the Arctic Ocean (SHEBA) field experiment, J. Geophys. Res.-Oceans, 110, C03002, doi:10.1029/2004JC002512, 2005.

Perovich, D. K. and Polashenski, C.: Albedo evolution of seasonal Arctic sea ice, Geophys. Res. Lett., 39, L08501, doi:10.1029/2012GL051432, 2012.

Perovich, D. K., Grenfell, T. C., Light, B., and Hobbs, P. V.: Seasonal evolution of the albedo of multiyear Arctic sea ice, J. Geophys. Res.-Oceans, 107, SHE 20-1–SHE 20-13, doi:10.1029/2000JC000438, 2002a.

Perovich, D. K., Tucker, W. B., and Ligett, K. A.: Aerial observations of the evolution of ice surface conditions during summer, J. Geophys. Res.-Oceans, 107, SHE 24-1–SHE 24-14, doi:10.1029/2000JC000449, 2002b.

Perovich, D. K., Grenfell, T. C., Light, B., Elder, B. C., Harbeck, J., Polashenski, C., Tucker, W. B., and Stelmach, C.: Transpolar observations of the morphological properties of Arctic sea ice, J. Geophys. Res.-Oceans, 114, C00A04, doi:10.1029/2008JC004892, 2009.

Petrich, C., Eicken, H., Polashenski, C. M., Sturm, M., Harbeck, J. P., Perovich, D. K., and Finnegan, D. C.: Snow dunes: A controlling factor of melt pond distribution on Arctic sea ice, J. Geophys. Res.-Oceans, 117, C09029, doi:10.1029/2012JC008192, 2012.

Polashenski, C., Perovich, D., and Courville, Z.: The mechanisms of sea ice melt pond formation and evolution, J. Geophys. Res.-Oceans, 117, C01001, doi:10.1029/2011JC007231, 2012.

Renner, A. H., Dumont, M., Beckers, J., Gerland, S., and Haas, C.: Improved characterisation of sea ice using simultaneous aerial photography and sea ice thickness measurements, Cold Reg. Sci. Technol., 92, 37–47, doi:10.1016/j.coldregions.2013.03.009, 2013.

Rösel, A., Kaleschke, L., and Birnbaum, G.: Melt ponds on Arctic sea ice determined from MODIS satellite data using an artificial neural network, The Cryosphere, 6, 431–446, doi:10.5194/tc-6-431-2012, 2012.

Schröder, D., Feltham, D. L., Flocco, D., and Tsamados, M.: September Arctic sea-ice minimum predicted by spring melt-pond fraction, Nat. Climate Change, 4, 353–357, doi:10.1038/nclimate2203, 2014.

Tschudi, M. A., Curry, J. A., and Maslanik, J. A.: Airborne observations of summertime surface features and their effect on surface albedo during FIRE/SHEBA, J. Geophys. Res, 106, 15335–15344, doi:10.1029/2000JD900275, 2001.

Tschudi, M. A., Maslanik, J. A., and Perovich, D. K.: Derivation of melt pond coverage on Arctic sea ice using MODIS observations, Remote Sens. Environ., 112, 2605–2614, doi:10.1016/j.rse.2007.12.009, 2008.

Untersteiner, N.: On the mass and heat budget of arctic sea ice, Archiv für Meteorologie, Geophysik und Bioklimatologie, Serie A, 12, 151–182, doi:10.1007/BF02247491, 1961.

World Meteorological Organization: WMO sea-ice nomenclature. Terminology, codes and illustrated glossary, Secretariat of the World Meteorological Organization Geneva, Switzerland, 147 pp., 1970.

Zubov, N.: Arctic Ice, Izd. Glavsevmorputi, Moscow, (translated by US Navy Oceanographic Office, Springfield, 1963), 360 pp., 1945.

2

Cyclone impact on sea ice in the central Arctic Ocean: a statistical study

A. Kriegsmann[1,*] **and B. Brümmer**[1]

[1]Meteorological Institute, University of Hamburg, Hamburg, Germany
[*]now at: Climate Service Center, Hamburg, Germany

Correspondence to: A. Kriegsmann (arne.kriegsmann@zmaw.de)

Abstract. This study investigates the impact of cyclones on the Arctic Ocean sea ice for the first time in a statistical manner. We apply the coupled ice–ocean model NAOSIM which is forced by the ECMWF analyses for the period 2006–2008. Cyclone position and radius detected in the ECMWF data are used to extract fields of wind, ice drift, and concentration from the ice–ocean model. Composite fields around the cyclone centre are calculated for different cyclone intensities, the four seasons, and different sub-regions of the Arctic Ocean. In total about 3500 cyclone events are analyzed. In general, cyclones reduce the ice concentration in the order of a few percent increasing towards the cyclone centre. This is confirmed by independent AMSR-E satellite data. The reduction increases with cyclone intensity and is most pronounced in summer and on the Siberian side of the Arctic Ocean. For the Arctic ice cover the cumulative impact of cyclones has climatologic consequences. In winter, the cyclone-induced openings refreeze so that the ice mass is increased. In summer, the openings remain open and the ice melt is accelerated via the positive albedo feedback. Strong summer storms on the Siberian side of the Arctic Ocean may have been important contributions to the recent ice extent minima in 2007 and 2012.

1 Introduction

Arctic sea-ice extent and thickness have undergone a significant decline in the last decade (Stroeve et al., 2012) with the latest record sea-ice extent minimum in September 2012. The reasons are manifold and cannot be assigned to one process only. However, it is generally accepted that the decline is also a signature of anthropogenic climate change (e.g. Notz and Marotzke, 2012).

Although sea ice is finally the result of local thermodynamic freezing and melting, there are many processes that significantly influence sea-ice formation and destruction. To this end, cyclones play an important role with both thermodynamic and dynamic aspects. With respect to the thermodynamics, cyclones contribute the major part of the sensible heat and moisture advection to the Arctic (e.g. Overland and Turet, 1994). In addition, the related higher amount of clouds enhances the surface radiation and energy budget in autumn through spring and reduces it in summer (e.g. Curry et al., 1993; Intrieri et al., 2002). With respect to the dynamics, the strong, inhomogeneous wind field in cyclones deforms the ice cover and causes cracks, leads and polynias. The effect of these openings is different in winter and summer. In winter, the openings are the places with the largest heat fluxes from the ocean to the atmosphere. The openings refreeze so that the heat fluxes decrease with time and salt is released from the newly formed ice to the ocean. In summer, refreezing does not occur, so that almost all downwelling short-wave radiation is absorbed. This leads to a local warming of the uppermost ocean layer and promotes further ice melt (e.g. Perovich et al., 2008).

The potential of cyclones to significantly affect the sea ice has been demonstrated in many case studies. Zwally and Walsh (1987) investigated by means of satellite images and model data the local change in ice concentration under the influence of a wintertime cyclone north of Alaska. Even though the impact of the wind field of the cyclone lasted only between 1 and 3 days, the reduction of the multiyear sea ice persisted for months. With satellite data and ice-drift buoy measurements Maslanik and Barry (1989) found for wintertime situations in the Canadian part of the central Arctic, that cyclones lead to reductions in sea-ice concentration and increased divergence of the sea-ice drift. Barry and Maslanik (1989) also reported a reduction in sea-ice concentration in summer in the Canadian part of the Arctic, caused by cyclonic conditions in the wind field and the associated divergent and sheared ice drift. In the western Arctic, a strong cyclone reduced the sea-ice concentration by 3 to 5 % in October 1991 (Maslanik et al., 1995). A measured increase in the fraction of first year ice was attributed to the freezing of the cyclone-induced open water areas.

Through the breaking of the ice cover, internal stress within the ice is reduced and the mobility of the ice floes is increased. The ratio of ice-drift speed and wind speed (the wind factor) is increased for a short time period after the cyclone passage (Brümmer and Hoeber, 1999). On the basis of satellite images, Holt and Martin (2001) investigated the passage of a cyclone through the Beaufort Sea, the Chukchi Sea, and the East Siberian Sea in August 1992. They reported a growth of the fraction with open water and a decrease in the size of floes. Using RADARSAT satellite images, Brümmer et al. (2008) analyzed the ice-drift field around a cyclone in March 2002. They found a divergent drift in the centre of the cyclone and a convergent drift at the margins.

On the basis of buoy measurements, Brümmer et al. (2003) found cyclone-induced inertial oscillations or tidal oscillations in the Fram Strait. They showed, that the oscillations are also detectable in the divergence, the vorticity and the deformation of the ice drift. Lammert et al. (2009) reported on a cyclone that caused a resonant forcing of inertial oscillations of the sea ice in the Fram Strait in March 2007. On the right side of the cyclone track (looking along the propagation vector) the wind direction changes were in phase with the drift direction changes of the inertial oscillation so that the ice concentration was reduced by up to 11 %.

More process-oriented, Haapala et al. (2005) used a multicategory sea-ice model in order to separate sea ice in "undeformed" and "deformed" ice classes. In model simulations with a prescribed cyclone propagating over ice, they investigated how the amount of the open water forming through the deformation influence of the cyclone depends on the chosen cyclone and ice parameters. Kawaguchi and Mitsudera (2008) used a Lagrangian ice model with an idealized stationary cyclone to examine mechanisms that cause ice divergence and its seasonal changes. They found, among other factors, that even the stratification of the ocean between winter and summer can change the ice divergence by a factor of 2. This is because the divergence of the Ekman flow in the shallow, stable ocean boundary layer below the ice in summer is larger than in the deeper and less stable boundary layer in winter.

Based on reanalysis data, Simmonds and Rudeva (2012) investigated a strong Arctic cyclone that occurred during August 2012. While their paper focuses mainly on the properties of the cyclone, they also note, that the cyclone led to a dispersion and separation of a significant amount of ice from the main Arctic ice shield.

Screen et al. (2011) investigated the relationship between sea-ice extent at the end of the Arctic summer, and cyclone activity in the preceding spring and summer. They found that fewer cyclones over the central Arctic Ocean in May, June, and July favour a low sea-ice extent at the end of the melt season. They argue that fewer storms are associated with above-average sea-level pressure, strengthened anticyclonic winds, intensification of the transpolar drift stream, and reduced cloud cover, which all favor ice melt.

In this paper, we use a statistical approach to quantify the local impact of cyclones on the sea ice in the Arctic Ocean, applying a combination of observations and modelling. We use the six-hourly operational analyses of the ECMWF (European Centre for Medium-Range Weather Forecasts) as atmospheric forcing of the coupled ice–ocean model NAOSIM (North Atlantic–Arctic Ocean Sea-Ice Model). Within a certain radius around the cyclone position the model-simulated impact on the sea ice is determined with respect to ice concentration and the wind–ice-drift relation. The model results for the individual cyclone cases are stratified by different cyclone intensities, different seasons, and different sub-regions of the Arctic Ocean. For all classes, the spatial distribution of wind, drift, and ice concentration within the cyclone radius and the time evolution from before, during, to after the cyclone event are investigated. The study is based on a three-year period 2006–2008, thus, including the time of the so-far second-lowest summer ice extent minimum in September 2007.

The paper is organized as follows: Sect. 2 describes the model, the model domain, and the study area. Section 3 details the method of cyclone detection. The results of the statistical analyses of the impact of the cyclones on sea ice in the central Arctic Ocean are presented in Sect. 4 with respect to the cyclone intensity, to seasonal differences, and to regional differences. Conclusions are given in Sect. 5.

2 The ice–ocean model

The model NAOSIM is a coupled sea-ice–ocean model. It has been applied in many research papers (e.g. Köberle and Gerdes, 2003; Karcher et al., 2007; Rozman et al., 2011) and also participated in the Arctic Ocean Model Intercomparison Project (e.g. Martin and Gerdes, 2007). The ocean part is based on the Modular Ocean Model 2 of the Geophysical Fluid Dynamics Laboratory. The sea-ice part is the dynamic–thermodynamic ice model based on the work of Hibler (1979), and was enhanced, for example, in the subgrid sea-ice thickness (Hibler, 1984) and in the rheology (Harder, 1996). The sea ice follows a viscous-plastic rheology. The prognostic variables of the ice model are ice concentration and thickness, snow height, and ice drift. On a subgrid scale,

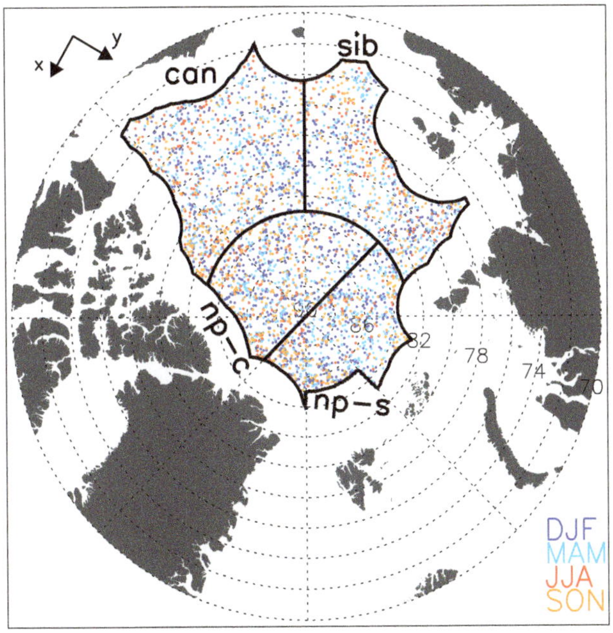

Fig. 1. Study area with spatial and seasonal distribution of the cyclone detections (dots) for the period 2006–2008. The orientation of the (x, y) coordinate axes is shown in the upper left corner.

the thermodynamic ice thickness growth is distinguished between seven ice thickness classes. The model domain covers the whole Arctic and the North Atlantic with a southern border at the 50° N latitude line. The x axis is parallel to the 30° W/150° E longitude line. Since the impact of the coast on the ice drift is detectable up to several hundred kilometres (Thorndike and Colony, 1982), we restrict our studies to those parts of the Arctic Ocean that are at least 300 km distant from the nearest coast (Fig. 1). The spatial resolution is ≈ 9 km. The model time step is 5 min and the prognostic ice variables were stored every 3 h. The model needs a prescribed atmosphere every 6 h. In this work, forcing data of the ECMWF operational analysis are used. These data have a spatial resolution of 25 km and are interpolated to the 9 km model grid. The forcing variables are temperature, dew point temperature, wind, cloud cover, and precipitation. Precipitation is taken from ECMWF 6 h forecast. The short-wave and long-wave up- and downwelling radiation fluxes are parameterized in the NAOSIM model as well as the turbulent fluxes of momentum and sensible and latent heat.

Since the cyclone impact on sea ice is taken here from the NAOSIM model, the degree of realism of the results depends essentially on the model quality. The quality of NAOSIM with respect to ice drift has been tested against satellite and in-situ data in several publications (e.g. Martin and Gerdes, 2007; Rozman et al., 2011). The results show a high correlation of at least 0.84 between the simulated drift and the in situ data. The standard deviation of the error for the model simulations validated with mooring observations amounts to $0.03\,\mathrm{m\,s^{-1}}$. Results from comparisons between the model simulations during the period 2006–2008 in this paper and simultaneous buoy drift data are presented in Kriegsmann (2011). The data were taken from two ice-drift experiments with 16 buoys in the region north of 85° N for the period April 2007–January 2008 and with eight buoys in the American sector of the Arctic for the period April 2008–February 2009. For both drift experiments, model–buoy correlations for both the u- and v-ice-drift component were between 0.82 and 0.85. In general, there were more model overestimations for ice drifts $< 0.2\,\mathrm{m\,s^{-1}}$ and underestimations for ice drifts $> 0.2\,\mathrm{m\,s^{-1}}$, that is, model variability was less than observed variability. The angle difference varies around zero, but with more positive than negative deviations. The angle difference for each buoy averaged over the lifetime of the buoy is between 3° and 19° to the right of the buoy drift. It should be mentioned that the model–buoy differences cannot solely be attributed to the model, but can result also from uncertainties of the ECMWF analysis and to a minor degree from measurement errors.

For our study, there might have been the option to use satellite-derived fields of ice motion. However, even the best time resolution of 1 day (mean over 1 day) and spatial resolution of 25 km appear not yet adequate for a cyclone impact study which requires simultaneous fields of atmosphere and ice with high time and spatial resolution as applied in this paper.

In order to compare the cyclone-induced changes in ice concentration between model and observations, we use the satellite-based AMSR-E (Advanced Microwave Scanning Radiometer - Earth Observing System) microwave (89 GHz) sea-ice concentration, retrieved with the ASI-algorithm. The satellite data were obtained from the Integrated Climate Data Center (ICDC, http://icdc.zmaw.de), University of Hamburg, Germany. The AMSR-E ice concentration data have a horizontal resolution of 6.5 km and a time resolution of 1 day. For the comparison, the AMSR-E data are interpolated to the 9 km model grid. AMSR-E has the tendency to overestimate the ice concentration (Spreen et al., 2008). This is mainly the case in summer so that the ice-covered area in the simulation and from the AMSR-E data do not match well (Fig. 2). Therefore, all cyclone cases with a difference in the mean ice concentration within the cyclone radius (for definition, see Sect. 3) of more than 10 % are neglected for the cyclone impact study (called ADiff-Filter). The ADiff-filter removes about 30 % of the cyclone detection cases (Table 1). It should be mentioned that in this study we are only interested in changes of ice concentration, not in the absolute values.

3 Cyclone detection

Cyclone position, radius, and intensity are determined from the 6-hourly sea level pressure field of the ECMWF analysis for the period 2006–2008 applying the algorithm of Murray

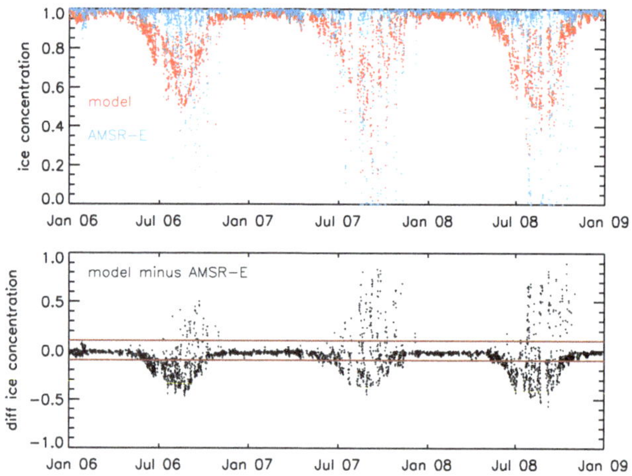

Fig. 2. Top: mean simulated and AMSR-E ice concentration within the cyclone radius for all 5057 detections. Bottom: difference between mean simulated and AMSR-E ice concentration. Red line marks range of Adiff-filter.

and Simmonds (1991). In the beginning, the first and the second derivatives of the two-dimensional field of sea level pressure p are calculated at all grid points. Grid points where the Laplace operator $\nabla^2 p = p_{xx} + p_{yy}$, which is proportional to the geostrophic vorticity, has a local maximum and is above a certain threshold are candidates for a cyclone. In the case that a pressure minimum in the surrounding of this point exists, the exact position of the cyclone core is calculated. The cyclone detection according to Murray and Simmonds (1991) depends on many settings, for example on the threshold of $\nabla^2 p$, on the search radius around the maximum of the Laplace operator, on the search area and the kind of search. For cyclone tracking, the detections are connected to cyclone tracks, based on probability calculations regarding cyclone track direction, cyclone speed, and core pressure among others.

In this work, two further quantities are used in addition to position and time of detection: the radius and the intensity of a cyclone. The radius is defined as a weighted distance (weighted over 12 directions) between core and margin of a cyclone, where margin is the place with $\nabla^2 p = 0$. The intensity of a cyclone is defined as the mean of $\nabla^2 p$ within the radius region.

The cyclone data set which we use was prepared by Haller (2011) and contains all detections for the area north of 60° N, for the period 2006–2008. This amounts to a total of 70 405 detections belonging to 7987 cyclones (Table 1). On average, each cyclone has 8.8 detections (i.e. a lifetime of about 53 h). Detections in the study area after applying the Adiff-filter amount to 3496 detections belonging to 692 cyclones.

The automated detection of cyclones is an often applied method. For example, Jahnke-Bornemann (2010) uses an automated procedure of Blender et al. (1997) for investigations of cyclones in the Norwegian Sea. Also with an automated

Table 1. Number of detections (cyclones).

	North of 60° N	Within study area	After ADiff-filter
2006	23 220 (2627)	1874 (316)	1245 (248)
2007	23 171 (2570)	1462 (248)	1100 (204)
2008	24 014 (2790)	1721 (304)	1151 (240)
Total	70 405 (7987)	5057 (868)	3496 (692)

procedure, Serreze and Barrett (2008) found in reanalyses of the National Centers for Environmental Prediction (NCEP) in the long-time mean a maximum of cyclones in the central Arctic during the summer. Affeld (2003) used automatically detected cyclones to investigate the impact of cyclones on the ice transport through the Fram Strait. A good review of the different approaches to detect (and track) cyclones is given in Ulbrich et al. (2009).

On the basis of the position of the cyclones the mean impact on ice concentration and ice drift was investigated. The cyclone detections were sorted into different classes in order to find out how the impact of the cyclone on the sea ice varies with cyclone intensity, season, and sub-region of the Arctic Ocean.

The study period 2006–2008 contains the time (September 2007) when the so-far second lowest summer ice-extent minimum occurred. The cyclone frequency during this period was at an almost constant level, embedded in an otherwise increasing trend of cyclone frequency since 1958 (Haller, 2011). In terms of general circulation, characterized by North Atlantic Oscillation and Arctic Oscillation indices (e.g.http://www.ersl.noaa.gov/psd/gcos_wgsp/Timeseries), this period was not particularly outstanding. So, our results can be regarded to represent Arctic conditions typical for the first decade of the 21th century.

4 Statistical analysis of cyclone impact on sea ice

In this section the impact of cyclones on the Arctic Ocean sea ice is presented and distinguished with respect to three categories. The first category is the strength of the atmospheric forcing which we assume to be represented by the cyclone intensity. The second category is the season because in addition to, for example, different air temperatures, a cyclone meets different sea-ice conditions such as concentration, thickness, or ice strength and, thus, the consequences on the different sea ice are expected to be different. The third category is the sub-region of the Arctic Ocean because the sea ice has a distinct gradient, for example, in thickness from the Siberian (thin) to the Canadian side (thick) and belongs to different drift regimes (Transpolar Drift, Beaufort Gyre).

Each category (intensity, season, region) is subdivided into four classes. For each class composite fields for wind, ice drift and concentration around the cyclone are calculated. This is performed in a relative horizontal coordinate frame which has its origin at the cyclone centre and is normalized with the cyclone radius. The composite averaging is made without any rotation of the frame (e.g. with respect to North). So, the relative x, y frame in the figures below is parallel to the axes of the model domain (see Fig. 1) and $x = y = 1$ is the normalized cyclone radius.

4.1 Cyclone intensity

The intensity $I = \nabla^2 p$ of all cyclone detections in the study area during the period 2006–2008 ranges from $I = 0$ to $I = I_{\max}$ ($= 7.31$ hPa (deg.lat.)$^{-2}$). We distinguish between four intensity classes, I_1 to I_4, belonging to the four quartiles of the frequency distribution. The intensity range for each class together with the average cyclone radius is given in Table 2.

For each class the mean wind vector field and the mean wind speed are shown in Fig. 3 (top). The cyclonic structure of the wind field does not end at the cyclone radius, but reaches to twice the cyclone radius. As expected, the wind speed increases with increasing intensity. The maximum mean wind speed increases from about $4\,\mathrm{m\,s^{-1}}$ in I_1 to almost $9.5\,\mathrm{m\,s^{-1}}$ in I_4. The wind speed slows down towards the cyclone centre where it is around $2.5\,\mathrm{m\,s^{-1}}$ in I_1 and $4.5\,\mathrm{m\,s^{-1}}$ in I_4.

The wind fields in Fig. 3 are not rotationally symmetric but show the highest winds and, thus, the strongest pressure gradients in the direction towards the negative x and y directions. This is (see Fig. 1) along the direction from the North Atlantic side to the Pacific side of the Arctic Ocean, and coincides with the gradient of the climatologically mean sea-level pressure distribution which exhibits lower pressure over the northeast Atlantic and higher pressure over the East Siberian Sea and Beaufort Sea (e.g. Serreze and Barry, 2005). If a symmetric cyclone pressure field is superposed to this climatological pressure distribution the resulting pressure gradient in the cyclone is stronger on the side towards the climatological high pressure than towards the low pressure.

The mean ice-drift field in Fig. 3 (bottom) has a similar cyclonic structure as the wind field. The ice drift decreases towards the centre in accordance with the decreasing wind. This is true for all intensity classes.

The relation between drift and wind is presented in Fig. 4 by means of the wind factor WF (ratio drift vs. wind) and the deviation angle DA (positive for ice drift to the right of the wind). WF and DA are relatively easily accessible from observations and are frequently applied in the literature because they are indirect indicators for the importance of the individual forces in the ice momentum balance equation. In case of a free ice drift (internal ice force negligible which is typically assumed for ice concentrations below 0.8, Leppäranta, 2005, p. 141) and thin ice (Coriolis force negligible), the WF is the square root of the ratio of the products of density and drag coefficient in atmosphere and ocean. If, in addition, the ocean current is zero the ice drifts along the wind. If we assume a wind-driven Ekman current in the ocean the ice drifts to the right of the wind and to the left of the current. With increasing ice thickness (i.e. increasing Coriolis force), the WF decreases and the DA increases (Leppäranta, 2005, p. 144). In case of ice concentrations above 0.8, the internal ice force has to be taken into account. It is about opposite to the drift and acts towards a decrease of WF and DA (Thorndike and Colony, 1982). The inertial force is typically small but can modify these relations for example if rapid wind changes occur.

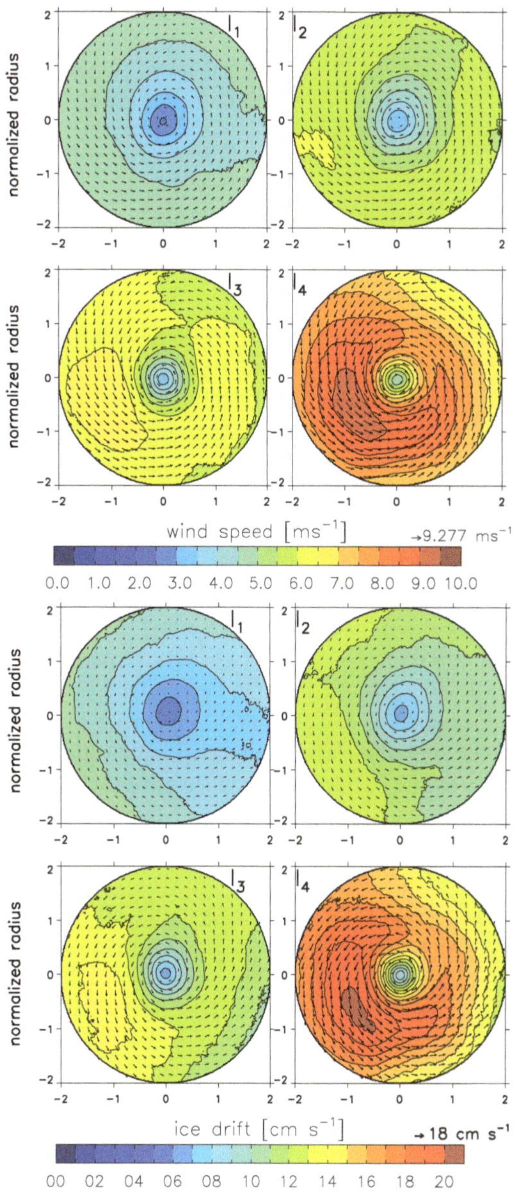

Fig. 3. Mean wind (top) and mean ice drift (bottom) for the intensity classes I_1 to I_4, centred on the location of the cyclone detection and normalized with the cyclone radius. Mean vectors as arrows, mean over absolute value of the speed as color. Isoline interval is $0.5\,\mathrm{m\,s^{-1}}$ for wind speed and $1\,\mathrm{cm\,s^{-1}}$ for ice drift.

Table 2. Number of cyclone detections with corresponding mean radius and intensity for intensity quartiles, seasons, and regions.

Intensity (% of I_{max})	No.	Radius [km]	Season	No.	Radius [km]	Intensity hPa (deg.lat.)$^{-2}$	Region	No.	Radius [km]	Intensity hPa (deg.lat.)$^{-2}$
I_1 (0–14.7)	873	266	winter	1360	315	1.76	np-s	782	298	1.95
I_2 (14.7–20.3)	875	306	spring	1067	320	1.73	np-c	844	309	1.69
I_3 (20.3–29.0)	874	327	summer	374	303	1.64	sib	789	316	1.70
I_4 (29.0–100)	874	344	autumn	694	293	1.75	can	1080	318	1.58

Averaging the WF and DA over the area within the double cyclone radius and over all cyclone detections and all intensities, they amount to about 2.0 % and 30°, respectively. This agrees with the values cited by Leppäranta (2005, p. 144) as typical for the Arctic Ocean. This is again an indirect indication for the validity of the NAOSIM model and that the above-mentioned (Sect. 2) differences found between model and drift buoys cannot be attributed to model alone. For all intensities, Fig. 4 exhibits a mean gradient in the WF and DA composite fields: WF (DA) decreases (increases) towards the positive x direction and weakly negative y direction. This direction points from the Siberian side to the Canadian side (i.e. from thinner to thicker ice). These mean gradients in Fig. 4 agree with the above-mentioned increasing influence of the Coriolis force with increasing ice thickness. These gradients remain, even in the case of ice concentrations above 0.8, when the internal ice stress has to be regarded in the balance of forces.

The WF and DA values deviate from this gradient around the cyclone centre in a region with a non-dimensional radius of 0.25 to 0.5. The DA minimum below 20° and WF maximum of up to 2.6 % are suggestive of a smaller ice thickness in the centre than on the perimeter. The WF maximum is almost absent in I_4 and increases with decreasing intensity. This may be related to the slowing down of the wind in the cyclone centre. At these weak winds the inertia of the ice drift and, in addition, the ocean current come into play and can make the WF particularly high (Leppäranta, 2005, p. 146).

Next we investigate the time development of wind and drift in the fixed circular area where the cyclone passes over (is detected) at $t = 0$. We calculate composite fields every 3 h from 2 days before until 5 days after the cyclone passage. The horizontal averages of the composite fields of wind, ice drift, and wind factor WF within the cyclone radius are calculated and displayed in Fig. 5. Hereby, the mean WF is not calculated as ratio of mean ice drift and mean wind but as the mean of the individual ratios at each grid point. The ratio ice drift / wind can be arbitrarily high for small wind speed; therefore WF was only taken into account when the wind stress was at least 0.01 N m^{-2}, that corresponds to a wind of about 1.8 m s^{-1}.

The time series of the horizontally averaged wind speed show an opposite time development for the intensity classes I_1 and I_4 with a minimum and a maximum near $t = 0$ and $t = -0.25$ d, respectively. The different behaviour can be understood from the different wind fields in Fig. 3. In I_1, the wind speed increases from the centre to the single radius and remains relatively constant further outside. If such a cyclone approaches, the mean wind speed within the fixed circular area decreases the closer the cyclone is and becomes minimal around $t = 0$, when the wind field including its minimum is centred over the fixed averaging area. The situation is different in I_4: the wind speed has a distinct maximum around the cyclone radius, centred in the negative (x, y) direction, and decreases towards both the centre and further outside. If such a cyclone approaches the fixed area, the mean wind first has to increase. I_2 and I_3 are transition classes between I_1 and I_4.

As expected, the time series of the ice drift are similar to those of the wind speed. The four time series of WF differ significantly only between the intensity classes I_1 and I_4 around $t = 0$. There, surprisingly, WF exhibits a minimum and a maximum, respectively. A possible reason may be related to regional differences, because I_1-cyclones occur most frequently over the Canadian sector (Can) and I_4-cyclones over the North Pole sector on the Siberian side (NP-S) (see Fig. 1 and Table 2). This is further outlined in Sect. 4.3 on the regional differences.

In the following, the changes in ice concentration within 24 h after cyclone passage ($t = 0$) are displayed in Fig. 6. For each intensity class a reduction in ice concentration around the cyclone centre is simulated. With increasing intensity, the area, where the ice concentration is reduced, becomes larger and the reduction becomes stronger. The change in the centre is at −0.7 % day^{-1} in I_1 and at −2.1 % day^{-1} in I_4.

For comparison, the 24 h ice concentration change based on AMSR-E is also shown in Fig. 6. It should be kept in mind that the time resolution of the AMSR-E ice concentration fields is 1 day and that a field is composed of several satellite orbits. For the calculation of the ice concentration change, we assume that the daily fields hold for 12 UTC and then interpolate the daily fields to the time of cyclone detection ($t = 0$) and to $t = 24$ h. The 24 h change in the AMSR-E ice concentration shows no major difference between the I_1 to I_3. The ice concentration decreases in the cyclone centre by up to −1 % day^{-1}. The single radius area is surrounded by an area with ice concentration increase. In I_4, the reduction

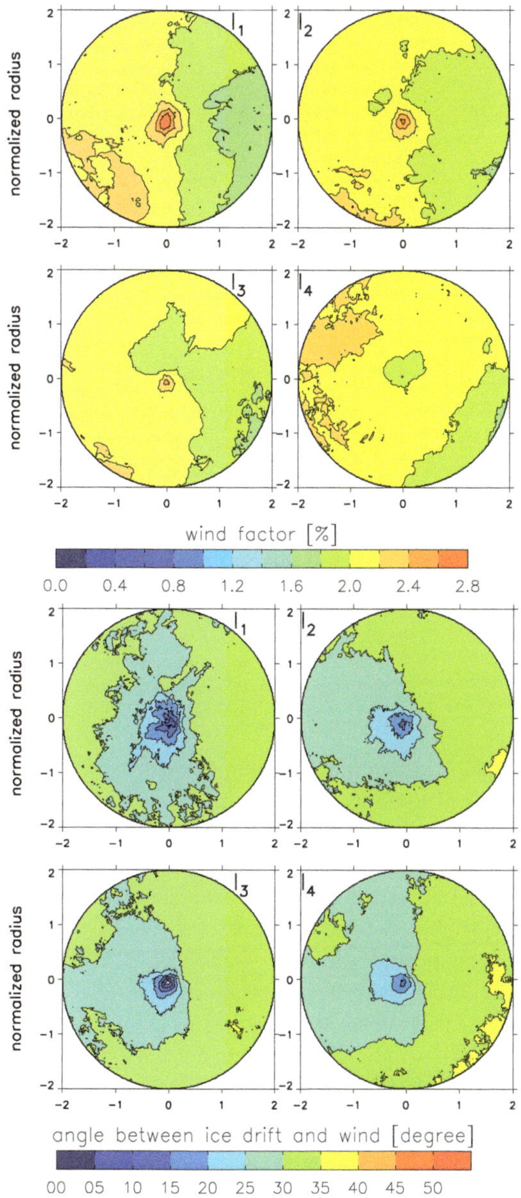

Fig. 4. Mean wind factor (top) and mean angle between wind forcing and ice drift (bottom) for the cyclone intensity classes I_1 to I_4. Isoline interval is 0.2 % for wind factor and 5° for deviation angle.

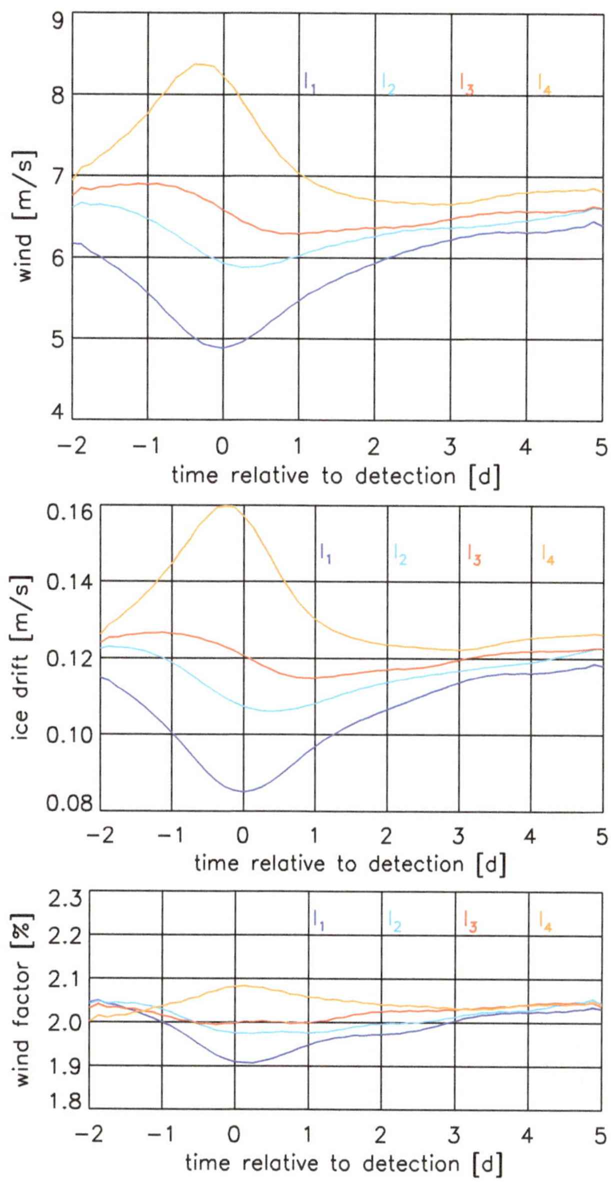

Fig. 5. Time series of wind (top), ice drift (middle), and wind factor (bottom) for different intensity classes. Shown are the mean values within the cyclone radius for the time period from 2 days before to 5 days after detection ($t = 0$).

in ice concentration is stronger than in the other three classes. Altogether, AMSR-E observations support the model results that the ice concentration is reduced within the radius of the detection and that the ice reduction increases with increasing cyclone intensity.

Ice concentration is changed by two processes: divergence and deformation. In Fig. 7 the composite fields of vorticity, divergence, and shear deformation for the weakest and the strongest intensity classes, I_1 and I_4, are displayed. Vorticity and divergence are about three times stronger for I_4-cyclones than for I_1-cyclones, with respect to both the mean values within the cyclone radius and the maximum values in the cyclone centre. The maximum vorticity in the centre of I_4-cyclones amounts to $22 \times 10^{-7}\,s^{-1}$ corresponding to a rotation of 11° per day. The maximum divergence in I_4-cyclones amounts to $7.5 \times 10^{-7}\,s^{-1}$ corresponding to an opening of the ice by 5 % per day. The shear deformation in I_1-cyclones is small whereas in I_4-cyclones maximum values of $15 \times 10^{-7}\,s^{-1}$ are reached. This means that a right angle opens by 7.5° per day. Of course, these maximum values are not reached in the course of a day, because the composite fields do not remain for 24 h over the same location. The

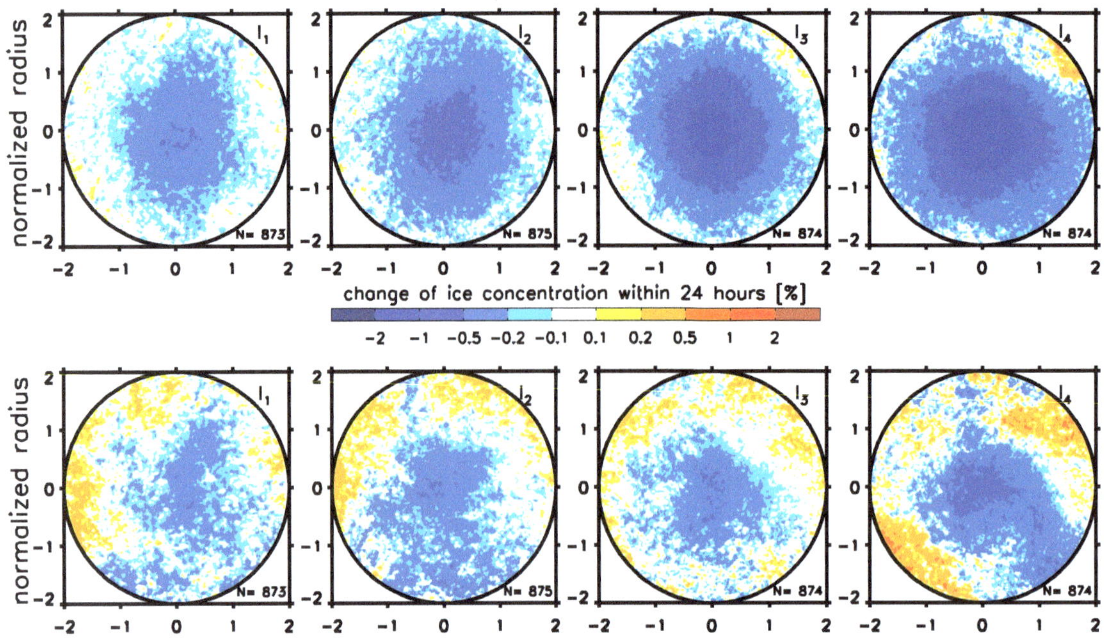

Fig. 6. Mean change of simulated (top) and AMSR-E (bottom) ice concentration within 24 h after cyclone passage ($t = 0$) for I_1 to I_4.

impact on vorticity, divergence, and deformation extends up to the cyclone radius for I_4-cyclones and is significant only in the very centre of the I_1-cyclones.

4.2 Seasonal differences of cyclone impact

The number of cyclone detections in the study area during the period 2006 to 2008 is almost the same in all seasons. Also in summer, the cyclone activity is high in the central Arctic. The maximum of cyclone detections is placed over the central Arctic Ocean in summer (Serreze and Barrett, 2008), while it is placed over the northeast Atlantic in winter (e.g. Sorteberg and Walsh, 2008). Due to the above-mentioned problem of ice concentration differences between AMSR-E and model in summer and autumn we use here only those detections where simulated and observed ice concentration are in the same range (Table 2). The frequency distribution of cyclone intensities in Fig. 8 shows a similar distribution in autumn (SON), winter (DJF), and spring (MAM), whereas the summer (JJA) cyclone intensity distribution has more weaker cyclones and less stronger cyclones. This ranking is also reflected in the season means in Table 2.

The composite cyclone wind fields for each season are presented in Fig. 9. As already mentioned for the intensity classes, the cyclonic wind field is not rotationally symmetric during autumn, winter, and spring but shows a maximum towards the negative x and y directions (i.e. towards the western Arctic because of the large-scale pressure increase from the Atlantic to the Pacific side of the Arctic). This is not the case in summer when the high-pressure zone over the western Arctic is absent. On the average, the wind is weaker in spring and summer than in autumn and winter as may be expected because of the more frequent weak summer cyclones. However, the differences are not large: the mean is $5.9\,\mathrm{m\,s^{-1}}$ in summer and $6.8\,\mathrm{m\,s^{-1}}$ in winter. This reflects the fact that the frequency distribution of cyclone intensities is not too much different between the seasons (Fig. 8). The ice-drift fields (not shown) exhibit an analogous pattern to the wind pattern, but have a different ranking of the seasons concerning mean drift speed. It is highest in autumn and lowest in summer and spring. This reflects a combination of the seasonal variations in wind speed and internal ice stress.

The composite fields of WF and DA around cyclones during different seasons are displayed in Fig. 10. The large-scale gradients with decreasing WF and increasing DA towards the positive x direction and slightly negative y direction (i.e. towards the Canadian side) are, as was shown for the different cyclone intensities, also present in each season. This reflects the increasing ice thickness and, thus, increasing role of the Coriolis force towards the Canadian side. Figure 10 further represents the well-known observationally based seasonal differences of DA and WF. DA is in winter (average 29°) smaller than in summer (average 37°) and also WF is in winter (average 2.0 %) smaller than in summer. However, the largest WF values occur after summer, in autumn (2.2 %). This underlines the seasonally different importance of the internal ice stress. We remind that, according to the above-mentioned model–buoy comparison, the DA values may be slightly overestimated. But this does not affect the general seasonal and spatial patterns. The superposed cyclone effect with small DA and large WF (in accordance with a smaller mean ice thickness around the cyclone centre)

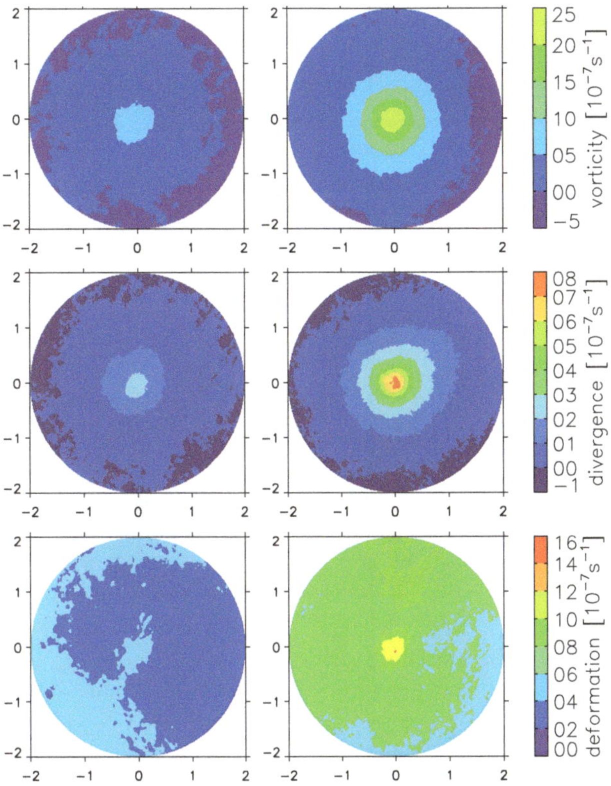

Fig. 7. Composite fields of vorticity (top), divergence (middle), and deformation (bottom) for intensity classes I_1 (left) and I_4 (right).

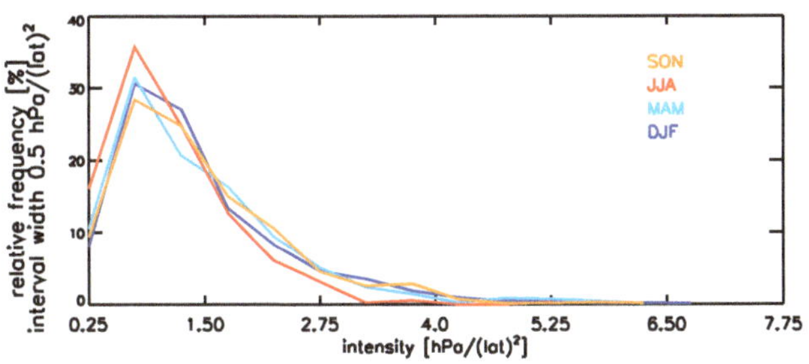

Fig. 8. Relative frequency of cyclone intensity per 0.5 hPa (deg.lat.)$^{-2}$ for each season.

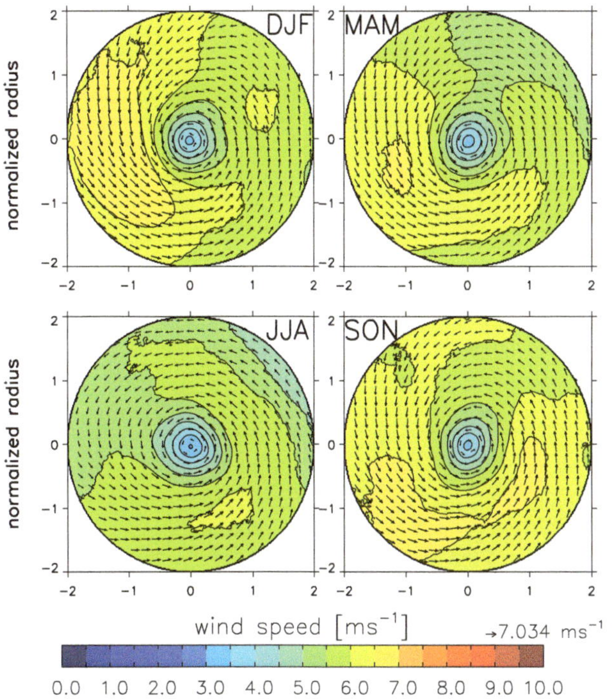

Fig. 9. Mean seasonal wind at the position of detection, normalized by radius. Shown are mean vector components and, as color, the mean value of wind speed. Isoline interval is 0.5 m s^{-1}.

is most obvious in winter and spring concerning both amplitude and extent. Though discernible, it is less distinct in summer. This results from the different mean ice concentrations in winter (about 98 %) and summer (about 77 %) (see Fig. 12). The change of the internal ice stress caused by the cyclone is less in case of a small rather than high ice concentration.

The seasonal impact of cyclones on the ice concentration during the first 24 h after the passage is shown in Fig. 11. In all seasons, a reduction is present which is most pronounced in spring and summer. In the centre of the cyclone the ice concentration is reduced within 24 h by 0.9 % in winter, 1.7 % in spring, 2.5 % in summer and 1.2 % in autumn.

For a closer investigation of the cyclone-induced part of the ice concentration changes, the annual cycle of ice concentration has to be considered. Melting in summer and freezing in winter result in a long-term (longer than the cyclone event) decrease and increase of the sea-ice concentration, respectively. The cyclone impact is then superimposed on these seasonal trends, producing less of a decrease in concentration in winter than in summer. This effect can be seen in Fig. 11. The outer area (beyond 1.5 times the cyclone radius) shows no change in ice concentration in winter, a slight decrease in spring, a stronger decrease in summer, and an increase in autumn. For comparison, the 24 h change in the observed AMRS-E ice concentration is also shown in Fig. 11. The main features simulated by the model are also present in the satellite observations: cyclones reduce the ice concentration in all seasons. Outside the cyclone radius the mean seasonal trends of ice concentration are present: constant in winter, variable (decrease as well as increase) in spring, decrease in summer, and increase in autumn.

Figure 12 displays the temporal evolution of ice concentration within the cyclone radius in comparison to the ice concentration trend in the whole study area. In winter, temperatures are far below the freezing point and the ice concentration is high (> 98 %). Open water, caused by cyclones, freezes quickly and the ice concentration has recovered ≈ 5 days after the passage of the cyclone. In spring, the mean ice concentration is around 97 %. However, areas of

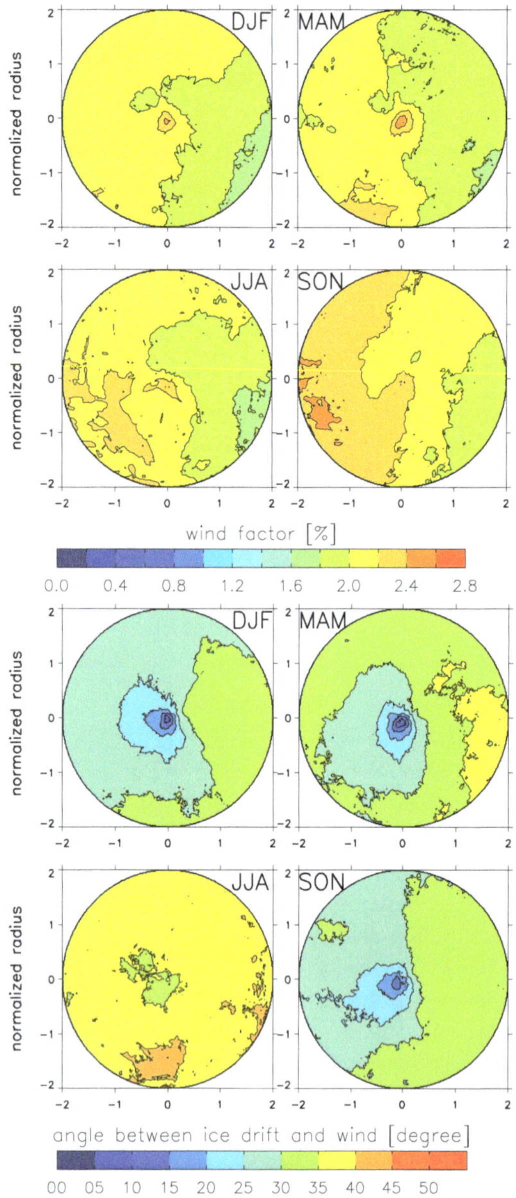

Fig. 10. Spatial distribution of the wind factor (top) and the angle between wind forcing and ice drift (bottom) for each season. Isoline interval is 0.2 % for wind factor and 5° for deviation angle.

open water, caused by cyclones, do not freeze as quickly as in winter. They exist even after a few days. The reason is the onset of the warming due to the solar radiation. In summer, the ice concentration is around 77 % and is reduced by 2 % per week even without cyclone impact. However, with cyclones the reduction is accelerated to 4 % per week. In autumn, the mean ice concentration is about 83.5 % and new ice is formed in large areas of the region. The ice concentration in the whole study area increases by about 2.5 % per week. Under the influence of a cylone, the sea-ice concentration grows only by 1.5 % per week. An important difference between the seasons is that the impact of a cyclone in winter is greatly diminished after about 5 days, whereas in the other seasons the impact remains visible. In summer, the difference between mean ice concentration in the whole study area and mean ice concentration within the area of detection is 2 % even after 5 days. The cyclone-induced change in ice concentration relative to the mean seasonal trend underlines the quantitatively stronger reduction in summer (Fig. 12 bottom).

The seasonal difference in the duration of a cyclone's impact may have important climatological consequences. Many studies have examined the change in Arctic cyclone activity over the past decades, some finding an increase in the number of cyclones in the Arctic (e.g. Sepp and Jaagus, 2011). Zhang et al. (2004) found that the number of summer cyclones are increasing. With an increasing number of cyclones in summer, the melting of sea ice could be accelerated. This conclusion appears to be in contrast to the results of Screen et al. (2011), who find a positive correlation between Arctic sea-ice extent at the end of the summer and cyclone frequency in the preceding spring and summer. However, their investigation is on an Arctic-wide scale where other Arctic-wide processes, like the strength of the Transpolar Drift, come into play, whereas our investigations are on the local cyclone scale. Of course, a cyclone in summer has two opposite effects. On one side, the cyclone breaks the ice and favors ice melt via the positive ice–albedo feedback. On the other side, a cyclone is related with a higher amount of clouds and reduces solar insolation and, thus, ice melting. However, both opposite effects are taken into account in the NAOSIM model, so the reason for the contradiction between our results and those of Screen et al. (2011) is not clear.

4.3 Cyclone impact in different regions of the Arctic Ocean

We divide the study area into four sub-regions: Siberian sector (Sib), Canadian sector (Can), North Pole region on the Siberian side (NP-S), and North Pole region on the Canadian side (NP-C) (Fig. 1). The regions were chosen in a way that all have similar numbers of detections (Table 2), but that the ice conditions differ. For example, in the Can sector thick and compact ice prevails moving with the Beaufort Gyre. The sea ice in the Sib sector is thinner and moves with the Transpolar Drift towards the Fram Strait. The mean intensity of cyclones is highest (see Table 2) in NP-S which is closest to the North Atlantic low-pressure zone (e.g. Jahnke-Bornemann and Brümmer, 2009), followed by Sib, NP-C, and finally Can which is close to the Arctic high pressure zone.

Figure 13 shows the frequency distribution of the wind factor WF at each model grid point within the cylone radius. In addition, the corresponding mean values of wind speed, ice concentration and thickness are listed. WF is highest in Sib where thin ice is present, followed by NP-S, NP-C,

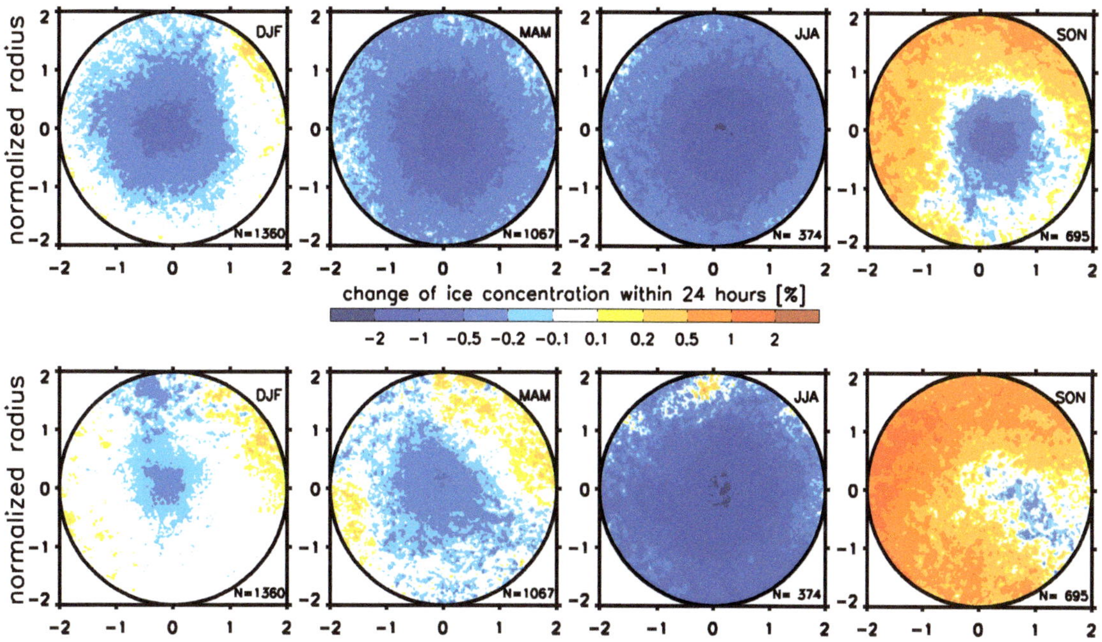

Fig. 11. Mean change of simulated (top) and AMSR-E (bottom) ice concentration within 24 h after cyclone passage ($t = 0$) for each season.

and Can where the thickest ice is present. Sectors Sib and NP-S are rather similar and so are the sectors Can and NP-C. For NP-C and Can, there are relatively more grid-points within the cyclone radius with a small WF below 1.75 %. This is attributed to the thick (mean values 3.3 m and 4.2 m, respectively) and compact ice (mean values 96 % and 95 %, respectively) that piles up ahead of the Canadian Arctic Archipelago and that drifts relatively slow.

The mean deviation angle DA within the cyclone radius shows smaller values around 28° in the Sib region and larger values around 35° in the Can region (not shown). Thicker ice (larger Coriolis force) causes an increase of DA and more compact ice (larger internal ice stress) a decrease of DA. Obviously, the effect of ice thickness dominates over the effect of ice concentration when looking at the Sib–Can differences.

Time variations of wind, drift, and wind factor WF in the cyclone radius during a 7 day period around the cyclone passage ($t = 0$) within the four sub-regions are shown in Fig. 14. Wind speeds in the regions Can and NP-C are nearly identical and are the smallest, compared to the other regions. Wind speed in the Sib sector is slightly higher. In these three regions, the wind speed varies similarly with time, whereas in NP-S not only the wind speed is highest, but also the wind extremes occur at other times and the minimum is less distinct. The mean ice drift during a cyclone event is slowest in the Can and NP-C regions. The fastest moving ice can be found in NP-S, where the wind is strongest. The different WF levels between the Can–NP-C sectors on one side and Sib–NP-S sectors on the other side are, as already mentioned above, due to the clearly different ice conditions (thickness, concentration). Furthermore striking is the opposite time variation of WF around the cyclone passage between Can–NP-C and Sib–NP-S. In Sib–NP-S, the passage is connected with an increase of WF by up to 0.1 %. In contrast to that, in Can–NP-C WF decreases by up to 0.2 %. We presume that this is a consequence of the different climatological ice drift and ocean current systems on the Can and Sib side. In the Can sector, mean drift and current rotate anticyclonically with the Beaufort Gyre, whereas a cyclone rotates cyclonically. Thus, the atmospheric cyclone impact on the ice drift is reduced. In contrast to that, mean drift and current in Sib follow with the weakly cyclonically curved Transpolar Drift Stream. Thus, the cyclone impact on ice drift is supported and increases WF.

The time series of ice concentration within the cyclone radius show that the mean ice concentration 2 days before the detection starts at different levels between 91.5 % in Sib and 97 % in NP-C (Fig. 15 top). In all regions, the ice concentration 5 days after detection is reduced, compared to the mean ice concentration in the whole study area, and is below the value it had before the cyclone passage. Relating the ice concentration to that before the cyclone passage ($t = -2$ d) (Fig. 15 bottom), there is a similar development of the ice concentration in all regions. The reduction amounts between 1 % and 1.5 %. The maximum cyclone-related ice reduction is reached about 1 day after the cyclone passage.

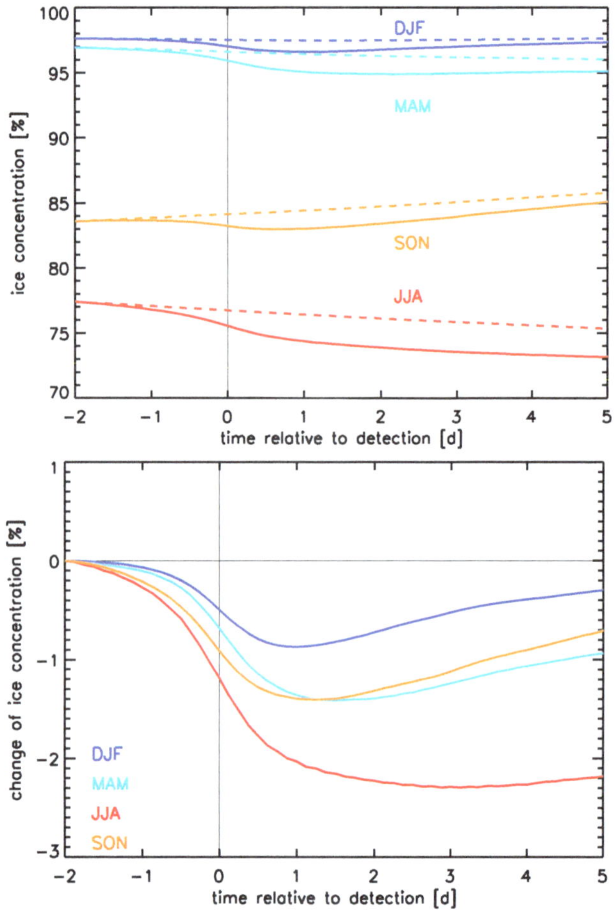

Fig. 12. Top: ice concentration in the period around the time of detection ($t = 0$). Solid line: mean over all detections within the cyclone radius, dashed line: simultaneous ice concentration in the whole study area (shifted to the starting value of the corresponding ice concentration within the cyclone radius). Bottom: detrended change in ice concentration, relatively to $t = -2$d.

5 Conclusions

In this study the impact of cyclones on the sea ice in the central Arctic Ocean was investigated, in contrast to many case studies, for the first time in a statistical manner. To this end we applied the coupled ice–ocean model NAOSIM and forced it with the 6-hourly ECMWF analysis data for the period 2006–2008. Cyclone position and radius detected from the ECMWF sea-level pressure were used to extract the local fields of wind, ice drift, and concentration within twice the cyclone radius for a time interval from 2 days before to 5 days after the cyclone passage. The cyclone impact is quantified by the drift-wind ratio, the deviation angle, and the change of ice concentration. Centred at the cyclone position and normalized by the cyclone radius, composite fields of these quantities are calculated for different cyclone intensities, the four seasons, and different sub-regions of the Arctic Ocean.

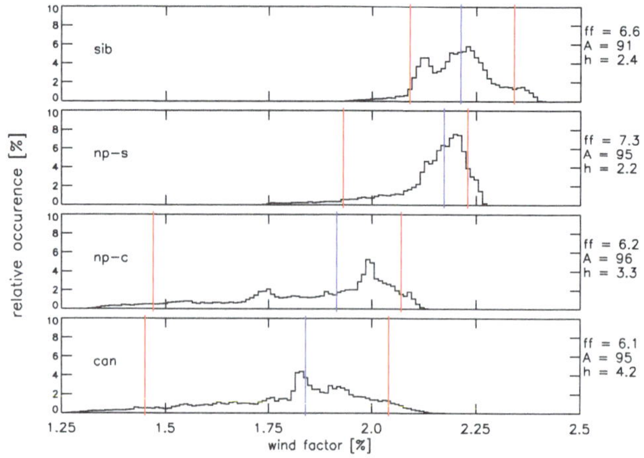

Fig. 13. Frequency distribution of the wind factor within the cyclone radius in four sub-regions of the Arctic Ocean. Red lines mark 5 and 95 percentiles, blue line marks the median. Also given is the mean wind speed ff [m/s], the mean ice concentration A [%] and the mean ice thickness h [m].

The classification of the cyclones according to their intensities shows that wind speed and ice drift increase with increasing intensity. In general, the cyclone passage leads to a reduction of the ice concentration. The reduction increases from the cyclone edge to the centre. In the centre, the reduction within 24 h after the passage is of the order of 0.5 to 2 % increasing with cyclone intensity. This spatial distribution and ranking of the ice concentration change is supported by independent AMSR-E satellite data.

Mean wind speed and ice drift within the cyclone radius show only small seasonal differences with a minimum in summer and a maximum in autumn/winter. The deviation angle between 10 m-wind and ice drift is largest in summer with up to 37° and smallest in autumn/winter with up to 29°. The wind factor has a minimum in spring and a maximum in autumn. Reductions in sea-ice concentration occur in all seasons, but with different magnitude and duration. The largest reduction of up to 4 % occurs in summer and is persistent. In winter, the sea-ice concentration returns to its initial value a few days after the passage of a cyclone, because areas with open water refreeze quickly.

The regional investigation of cyclone impact reveals that wind speed and ice drift are strongest in the NP-S region, which is close to the North Atlantic cyclone track, and are weakest in the Can region, which is close to the high pressure zone over the western Arctic Ocean. The wind factor is smallest in the Can region with 1.8 % and largest in the Sib region with 2.2 %. The main reason for the regional differences of cyclone impact is the different ice thickness and concentration in these sectors.

On the short timescale (less than 12 h) of a passing cyclone, freezing and melting play a minor role in the change of ice concentration. Thus, the cyclone-induced reduction of

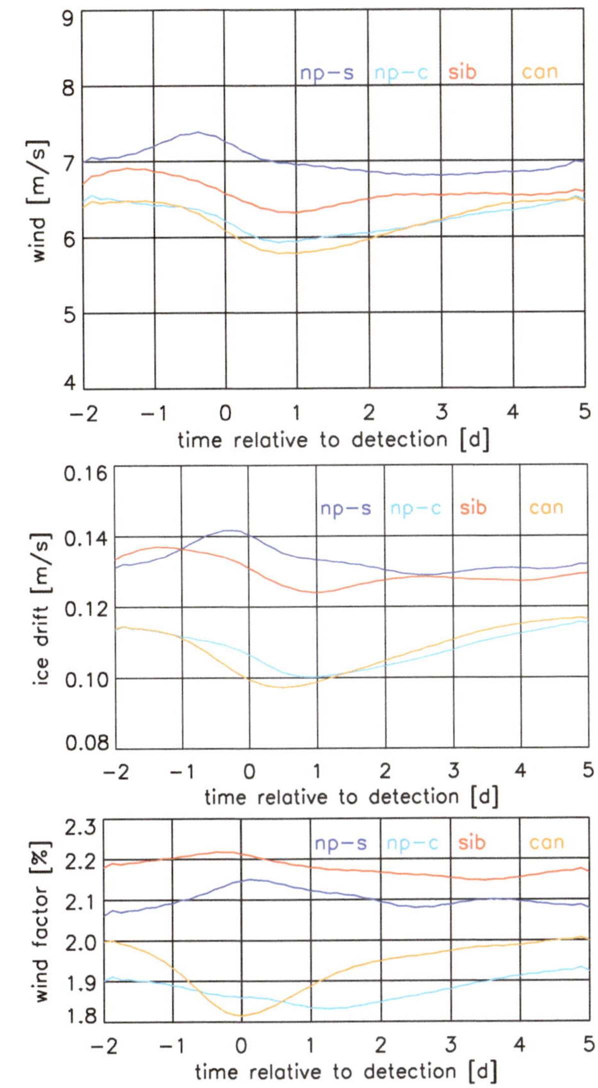

Fig. 14. As in Fig. 5, but for different sub-regions of the Arctic Ocean.

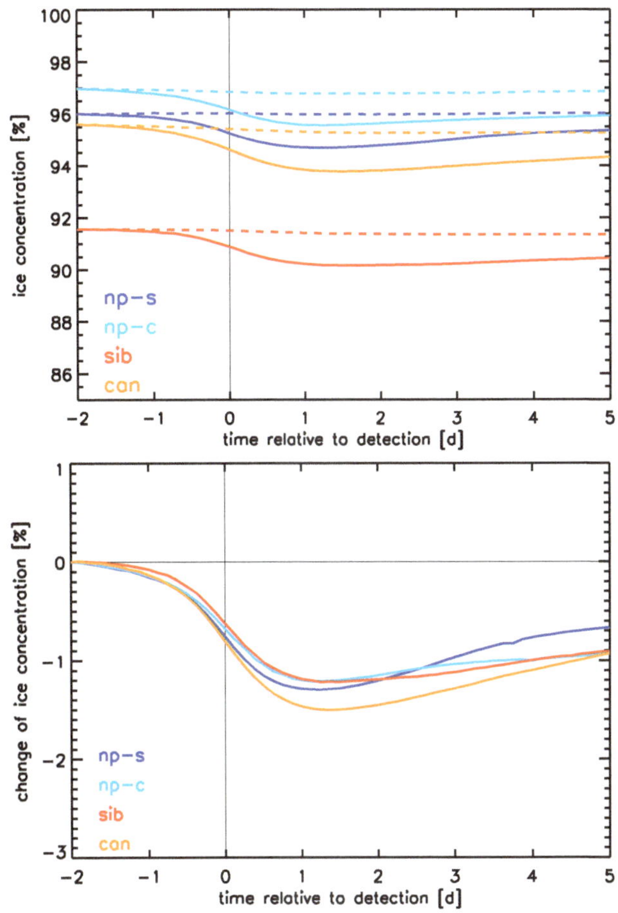

Fig. 15. As in Fig. 12, but for different sub-regions of the Arctic Ocean.

ice concentration is almost solely due to the dynamics (i.e. ice-drift divergence and ice sheet deformation). This means that there is no loss of ice mass. The subsequent processes of freezing and melting have different longer term or even climatologic consequences, especially in winter and summer. In winter, the heat flux between ocean and atmosphere over the cyclone-induced open water areas is increased for a few days. This heats and moistens the shallow Arctic boundary layer. At the same time the freezing of the open water areas leads to the formation of new sea ice, so that a further important impact of the wintertime cyclones is an increase of the Arctic ice mass. In summer, the cyclone-induced reduction of sea-ice concentration is largest and the open water areas remain open, decreasing the area average surface albedo and increasing the solar absorption. This summertime impact is expected to be especially large in areas with thinner ice as, for example, on the Siberian side and in areas with smaller ice floes as, for example, in the marginal ice areas of the Arctic Ocean. Strong summer storms in those areas can lead to increased exceptional reduction of the sea-ice concentration, because they can separate and disperse large ice fields from the main Arctic ice cover. The strong summer cyclone between 4 and 8 August 2012 is believed to be one reason for the following record ice extent minimum in September 2012 (Simmonds and Rudeva, 2012). Based on model simulations, Zhang et al. (2013) hint at another impact of the strong 2012 August storm. They found an exceptionally strong bottom melt of the sea ice resulting from the cyclone-induced enhanced upward mixing of water from the near sub-surface ocean maximum temperature layer.

In this paper, we distinguished between three categories (intensity, season, region). This stratification can only show tendencies. A combined stratification would have been better and more instructive, but was not made here because of the only 3 year-long investigation period and the consequently limited number of cases. This could be improved with a longer investigation period. The cyclone radius may be a further classification parameter. Uotila et al. (2011) studied the relationship between cyclone properties and surface properties in the Antarctic. They distinguished between mesoscale ($r \leq 500$ km) and synoptic-scale cyclones ($r >$ 500 km) and found that cyclones impacting the surface are more synoptic-scale and cyclones being impacted by the surface are more mesoscale. A further stratification criterion when applying the compositing method should be the propagation direction of the cyclone. Haapala et al. (2005) and Kriegsmann (2011) demonstrated that the cyclone impact is not symmetrical on the left and right side with regard to the propagation direction. On the right side, the cyclone presses the ice towards the not yet impacted ice cover, and away from it on the left side. All these additional stratification criteria would help to increase our knowledge of cyclone impact on sea ice, however, this requires a far larger number of cases.

Acknowledgements. This work was mainly supported by the European Union through the 6th framework programme with the Integrated Project DAMOCLES. Some support was obtained through the Cluster of Excellence "CliSAP" (EXC177), established at the University of Hamburg and funded by the German Science Foundation (DFG). We thank the editor of TC, O. Persson, for his constructive comments.

Edited by: O. Persson

References

Affeld, B.: Zyklonen in der Arktis und ihre Bedeutung für den Eistransport durch die Framstraße, Ph.D. thesis, Meteorologisches Institut, Universität Hamburg, available at: http://ediss.sub.uni-hamburg.de/volltexte/2003/1009 (last access: 26 February 2014), 2003.

Barry, R. G. and Maslanik, J.: Arctic sea ice characteristics and associated atmosphere – ice interactions in summer inferred from SMMR sata and drifting buoys: 1979–1984, GeoJournal, 18, 35–44, 1989.

Blender, R., Fraedrich, K., and Lunkeit, F.: Identification of cyclone-track regimes in the North Atlantic, Q. J. Roy. Meteor. Soc., 123, 727–741, 1997.

Brümmer, B. and Hoeber, H.: A mesoscale cyclone over the Fram Strait and its effects on sea ice, J. Geophys. Res., 104, 19085–19098, 1999.

Brümmer, B., Müller, G., and Hoeber, H.: A Fram Strait cyclone: Properties and impact on ice drift as measured by aircraft and buoys, J. Geophys. Res., 108, 4217, doi:10.1029/2002JD002638, 2003.

Brümmer, B., Schröder, D., Müller, G., Spreen, G., Jahnke-Bornemann, A., and Launiainen, J.: Impact of a Fram Strait cyclone on ice edge, drift, divergence, and concentration: Possibilities and limits of an observational analysis, J. Geophys. Res., 113, C12003, doi:10.1029/2007JC004149, 2008.

Curry, J., Schramm, J., and Ebert, E.: Impact of clouds on the surface radiation balance of the Arctic Ocean, Meteorol. Atmos. Phys., 51, 197–217, doi:10.1007/BF01030494, 1993.

Haapala, J., Lönnroth, N., and Stössel, A.: A numerical study of open water formation in sea ice, J. Geophys. Res., 110, C09011, doi:10.1029/2003JC002200, 2005.

Haller, M.: Arktische Zyklonen und Meereisdrift während der DAMOCLES-Bojenexperimente 2007–2010, Ph.D. thesis, Meteorologisches Institut, Universität Hamburg, 212 pp., available at: http://ediss.sub.uni-hamburg.de/volltexte/2011/5283 (last access: 26 February 2014), 2011.

Harder, M.: Dynamik, Rauhigkeit und Alter des Meereises in der Arktis – numerische Untersuchungen mit einem großskaligen Modell, Ph.D. thesis, Berichte zur Polarforschung, 203, Univ. Bremen, 129 pp., 1996.

Hibler, III., W.: A dynamic thermodynamic sea ice model, J. Phys. Oceanogr., 9, 815–846, 1979.

Hibler III., W.: The role of sea ice dynamics in modeling CO_2 increases., Climate Processes and Climate Sensitivity, Geoph. Monog. Series, 29, 238–253, 1984.

Holt, B. and Martin, S.: The effect of a storm on the 1992 summer sea ice cover of the Beaufort, Chukchi, and East Siberian Seas, J. Geophys. Res., 106, 1017–1032, 2001.

Intrieri, J. M., Fairall, C. W., Shupe, M. D., Persson, P. O. G., Andreas, E. L., Guest, P. S., and Moritz, R. E.: An annual cycle of Arctic surface cloud forcing at SHEBA, J. Geophys. Res.-Oceans, 107, 8039, doi:10.1029/2000JC000439, 2002.

Jahnke-Bornemann, A.: Zyklonen im Europäischen Nordmeer und Zusammenhänge mit der nordatlantischen Tiefdruckrinne, Ph.D. thesis, Meteorologisches Institut, Universität Hamburg, available at: http://ediss.sub.uni-hamburg.de/volltexte/2010/4494 (last access: 26 February 2014), 2010.

Jahnke-Bornemann, A. and Brümmer, B.: The Iceland-Lofotes pressure difference: different states of the North Atlantic low-pressure zone, Tellus A, 61, 466–475, 10.1111/j.1600-0870.2009.00401.x, 2009.

Karcher, M., Kauker, F., Gerdes, R., Hunke, E., and Zhang, J.: On the dynamics of Atlantic Water circulation in the Arctic Ocean., J. Geophys. Res., 112, C04S02, doi:10.1029/2006JC003630, 2007.

Kawaguchi, Y. and Mitsudera, H.: A numerical study of ice-drift divergence by cyclonic wind with a Lagrangian ice model, Tellus A, 60, 789–802, doi:10.1111/j.1600-0870.2008.00321.x, 2008.

Köberle, C. and Gerdes, R.: Mechanisms Determining the Variability of Arctic Sea Ice Conditions and Export, J. Climate, 16, 2843–2858, doi:10.1175/1520-0442(2003)016<2843:MDTVOA>2.0.CO;2, 2003.

Kriegsmann, A.: Einfluss von Zyklonen auf das Meereis in der zentralen Arktis: Modellstudien und Beobachtungen, Ph.D. thesis, Universität Hamburg, Von-Melle-Park 3, 20146 Hamburg, available at: http://ediss.sub.uni-hamburg.de/volltexte/2011/5257 (last access: 16 February 2014), 2011.

Lammert, A., Brümmer, B., and Kaleschke, L.: Observation of cyclone-induced inertial sea-ice oscillation in Fram Strait, Geophys. Res. Lett., 36, L10503, doi:10.1029/2009GL037197, 2009.

Leppäranta, M.: The Drift of Sea Ice, Springer Verlag, Berlin, Germany, 266 pp., 2005.

Martin, T. and Gerdes, R.: Sea ice drift variability in Arctic Ocean Model Intercomparison Project models and observations., J. Geophys. Res., 112, C04S10, doi:10.1029/2006JC003617, 2007.

Maslanik, J. A. and Barry, R.: Short-term interactions between atmospheric synoptic conditions and sea-ice behaviour in the Arctic, Ann. Glaciol., 12, 113–117, 1989.

Maslanik, J. A., Fowler, C., Heinrichs, J., Barry, R., and Emery, W. J.: Remotely-sensed and simulated variability of Arctic sea-ice concentrations in response to atmospheric synoptic systems, Int. J. Remote Sens., 16, 3325–3342, 1995.

Murray, R. J. and Simmonds, I.: A numerical scheme for tracking cyclone centers from digital data. Part I: Development and operation of the scheme., Aust. Meteorol. Mag., 39, 155–166, 1991.

Notz, D. and Marotzke, J.: Observations reveal external driver for Arctic sea-ice retreat, Geophys. Res. Lett., 39, L08502, doi:10.1029/2012GL051094, 2012.

Overland, J. E. and Turet, P.: Variability of the atmospheric energy flux across 70° N computed from the GFDL data set, in: Nansen Centennial Volume, Geophys. Monogr., 84, 313–325, American Geophysical Union, 1994.

Perovich, D. K., Richter-Menge, J. A., Jones, K. F., and Light, B.: Sunlight, water, and ice: Extreme Arctic sea ice melt during the summer of 2007, Geophys. Res. Lett., 35, L11501, doi:10.1029/2008GL034007, 2008.

Rozman, P., Hölemann, J. A., Krumpen, T., Gerdes, R., Köberle, C., Lavergne, T., Adams, S., and Girard-Ardhuin, F.: Validating satellite derived and modelled sea-ice drift in the Laptev Sea with in situ measurements from the winter of 2007/08, Polar Res., 30, 7218, doi:10.3402/polar.v30i0.7218, 2011.

Screen, J. A., Simmonds, I., and Keay, K.: Dramatic interannual changes of perennial Arctic sea ice linked to abnormal summer storm activity, J. Geophys. Res., 116, D15105, doi:10.1029/2011JD015847, 2011.

Sepp, M. and Jaagus, J.: Changes in the activity and tracks of Arctic cyclones, Clim. Change, 577–595, 2011.

Serreze, M. and Barry, R. G.: The Arctic Climate System, Cambridge University Press, 385 pp., 2005.

Serreze, M. C. and Barrett, A. P.: The Summer Cyclone Maximum over the Central Arctic Ocean, J. Climate, 21, 1048–1065, 2008.

Simmonds, I. and Rudeva, I.: The great Arctic cyclone of August 2012, Geophys. Res. Lett., 39, L23709, doi:10.1029/2012GL054259, 2012.

Sorteberg, A. and Walsh, J.: Seasonal cyclone variability at 70° N and its impact on moisture transport into the Arctic, Tellus A, 60, 570–586, doi:10.1111/j.1600-0870.2008.00314.x, 2008.

Spreen, G., Kaleschke, L., and Heygster, G.: Sea ice remote sensing using AMSR-E 89-GHz channels, J. Geophys. Res., 113, C02S03, doi:10.1029/2005JC003384, 2008.

Stroeve, J. C., Kattsov, V., Barrett, A., Serreze, M., Pavlova, T., Holland, M., and Meier, W. N.: Trends in Arctic sea ice extent from CMIP5, CMIP3 and observations, Geophys. Res. Lett., 39, L16502, doi:10.1029/2012GL052676, 2012.

Thorndike, A. and Colony, R.: Sea Ice Motion in Response to Geostrophic Winds, J. Geophys. Res., 87, 5845–5852, 1982.

Ulbrich, U., Leckebusch, G., and Pinto, J.: Extra-tropical cyclones in the present and future climate: a review, Theor. Appl. Climatol., 96, 117–131, 2009.

Uotila, P., Vihma, T., Pezza, A. B., Simmonds, I., Keay, K., and Lynch, A. H.: Relationships between Antarctic cyclones and surface conditions as derived from high-resolution numerical weather prediction data, J. Geophys. Res., 116, D07109, doi:10.1029/2010JD015358, 2011.

Zhang, J., Lindsay, R., Schweiger, A., and Steele, M.: The impact of an intense summer cyclone on 2012 Arctic sea ice retreat, Geophys. Res. Lett., 40, 720–726, doi:10.1002/grl.50190, 2013.

Zhang, X., Walsh, J., Zhang, J., Bhatt, U., and Ikeda, M.: Climatology and Interannual Variability of Arctic Cyclone Activity: 1948–2002, J. Climate, 17, 2300–2317, 2004.

Zwally, H. J. and Walsh, J. E.: Comparison of observed and modeled ice motion in the Arctic Ocean, Ann. Glaciol., 9, 136–144, 1987.

3

Brief Communication: Further summer speedup of Jakobshavn Isbræ

I. Joughin[1], B. E. Smith[1], D. E. Shean[1,2], and D. Floricioiu[3]

[1]Polar Science Center, Applied Physics Lab, University of Washington, 1013 NE 40th St., Seattle, WA 98105-6698, USA
[2]Earth & Space Sciences, University of Washington, Seattle, WA 98195, USA
[3]German Aerospace Center (DLR), Remote Sensing Technology Institute, SAR Signal Processing, Muenchenerstr. 20, 82230 Wessling, Germany

Correspondence to: I. Joughin (ian@apl.washington.edu)

Abstract. We have extended the record of flow speed on Jakobshavn Isbræ through the summer of 2013. These new data reveal large seasonal speedups, 30 to 50 % greater than previous summers. At a point a few kilometres inland from the terminus, the mean annual speed for 2012 is nearly three times as great as that in the mid-1990s, while the peak summer speeds are more than a factor of four greater. These speeds were achieved as the glacier terminus appears to have retreated to the bottom of an over-deepened basin with a depth of $\sim$ 1300 m below sea level. The terminus is likely to reach the deepest section of the trough within a few decades, after which it could rapidly retreat to the shallower regions $\sim$ 50 km farther upstream, potentially by the end of this century.

1 Introduction

The speeds of many of Greenland glaciers have varied dramatically over the last two decades (Howat et al., 2008; Moon et al., 2012), which has contributed to the ice sheet's increasingly negative mass imbalance (Rignot and Kanagaratnam, 2006; Shepherd et al., 2012; van den Broeke et al., 2009). Nowhere are such changes more evident than on Greenland's fastest glacier, Jakobshavn Isbræ (Fig. 1), which sped up more than twofold over the last decade and a half (Joughin et al., 2012a). After a period of multi-decadal relative stability (Pelto et al., 1989), speedup began in the late 1990s when Jakobshavn Isbræ's floating ice tongue began to weaken and break up (Joughin et al., 2004; Luckman and Murray, 2005; Thomas et al., 2003), likely in response to increased basal melting (Holland et al., 2008; Motyka et al., 2011) and to weakened ice mélange in the fjord (Amundson et al., 2010; Joughin et al., 2008). Since the loss of this ice tongue, the glacier's speed has varied, seasonally slowing down with terminus advance in winter and speeding up with terminus retreat in summer (Joughin et al., 2012a). This seasonal variation of the terminus position may be driven by seasonal changes in the rigidity of the ice mélange, which appear to reduce winter calving (Amundson et al., 2010). Over the past several years, the speedup has gradually increased and migrated inland due to a number of feedbacks as the glacier has thinned and retreated (Joughin et al., 2012a; Van der Veen et al., 2011). Largely as a consequence of this speedup, Jakobshavn Isbræ alone has contributed nearly 1 mm to global sea level over the period from 2000 to 2011 (Howat et al., 2011).

A record of Jakobshavn Isbræ's variation in speed from the mid-1990s through mid-2011 was published recently (Joughin et al., 2012a). Since then, the summer speedups in 2012 and 2013 were in excess of those observed in previous summers. Thus, here we provide an extension to the record of flow speed to include events through 2013.

2 Results

Since 2009, we have mapped the speed of Jakobshavn Isbræ regularly using data from the German Space Agency's (DLR) TerraSAR-X synthetic aperture radar (SAR). To do this, we applied a set of well-established speckle-tracking techniques

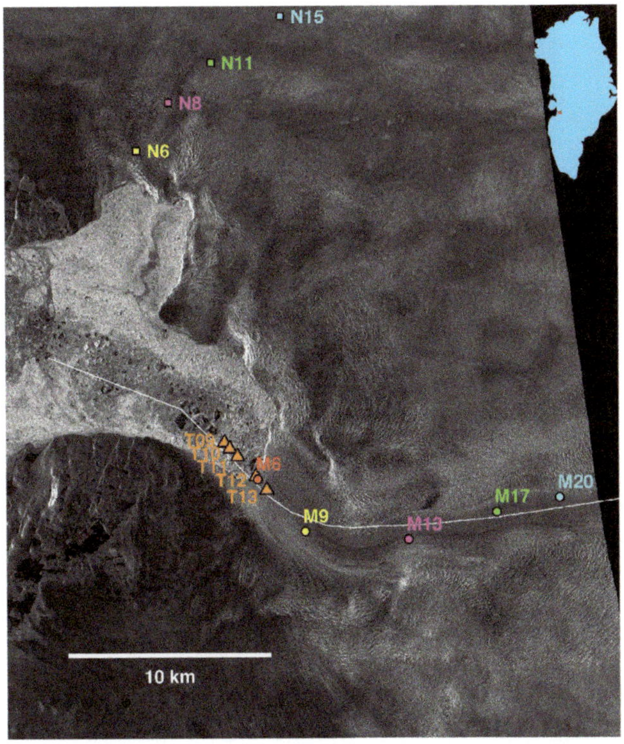

Fig. 1. TerraSAR-X image acquired 20 September when the terminus was near the point of maximum retreat in the summer of 2013. Markers M6–M20 and T09–T13 show the locations of points plotted in Fig. 2. The white profile indicates the location of the profile plotted in Fig. 3. TerraSAR-X image copyright DLR, 2013.

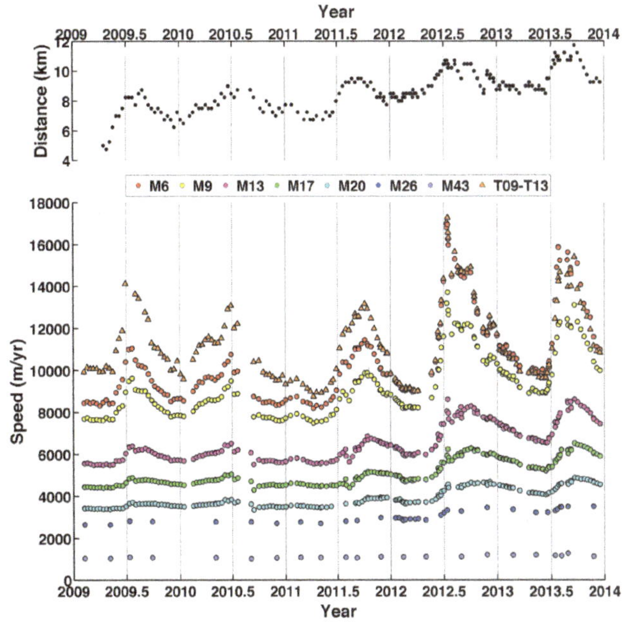

Fig. 2. Plots of (top) terminus position and (bottom) speed through time for Jakobshavn Isbræ determined from TerraSAR-X data collected from 2009 to 2013. Terminus position was digitized where it intersects the white profile shown in Fig. 1. The colour circles (M6–M43) show the speed at several points along the glacier's main trunk at the locations shown in Fig. 1 (M26 and M43 locations not shown; see Joughin et al., 2008). Each point's numerical designation (e.g. M6) gives the approximate distance in kilometres from glacier terminus in late summer 2003 and these points are used for consistency with earlier records (Joughin et al., 2008, 2012b). Additional markers, T09–T13 (orange triangles) (locations shown in Fig. 1), are each situated 1 km upstream of the terminus at its position of maximum retreat for the years 2009–2013. Each year, speeds are plotted for the corresponding point (T09–T13).

(Joughin, 2002) to pairs of TerraSAR-X images separated by 11 days. Figure 2 shows a time series of speeds extracted from these velocity maps. These speeds typically have slope-dependent errors of up to $\sim 3\,\%$, which are also a function of the particular imaging geometry. One complication with Jakobshavn Isbræ is that surface elevations near the terminus are lowering rapidly ($\sim 15\,\mathrm{m\,a^{-1}}$) (Joughin et al., 2008; Krabill et al., 2004), which can increase slope errors and introduce geolocation errors. To reduce such errors, we used an updated digital elevation model (DEM) for each calendar year determined using a combination of ASTER-stereo (Howat et al., 2014), WorldView-stereo, and TanDEM-X elevation data.

The speeds plotted in Fig. 2 are from the locations along the glacier's main trunk that are shown in Fig. 1 (M26 and M43 locations not shown). The coloured circles correspond to fixed points such that the name (e.g. M6) indicates the distance from the late 2003 ice front. Because these points are fixed in space, the changes in speed reflect both the influence of proximity to the terminus and variation in terminus thickness, both of which vary with terminus advance and retreat. To help separate these effects, we also plotted the speed at a point (T09–T13) 1 km behind the location of where the terminus reaches its point of maximum summer retreat for the corresponding calendar year. As a result, speed at these points is largely influenced by near-terminus conditions (e.g. thickness) rather than by diminishing proximity to the terminus.

Figure 2 shows the change in speed on Jakobshavn Isbræ since 2009, extending a satellite-derived record that reaches back to 1992 (see Fig. S1 in the Supplement) (Joughin et al., 2012a). Although our focus is on the main branch, for completeness, Fig. S2 provides a record of variation in speed along the north branch (points N6–N15 in Fig. 1). Following the near doubling of speed near the terminus of the main branch in the late 1990s to early 2000s (Joughin et al., 2004; Luckman and Murray, 2005), Jakobshavn Isbræ sped up more moderately at rates of 2.6–4.4 % per year from 2004 to 2011, coincident with a strong seasonal variation in speed (Joughin et al., 2012a). Our data show that in the last two years this pattern has altered, beginning with the increase in the peak summer speed at M6 by 50 % from 2011 to 2012 (11 300 to $17\,000\,\mathrm{m\,a^{-1}}$). Some of this change can be attributed to the terminus having a greater influence on speed as it moves increasingly close to M6 each summer (< 1 km in 2013). Just above the terminus (orange triangles Fig. 2), peak speeds increased by 31 % from summer 2011 to 2012 (13 300 to 17 100 m a^{-1}). From 2012 to 2013, peak summer speeds near the terminus (M6 & M9) appear to have declined slightly, but increased at points farther inland (M13–M46). The TerraSAR-X satellite was inoperative for a brief period in early August 2013, so that we missed acquisitions near the time of the 2013 peak. As a result and because the peak in 2012 was brief, we cannot rule out the possibility of a similarly brief peak in 2013 with a similar or even greater magnitude than 2012. The summer 2012 peak at M6 represents a 420 % increase in speed relative to the 1992 value, which corresponds to a period with little observed seasonal variation (Echelmeyer and Harrison, 1990; Pelto et al., 1989). Thus, a more direct comparison is that of the mean annual speed at M6 in 2012 (11 600 m a^{-1}), which yields a 1992 to 2012 speedup of 286 %. Winter near terminus speeds (T9–T13) show little increase.

To evaluate terminus retreat, for each TerraSAR-X image, we digitized the location where the terminus intersects the white profile shown in Fig. 1 and plotted the results in Fig. 2 (top). Since there are geolocation errors associated with rapidly changing topography as described above, this yields position errors of ±100 m. While terminus position often appears to influence terminus speed (Howat et al., 2008), of greater importance is whether the terminus is retreating into deeper water (Howat et al., 2005; Thomas, 2004). To examine the relationship of retreat to surface and bed geometry, Fig. 3 illustrates the glacier geometry, corresponding to the white profile shown in Fig. 1, along with the position of the terminus through time. We have aligned this profile to follow the deepest part of the gridded bed map, which differs from the points where we have plotted speeds.

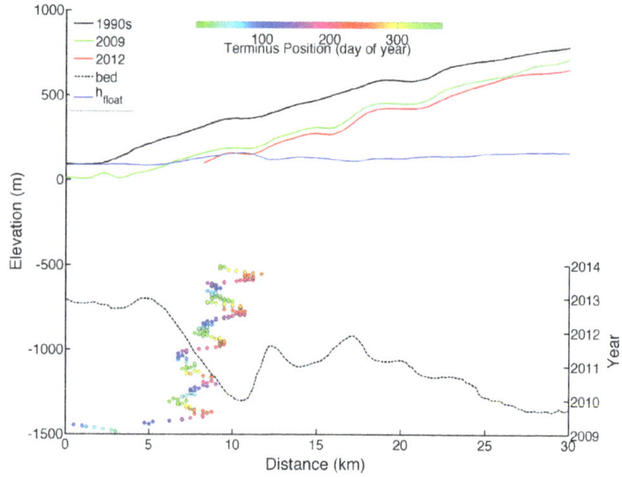

Fig. 3. Surface and bed elevations in the near-terminus region of Jakobshavn Isbræ along the profile shown in Fig. 1. Terminus position (x axis) is shown as a function of time (right y axis) with colour to indicate day of year (see also same data in Fig. 2). Surface elevations were determined (Joughin et al., 2012a) by interpolating data collected by NASA's Airborne Topographic Mapper (ATM) in the 1990s, 2009, and 2012 as part of Operation IceBridge and its predecessor missions (Krabill et al., 2004). Bed elevations were interpolated from a gridded map of radar depth soundings produced by the Center for Remote Sensing of Ice Sheets (CReSIS)(Li, 2009; Van der Veen et al., 2011). Multiple versions of the DEM exist, but based on comparison with other data sets our preferred version is the one located at (ftp://data.cresis.ku.edu/data/grids/old_format/2008_Jakobshavn.zip).

3 Discussion

As a terminus with little or no floating extension advances and retreats, its thickness and height above flotation vary. Several studies indicate that this variation exerts a strong influence on speed within several ice thicknesses of the calving front at time scales ranging from minutes to years (Thomas, 2004; Howat et al., 2005; Nettles et al., 2008; Nick et al., 2009; Joughin et al., 2012b). This means that as the terminus retreats into deeper water, the pressure boundary condition at the near-vertical terminus face produces a force that must be balanced upstream by longitudinal stress gradients, which are produced through increased stretching (i.e. speedup). This stretching should produce thinning that contributes to further retreat. In response, additional feedbacks (e.g. evolving driving stress, changes in basal water pressure, and margin softening) contribute to the overall variation in a glacier's speed as its geometry evolves in response (Joughin et al., 2012a; Van der Veen et al., 2011).

Analysis of data from 2009 indicates that the forces associated with the terminus depth variation and height above flotation account for most of Jakobshavn Isbræ's seasonal flow variation (Joughin et al., 2012a). Such results are consistent with the large summer speedups in 2012 and 2013 when the terminus appears to have reached the bottom of an overdeepened basin (Fig. 3), which occurred after the terminus retreated more than a kilometre farther inland than previous summers. While the correspondence between seasonally varying terminus position and speed is relatively strong, there are some notable differences. For instance, peaks in speed tend to be sharper in time than corresponding peaks in terminus retreat, which, in addition to the nonlinearity of the ice dynamics, likely reflects the glacier's rapid evolution. As an example, if the terminus maintains roughly the same location for a period of weeks (e.g. relatively flat peak in late summer 2011), then the terminus should continue to thin by several metres over the same period. Since speed is sensitive to height above flotation, this thinning is likely to reduce the peak speed even with no change in terminus position.

Over the 5 yr period of our observations, the correspondence between terminus depth and speed is less clear. The greatest mean annual speeds do occur in 2012 when the terminus is near what appears to be an overdeepening and subsequently decline in 2013 as the terminus retreats to higher ground. By contrast, the 2009 mean speed was faster (11 130 m a^{-1}) than the 2011 (10 590 m a^{-1}), despite a more advanced 2009 terminus position. This difference likely reflects the fact that at these time scales, the other processes and feedbacks mentioned above have a substantial influence on flow (Joughin et al., 2012a; Van der Veen et al., 2011).

If Jakobshavn Isbræ's terminus has reached the bottom of an overdeepened region, then the terminus may be able to find a position of transient stability on the high spot farther upstream (located at ~ 12 to 17 km in Fig. 3) as retreat to shallower depths yields slower speeds (Joughin et al., 2012a). The relatively high surface slope region above the basin, where the present heights are tens to hundreds of metres above flotation, may further slow retreat, since it should take more time to thin to near flotation. By contrast, low surface slopes and heights near flotation (Fig. 3) likely facilitated the rapid retreat since 2009. While the high spot above the basin may slow flow, the terminus would still be grounded on a bed at least 900 m below sea level, likely yielding speeds well above balance that would maintain strong, although potentially diminished, thinning. As a consequence, the terminus likely will continue to retreat, albeit perhaps more slowly in the near term, until it again reaches bed depths similar to summer 2012 (~ 1300 m below sea level) at ~ 15 km farther upstream (Fig. 3). As the large drawdown since the 1990s indicates, such a retreat could happen over the span of a few decades or less (Joughin et al., 2012a). At the current stage, further retreat may largely be driven by ice dynamics with only weak coupling to climate forcing. Thus, a re-advance rather than retreat likely would require a period of extended (several years to decades) cooling.

Once past the high spot, the trough extends roughly 50 km farther inland at depths below sea level of ~ 1200 m and greater before eventually reaching shallower depths. Thus, although there is some variation in trough width (~ 3.5 to 5.5 km) that could modulate the rate of flow, once into this deepest part of the trough, extreme velocities (> 12 000 m a^{-1}) are likely to persist as the terminus rapidly retreats (Thomas et al., 2011). Furthermore, without the ability to seasonally advance up a relatively steep bed slope as in the past several winters, such high speeds may be sustained year round. Because the deep trough of Jakobshavn is extremely difficult to measure with conventional techniques (Li, 2009), we cannot rule out that some or all of the high spot might be a gridding artifact, in which case retreat may occur even more rapidly. Similarly, we cannot rule out that some bed highs may not have been resolved that could provide additional points of transient stability.

The transient summer speeds we observe for 2012 (> 17 000 m a^{-1}) appear to represent the fastest observed speed for any outlet glacier or ice stream in Greenland or Antarctica. This yields a transient peak speed a factor of four greater than the speeds observed in the 1990s, while the mean annual speedup is by just under a factor of three. If, as the glacier recedes up the trough, it is able to maintain the peak speeds year round, then a sustained speedup of the terminus by a factor of four or five is conceivable based on recent behaviour, which is about half of the nominal tenfold upper limit on the amount a glacier could speed up by that was proposed by Pfeffer et al. (2008). Nevertheless, these speeds would occur in a trough roughly twice as deep as prior to the speedup. Hence, a tenfold increase in ice flux may be possible for Jakobshavn Isbræ if the trough does not narrow substantially with distance upstream. Equivalently, while the increase in terminus speed and the glacier's overall maximum speed may remain under a factor of five, as the terminus retreats farther inland where the speeds now are comparatively slow, the relative speedup is much greater (e.g. if the terminus retreated to M26 with a speed of 16 000 m a^{-1}, this would represent a twelvefold speedup). Thinning by hundreds of metres to a terminus near flotation, however, would yield something closer to a tenfold flux increase (~ 8.5 based on simple scaling by the ratio of thickness at flotation to thickness in 1990s). It is unlikely that such retreat could be sustained for more than a few decades because the terminus would rapidly retreat ~ 50 km to shallower depths (Joughin et al., 2012a).

4 Conclusions

Our results show that Jakobshavn Isbræ has accelerated to speeds unprecedented in its observational record as its terminus has retreated to a region where the bed is ~ 1300 m below sea level. While the current increase in annual discharge flux remains less than a factor of three, the increase plausibly could reach or exceed a factor of 10 within decades. This is a consequence of the fact that retreat into deeper water increases both speed and thickness of the terminus. Conversely, where retreat to shallower depths occurs, losses will be far more moderate. Hence, a tenfold increase in discharge is likely only to be sustained in the few decades before rapid thinning would cause the terminus to retreat out of the deep trough. Thus, the potential for large losses from Greenland is likely to be determined by the depth and inland extent of the troughs through which its outlet glaciers drain. These features are only beginning to be well resolved by international efforts such as NASA's Operation IceBridge. The relatively sparse data collected thus far indicate that, with its great depths and inland extent, Jakobshavn's Isbræ is somewhat unique (Bamber et al., 2013), suggesting that it may be difficult for the majority of Greenland's outlet glaciers to produce or to sustain such large increases in ice discharge.

Acknowledgements. The contribution to this work by I. Joughin was supported National Science Foundations Office of Polar Programs (NSF-OPP) through the Center for Remote Sensing of Ice Sheets (CReSIS) (NSF ANT-0424589). The National Aeronautics and Space Administration's (NASA) Earth Science Division (NNX09AE47G) supported the contribution to this work by B. E. Smith. Support for the SAR data processing was from NASA's MEaSUREs Program (NNX08AL98A and NNX13AI21A). The TerraSAR-X and TanDEM-X data were provided by the German Space Agency (DLR), project HYD0754 and XTI_GLAC0400. The bed elevation data were provided by CRESIS and the surface elevation data were provided by NASA's Airborne Topographic Mapper (ATM) Program. NASA's Operation IceBridge and its predecessor campaigns provided the instrument platform for both the bed and elevation data collection. Comments by J. Bamber, P. Christoffersen, A. Luckman, M. Maki, M. Pelto, and an anonymous reviewer improved the revised manuscript.

Edited by: J. L. Bamber

References

Amundson, J. M., Fahnestock, M., Truffer, M., Brown, J., Luethi, M. P., and Motyka, R. J.: Ice melange dynamics and implications for terminus stability, Jakobshavn Isbrae Greenland, J. Geophys. Res.-Earth, 115, F01005, doi:10.1029/2009JF001405, 2010.

Bamber, J. L., Griggs, J. A., Hurkmans, R. T. W. L., Dowdeswell, J. A., Gogineni, S. P., Howat, I., Mouginot, J., Paden, J., Palmer, S., Rignot, E., and Steinhage, D.: A new bed elevation dataset for Greenland, The Cryosphere, 7, 499–510, doi:10.5194/tc-7-499-2013, 2013.

Echelmeyer, K. and Harrison, W. D.: Jakobshavns Isbræ, West Greenland: Seasonal variations in velocity-or lack thereof, J. Glaciol., 36, 82–88, 1990.

Holland, D. M., Thomas, R. H., De Young, B., Ribergaard, M. H., and Lyberth, B.: Acceleration of Jakobshavn Isbrae triggered by warm subsurface ocean waters, Nat. Geosci., 1, 659–664, doi:10.1038/ngeo316, 2008.

Howat, I. M., Joughin, I., Tulaczyk, S., and Gogineni, S.: Rapid retreat and acceleration of Helheim Glacier, east Greenland, Geophys. Res. Lett., 32, L22502, doi:10.1029/2005GL024737, 2005.

Howat, I. M., Joughin, I., Fahnestock, M., Smith, B. E., and Scambos, T. A.: Synchronous retreat and acceleration of southeast Greenland outlet glaciers 2000–06: ice dynamics and coupling to climate, J. Glaciol., 54, 646–660, doi:10.3189/002214308786570908, 2008.

Howat, I. M., Ahn, Y., Joughin, I., Van Den Broeke, M. R., Lenaerts, J. T. M., and Smith, B.: Mass balance of Greenland's three largest outlet glaciers, 2000–2010, Geophys. Res. Lett., 38, L12501, doi:10.1029/2011GL047565, 2011.

Howat, I. M., Negrete, A., and Smith, B. E.: The Greenland Ice Mapping Project (GIMP) land classification and surface elevation datasets, The Cryosphere Discuss., 8, 453–478, doi:10.5194/tcd-8-453-2014, 2014.

Joughin, I.: Ice-sheet velocity mapping: A combined interferometric and speckle-tracking approach, Ann. Glaciol., 34, 195–201, 2002.

Joughin, I., Abdalati, W., and Fahnestock, M.: Large fluctuations in speed on Greenland's Jakobshavn Isbræ glacier, Nature, 432, 608–610, doi:10.1038/nature03130, 2004.

Joughin, I., Howat, I. M., Fahnestock, M., Smith, B., Krabill, W., Alley, R. B., Stern, H., and Truffer, M.: Continued evolution of Jakobshavn Isbrae following its rapid speedup, J. Geophys. Res.-Earth, 113, F04006, doi:10.1029/2008JF001023, 2008.

Joughin, I., Smith, B. E., Howat, I. M., Floricioiu, D., Alley, R. B., Truffer, M., and Fahnestock, M.: Seasonal to decadal scale variations in the surface velocity of Jakobshavn Isbrae, Greenland: Observation and model-based analysis, J. Geophys. Res., 117, F02030, doi:10.1029/2011JF002110, 2012a.

Joughin, I., Alley, R. B., and Holland, D. M.: Ice-sheet response to oceanic forcing, Science, 338, 1172–1176, doi:10.1126/science.1226481, 2012b.

Krabill, W., Hanna, E., Huybrechts, P., Abdalati, W., Cappelen, J., Csatho, B., Frederick, E., Manizade, S., Martin, C., Sonntag, J., Swift, R., Thomas, R., and Yungel, J.: Greenland Ice Sheet: Increased coastal thinning, Geophys. Res. Lett., 31, L24402, doi:10.1029/2004GL021533, 2004.

Li, J.: Mapping of ice sheet deep layers and fast outlet glaciers with multi-channel-high-sensitivity radar, University of Kansas, Lawrence, 2009.

Luckman, A. and Murray, T.: Seasonal variation in velocity before retreat of Jakobshavn Isbrae, Greenland, Geophys Res Lett, 32, L08501, doi:10.1029/2005GL022519, 2005.

Moon, T., Joughin, I., Smith, B., and Howat, I.: 21st-Century evolution of Greenland outlet glacier velocities, Science, 336, 576–578, doi:10.1126/science.1219985, 2012.

Motyka, R. J., Truffer, M., Fahnestock, M., Mortensen, J., Rysgaard, S., and Howat, I.: Submarine melting of the 1985 Jakobshavn Isbrae floating tongue and the triggering of the current retreat, J. Geophys. Res.-Earth, 116, F01007, doi:10.1029/2009JF001632, 2011.

Nettles, M., Larsen, T. B., Elósegui, P., and Hamilton, G. S.: Step-wise changes in glacier flow speed coincide with calving and glacial earthquakes at Helheim Glacier, Greenland, Geophys. Res. Lett, 35, L24503, doi:10.1029/2008GL036127, 2008.

Nick, F. M., Vieli, A., Howat, I. M., and Joughin, I.: Large-scale changes in Greenland outlet glacier dynamics triggered at the terminus, Nat. Geosci., 2, 110–114, doi:10.1038/NGEO394, 2009.

Pelto, M. S., Hughes, T. J., and Brecher, H. H.: Equilibrium state of Jakobshavns Isbræ, West Greenland, Ann. Glaciol., 1989.

Pfeffer, W. T., Harper, J. T., and O'Neel, S.: Kinematic constraints on glacier contributions to 21st-century sea-level rise, Science, 321, 1340–1343, doi:10.1126/science.1159099, 2008.

Rignot, E. and Kanagaratnam, P.: Changes in the velocity structure of the Greenland ice sheet, Science, 311, 986–990, doi:10.1126/science.1121381, 2006.

Shepherd, A., Ivins, E. R., Geruo, A., Barletta, V. R., Bentley, M. J., Bettadpur, S., Briggs, K. H., Bromwich, D. H., Forsberg, R., Galin, N., Horwath, M., Jacobs, S. S., Joughin, I., King, M. A., Lenaerts, J. T. M., Li, J., Ligtenberg, S. R. M., Luckman, A., Luthcke, S. B., McMillan, M., Meister, R., Milne, G., Mouginot, J., Muir, A., Nicolas, J. P., Paden, J., Payne, A. J., Pritchard, H., Rignot, E., Rott, H., Sørensen, L. S., Scambos, T. A., Scheuchl, B., Schrama, E. J. O., Smith, B., Sundal, A. V., van Angelen, J. H., van de Berg, W. J., van den Broeke, M. R., Vaughan, D. G., Velicogna, I., Wahr, J., Whitehouse, P. L., Wingham, D. J., Yi, D., Young, D., and Zwally, H. J.: A reconciled estimate of ice-sheet mass balance, Science, 338, 1183–1189, doi:10.1126/science.1228102, 2012.

Thomas, R. H.: Force-perturbation analysis of recent thinning and acceleration of Jakobshavn Isbræ, Greenland, J. Glaciol., 50, 57–66, 2004.

Thomas, R. H., Abdalati, W., Frederick, E., Krabill, W., Manizade, S., and Steffen, K.: Investigation of surface melting and dynamic thinning on Jakobshavn Isbrae, Greenland, J. Glaciol., 49, 231–239, 2003.

Thomas, R., Frederick, E., Li, J., Krabill, W., Manizade, S., Paden, J., Sonntag, J., Swift, R., and Yungel, J.: Accelerating ice loss from the fastest Greenland and Antarctic glaciers, Geophys. Res. Lett., 38, L10502, doi:10.1029/2011GL047304, 2011.

van den Broeke, M., Bamber, J., Ettema, J., Rignot, E., Schrama, E., van de Berg, W. J., van Meijgaard, E., Velicogna, I., and Wouters, B.: Partitioning recent Greenland mass loss, Science, 326, 984–986, doi:10.1126/science.1178176, 2009.

Van der Veen, C. J., Plummer, J. C., and Stearns, L. A.: Controls on the recent speed-up of Jakobshavn Isbrae, West Greenland, J. Glaciol., 57, 770–782, 2011.

Weertman, J.: Stability of the junction of an ice sheet and an ice shelf, J. Glaciol., 13, 3–11, 1974.

4

Recent summer Arctic atmospheric circulation anomalies in a historical perspective

A. Belleflamme, X. Fettweis, and M. Erpicum

Laboratory of Climatology, Department of Geography, University of Liège, Allée du 6 Août, 2, 4000 Liège, Belgium

Correspondence to: A. Belleflamme (a.belleflamme@ulg.ac.be)

Abstract. A significant increase in the summertime occurrence of a high pressure area over the Beaufort Sea, the Canadian Arctic Archipelago, and Greenland has been observed since the beginning of the 2000s, and particularly between 2007 and 2012. These circulation anomalies are likely partly responsible for the enhanced Greenland ice sheet melt as well as the Arctic sea ice loss observed since 2007. Therefore, it is interesting to analyse whether similar conditions might have happened since the late 19th century over the Arctic region. We have used an atmospheric circulation type classification based on daily mean sea level pressure and 500 hPa geopotential height data from five reanalysis data sets (ERA-Interim, ERA-40, NCEP/NCAR, ERA-20C, and 20CRv2) to put the recent circulation anomalies in perspective with the atmospheric circulation variability since 1871. We found that circulation conditions similar to 2007–2012 have occurred in the past, despite a higher uncertainty of the reconstructed circulation before 1940. For example, only ERA-20C shows circulation anomalies that could explain the 1920–1930 summertime Greenland warming, in contrast to 20CRv2. While the recent anomalies exceed by a factor of 2 the interannual variability of the atmospheric circulation of the Arctic region, their origin (natural variability or global warming) remains debatable.

1 Introduction

In recent years, particularly since 2007, significant atmospheric circulation anomalies have been observed over different parts of the Arctic. Based on 500 hPa geopotential height (Z500), Fettweis et al. (2013) reported a doubled frequency of summertime anticyclones centred over Greenland, representing an increased frequency of negative NAO (North Atlantic Oscillation) conditions. This circulation anomaly impacts the climate of a major part of the Arctic region by favouring warm southerly air advection over western Greenland and the Canadian Arctic Archipelago, and rather cold polar flow over Svalbard and the Barents Sea. This circulation anomaly partly explains the sharply enhanced melt of the Greenland ice sheet (Tedesco et al., 2008; Hanna et al., 2009; Fettweis et al., 2013; Rajewicz and Marshall, 2014) and the stabilisation of the melt rate of the Svalbard glaciers (Moholdt et al., 2010) despite Arctic warming (Serreze et al., 2009). In the same way, Bezeau et al. (2014) have shown that the increased melt of glaciers and ice caps in the Canadian Arctic Archipelago is related to the increased occurrence of high pressure systems over this region over 2007–2012. Ballinger et al. (2014) highlighted an increase in the summertime frequency of the Beaufort Sea High over the last decade, on the basis of the mean sea level pressure (SLP). They found that this circulation anomaly is significantly anticorrelated with the Arctic Oscillation (AO) index, which has decreased over the same period (Hanna et al., 2014a). Moreover, this anomaly is simultaneous with the increased frequency of the Greenland High described above suggesting that both anomalies are linked. Finally, these circulation anomalies have been implicated in the recent Arctic sea ice extent (SIE) decrease (Wang et al., 2009; Overland et al., 2012; Matsumura et al., 2014; Stroeve et al., 2014; Simmonds, 2015). High pressure systems over the Canadian Arctic Archipelago and Greenland favour sea ice export from the Arctic Basin through the Fram Strait and the Barents Sea, which is particularly effective for sea ice loss during summer (Wang et al., 2009).

It is interesting to study whether the recent circulation anomalies are unique (and potentially caused by global warming) or if similar anomalies have already occurred during the instrumental period (since the late 19th century) due to the natural variability of the climatic system. We have put the 2007–2012 summertime atmospheric circulation anomaly over the Arctic region in perspective with the reconstructed circulation over the instrumental period. To achieve this, we have used an atmospheric circulation type classification (CTC) to distinguish the main circulation types over the Arctic region and to analyse their frequency changes over time, as done by Ballinger et al. (2014) over the Beaufort Sea, Bezeau et al. (2014) over the Canadian Arctic Archipelago, and Fettweis et al. (2013) over Greenland. Since the aim of CTCs is to group similar circulation situations together (Huth et al., 2008; Philipp et al., 2010; Käsmacher and Schneider, 2011), this methodology allows for a synthetic analysis of the atmospheric circulation over a given region at a daily scale (i.e. the characteristic timescale of synoptic circulation patterns like high pressure systems). CTCs are widely used to compare data sets (e.g. reanalyses, general circulation model outputs), to evaluate their ability to reproduce the observed atmospheric circulation, and to detect changes in the observed and projected atmospheric circulation (Bardossy and Caspary, 1990; Kyselý and Huth, 2006; Philipp et al., 2007; Anagnostopoulou et al., 2009; Demuzere et al., 2009; Pastor and Casado, 2012; Fettweis et al., 2013; Belleflamme et al., 2013, 2014). While a wide range of classifications has been developed to study the atmospheric circulation (e.g. leader-algorithm approaches, Fettweis et al., 2011; principal component analyses, Huth, 2000; optimisation algorithms, Philipp et al., 2007, including self-organising maps, Käsmacher and Schneider, 2011; Bezeau et al., 2014; Hope et al., 2014), no method can be considered as being overall better than the others (Philipp et al., 2010). Thus, we use our CTC that has been developed for the Arctic region and especially for Greenland (Fettweis et al., 2011). This CTC has already been used to compare reanalysis data sets and general circulation model outputs over Greenland with the aim of detecting circulation changes (Belleflamme et al., 2013) and to analyse temperature-related flow analogues over the Greenland ice sheet (Fettweis et al., 2013).

In this study, we apply the CTC developed by Fettweis et al. (2011) (described in Sect. 3) to daily SLP and Z500 fields from different reanalysis data sets (detailed in Sect. 2). In Sect. 4.1.1, we put in perspective the summertime circulation of 2007–2012 with the circulation variability observed since 1871. The influence of the uncertainties of the past circulation on our results is discussed in Sect. 4.1.2. After a comparison between the SLP- and the Z500-based results in Sect. 4.2, we analyse the links between the circulation type frequencies and NAO and SIE in Sect. 4.3.

2 Data

We used daily SLP and Z500 data for the summer months (JJA – June, July, and August) of five reanalysis data sets.

- The ERA-Interim reanalysis (Dee et al., 2011) from the European Centre for Medium-Range Weather Forecasts (ECMWF) (spatial resolution: $0.75^\circ \times 0.75^\circ$) over the period 1979–2014.
- The ERA-40 reanalysis from the ECMWF (Uppala et al., 2005) (spatial resolution: $1.125^\circ \times 1.125^\circ$) over the period 1958–1978 used to extend ERA-Interim. It should be noted that ERA-40 is known to have significant biases in its vertical temperature profile (Screen and Simmonds, 2011), which is used in the geopotential height calculation. However, the impact of these biases on our Z500-based results should be limited, since the most problematic year (i.e. 1997) is not included in the ERA-40 period considered here.
- The NCEP/NCAR reanalysis from the National Centers for Environmental Prediction – National Center for Atmospheric Research (Kalnay et al., 1996) (spatial resolution: $2.5^\circ \times 2.5^\circ$) over the period 1948–2014.
- The ERA-20C reanalysis from the ECMWF (Poli et al., 2013) (spatial resolution: $1.125^\circ \times 1.125^\circ$) over the period 1900–2010. The spread evaluating the uncertainty of the ERA-20C data was not yet available when conducting this study.
- The Twentieth Century Reanalysis version 2 (20CRv2) (Compo et al., 2011) from the NOAA ESRL/PSD (National Oceanic and Atmospheric Administration Earth System Research Laboratory/Physical Sciences Division) (spatial resolution: $2^\circ \times 2^\circ$) over the period 1871–2012. The 20CRv2 data are constructed as the ensemble mean of 56 runs. The standard deviation (called spread) of this ensemble mean is also given for each variable (in our case SLP and Z500). We used it to estimate the uncertainty of our 20CRv2-based results, in particular before the overlapping period with the other reanalysis data sets when the assimilated observations are sparse. In fact, the spread, and thus the uncertainty of the reconstructed atmospheric circulation in 20CRv2, strongly depends on the number of pressure observations, which is low before 1940 (Compo et al., 2011).

It is important to note that only SLP, sea surface temperature (SST), and sea ice are assimilated into 20CRv2, and SLP, SST, and oceanic near-surface air temperature and wind into ERA-20C. The other reanalyses also assimilate satellite and upper air data every 6 h. Therefore, 20CRv2 and ERA-20C are a priori less reliable than the other more constrained reanalyses.

Additionally, daily sea ice cover data from the ERA-Interim are also used over the 1980–2014 JJA period.

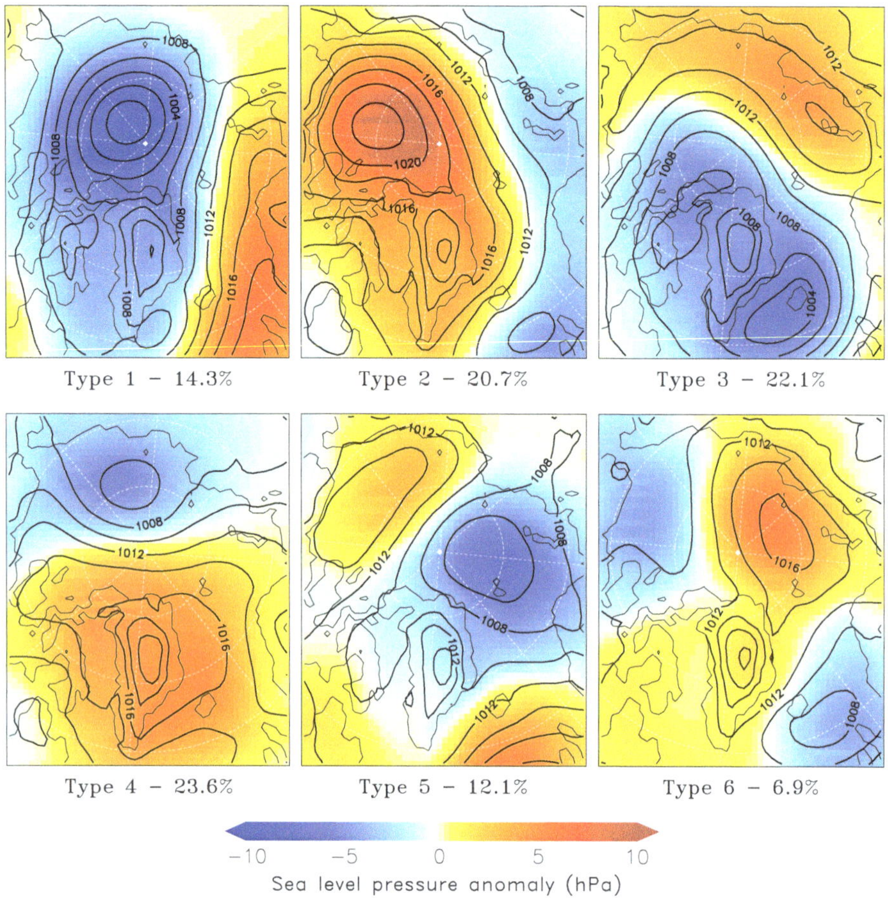

Figure 1. The SLP-based reference circulation types over the 1980–2012 (JJA) period for the ERA-Interim are represented by the solid black isobars (in hPa). The SLP anomaly (in colours) is calculated as the difference between the class mean SLP and the seasonal mean SLP over 1980–2012. The average frequency of each type is also given.

Since the reanalyses have different spatial resolutions, and to avoid the problem of decreasing pixel area near the pole when using geographic coordinates, all reanalysis outputs have been linearly interpolated to a regular grid with a spatial resolution of 100 km. Our integration domain has a size of 5000 × 6000 km and covers the whole Arctic Ocean, Greenland, and the northern part of the Atlantic Ocean (Fig. 1).

Finally, monthly NAO data over the period 1871–2013 were obtained from the Climatic Research Unit (CRU). This NAO index is defined as the normalised difference between the SLP measured in the Azores (Ponta Delgada) and Iceland (Reykjavik).

3 Method

The SLP data from the different reanalyses were compared using the automatic circulation type classification developed by Fettweis et al. (2011) and used over Greenland by Belleflamme et al. (2013) and Fettweis et al. (2013), and over Europe by Belleflamme et al. (2014). This CTC is considered a leader-algorithm method (Philipp et al., 2010), because each class is defined by a reference day and a similarity threshold. After having calculated the similarity index (see below) between all pairs of days of the data set, the day counting the most similar days (i.e. with a similarity index value above the similarity threshold) is selected as the reference day for the first type. All days considered as similar to this reference day are grouped into this type. The same procedure is repeated type by type over the remaining days of the data set. This whole process is repeated many times for various similarity thresholds in order to optimise the classification. The similarity between the days is gauged by the Spearman rank correlation coefficient. The key feature of using correlation-based similarity indices is that they are not influenced by the average SLP of a day, but only by its spatial pattern (Philipp et al., 2007). Thus, in contrary to Fettweis et al. (2011, 2013), who used the Euclidean distance as similarity index and Z500 to take into account the influence of the temperature on the upper level circulation, we used the Spearman rank correlation and SLP to focus exclusively on the circulation pattern. In order to minimise the influence of eventual temperature biases into the SLP retrieving computation, especially over elevated regions like Greenland (Lindsay et al., 2014), we only considered oceanic pixels when performing the SLP-based classification. For comparison, the same procedure was done using Z500, but all pixels of the domain were taken into account, since there is much less influence of the surface and its elevation at this level.

This CTC is automatic, meaning that the circulation types are built by the algorithm and not predefined by the user. This implies that the circulation types obtained using different data sets will be different and thus difficult to compare. To overcome this problem, we "projected" the types of a reference data set onto the other data sets, i.e. the types obtained for the reference data set were imposed as predefined types for the other data sets, as proposed by Huth (2000) and implemented by Belleflamme et al. (2013, 2014). Since the types are now the same for all data sets, they can easily be compared. Lindsay et al. (2014) compared seven reanalysis data sets (including ERA-Interim, NCEP/NCAR, and 20CRv2) over the Arctic for the 1980–2009 period and they conclude that the ERA-Interim gives the best results for various variables (e.g. SLP, T2M, wind speed) compared with observations. Thus, we used the ERA-Interim data set over the 1980–2012 period, which is common to all reanalyses used here (except ERA-20C) and includes the 2007–2012 circulation anomaly as the reference data set.

As said above, the 20CRv2 reanalysis SLP data are given as an ensemble mean of 56 runs and the spread around this ensemble mean. To evaluate the uncertainty from the 20CRv2-based data estimated by this spread, we have performed 20 000 classification runs (note that using 5000 or 10 000 runs does not affect the results). For each run, the daily spread, multiplied by a factor varying randomly between −1 and 1, is added to the daily SLP. Due to the high number of runs, all multiplying factor values have equal likelihood and their average tends to zero. Thus, no systematic SLP bias is introduced in the 20 000-run ensemble. If adding or subtracting the spread implies a sufficient alteration in the SLP pattern of a given day, this day could be classified into another circulation type, compared to the run using the 20CRv2 SLP ensemble mean (called hereafter 20CRv2 reference run). The same procedure has been done for Z500.

In our CTC, the number of classes is fixed by the user. On the basis of the ERA-Interim SLP over the 1980–2012 summers (JJA), six circulation types were retained. This is the lowest number of classes needed to obtain the two patterns in which we are the most interested in (i.e. the Beaufort Sea High and the Greenland High) as well marked types. Minimising the number of types allows for a more concise and synthetic analysis. Nevertheless, the main conclusions of this study are the same when using 5, 8, or 10 types. The six obtained circulation types can be described as follows (Fig. 1): type 1 is characterised by a low pressure system centred over the Arctic Ocean and represents about 14 % of the classified days during 1980–2012. However, type 2 is marked by a high pressure located over the Beaufort Sea and the western part of the Arctic Ocean. Type 3 presents a strong Icelandic Low and higher pressure along the Russian coast. Type 4 is the Greenland High type and accounts for 24 % of the classified days. Types 5 and 6 show opposite patterns, with a low (respectively, high) pressure east of Svalbard surrounded by high (respectively, low) pressure systems. It is important to note that type 6 also contains the unclassified days, i.e. the days for which the similarity index values with regard to the reference days of all types lie below the similarity thresholds of these types (< 1 % of the classified days for the ERA-Interim reference classification).

Despite the difference between the SLP- and Z500-based circulation types, six types were also retained for Z500 (Fig. S1 in the Supplement). Type 1 is characterised by a strong depression centred over the Arctic Ocean. This depression is located further to the east in type 2, which also presents a slight ridge over Greenland. In type 3, the depression is situated over the Greenland and Svalbard region. Type 4 is marked by two depressions, the Icelandic Low and a low over the Chukchi Sea, and a ridge over the Barents and Kara seas. Type 5 combines the Greenland High and the Beaufort Sea High. Finally, type 6 shows a high pressure system over the Arctic Ocean, while the depression is split into three parts (i.e. the Icelandic Low, a low over the Kara Sea, and a low over the Canadian Arctic Archipelago).

4 Results

There is a very good agreement between the frequencies of the circulation types and their evolution over time for all reanalyses over 1958–2012, as well SLP (Fig. 2) and Z500 (Fig. S2). Nevertheless, two notable differences have to be pointed out. The first difference is a systematic overestimation of about 4–6 % of the frequency of type 3 at the expense of type 2 by 20CRv2 compared to the full constrained reanalyses (ERA-40/ERA-Interim and NCEP/NCAR) for SLP. This bias is in agreement with the findings of Lindsay et al. (2014), who report a positive SLP bias over Asia for 20CRv2 (using monthly data), since type 3 is characterised by an anticyclone over the Asian part of the domain, in contrary to type 2. The second difference is an overestimation of type 2 at the expense of type 1 and to a lesser extent of type 3 by ERA-20C over its whole period (1900–2010) compared to all the other reanalyses used here, for the SLP-based classification. This frequency bias is particularly important before 1950 compared to 20CRv2. In fact, ERA-20C overestimates SLP over the whole Arctic Ocean, and especially over the Beaufort Sea, compared to the other reanalyses (not shown). This implies that more ERA-20C days are considered as similar to type 2 (Beaufort Sea–Arctic Ocean High) at the expense of type 1 (low pressure over the Arctic Ocean) and type 3 (low pressure over Greenland and the Canadian Arctic Archipelago). The ERA-20C SLP bias is particularly impor-

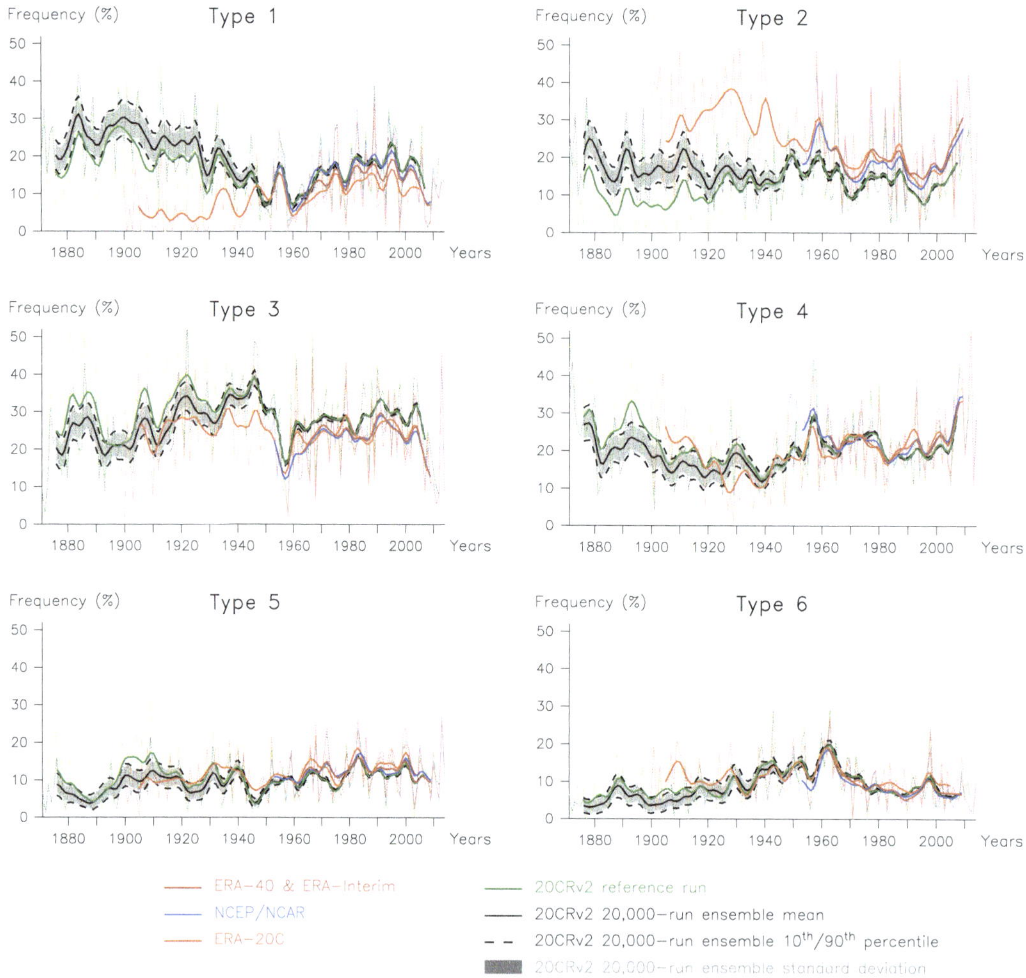

Figure 2. The dotted light lines represent the annual (JJA) SLP-based circulation type frequencies for ERA-Interim (1979–2014) and ERA-40 (1958–1978), NCEP/NCAR (1948–2014), ERA-20C (1900–2010), and the 20CRv2 reference run and the 20CRv2 20 000-run ensemble mean (1871–2012). The corresponding solid lines represent the 10-year binomial running mean frequencies. For the 20CRv2 20 000-run ensemble, the 10th and the 90th percentiles as well as the 1 standard deviation interval around the mean are also given.

tant before 1950 compared to 20CRv2, but it is still present over the last decades compared to ERA-40, ERA-Interim, and NCEP/NCAR. However, since 20CRv2 also shows systematic biases, it is not possible to consider one of these two reanalyses as more reliable than the other.

4.1 Sea level pressure

4.1.1 Circulation type frequency evolution

The frequencies of type 2 (Beaufort Sea–Arctic Ocean High) and type 4 (Greenland High) are almost twice as large during 2007–2012 compared with the 1871–2014 average (Fig. 2). These frequency anomalies are similar to those found by Ballinger and Sheridan (2014) and Ballinger et al. (2014) for the Beaufort Sea, by Fettweis et al. (2013) for Greenland, and by Bezeau et al. (2014) for the Canadian Arctic Archipelago. They are compensated by a decrease in frequency of type 3 by a factor of 2, and to a lesser extent of type 1. Both types are characterised by a low pressure system over the Arctic. Over the record, types 2 and 4 never experienced such high frequencies over several consecutive summers since 1871. Between 2007 and 2012, two summers for type 2 and five summers for type 4 presented a higher frequency than the 90th percentile frequency, meaning a return period of about 10 years (Table 1). However, 20CRv2 suggests that similar circulation type frequencies were observed before 1880; although, the uncertainty is very high over that period. Moreover, type 4 shows some summers with anomalously high frequencies between 1891 and 1896. Despite the frequency biases described above, ERA-20C shows many summers with exceptionally high frequencies of type 2 between 1923 and 1931. Thus, on the basis of ERA-20C, the anomalously warm conditions and the associated high surface mass loss rates observed over the Green-

Table 1. For the SLP-based circulation types, which show a frequency increase over the 2007–2012 period, the summers presenting a higher frequency than the 90th percentile frequency (i.e. a 10-year return period) on the basis of the 20CRv2 reference run over 1871–2012 (JJA) are listed chronologically.

Type 2	Type 4
1873	1871
1877	1872
1879	1880
1891	1928
1897	1957
1911	1958
1912	1971
1923	1977
1948	1980
1960	1993
1965	2007
1982	2008
1987	2009
2007	2011
2011	2012

land ice sheet over that period (Chylek et al., 2006; Fettweis et al., 2008) could be attributed to atmospheric circulation anomalies. In contrast, the 20CRv2 circulation type frequencies do not present any anomalies over the 1923–1931 period. Finally, atmospheric circulation conditions similar to 2007–2012 are observed around 1957–1960 for all reanalyses used here. Nevertheless, these four periods were shorter than 2007–2012 and not marked by as many anomalous summers, except for the 1891–1896 period for 20CRv2 type 4 and the 1923–1931 period for ERA-20C type 2. The anomalies of the frequencies of types 2 and 4 (+20 %), and type 3 (−15 to −20 %), are much higher than their interannual frequency variability (with a standard deviation over the 1871–2012 period for the 20CRv2 20 000-run ensemble mean of about 7.7, 9.7, and 8.2 % for types 2, 3, and 4, respectively). The frequency anomalies of the other types are of the same order than their interannual variability (with a standard deviation of about 9.4, 5, and 5.6 % for types 1, 5, and 6, respectively). The exceptional frequency anomalies of 2007–2012 could suggest that they are related to global warming. However, the 2013 summer shows opposite extremes. Nevertheless, the circulation type frequencies of the 2014 summer are of the same order as the 2007–2012 average. This suggests that, even if the 2007–2012 circulation anomalies might be related to global warming, this link is not straightforward, and the natural variability could largely exceed the global warming induced signal.

The circulation type frequency anomalies are not due to changes in the persistence (i.e. the duration of consecutive days grouped in the same type). In fact, there is a persistence increase for types 2 and 4, and a decrease for type 3 over 2007–2012 with regard to the overall average (not shown). But a more detailed analysis shows that these persistence changes are artefacts due to the frequency anomalies. Note that the 20CRv2 20 000-run ensemble persistence cannot be used for a persistence analysis. Since the spread is added with a multiplying factor determined randomly for each day, the continuity of the atmospheric circulation over time, i.e. the transitions between the circulation types and the succession of the types themselves, is not preserved.

The analysis of the 20CRv2 reference run monthly circulation type frequencies shows that the 2007–2012 frequency anomalies affect all three months (JJA). In this way, the 2007–2012 period differs from the other anomalous periods (1871–1880, 1891–1896, 1923–1931, and 1957–1960). For example, the positive frequency anomaly of type 2 over 1958–1960 is due to high frequencies during August and to a lesser extent during June. For the year 1957, type 4 shows frequencies far above normal for June and July, but not for August. The 1871–1880 period has many summers with above normal frequencies for type 4, but no systematic frequency anomaly appearing in several summers can be detected for one particular month.

4.1.2 20CRv2 frequency uncertainty

The uncertainty of the 20CRv2 frequencies strongly decreases between 1930 and 1950 to become insignificant over the 4 or 5 last decades for all types (Fig. 2). It is interesting to observe that this uncertainty remains relatively constant over time before 1940, at a level of around 7–11 % for the first four types, and around 4–6 % for types 5 and 6. Moreover, the 20 000-run ensemble mean frequency, its standard deviation, and its 10th and 90th percentiles show an evolution over time that is almost parallel to the 20CRv2 reference run. There is no smoothing of the interannual variability, which remains similar to the variability of the last decades when going back in time. Finally, the last class, which groups the unclassified days, does not show any increase towards the beginning of the 20CRv2 period, meaning that there are not more days that do not correspond to the main types before 1940 than over the last 3 decades (1980–2012). Thus, even if there is some uncertainty about the exact frequencies before 1940, there is high confidence in the magnitude and the time evolution of the circulation type frequencies.

There are significant circulation type frequency differences between the 20 000-run ensemble mean and the 20CRv2 reference run before 1940. In particular, the frequencies of types 1 and 2 are strongly overestimated by the 20 000-run ensemble compared to the 20CRv2 reference run. The 20CRv2 reference run annual frequencies approximate the 10th percentile of the 20 000-run ensemble for these two types. This is compensated by an underestimation of the frequencies of the other types by the 20 000-run ensemble whose 90th percentile frequencies are of the same order as the 20CRv2 reference run frequencies. These frequency shifts are due to the pattern of the SLP spread, which is much higher over the Arctic Ocean than over the rest of the domain for the 1871–1930 period (Fig. 3, bottom panels). Consequently, when adding (with a multiplying factor of between 0 and 1) the SLP spread, the pattern of the SLP daily mean is changed towards a more anticyclonic pattern over the Arctic Ocean, making it similar to type 2. In the same way, when subtracting (with a multiplying factor of between −1 and 0) the SLP spread, the SLP daily mean becomes more similar to type 1, which presents a low pressure over the Arctic Ocean. Furthermore, since the Spearman rank correlation coefficient is not sensitive to the average SLP, but only to the SLP pattern, the evolution of the frequency uncertainty over time is rather more related to the spatial maximum and the standard deviation of the SLP spread than to its average. As shown in Fig. 3 (top panel), the average SLP spread over the Arctic region decreases already at the beginning of the 20CRv2 era, while the spatial maximum and the standard deviation of the SLP spread remain high until around 1940.

4.2 Geopotential height at 500 hPa

The detected frequency changes are similar to those of SLP. The Greenland High and the Beaufort Sea High (types 2 and 5) were almost twice as frequent over 2007–2012 than over the whole 1871–2014 period (Fig. S2, Table S1 in the Supplement). For type 2 (Greenland High), two of the four other high frequency periods found for SLP are also detected: 1871–1880 and 1891–1896. The 1957–1960 period is not exceptional on the basis of Z500. This is in agreement with the findings of Bezeau et al. (2014), who showed on the basis of NCEP/NCAR Z500 that the 2007–2012 period reached record values since 1948, despite more frequent anticyclones over the Canadian Arctic Archipelago before 1960. The 1923–1931 period, which is only exceptional on the basis of ERA-20C SLP, is not anomalous at the Z500 level.

As for SLP, the Z500 spread plays an important role in the frequency distribution before 1940, when it is the highest (Fig. S3, top panel). Before 1940, the frequencies of the 20CRv2 reference run and to a lesser extent of the 20 000-run ensemble mean are much higher for type 1 (about 20 %) and type 2 (about 10 %) with respect to the second half of the 20th century. This is compensated by particularly low frequencies of types 3, 5, and 6, and to a lesser extent of type 4. These frequency shifts are probably due to the uncertainties in the 20CRv2 data before 1940 since they are lower for the 20 000-run ensemble compared to the 20CRv2 reference run. Moreover, as for SLP, the frequency differences between the 20CRv2 reference run and the 20 000-run ensemble mean can be explained by the pattern of the Z500 spread, which is very close to the SLP spread pattern (Fig. S3, bottom panels). When subtracting the spread, the circulation tends to become more cyclonic over the Arctic Ocean, favouring the shift of days into types 1, 2, and 4. In the same way, adding the spread gives a more anticyclonic character to the circulation.

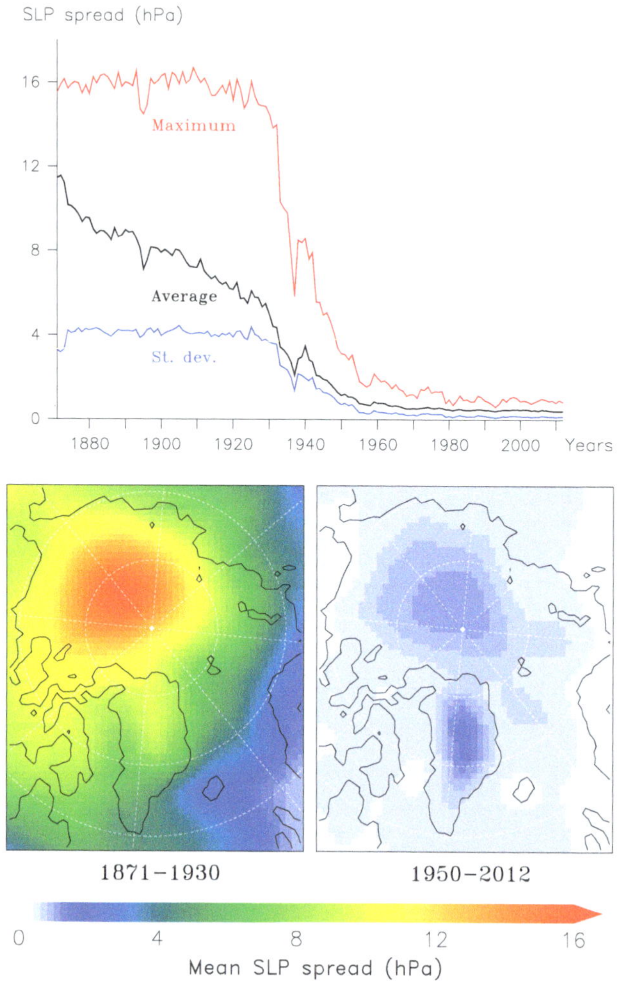

Figure 3. Top: the average SLP spread and its standard deviation are calculated as the seasonal (JJA) average 20CRv2 spread and its standard deviation over the oceanic pixels of our domain. The maximum SLP spread is the value of the oceanic pixel showing the highest seasonal (JJA) average spread of each year. Bottom: the SLP spread is calculated as the average 20CRv2 spread over the 1871–1930 summers (JJA), left, and over the 1950–2012 summers, right.

This favours types 3, 5, and 6. But types 1 and 2 count for about 80 % of all days before 1940. Therefore, subtracting the spread has only a limited impact on the frequency distribution, since it favours types 1 and 2, which already contain most of the days. However, adding the spread at the expense of types 1 and 2 induces many more frequency changes, since more days can be shifted into another type.

4.3 Links with other variables

4.3.1 North Atlantic Oscillation

The 30-year running correlation between the circulation type frequencies and the JJA CRU NAO index (calculated as the average of the JJA monthly CRU NAO index values) shows

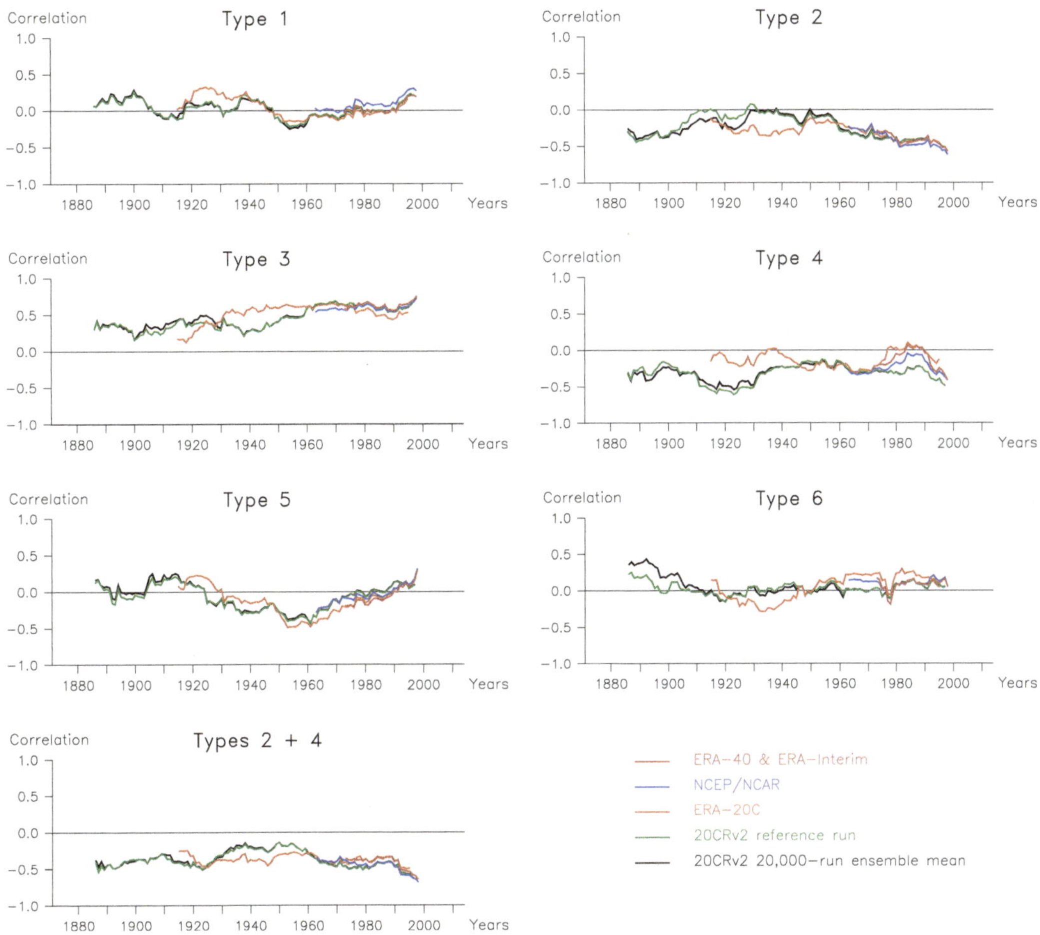

Figure 4. The 30-year running correlation is calculated between the annual (JJA) SLP-based frequencies of each type and the JJA CRU NAO index. For the "Types 2 + 4" correlation, the frequencies of types 2 and 4 have been summed before computing the correlation.

a very good agreement between the different reanalyses on the basis of SLP and Z500. Furthermore, the almost similar evolution of this correlation for the 20CRv2 reference run and the 20 000-run ensemble mean confirms that taking into account the spread does not impact the frequency variations over time.

For the SLP-based classification, types 2 and 4 both show negative correlations with NAO, while only the correlation of type 3 is always positive (Fig. 4). The link between NAO and the three other types is not as clear. The association of negative NAO phases with high frequencies of types 2 and 4 is even more evident when adding both frequencies together, the average correlation between NAO and the 20CRv2 reference run frequencies over 1871–2012 being of −0.24 for type 2, −0.32 for type 4, and −0.38 for types 2 + 4. Moreover, this association becomes more marked when the 2007–2012 anomalous summers are integrated into the calculations (i.e. starting in 1992), suggesting a stronger link over that period. This is in agreement with Fettweis et al. (2013) and Hanna et al. (2014a), who linked the recent circulation anomalies and the negative NAO anomalies. For most other periods with circulation anomalies (1871–1880, 1891–1896, and 1957–1960) there is no clear link between NAO, which is less exceptional than over 2007–2012, and the frequencies of types 2 and 4. Only the 1923–1931 period shows a clear negative correlation between NAO and the frequency of type 2 for ERA-20C.

The Z500-based results are basically the same as for SLP. Type 2 shows a negative correlation, while the correlation between NAO and the frequency of type 1 is always positive (Fig. S4). Nevertheless, the correlation of type 5 is not as clear. While it is negative since 1960, the 20CRv2 reference run and the 20 000-run ensemble mean show divergent values before 1940, suggesting that the uncertainty due to the 20CRv2 spread has more impact on our results for Z500 than for SLP. This is certainly reinforced by the underestimation of the frequency of type 5 before 1940. Again, the link between NAO and types 2 and 5 is stronger since 1992, when the 2007–2012 period is integrated into the 30-year running correlation.

4.3.2 Sea ice extent

There is a link between the circulation anomalies and the summertime SIE loss. As indicated by the correlation between the ERA-Interim SIE loss and the ERA-Interim SLP-based circulation type frequencies over 1980–2014, SIE loss is only favoured by types 2 ($r = -0.47$) and 4 ($r = -0.52$), while types 1 ($r = 0.50$) and 3 ($r = 0.48$) tend to mitigate it. Types 5 ($r = 0.17$) and 6 ($r = 0.22$) do not have important impacts on the SIE loss. When considering the sum of the frequencies of types 2 and 4, the relation appears to be even clearer with a correlation of −0.65. This means that the frequency increase of types 2 and 4 could partly explain the summertime record Arctic SIE loss observed over the last decade.

For the Z500-based classification, only type 5 ($r = -0.54$) can be clearly related to enhanced SIE loss and type 1 ($r = 0.57$) to mitigated SIE loss. For the remaining types, the correlation varies between −0.20 and −0.04. As said above, type 5 combines the Beaufort Sea High and the Greenland High. Type 2 shows only a slight ridge over Greenland and a depression centred over the Arctic Ocean, far away from the Russian coast. Thus, the conditions favouring sea ice export through the Fram Strait and the Barents Sea are not met for this type, as confirmed by its poor correlation ($r = -0.20$) with SIE.

Our results seem to confirm those of Wang et al. (2009) and Overland et al. (2012), who showed that the record Arctic sea ice loss observed over the last years can partly be attributed to more frequent positive Arctic dipole anomaly (DA) phases. In fact, positive DA phases are characterised by a higher occurrence of a high pressure system over the Canadian Arctic Archipelago and Greenland and a low pressure system over the Kara and Laptev seas (Wu et al., 2006). Thus, at first glance, the SLP-based types 2 and 4 can both be associated with a positive DA phase, while the other types, particularly types 1 and 3, can be related to a negative DA phase. During positive DA phases, the sea ice export from the Arctic Basin through the Fram Strait and the Barents Sea is strongly enhanced, which is particularly effective for important sea ice loss during summer (Wang et al., 2009). Furthermore, our results agree with those of Simmonds and Keay (2009) and Screen et al. (2011), who have shown that SIE in September is lower in years characterised by a weaker than normal summertime Arctic cyclonic activity, which induces a higher average SLP over the region.

5 Conclusions

We have used an automatic circulation type classification to study the anomalies in the summertime (JJA) atmospheric circulation based on (i) the sea level pressure and (ii) the 500 hPa geopotential height over the Arctic region over the 1871–2014 period. Three reanalysis data sets were used over the second half of the 20th century (ERA-Interim as reference, ERA-40, and NCEP/NCAR). The 20CRv2 and ERA-20C reanalyses were used over the 1871–2012 and 1900–2010 periods, respectively, to evaluate if circulation anomalies similar to 2007–2012 could already have occurred. Furthermore, since 20CRv2 data are given as a 56-member ensemble mean with its standard deviation (spread), 20 000 runs have been done to take into account the 20CRv2 uncertainty. For these runs, the spread multiplied by a factor varying randomly between −1 and 1 has been added to the daily mean.

Despite an uncertainty of about 5–11 % for the circulation type frequencies before 1930, the magnitude and the time evolution of the frequency anomalies can be reasonably well estimated using 20CRv2 SLP. Furthermore, this uncertainty becomes less significant after 1950, due to improved assimilated data availability and reliability. The strong impact of the number and quality of observational data on the reliability of reanalysis data sets is also highlighted by the important discrepancies between 20CRv2 and ERA-20C during the first half of the 20th century. These discrepancies can have strong impacts on the interpretation of the results. For example, the 1923–1931 warmer summers over Greenland (Chylek et al., 2006) could be attributed to anomalous atmospheric circulation conditions according to ERA-20C but not to 20CRv2. The particular spatial pattern of the 20CRv2 spread, i.e. highest over the Arctic Ocean, causes a strong overestimation of type 1 (low pressure over the Arctic Ocean) and type 2 (high pressure over the Arctic Ocean) at the expense of all other types for the SLP-based 20 000-run ensemble compared to the 20CRv2 reference run. Thus, it is interesting to note that, although no systematic SLP bias is introduced through adding the spread, systematic circulation type frequency shifts appear. In a similar way, the Z500 spread also introduces artefacts in the 20 000-run ensemble. This shows the importance of accounting for the spread of the 20CRv2 data to get an estimation of the range of plausible results.

We have found the same summertime circulation anomalies as described by other authors (Fettweis et al., 2013; Ballinger et al., 2014), i.e. a doubling in frequency of the Beaufort Sea–Arctic Ocean High and of the Greenland High over 2007–2012. Only four other periods (1871–1880, 1891–1896, 1923–1931, and 1957–1960) of similar circulation anomalies were detected but the successions of summers with such anomalies are shorter than in the 2000s. These anomalies all largely exceed the interannual variability of the circulation type frequencies. Nevertheless, it is not possible to attribute the circulation anomalies over 2007–2012 to global warming. First, these anomalies are observed over a too short period, so that they could simply be an exceptionally strong deviance from the average circulation. In this way, the 2013 summer was marked by opposite frequency extremes, positive NAO index values, low melting of the Greenland ice sheet, and lower Arctic sea ice decline compared to 2007–2012. Our findings corroborate those of Rajewicz and Marshall (2014), who state that the 2013 JJA mean Z500 over Greenland was significantly lower than the average over the last 7 decades, which contrasts with the strong positive anomaly of the preceding summers. The opposite extreme anomalies between 2012 (positive anomaly) and 2013 (negative anomaly) have also been highlighted by Hanna et al. (2014b) on the basis of the Greenland Blocking Index. Secondly, as said above, similar circulation conditions were observed before 1880 and around 1891–1896, when the Arctic climate was likely to have been much colder than now. Furthermore, Ding et al. (2014) suggest that the geopotential height increase observed over north-eastern Canada and Greenland, as well as the negative NAO trend, could be due to SST changes in the tropical Pacific that induce changes in the Rossby wave train affecting the North American region. Since the tropical SST changes are not reproduced by general circulation models under current greenhouse gas concentrations, Ding et al. (2014) conclude that these changes are due to the natural variability of the climatic system. However, Wu et al. (2014) showed that the progressive intensification of the Beaufort Sea High over 1979–2005 can only be reproduced by climate models by including the observed greenhouse gas concentration increase. Moreover, Screen et al. (2012) have shown that various forcings are needed to explain the observed Arctic warming: while Arctic sea ice and associated SST changes, as well as remote SST changes (corroborating the conclusions of Ding et al., 2014) are the main drivers of the winter warming, the summertime temperature increase could mainly be due to increased radiative forcing, suggesting a role of global warming. Matsumura et al. (2014) have found a significant relation between the earlier spring snowmelt over the Eurasian continent and the enhanced summertime Arctic anticyclonic circulation. The earlier snowmelt could induce a negative SLP anomaly over Eurasia, which is compensated by an SLP increase over the western part of the Arctic region. Finally, Bezeau et al. (2014) conclude that the anomalous anticyclonic patterns over the Arctic over 2007–2012 are due to combined effects of sea ice loss, snow extent reduction, and enhanced meridional heat advection. Thus, while it is widely admitted that the Arctic region experiences a strong warming since some years (Screen et al., 2012), the complexity of the climate of this region due to its multiple internal and external forcings and feedbacks does not allow us to solve the question of whether the 2007–2012 circulation anomaly is (mainly) due to global warming or to natural variability.

Our findings corroborate those of Ballinger et al. (2014) and Overland et al. (2012), who found that the Beaufort Sea High is associated with anticyclonic conditions over Greenland. This is particularly clear for the Z500-based classification, where type 5 combines both the Beaufort Sea High and the Greenland High. Moreover, the circulation type frequency anomalies observed over the 2007–2012 period on the basis of SLP and Z500 are linked with the observed negative NAO trend (Hanna et al., 2014a). Thus, our results seem to be in agreement with the hypothesis of Overland et al. (2012), who suggest that recently more frequent Beaufort Sea High and Greenland High pressure systems might be part of an enhanced North American blocking mechanism. Additionally, we have shown that the 2007–2012 circulation anomaly affects the whole tropospheric circulation, from the surface (SLP) to upper levels (Z500). In this way, Mahieu et al. (2014) have shown that increased HCl concentrations observed since 2007 in the lower stratosphere of the Northern Hemisphere can be attributed to atmospheric circulation anomalies. This confirms once more that the 2007–2012 summertime Arctic circulation anomaly analysed here could be part of a major climatic anomaly extending beyond the Arctic region.

Finally, the observed summertime decrease in sea ice extent (SIE) between 1980 and 2014 seems to be partly due to the higher occurrence of the Beaufort Sea High and the Greenland High. This means that, in addition to the factors influencing the Arctic sea ice melt cited by Stroeve et al. (2014) and Parkinson (2014) (e.g. generalised warming over the Arctic, earlier melt onset, enhanced ice–albedo feedback, increased SST, and a delayed autumn freeze-up), the JJA atmospheric circulation could also play a significant role in the sea ice melt. However, some studies (e.g. Petoukhov and Semenov, 2010; Inoue et al., 2012; Bezeau et al., 2014) suggest that atmospheric circulation changes can be induced by SIE anomalies. Therefore, the recent SIE decrease could be a trigger of the recent atmospheric circulation change inducing in turn a SIE decrease, suggesting a positive feedback.

Acknowledgements. NCEP/NCAR reanalysis data were provided by the NOAA/OAR/ESRL PSD, Boulder, Colorado, USA, from their website at http://www.esrl.noaa.gov/psd/.

The ECMWF ERA-40, ERA-Interim, and ERA-20C reanalysis data used in this study were obtained from the ECMWF Data Server (http://www.ecmwf.int).

Support for the Twentieth Century Reanalysis (20CR) Project data set was provided by the US Department of Energy, Office of Science Innovative and Novel Computational Impact on Theory and Experiment (DOE INCITE) program, by the Office of Biological and Environmental Research (BER), and by the National Oceanic and Atmospheric Administration (NOAA) Climate Program Office (http://www.esrl.noaa.gov/psd/data/gridded/data.20thC_ReanV2.html).

The monthly NAO data were obtained from the University of East Anglia (UEA) Climatic Research Unit (CRU) website (http://www.cru.uea.ac.uk/cru/data/nao/).

This paper is published with the support of the Belgian University Foundation. Cet article est publié avec le concours de la Fondation Universitaire de Belgique.

Edited by: M. van den Broeke

References

Anagnostopoulou, C., Tolika, K., and Maheras, P.: Classification of circulation types: a new flexible automated approach applicable to NCEP and GCM datasets, Theor. Appl. Climatol., 96, 3–15, doi:10.1007/s00704-008-0032-6, 2009.

Ballinger, T., and Sheridan, S.: Associations between circulation pattern frequencies and sea ice minima in the western Arctic, Int. J. Climatol., 34, 1385–1394, doi:10.1002/joc.3767, 2014.

Ballinger, T., Sheridan, S., and Hanna, E.: Short Communication "Resolving the Beaufort Sea High using synoptic climatological methods", Int. J. Climatol., 34, 3312–3319, doi:10.1002/joc.3907, 2014.

Bardossy, A. and Caspary, H.-J.: Detection of climate change in Europe by analyzing European atmospheric circulation patterns from 1881 to 1989, Theor. Appl. Climatol., 42, 155–167, doi:10.1007/BF00866871, 1990.

Belleflamme, A., Fettweis, X., Lang, C., and Erpicum, M.: Current and future atmospheric circulation at 500 hPa over Greenland simulated by the CMIP3 and CMIP5 global models, Clim. Dynam., 41, 2061–2080, doi:10.1007/s00382-012-1538-2, 2013.

Belleflamme, A., Fettweis, X., and Erpicum, M.: Do global warming-induced circulation pattern changes affect temperature and precipitation over Europe during summer?, Int. J. Climatol., doi:10.1002/joc.4070, in press, 2014.

Bezeau, P., Sharp, M., and Gascon, G.: Variability in summer anticyclonic circulation over the Canadian Arctic Archipelago and west Greenland in the late 20th/early 21st centuries and its effect on glacier mass balance, Int. J. Climatol., doi:10.1002/joc.4000, 2014.

Chylek, P., Dubey, M. K., and Lesins, G.: Greenland warming of 1920–1930 and 1995–2005, Geophys. Res. Lett., 33, L11707, doi:10.1029/2006GL026510, 2006.

Compo, G., Whitaker, J., Sardeshmukh, P., Matsui, N., Allan, R., Yin, X., Gleason, B., Vose, R., Rutledge, G., Bessemoulin, P., Brönnimann, S., Brunet, M., Crouthamel, R., Grant, A., Groisman, P., Jones, P., Kruk, M., Kruger, A., Marshall, G., Maugeri, M., Mok, H., Nordli, Ø., Ross, T., Trigo, R., Wang, X., Woodruff, S., and Worley, S.: The twentieth century reanalysis project, Q. J. Roy. Meteorol. Soc., 137, 1–28, doi:10.1002/qj.776, 2011.

Dee, D., Uppala, S., Simmons, A., Berrisford, P., Poli, P., Kobayashi, S., Andrae, U., Balmaseda, M., Balsamo, G., Bauer, P., Bechtold, P., Beljaars, A., van de Berg, L., Bidlot, J., Bormann, N., Delsol, C., Dragani, R., Fuentes, M., Geer, A., Haimberger, L., Healy, S., Hersbach, H., Hólm, E., Isaksen, L., Kållberg, P., Köhler, M., Matricardi, M., McNally, A., Monge-Sanz, B., Morcrette, J.-J., Park, B.-K., Peubey, C., de Rosnay, P., Tavolato, C., Thépaut, J.-N., and Vitart, F.: The ERA-Interim reanalysis: configuration and performance of the data assimilation system, Q. J. Roy. Meteorol. Soc., 137, 553–597, doi:10.1002/qj.828, 2011.

Demuzere, M., Werner, M., van Lipzig, N., and Roeckner, E.: An analysis of present and future ECHAM5 pressure fields using a classification of circulation patterns, Int. J. Climatol., 29, 1796–1810, doi:10.1002/joc.1821, 2009.

Ding, Q., Wallace, J. M., Battisti, D. S., Steig, E. J., Gallant, A. J. E., Kim, H.-J., and Geng, L.: Tropical forcing of the recent rapid Arctic warming in northeastern Canada and Greenland, Nature, 509, 209–212, doi:10.1038/nature13260, 2014.

Fettweis, X., Hanna, E., Gallée, H., Huybrechts, P., and Erpicum, M.: Estimation of the Greenland ice sheet surface mass balance for the 20th and 21st centuries, The Cryosphere, 2, 117–129, doi:10.5194/tc-2-117-2008, 2008.

Fettweis, X., Mabille, G., Erpicum, M., Nicolay, S., and Van den Broeke, M.: The 1958–2009 Greenland ice sheet surface melt and the mid-tropospheric atmospheric circulation, Clim. Dynam., 36, 139–159, doi:10.1007/s00382-010-0772-8, 2011.

Fettweis, X., Hanna, E., Lang, C., Belleflamme, A., Erpicum, M., and Gallée, H.: *Brief communication* "Important role of the mid-tropospheric atmospheric circulation in the recent surface melt increase over the Greenland ice sheet", The Cryosphere, 7, 241–248, doi:10.5194/tc-7-241-2013, 2013.

Hanna, E., Cappelen, J., Fettweis, X., Huybrechts, P., Luckman, A., and Ribergaard, M. H.: Hydrologic response of the Greenland ice sheet: the role of oceanographic warming, Hydrol. Process. (Special issue: Hydrol. Effect Shrink Cryosphere), 23, 7–30, doi:10.1002/hyp.7090, 2009.

Hanna, E., Cropper, T. E., Jones, P. D., Scaife, A. A., and Allan, R.: Recent seasonal asymmetric changes in the NAO (a marked summer decline and increased winter variability) and associated changes in the AO and Greenland Blocking Index, Int. J. Climatol., doi:10.1002/joc.4157, in press, 2014a.

Hanna, E., Fettweis, X., Mernild, S. H., Cappelen, J., Ribergaard, M. H., Shuman, C. A., Steffen, K., Wood, L., and Mote, T. L.: Atmospheric and oceanic climate forcing of the exceptional Greenland ice sheet surface melt in summer 2012, Int. J. Climatol., 34, 1022–1037, doi:10.1002/joc.3743, 2014b.

Hope, P., Keay, K., Pook, M., Catto, J., Simmonds, I., Mills, G., McIntosh, P., Risbey, J., and Berry, G.: A comparison of automated methods of front recognition for climate studies: A case study in southwest Western Australia, Mon. Weather Rev., 142, 343–363, doi:10.1175/MWR-D-12-00252.1, 2014.

Huth, R.: A circulation classification scheme applicable in GCM studies, Theor. Appl. Climatol., 67, 1–18, doi:10.1007/s007040070012, 2000.

Huth, R., Beck, C., Philipp, A., Demuzere, M., Ustrnul, Z., Cahynová, M., Kyselý, J., and Tveito, O. E.: Classifications of atmospheric circulation patterns: Recent advances and applications, Ann. NY. Acad. Sci., 1146, 105–152, doi:10.1196/annals.1446.019, 2008.

Inoue, J., Hori, M., and Takaya, K.: The role of Barents sea ice in the wintertime cyclone track and emergence of a warm-Arctic cold-Siberian Anomaly, J. Climate., 25, 2561–2568, doi:10.1175/JCLI-D-11-00449.1, 2012.

Kalnay, E., Kanamitsu, M., Kistler, R., Collins, W., Deaven, D., Gandin, L., Iredell, M., Saha, S., White, G., Woollen, J., Zhu, Y., Leetmaa, A., Reynolds, B., Chelliah, M., Ebisuzaki, W., Higgins, W., Janowiak, J., Mo, K., Ropelewski, C., Wang, J., Jenne, R., and Joseph, D.: The NCEP/NCAR 40-Year Reanalysis Project, B. Am. Meteorol. Soc., 77, 437–471, doi:10.1175/1520-0477(1996)077<0437:TNYRP>2.0.CO;2, 1996.

Käsmacher, O. and Schneider, C.: An objective circulation pattern classification for the region of Svalbard, Geogr. Ann. A., 93, 259–271, doi:10.1111/j.1468-0459.2011.00431.x, 2011.

Kyselý, J. and Huth, R.: Changes in atmospheric circulation over Europe detected by objective and subjective methods, Theor. Appl. Climatol., 85, 19–36, doi:10.1007/s00704-005-0164-x, 2006.

Lindsay, R., Wensnahan, M., Schweiger, A., and Zhang, J.: Evaluation of seven different atmospheric reanalysis products in the Arctic, J. Climate., 27, 2588–2606, doi:10.1175/JCLI-D-13-00014.1, 2014.

Mahieu, E., Chipperfield, M. P., Notholt, J., Reddmann, T., Anderson, J., Bernath, P. F., Blumenstock, T., Coffey, M. T., Dhomse, S. S., Feng, W., Franco, B., Froidevaux, L., Griffith, D. W. T., Hannigan, J. W., Hase, F., Hossaini, R., Jones, N. B., Morino, I., Murata, I., Nakajima, H., Palm, M., Paton-Walsh, C., Russell III, J. M., Schneider, M., Servais, C., Smale, D., and Walker, K. A.: Recent Northern Hemisphere stratospheric HCl increase due to atmospheric circulation changes, Nature, 515, 104–107, doi:10.1038/nature13857, 2014.

Matsumura, S., Zhang, X., and Yamazaki, K.: Summer Arctic atmospheric circulation response to spring Eurasian snow cover and its possible linkage to accelerated sea ice decrease, J. Climate., 27, 6551–6558, doi:10.1175/JCLI-D-13-00549.1, 2014.

Moholdt, G., Nuth, C., Hagen, J., and Köhler, J.: Recent elevation changes of Svalbard glaciers derived from ICESat laser altimetry, Remote Sens. Environ., 114, 2756–2767, doi:10.1016/j.rse.2010.06.008, 2010.

Overland, J., Francis, J., Hanna, E., and Wang, M.: The recent shift in early summer Arctic atmospheric circulation, Geophys. Res. Lett., 39, L19804, doi:10.1029/2012GL053268, 2012.

Parkinson, C. L.: Spatially mapped reductions in the length of the Arctic sea ice season, Geophys. Res. Lett., 41, 4316–4322, doi:10.1002/2014GL060434, 2014.

Pastor, M. and Casado, M.: Use of circulation types classifications to evaluate AR4 climate models over the Euro-Atlantic region, Clim. Dynam., 39, 2059–2077, doi:10.1007/s00382-012-1449-2, 2012.

Petoukhov, V. and Semenov, V.: A link between reduced Barents-Kara sea ice and cold winter extremes over northern continents, J. Geophys. Res., 115, D21111, doi:10.1029/2009JD013568, 2010.

Philipp, A., Della-Marta, P., Jacobeit, J., Fereday, D., Jones, P., Moberg, A., and Wanner, H.: Long-term variability of daily North Atlantic-European pressure patterns since 1850 classified by simulated annealing clustering, J. Climate, 20, 4065–4095, doi:10.1175/JCLI4175.1, 2007.

Philipp, A., Bartholy, J., Beck, C., Erpicum, M., Esteban, P., Fettweis, X., Huth, R., James, P., Jourdain, S., Kreienkamp, F., Krennert, T., Lykoudis, S., Michalides, S., Pianko, K., Post, P., Rassilla Álvarez, D., Schiemann, R., Spekat, A., and Tymvios, F. S.: COST733CAT – a database of weather and circulation type classifications, Phys. Chem. Earth, 35, 360–373, doi:10.1016/j.pce.2009.12.010, 2010.

Poli, P., Hersbach, H., Tan, D., Dee, D., Thépaut, J.-N., Simmons, A., Peubey, C., Laloyaux, P., Komori, T., Berrisford, P., Dragani, R., Trémolet, Y., Hólm, E., Bonavita, M., Isaksen, L., and Fisher, M.: The data assimilation system and initial performance evaluation of the ECMWF pilot reanalysis of the 20th-century assimilating surface observations only (ERA-20C), ERA Report Series, 14, 59 pp., available at: http://old.ecmwf.int/publications/library/ecpublications/_pdf/era/era_report_series/RS_14.pdf (last access: 17 November 2014), 2013.

Rajewicz, J. and Marshall, S.: Variability and trends in anticyclonic circulation over the Greenland ice sheet, 1948–2013, Geophys. Res. Lett., 41, 2842–2850, doi:10.1002/2014GL059255, 2014.

Screen, J. A. and Simmonds, I.: Erroneous Arctic temperature trends in the ERA-40 reanalysis: A closer look, J. Climate., 24, 2620–2627, doi:10.1175/2010JCLI4054.1, 2011.

Screen, J. A., Simmonds, I., and Keay, K.: Dramatic interannual changes of perennial Arctic sea ice linked to abnormal summer storm activity, J. Geophys. Res., 116, D15105, doi:10.1029/2011JD015847, 2011.

Screen, J. A., Deser, C., and Simmonds, I.: Local and remote controls on observed Arctic warming, Geophys. Res. Lett., 39, L10709, doi:10.1029/2012GL051598, 2012.

Serreze, M. C., Barrett, A. P., Stroeve, J. C., Kindig, D. N., and Holland, M. M.: The emergence of surface-based Arctic amplification, The Cryosphere, 3, 11–19, doi:10.5194/tc-3-11-2009, 2009.

Simmonds, I.: Comparing and contrasting the behaviour of Arctic and Antarctic sea ice over the 35 year period 1979–2013, Ann. Glaciol., 56, 18–28, doi:10.3189/2015AoG69A909, 2015.

Simmonds, I. and Keay, K.: Extraordinary September Arctic sea ice reductions and their relationships with storm behavior over 1979–2008, Geophys. Res. Lett., 36, L19715, doi:10.1029/2009GL039810, 2009.

Stroeve, J., Markus, T., Boisvert, L., Miller, J., and Barrett, A.: Changes in Arctic melt season and implications for sea ice loss, Geophys. Res. Lett., 41, 1216–1225, doi:10.1002/2013GL058951, 2014.

Tedesco, M., Serreze, M., and Fettweis, X.: Diagnosing the extreme surface melt event over southwestern Greenland in 2007, The Cryosphere, 2, 159–166, doi:10.5194/tc-2-159-2008, 2008.

Uppala, S. M., Kållberg, P. W., Simmons, A. J., Andrae, U., da Costa Bechtold, V., Fiorino, M., Gibson, J. K., Haseler, J., Hernandez, A., Kelly, G. A., Li, X., Onogi, K., Saarinen, S., Sokka, N., Allan, R. P., Andersson, E., Arpe, K., Balmaseda, M. A., Beljaars, A. C. M., van de Berg, L., Bidlot, J., Bormann, N., Caires, S., Chevallier, F., Dethof, A., Dragosavac, M., Fisher, M., Fuentes, M., Hagemann, S., Hólm, E., Hoskins, B., Isaksen, L., Janssen, P. A. E. M., Jenne, R., McNally, A. P., Mahfouf, J.-F., Morcrette, J.-J., Rayner, N. A., Saunders, R. W., Simon, P., Sterl, A., Trenberth, K. E., Untch, A., Vasiljevic, D., Viterbo, P., and Woollen, J.: The ECMWF re-analysis, Q. J. Roy. Meteorol. Soc., 131, 2961–3012, doi:10.1256/qj.04.176, 2005.

Wang, J., Zhang, J., Watanabe, E., Ikeda, M., Mizobata, K., Walsh, J., Bai, X., and Wu, B.: Is the Dipole Anomaly a major driver to record lows in Arctic summer sea ice extent?, Geophys. Res. Lett., 36, L05706, doi:10.1029/2008GL036706, 2009.

Wu, B., Wang, J., and Walsh, J.: Dipole anomaly in the winter Arctic atmosphere and its association with sea ice motion, J. Climate, 19, 210–225, doi:10.1175/JCLI3619.1, 2006.

Wu, Q., Zhang, J., Zhang, X., and Tao, W.: Interannual variability and long-term changes of atmospheric circulation over the Chukchi and Beaufort Seas, J. Climate, 27, 4871–4889, doi:10.1175/JCLI-D-13-00610.1, 2014.

5

Surface elevation and mass changes of all Swiss glaciers 1980–2010

M. Fischer[1]**, M. Huss**[1,2]**, and M. Hoelzle**[1]

[1]Department of Geosciences, University of Fribourg, 1700 Fribourg, Switzerland
[2]Laboratory of Hydraulics, Hydrology and Glaciology (VAW), ETH Zurich, 8093 Zurich, Switzerland

Correspondence to: M. Fischer (mauro.fischer@unifr.ch)

Abstract. Since the mid-1980s, glaciers in the European Alps have shown widespread and accelerating mass losses. This article presents glacier-specific changes in surface elevation, volume and mass balance for all glaciers in the Swiss Alps from 1980 to 2010. Together with glacier outlines from the 1973 inventory, the DHM25 Level 1 digital elevation models (DEMs) for which the source data over glacierized areas were acquired from 1961 to 1991 are compared to the swissALTI^{3D} DEMs from 2008 to 2011 combined with the new Swiss Glacier Inventory SGI2010. Due to the significant differences in acquisition dates of the source data used, mass changes are temporally homogenized to directly compare individual glaciers or glacierized catchments. Along with an in-depth accuracy assessment, results are validated against volume changes from independent photogrammetrically derived DEMs of single glaciers. Observed volume changes are largest between 2700 and 2800 m a.s.l. and remarkable even above 3500 m a.s.l. The mean geodetic mass balance is -0.62 ± 0.07 m w.e. yr^{-1} for the entire Swiss Alps over the reference period 1980–2010. For the main hydrological catchments, it ranges from -0.52 to -1.07 m w.e. yr^{-1}. The overall volume loss calculated from the DEM differencing is -22.51 ± 1.76 km^3.

1 Introduction

Fluctuations of mountain glaciers are known as a sensitive indicator for climatic changes (e.g. IPCC, 2013). Recent atmospheric warming has caused increased mass loss of glaciers all over the world (e.g. Zemp et al., 2009; Radić and Hock, 2014), which significantly contributes to present sea-level rise (e.g. Marzeion et al., 2012; Gardner et al., 2013) and affects the runoff regimes of glacierized catchments in different regions around the globe (e.g. Kaser et al., 2010; Huss, 2011; Sorg et al., 2012).

Glaciers of the European Alps showed general mass loss and shrinkage since the Little Ice Age maximum around 1850 (Zemp et al., 2008), despite intermittent phases of positive mass and area changes around 1890, during the 1910s and from the late 1970s to the mid-1980s (Glaciological Reports, 1960–2015). Since then, pronounced glacier retreat has been reported again (Paul et al., 2011; Huss, 2012). Glacier area changes are documented by the comparison of consecutive inventories (e.g. Lambrecht and Kuhn, 2007; Diolaiuti et al., 2012; Fischer et al., 2014; Gardent et al., 2014). Mass balance data are available either from annual field measurements on individual glaciers using the direct glaciological method (e.g. WGMS, 2012) or from the comparison of the glacier surface topography of different years and a density assumption for converting volume to mass change (e.g. Abermann et al., 2009; Carturan et al., 2013a). Together with the increasing number of digital elevation models (DEMs) available worldwide and the fact that inaccessible areas and entire glacier systems can be measured, this so-called geodetic method has become a popular approach to derive surface elevation and mass changes for a large number of glaciers (e.g. Rignot et al., 2003; Larsen et al., 2007; Bolch et al., 2008; Berthier et al., 2010; Nuth et al., 2010; Gardelle et al., 2012a).

Paul and Haeberli (2008) analyzed the spatial variability of glacier elevation changes in the Swiss Alps between 1985 and 1999 by comparing the DHM25 Level 1 DEMs (25 m resolution) created from topographic maps by the Swiss Federal Office of Topography (swisstopo) with the medium-resolution (90 m) Shuttle Radar Topography Mission (SRTM) DEM. Several factors that might have an important influence on the accuracy of glacier elevation changes derived from DEM differencing have, however, not been conclusively assessed in their study: differences in the reference years of the surface elevation information used for individual regions, the problem of radar penetration into snow and ice (Dall et al., 2001; Gardelle et al., 2012b) and/or impacts of downscaling DEMs to higher resolution (Gardelle et al., 2012b; Carturan et al., 2013a). Furthermore, applying the medium-resolution SRTM DEM to high-mountain areas is problematic due to the systematic biases reported for high elevations (Berthier et al., 2006). Numerically, these regions are generally dominated by very small glaciers, hereafter defined as being smaller than 0.5 km^2. Abermann et al. (2010) and Fischer et al. (2014) show that use of the most accurate and highest resolution source data available is of particular importance for change assessments of these smallest glacier size classes.

Due to the recent compilation of more detailed source data, we are able to present an extended re-evaluation of glacier-specific changes in surface elevation, volume and geodetic mass balance for every single glacier in Switzerland. We compare the DHM25 Level 1 DEMs dating from between 1961 and 1991 over glacierized areas with the swissALTI^{3D} DEMs from 2008–2011, and we combine the former with the 1973 inventory (Müller et al., 1976) and the latter with the new Swiss Glacier Inventory SGI2010 (Fischer et al., 2014). For direct comparison between individual glaciers or glacierized catchments, we temporally homogenize resulting mass changes to a consistent period, 1980–2010. This is necessary due to significant differences in the acquisition dates of the source data. We discuss various sources of possible error, perform an in-depth accuracy assessment of our results and validate them using independent volume changes from photogrammetrically derived DEMs. Furthermore, we comment on the controlling factors and the spatial variability of observed glacier mass changes.

2 Study region and data sets

2.1 Study region

The study area covers the entire Swiss Alps, where glaciers have generally shown rapid mass loss after a short period of mass gain between the late 1970s and the mid-1980s (Huss et al., 2010a). Overall, small, thin and rather steep glaciers dominate. Almost 90 % of all glaciers were smaller than 1.0 km^2 in 2010. At that time, the total glacierized area was $944.3 \pm 24.1\,km^2$, corresponding to an area change of $-362.6\,km^2$ (-27.7 or $-0.75\,\%\,yr^{-1}$) since 1973 (Fischer et al., 2014). If we apply the approach of Huss and Farinotti (2012) to all Swiss glaciers, the average estimated mean glacier thickness was 63 m in 2010, and 92 % of the estimated total glacier volume of 59.9 km^3 was stored within only 10 % of the 1420 glaciers comprised in the SGI2010.

2.2 Digital elevation models and glacier outlines

The initial glacier surface topography at the beginning of the observation period (hereafter referred to as t_1) is given by the DHM25 Level 1 DEMs from swisstopo, for which digitized contour lines and spot heights from the Swiss national topographic maps 1 : 25 000 were interpolated to a regular grid with 25 m spacing. The positional accuracy is reported to range between 2.5 and 7.5 m. The vertical accuracy was estimated by comparison of known spot heights on stable terrain with corresponding cell values of the DHM25 Level 1 DEMs and ranges between 3.7 and 8.2 m on average for rugged high-mountain topography depending on individual map sheets (Rickenbacher, 1999; swisstopo, 2000). Because spot heights are typically located at topographical extreme points like hilltops or depressions, the actual vertical accuracy over "average terrain" such as glacier surface topography is probably considerably higher. For glacierized areas, the dating of contour lines used for the creation of the DHM25 Level 1 DEMs is not consistent with corresponding specifications given in the product information (swisstopo, 2000). Therefore, we manually reconstructed the individual reference years of the surface topography at t_1 for every glacier by comparison of the DHM25 Level 1 contour lines with those from repeated updates of the 1 : 25 000 topographic maps with known reference years (cf. "journey through time" maps available through map.geo.admin.ch). In addition to the obvious regional differences in t_1, there is a certain trend towards earlier t_1 for small glaciers, for which surface contour lines were less frequently updated (Fig. 1).

Recent glacier surface topography, i.e. at the end of the observation period (hereafter referred to as t_2), is provided by the new 2 m resolution swissALTI^{3D} DEMs. For areas above 2000 m a.s.l., these were created by stereocorrelation of high-resolution (25 cm) 2008–2011 SWISSIMAGE Level 2 aerial orthophotographs. For these areas, the vertical accuracy is ± 1 to 3 m on average. For areas below 2000 m a.s.l., the swissALTI^{3D} DEMs are more accurate (± 0.5 m at the 1σ level) since they were created using airborne laser scanning data (swisstopo, 2013).

The surface area of individual glaciers at t_1 is derived from the finalized digital version of the Swiss Glacier Inventory 1973 (SGI1973; Maisch et al., 2000; Paul, 2004), which was originally compiled by Müller et al. (1976) from stereophotogrammetry-based interpretation of aerial photographs acquired in early September 1973. The considerable time difference between the acquisition of the SGI1973 source data and individual DHM25 Level 1 DEMs used for t_1 (Fig. 1) is acceptable as only small area changes and an almost balanced mass budget of glaciers were reported for the European Alps between 1960 and the mid-1980s (Glaciological Reports, 1960–2015; Paul et al., 2004; Huss, 2012). For t_2, the glacier outlines originate from the latest Swiss Glacier Inventory SGI2010 derived by manual digitization from high-resolution (25 cm) aerial orthophotographs ac-

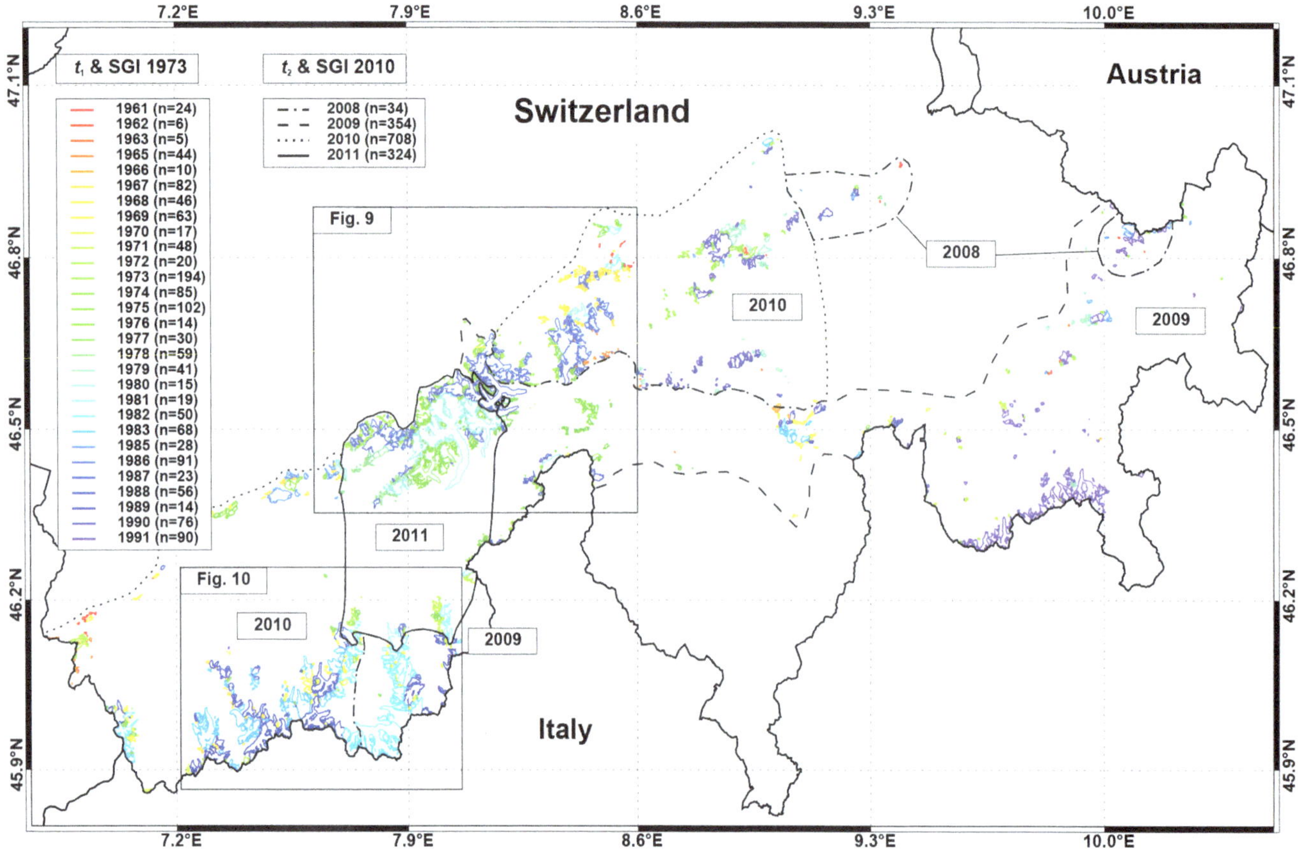

Figure 1. Glaciers in the Swiss Alps. Colours refer to the acquisition date of the elevation information (DHM25 Level 1) defining the beginning of the investigated period (t_1). Black lines delimit areas of equal acquisition dates of both glacier outlines and surface topography at the end of the investigated period (t_2). Two black rectangles show the perimeters of close-ups with resulting mean geodetic mass balance.

quired between 2008 and 2011 (Fischer et al., 2014). Applying the SGI2010 and the swissALTI^{3D} DEMs as source data for t_2 is ideal because their acquisition dates are mostly identical for individual glaciers.

2.3 Validation data

Time series of surface mass balance for glaciers of different type and size class covering the entire Swiss Alps over the last few decades (Huss et al., 2010a, b) are used to validate the geodetic mass balances presented here. These series rely on ice volume changes derived from high-accuracy photogrammetrical DEMs for sub-decadal to multi-decadal time intervals (Bauder et al., 2007). By using a distributed mass balance modelling approach including comprehensive field data sets (winter accumulation, summer ablation and discharge measurements), annual mass balance series were calculated that agree with the observed geodetic mass changes.

3 Methods

3.1 Calculation of glacier volume change and average mass balance

Prior to the calculation of surface elevation changes, the swissALTI^{3D} DEMs were resampled to a grid cell size of 25 m (i.e. equal to the resolution of the DHM25 Level 1 DEMs). Because glacier polygons of the SGI2010 were coded and named according to the 1973 outlines they fell into or overlapped with (Fischer et al., 2014), elevation changes could be calculated for individual glacier entities as a next step by subtracting the DHM25 Level 1 DEMs from the swissALTI^{3D} DEMs (Fig. 2).

The glacier-individual total volume change ΔV (m^3) for the respective survey period was calculated as follows:

$$\Delta V = \overline{\Delta z} \cdot A_{1973}, \qquad (1)$$

where $\overline{\Delta z}$ (m) is the average elevation change calculated from the difference between the swissALTI^{3D} and the DHM25 Level 1 DEMs within the perimeter covered by the

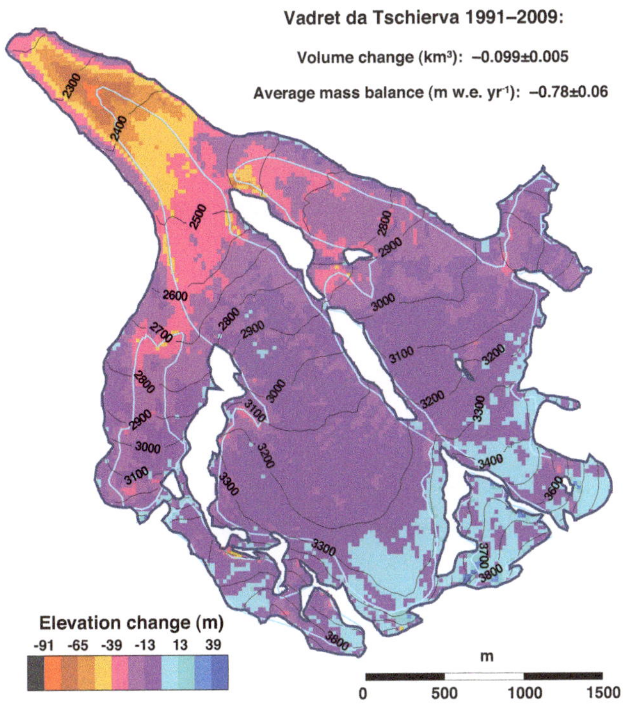

Figure 2. Spatial distribution of elevation changes for Vadret da Tschierva 1991–2009 within the digital glacier outlines from both the 1973 (dark blue) and the 2010 (light blue) inventories.

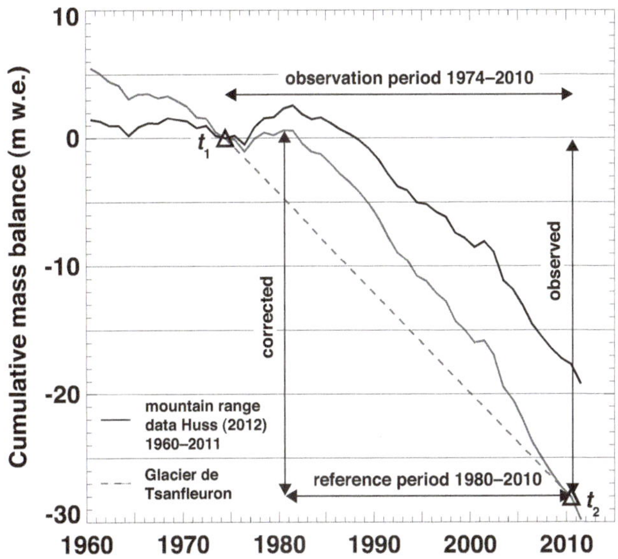

Figure 3. Temporal homogenization of cumulative average mass balance from DEM differencing for Glacier de Tsanfleuron (dark grey) based on cumulative mean annual mountain-range mass balance from Huss (2012).

glacier in 1973 (A_{1973} (m^2)). Then the area-averaged specific geodetic mass balance rate (m w.e. yr^{-1}) was calculated with

$$\dot{B} = \frac{\Delta V \cdot f_{\Delta V}}{\overline{A} \cdot \Delta t}, \tag{2}$$

where $f_{\Delta V}$ is a conversion factor (without unit) used to transform ΔV into a mass change, $\overline{A}$ (m^2) the average area between 1973 and 2010 calculated as $(A_{1973} + A_{2010})/2$ and Δt the length of the observation period $(t_2 - t_1)$ in years. We hereafter refer to the area-averaged specific geodetic mass balance rate as “average mass balance”. The conversion factor $f_{\Delta V}$ might vary from glacier to glacier depending on the length of the observation period, the respective mass balance and the firn compaction history. Due to the fairly long observation periods, $f_{\Delta V}$ is set as a constant of 0.85 corresponding to a density of volume change of $850 \pm 60\,\mathrm{kg\,m^{-3}}$ (Huss, 2013), which is consistent with other studies (Sapiano et al., 1998; Fischer, 2011; Zemp et al., 2013).

The significant regional differences in the length of the observation period (Fig. 1) imply that glacier-individual average mass balances $\dot{B}_g$ derived from the DEM differencing can not be directly compared to each other. In order to homogenize the glacier-individual observation periods to one comparable time interval, we make use of the data set of Huss (2012) that provides annual mass balance variability extrapolated to the entire European Alps based on a combination of all available mass balance data covering our period of interest. The deviation of average mass balance $\dot{B}_g$ for glacier g (dashed grey line in Fig. 3) from the mountain-range mean (black line in Fig. 3) over the respective observation period $\overline{B_{t_2-t_1}}$ is used as a scaling factor to account for the variability in glacier-wide mass balance (Kuhn et al., 1985). The mean mountain-range mass balance from Huss (2012) for an individual year i, $B_{i,\mathrm{mr}}$, accounts for the temporal mass balance variability. The annual mass balance $B_{i,g}$ for year i and every glacier g is thus calculated with

$$B_{i,g} = B_{i,\mathrm{mr}} + \dot{B}_g - \overline{B_{t_2-t_1}}. \tag{3}$$

Because 2010 is the reference year t_2 for most of the investigated glacier entities and the mean observation period is ≈ 30 years (Fig. 1), the hydrological years 1980/81–2009/10 are defined as the reference observation period over the entire Swiss Alps over which annual mass balances for individual glaciers $B_{i,g}$ are cumulated (grey line in Fig. 3). Using this approach, mass changes are temporally homogenized and can be compared and further analyzed.

3.2 Analysis of controls

Averaged over representative samples (number of glaciers) and observation periods (number of years), glacier area and elevation changes are usually in agreement with changes in air temperature and precipitation recorded over the investigated areas and time intervals (e.g. Abermann et al., 2009; VanLooy and Forster, 2011; Carturan et al., 2013a). Within a mountain range and despite similar climatic changes, the differences in long-term mass balance can, however, be significant between individual – and even adjacent – glaciers (e.g. Kuhn et al., 1985; Vincent, 2002; Larsen et al., 2007; Abermann et al., 2011). Different factors have been identified which can explain this variability to a certain extent. For instance, the glacier hypsometry, i.e. the distribution of glacier area and volume with altitude, plays an important role (Furbish and Andrews, 1984; Benn and Evans, 2010). Also, the characteristic glacier response time and dynamic adjustment to a certain climatic forcing varies with glacier size and affects the specific mass balance (Jóhannesson et al., 1989; Huss et al., 2012). Larger (and more gently sloping) glaciers are expected to lag behind the current climatic forcing and to show more negative mass balances than smaller (and steeper) glaciers (Hoelzle et al., 2003). Very small glaciers situated in cirques, niches and below headwalls generally react faster to climatic changes (Kuhn, 1995), but their individual response to the latter is – in relative terms – even more variable than for larger glaciers (Carturan et al., 2013b). Furthermore, local topographic and microclimatic factors are important for the behaviour and dynamics of mountain glaciers in general (e.g. Benn and Lehmkuhl, 2000; Oerlemans, 2010) and for very small glaciers in particular (e.g. DeBeer and Sharp, 2009; Hughes, 2009). These factors can be parameterized with geometrical indices which are related to the observed variability in long-term mass changes (Huss et al., 2012).

In order to identify the controlling factors and to better understand the spatial variability of the observed surface elevation and mass changes, a correlation analysis between the average mass balance over the reference period 1980–2010 and classes of (1) mean area, (2) median elevation, (3) surface slope of the glacier terminus and (4) dominant aspect, hereafter referred to as mean aspect, was performed. Huss (2012) showed that these four geometrical indices can explain some of the variability in observed long-term mass balances. For the surface slope of the glacier terminus, the testing of different values indicated that the average surface slope over the lowermost 25 % of the glacier at t_1 resulted in the highest correlation. For mean aspect, the initially eight classes were rearranged into five classes of equivalent potential clear-sky radiation (N, NW/NE, W/E, SW/SE, S) prior to the correlation analysis.

4 Accuracy assessment and validation

4.1 Quality of input DEMs over selected glaciers

Photogrammetrically derived DEMs of the same acquisition dates as the DHM25 Level 1 DEMs for t_1 and the swissALTI^{3D} DEMs for t_2 are available for 13 glaciers covering 207.3 km^2 (15.9 % of the total glacierized surface area) in 1973 and 12 glaciers covering 88.2 km^2 (9.3 %) in 2010 (Bauder et al., 2007, updated). Statistical comparisons of these DEMs show that for glacierized surface topography, the actual vertical accuracies of both the DHM25 Level 1 and the swissALTI^{3D} DEMs are likely much better in the glacier-wide mean than assumed by the average error values taken from literature. Both mean ($\mu = -1.19$ m) and median ($\widetilde{x} = -1.36$ m) elevation differences are slightly negative when subtracting the DHM25 Level 1 DEMs from the photogrammetrical DEMs at t_1. For the swissALTI^{3D} DEMs at t_2, mean ($\mu = 0.50$ m) and median ($\widetilde{x} = 0.21$ m) elevation differences are slightly positive and show even closer match with the photogrammetrical DEMs. The standard deviations (1σ level) and interquartile ranges of the elevation differences are comparable for both the DHM25 Level 1 DEMs and the swissALTI^{3D} DEMs. Hence, the uncertainties in glacier surface elevation are probably very similar for individual grid cells both at t_1 and t_2 (Fig. 4).

4.2 DEM comparison over stable terrain

An analysis of possible elevation-, slope- and aspect-dependent biases in DEM differencing is carried out by comparing the DHM25 Level 1 DEMs with the swissALTI^{3D} DEMs over stable terrain. The spatial distribution of surface elevation changes outside the glaciers is calculated within a mask around every entity (Fig. 5a) and over about twice the area glacierized at t_1. The mean offset between both DEMs, which we do not correct for, is -1.47 ± 6.82 (1σ) m (Fig. 5b). Thus, on average, the DHM25 Level 1 DEMs are slightly higher over stable terrain than the recent swissALTI^{3D} DEMs. The mean standard deviation over all stable terrain close to glacierized areas of ± 6.82 m agrees with the nominal uncertainty, $\sigma_{\Delta z}$, assessed with literature-based uncertainty estimates (Rickenbacher, 1999; swisstopo, 2000).

We find an almost linear but comparatively moderate increase of the elevation difference between both input DEMs with elevation (0.03 m between 1300 and 2000 m a.s.l. to -2.13 m above 3500 m a.s.l. on average) (Fig. 5c). Together with the scatter of the bias, which also increases towards higher elevations, this is probably due to the general occurrence of more rugged and steeper terrain with increasing elevation. No dependency is found when relating the mean offset to classes of slope. However, the scatter significantly increases towards steep slopes. The bias is slightly dependent on aspect. While areas with a mean aspect NE and SW show the same mean offset as the overall value, positive values result for pixels with a mean aspect W–NW–N and negative values for pixels facing E–SE–S, i.e. exposed to the opposite direction (Fig. 5a). This points to a slight horizontal shift in the NW–SE direction of the elevation information included in both input DEMs. We assume this shift to originate from the creation of the DHM25 Level 1 source data and therefore calculate the influence of its correction via co-registration according to Nuth and Kääb (2011) for the 45 largest glaciers spread over the entire Swiss Alps with a surface area of 650 km^2 at t_1. The effect of this correction on

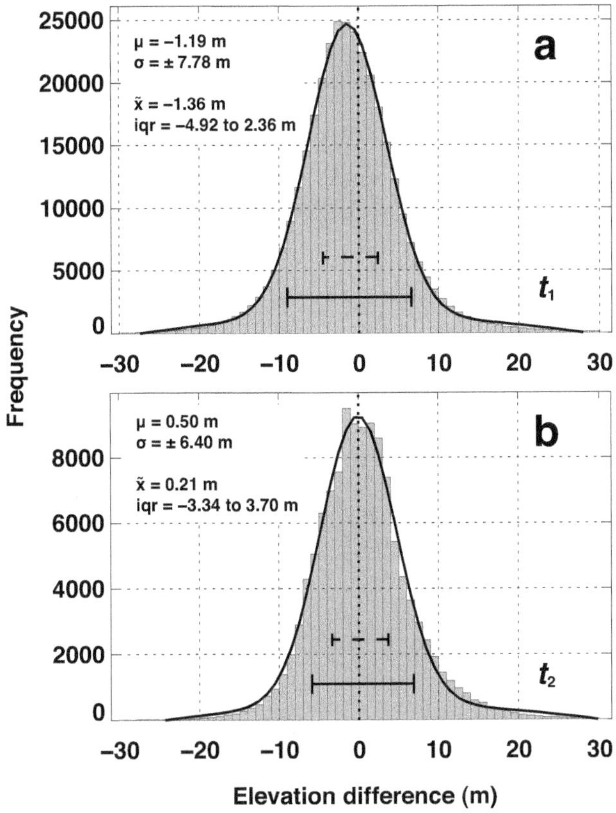

Figure 4. Distribution of **(a)** DHM25 Level 1 minus photogrammetrical DEM differences at t_1 and **(b)** swissALTI^{3D} minus photogrammetrical DEM differences at t_2 over selected glaciers with available independent, photogrammetrically derived DEMs. The black curves are normal fits over the data, the solid bars mark the standard deviation at the 1σ level and the dashed bars the interquartile range (iqr). In addition, the mean (μ) and median ($\tilde{x}$) elevation differences are given.

the average mass balance of individual glaciers turns out to be on the order of $\pm 10^{-4}$ to 10^{-2} m w.e. yr^{-1}. Therefore, we consider the impact of the detected DEM shifts on calculated surface elevation, volume and mass changes as negligible for the entire Swiss Alps and do not co-register the source DEMs prior to differencing.

4.3 Uncertainty

Two "extreme approaches" have been applied in literature to assess the uncertainty in glacier volume changes derived by DEM differencing. For individual grid cells, the uncertainty in surface elevation changes is either assumed to be totally correlated or totally uncorrelated in space. In the former case, the uncertainty in the glacier-individual volume change is given by the standard deviation of measured point elevation errors (e.g. Cox and March, 2004; Larsen et al., 2007). In the latter case, errors are considered as totally uncorrelated (totally random) in space (e.g. Thibert et al., 2008), and the uncertainty is smaller by a factor of $\sqrt{n}$ compared to totally correlated errors, where n is the number of grid cells over which the DEM differencing is carried out. By determining the spatial correlation lengths of elevation differences from the comparison of input DEMs over stable terrain, Rolstad et al. (2009) recommended a compromise between these two extreme approaches. Motyka et al. (2010) and Trüssel et al. (2013), for instance, also applied this third approach and showed that the degree of the spatial correlation of DEM errors influences the overall uncertainty in geodetic volume and mass changes.

The uncertainty in surface elevation, volume and mass changes presented in this study is mainly given by the uncertainty related to the two DEMs used. Hereafter, a simple implementation of Rolstad et al. (2009) is applied to estimate the uncertainty in the glacier-individual average elevation change $\sigma_{\Delta z,g}$ (m). This approach does not account for possible multi- or large-scale variance in the uncertainty. It can be regarded as a "rule of thumb" to assess the accuracy of the spatial average of elevation changes if the area to be averaged over, A_{1973} here, is larger than an area A_{cor}, within which errors in DEM differencing are considered as effectively correlated. A_{cor} (m^2) is defined as

$$A_{\text{cor}} = \pi \cdot L^2. \tag{4}$$

The radius L of this circular area is equal to the spatial correlation length. The variances of the elevation differences of both input DEMs over stable terrain were binned at 50 m intervals for scales < 1500 m and semivariograms produced. Analysing the latter, we find a mean correlation length of 400 m for the entire Swiss Alps. For glaciers with $A_{1973} > A_{\text{cor}}$, the uncertainty in the glacier-individual average elevation change $\sigma_{\Delta z,g}$ is calculated following Rolstad et al. (2009) as

$$\sigma_{\Delta z,g} = \pm\sqrt{\sigma^2_{\Delta h,g} \cdot \frac{A_{\text{cor}}}{5 \cdot A_{1973}}}, \tag{5}$$

where $\sigma_{\Delta h,g}$ (m) is the standard deviation of the elevation error over stable terrain derived within a mask around individual glaciers (Fig.5a). For glaciers with $A_{1973} < A_{\text{cor}}$, elevation errors are treated as totally correlated in space, and $\sigma_{\Delta z,g}$ is equal to $\sigma_{\Delta h,g}$.

The glacier-individual uncertainty in volume change $\sigma_{\Delta V,g}$ (m^3) is obtained by multiplying $\sigma_{\Delta z,g}$ with the initial glacier area A_{1973}. A conservative estimate of the uncertainty in the total volume change over the entire Swiss Alps is derived by

$$\sigma_{\Delta V,\text{tot}} = \pm\sum_{g=1}^{n} \sigma_{\Delta V,g} \tag{6}$$

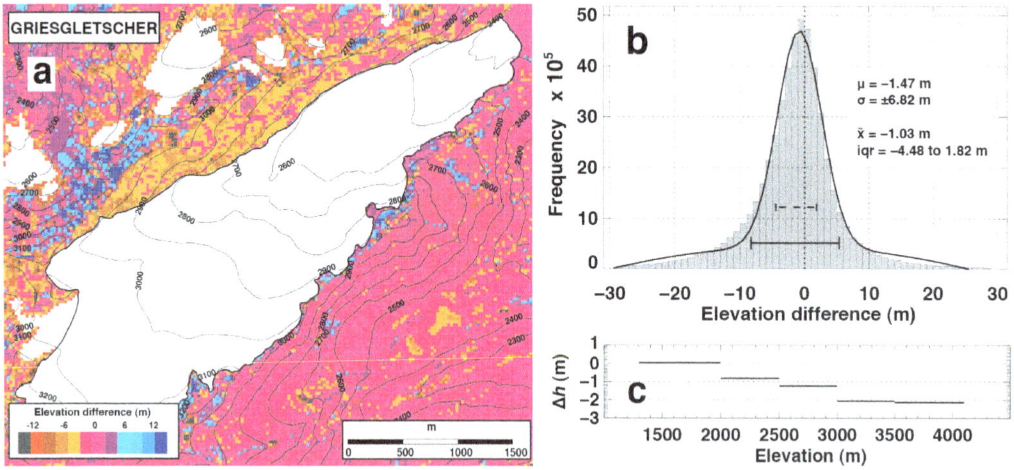

Figure 5. Comparison of input DEMs over stable terrain. (**a**) Spatial distribution of DEM differences outside the 1973 extents of Griesgletscher and adjacent smaller glaciers (masked out in white). (**b**) Frequency distribution of DEM differences over stable terrain around all glaciers. The black curve is a normal fit over the data, the solid bar marks standard deviation at the 1σ level and the dashed bar marks the interquartile range (iqr). In addition, the mean (μ) and median ($\tilde{x}$) elevation differences are given. (**c**) DEM differences over stable terrain around all glaciers averaged over 500 m elevation bands.

and results in $\pm 1.76\,\mathrm{km}^3$. The uncertainty in the geodetic mass balance of individual glacier entities $\sigma_{\mathrm{geod},g}$ (m w.e.) is calculated according to Huss et al. (2009) as

$$\sigma_{\mathrm{geod},g} = \pm\sqrt{(\Delta z_g \cdot \sigma_{f_{\Delta V}})^2 + (f_{\Delta V} \cdot \sigma_{\Delta z,g})^2}, \quad (7)$$

with a conversion factor $f_{\Delta V}$ of 0.85 and a corresponding uncertainty $\sigma_{f_{\Delta V}}$ of 0.06, derived from the assumed mean density of the total volume change of $850\,\mathrm{kg\,m}^{-3}$, and a corresponding uncertainty of $\pm 60\,\mathrm{kg\,m}^{-3}$ (Huss, 2013). The average geodetic elevation change Δz is assumed to be uncorrelated to $f_{\Delta V}$. The resulting values for $\sigma_{\mathrm{geod},g}$ are then divided by the glacier-individual observation period length (Δt) and range between ± 0.02 and $\pm 0.36\,\mathrm{m\,w.e.\,yr}^{-1}$. After the glacier-individual reconstruction of the t_1 values, we consider Δt as robust. Some uncertainty in t_2, and hence also in Δt, of maximum 2 years applies for only a few glaciers for which aerial orthophotographs used as source data for the creation of both the glacier outlines and DEMs at t_2 were acquired during 2 or more survey years (glaciers overlapping two or more areas of t_2 in Fig. 1).

Finally, the uncertainty in the average mass balance of all 1420 Swiss glaciers of the SGI2010 over the reference period 1980–2010 $\sigma_{\overline{B_{\mathrm{ref}}}}$ (m w.e. yr^{-1}) is derived with

$$\sigma_{\overline{B_{\mathrm{ref}}}} = \pm\frac{\sum_{g=1}^{n}((\sigma_{\mathrm{geod},g}/\Delta t)\cdot\overline{A_g})}{\sum_{g=1}^{n}\overline{A_g}}, \quad (8)$$

where $\overline{A_g}$ is the average area of glacier g between 1973 and 2010 and results in $\pm 0.07\,\mathrm{m\,w.e.\,yr}^{-1}$.

4.4 Validation

For validation of average mass balances between 1980 and 2010 we choose 31 glaciers from the data sets of Huss et al. (2010a, b), for which volume changes based on the independent, photogrammetrically derived DEMs show closest temporal accord with our respective observation period.

For individual glaciers, mean mass balance from Huss et al. (2010a, b) sometimes differs considerably from our results over the same period. Nevertheless, these differences do not indicate a systematic error and the mean difference is almost zero (Fig. 6). Hence, when analysing individual glaciers, the uncertainty in geodetic mass balance presented in this study has to be considered. Averaged over subsamples of glaciers or individual catchments though, the accuracy of the average geodetic mass balance is likely of the same order of magnitude as if derived with more precise source DEMs created, for instance, by photogrammetrical techniques.

5 Results

5.1 Changes in surface elevation and area-altitude distribution

The recent (t_2) area-altitude distribution was derived from the combination of the SGI2010 with the swissALTI^{3D} DEMs. For t_1 the SGI1973 and the DHM25 Level 1 DEMs were used. Averaged over the entire Swiss Alps, observed area changes between 1973 and 2010 were largest between 2800 and 2900 m a.s.l. The most heavily glacierized areas were located almost 200 m higher in 2010 (Fig. 7a). For the observation period, the overall volume loss for the entire Swiss

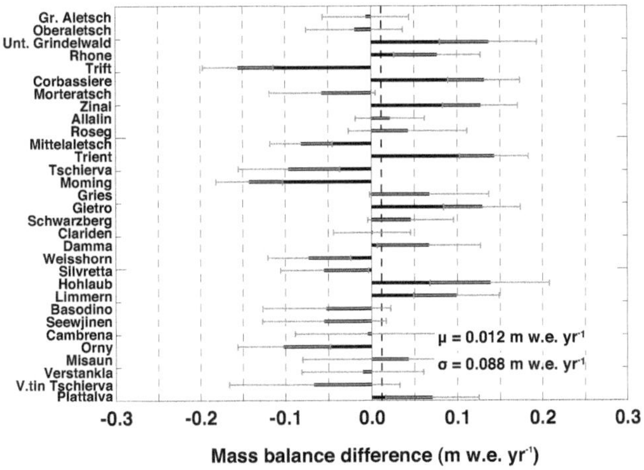

Figure 6. Validation of average geodetic mass balance with mass balance data derived from independent DEMs created using photogrammetry (Huss et al., 2010a, b). The mean difference μ, for which the standard deviation at the 1σ level is also given, is indicated with the bold dashed line. Individual glaciers are sorted according to their area. Error bars indicate the glacier-individual uncertainty in the geodetic mass balance ($\pm\sigma_{\mathrm{geod},g}/\Delta t$).

Table 1. Mean geodetic mass balance 1980–2010 for the main hydrological catchments of Switzerland shown in Fig. 8, listed in alphabetic order.

Catchment	Glacierized Area (km²) 1973	Glacierized Area (km²) 2010	Glacierization (%) 1973	Glacierization (%) 2010	$\overline{B_{\mathrm{ref}}}$ (m w.e. yr^{-1})
Alpenrhein	57.2	27.3	1.78	0.84	−0.69
Linth	23.3	16.0	1.09	0.75	−0.60
Reuss	102.5	75.7	3.03	2.23	−0.62
Aare b. Brugg	229.8	174.5	1.97	1.50	−0.64
Rhone	721.5	569.2	13.68	10.79	−0.59
Doveria	14.1	8.8	8.24	5.12	−0.52
Maggia	5.0	3.0	0.39	0.23	−0.83
Ticino	5.9	2.2	0.39	0.14	−0.67
Maira	20.4	13.0	10.76	6.86	−1.07
Poschiavino	9.5	6.9	5.60	4.06	−0.74
Inn	71.9	46.4	3.71	2.39	−0.81
All catchments	1261.2	942.8	3.89	3.04	−0.62

Alps is $-22.51 \pm 1.76\,\mathrm{km}^3$, of which glaciers still present in 2010 account for $-22.37\,\mathrm{km}^3$. Averaged within 100 m elevation bands, volume loss was strongest between 2700 and 2800 m a.s.l. Corresponding average elevation changes were in good agreement with theoretical considerations by Schwitter and Raymond (1993) and continuously decreased from largest changes nearly at lowermost elevations (terminus of valley glaciers) towards zero in the accumulation area. No elevation bands with positive volume changes were detected (Fig. 7b). Both surface elevation and area changes were remarkable even above 3500 m a.s.l. (Fig. 7). The observed thinning at high altitudes and over the accumulation areas of glaciers results from a combination of ice flow dynamics and reduced accumulation and emphasizes the current state of disequilibrium of glaciers in the Swiss Alps.

5.2 Average mass balance

For the entire Swiss Alps, the area-weighted average mass balance of all 1420 glaciers included in the SGI2010 was $-0.62 \pm 0.07\,\mathrm{m\,w.e.\,yr^{-1}}$ during our reference period 1980–2010. For the main hydrological catchments, it ranged between -0.52 and $-1.07\,\mathrm{m\,w.e.\,yr^{-1}}$ (Fig. 8, $\overline{B_{\mathrm{ref}}}$ in Table 1). Catchments along the north side of the Alps (Aare, Reuss, Linth) showed nearly the same mass changes ($-0.63\,\mathrm{m\,w.e.\,yr^{-1}}$ on average) as for all of Switzerland. In general, glaciers in the Valais Alps (Rhone, Doveria) lost comparatively less ($-0.59\,\mathrm{m\,w.e.\,yr^{-1}}$) and glaciers south of the main Alpine crest (Maggia, Ticino, Maira, Poschiavino) and of the inner-Alpine Inn valley comparatively more ($-0.84\,\mathrm{m\,w.e.\,yr^{-1}}$) mass than on average over the entire Swiss Alps. These differences might be attributed to regionally variable changes in the climatic forcing, glacier sensitivity and/or topographic effects.

The close-ups of the eastern Bernese Alps/western Alps of Central Switzerland (Fig. 9) and the central/eastern Valais Alps (Fig. 10) show a high spatial variability in the temporally homogenized geodetic mass balance 1980–2010. Individual glaciers showed strongly differing responses to a similar change in regional climate forcing. For most of the largest valley glaciers with flat termini (e.g. Unteraargletscher (UAR), Oberaletschgletscher (OAL) and Grosser Aletschgletscher (ALE) in Fig. 9 or Gornergletscher (GOR) and Glacier d'Otemma (OTE) in Fig. 10), mass changes were particularly high. In contrast, Fieschergletscher (FIE in Fig. 9) or Findelengletscher (FIN in Fig. 10) showed smaller mass loss. For small and very small glaciers, the scatter of resulting changes was maximal. Neighbouring glaciers sometimes exhibited a high spatial variability in observed geodetic mass changes (e.g. glaciers in the vicinity of ALE and UAR in Fig. 9 or glaciers close to OTE or within the Weissmiesgruppe in Fig. 10), while for other regions the response was quite uniform (e.g. Rotondogruppe or Blüemlisalp near Kanderfirn (KAN) in Fig. 9). Mass losses of the mostly medium-sized mountain and valley glaciers of the Mischabel- and Weisshorngruppe were also very uniform and comparatively moderate (Fig. 10). This is likely due in a major part to the continentality of these high-lying areas, influenced by one of the coldest and driest climate regimes in the entire Swiss Alps (e.g. Frei and Schär, 1998; Auer et al., 2007). The equilibrium line altitudes are highest for these glaciers (Maisch et al., 2000) and their sensitivity to changes in air temperature and precipitation is expected to be lowest (Oerlemans and Reichert, 2000).

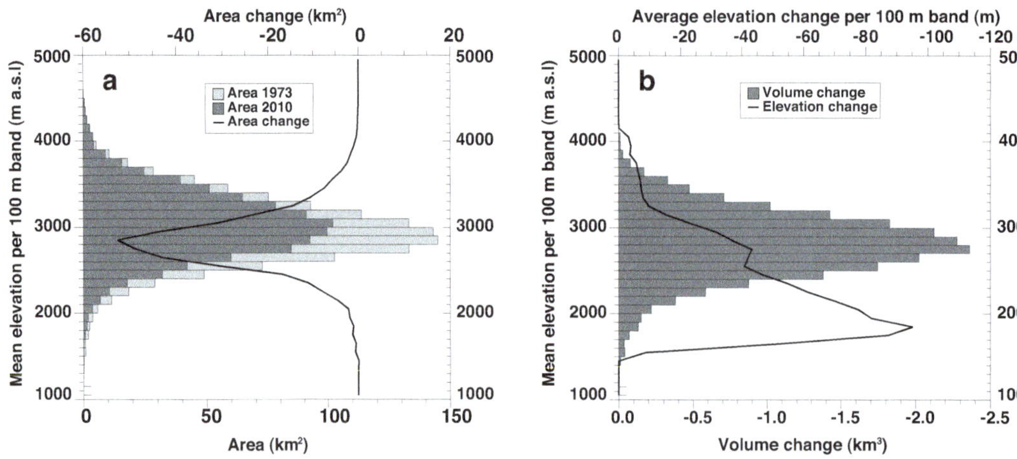

Figure 7. **(a)** Area-elevation distribution in 1973 and 2010 in 100 m elevation bands and corresponding area changes between the two inventories. **(b)** Elevation distribution of the observed volume changes within the observation period as well as resulting average elevation changes per 100 m elevation bands.

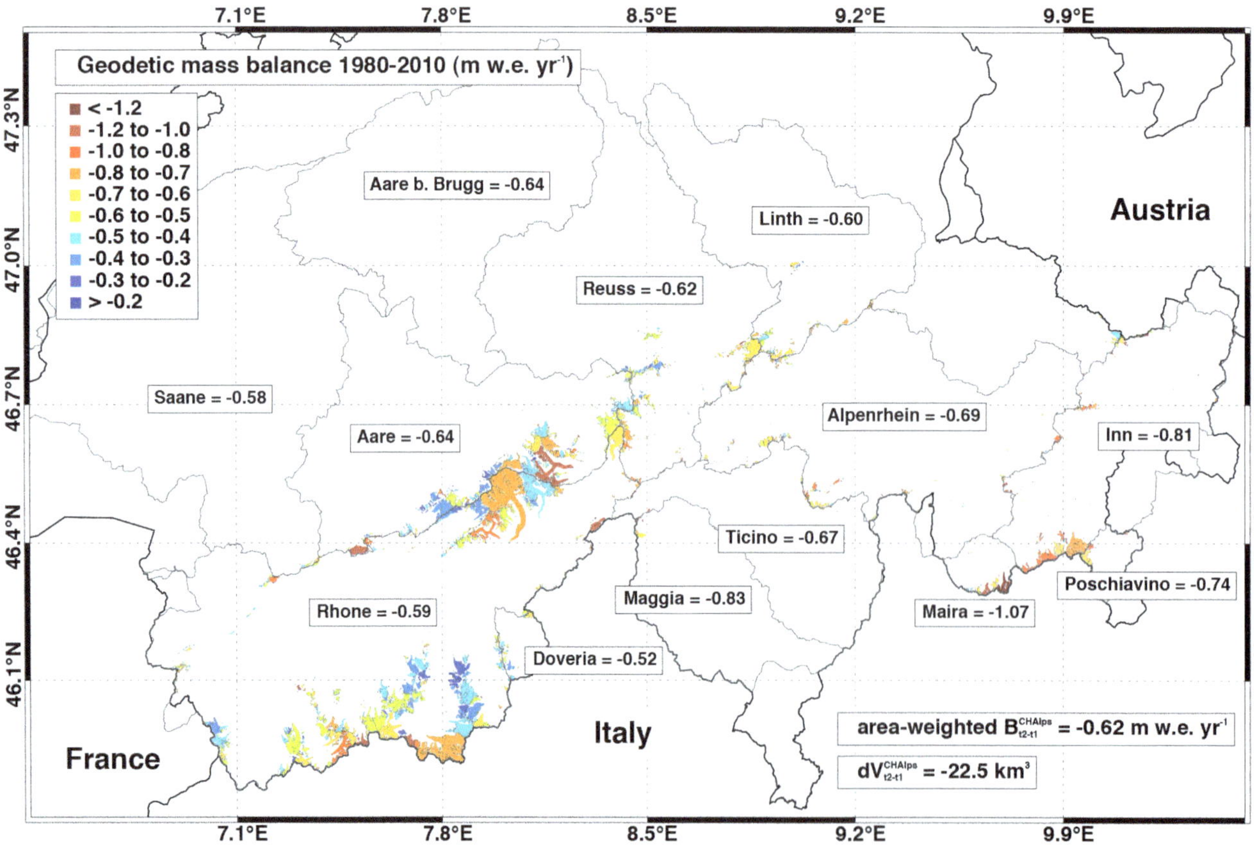

Figure 8. Mean area-weighted geodetic mass balance 1980–2010 for the main hydrological catchments and the entire Swiss Alps.

5.3 Geometrical indices and long-term mass balance variability

For individual glaciers, the relation between the observed geodetic mass changes of 1980–2010 and the explaining geometrical indices of (1) glacier area, (2) median elevation, (3) mean slope of the glacier terminus and (4) mean aspect were not straightforward. Neither average area during 1973–2010 ($r = -0.09$) nor mean aspect ($r = 0.03$) were significantly correlated with the geodetic mass balance (Fig. 11a,

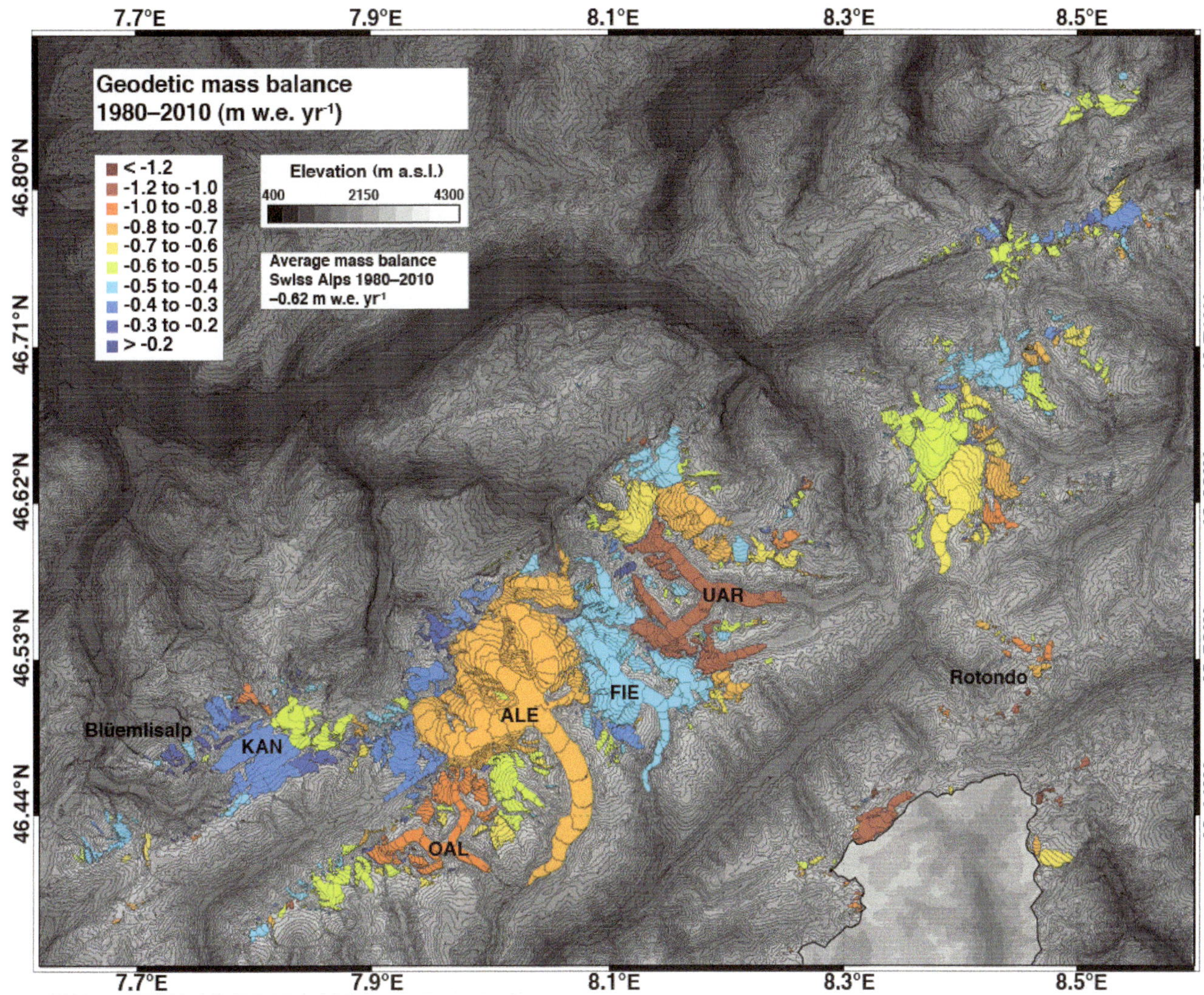

Figure 9. Spatial distribution of the temporally homogenized geodetic mass balance 1980–2010 for the eastern Bernese Alps/western Alps of Central Switzerland. The SGI2010 outlines are taken to illustrate glacier surface area.

d). A weak correlation ($r = 0.22$) was found for median elevation at t_1 (Fig. 11b) and a stronger one ($r = 0.42$) for mean slope over the lowermost 25 % of the glacier surface at t_1 (Fig. 11c). Because part of the significant scatter in Fig. 11a–c is likely caused by glacier-specific uncertainties and local effects, we also calculated the respective mean values for 5 % quantiles of the data (red triangles in Fig. 11a–c) by computing the mean average mass balance for 20 classes of equal sample size. Then, a more structured relation between the selected geometrical indices and long-term mass balance variability emerges. For average area of 1973–2010, the correlation is negative. However, if the smallest glaciers ($< 0.1\,km^2$) are neglected, mean average mass balances did not vary considerably for different size classes (Fig. 11a). For median elevation, the correlation is positive. The higher the median elevation of a glacier at t_1, the less negative the average mass budget tends to be (Fig. 11b). Median elevation is a proxy for the balanced-budget equilibrium line altitude (ELA_0) (Braithwaite et al., 2013), which in turn depends on continentality. The latter can be approximated as a function of mean annual air temperature and precipitation at the ELA_0 (Shumsky, 1964; Haeberli et al., 1989). Glaciers influenced by more maritime climatic conditions react more sensitively to changes in air temperature and precipitation than more continental glaciers do (Oerlemans and Reichert, 2000) and are thus expected to show more negative mass balance in a period of global warming. For mean slope of the glacier terminus, the correlation is again positive. The more gently sloping the glacier's ablation area at t_1, here approximated with the glacier surface over the lowermost 25 % at t_1, the more negative the average mass budget (Fig. 11c). The flattest glacier termini are typically those of larger valley glaciers, which also show comparatively more negative mass balances due to their longer response times (Hoelzle

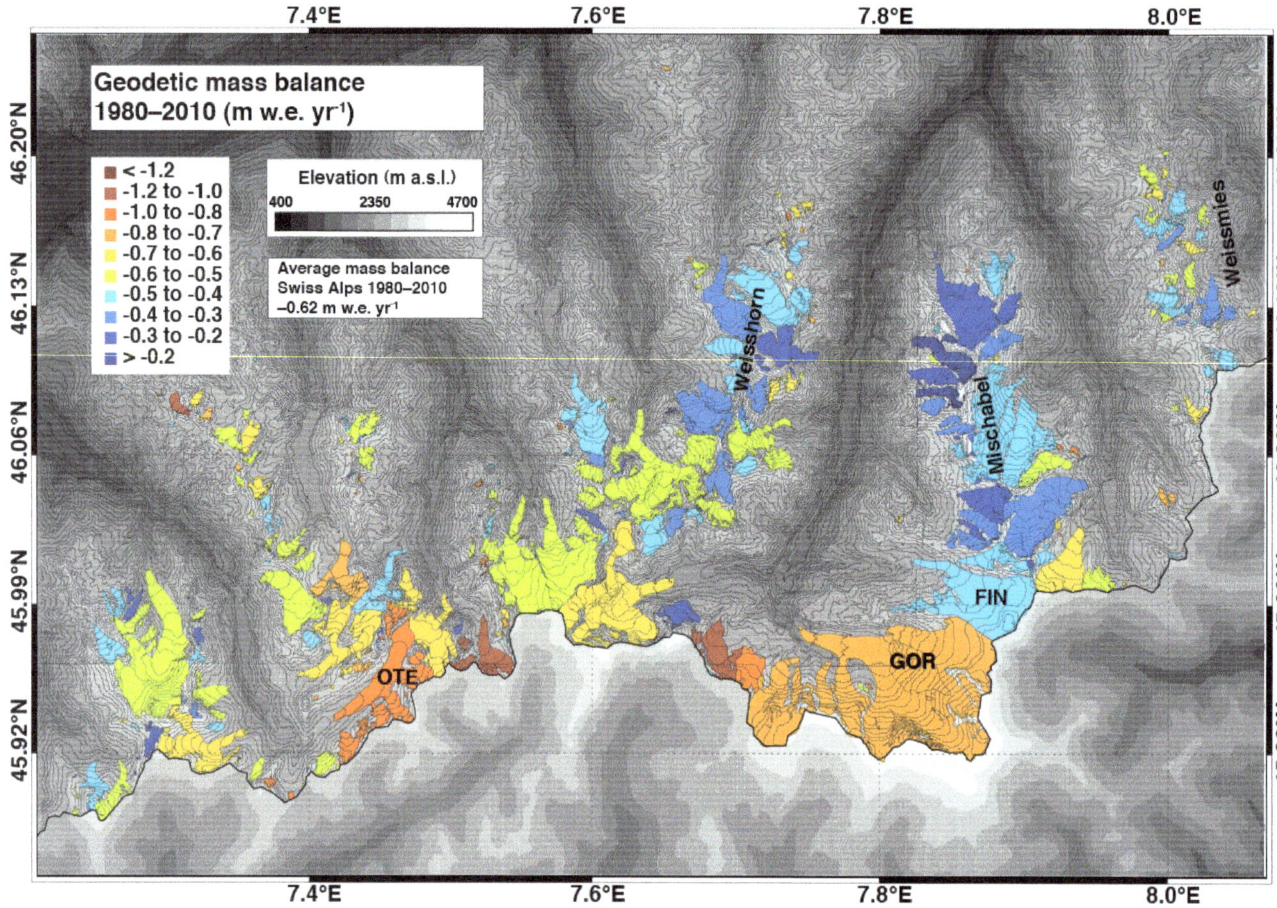

Figure 10. Spatial distribution of the temporally homogenized geodetic mass balance 1980–2010 for the central/eastern Valais Alps. The SGI2010 outlines are taken to illustrate glacier surface area.

et al., 2003). Furthermore, they usually flow down into narrow valley bottoms at low altitudes (Maisch et al., 2000). There, higher air temperatures are in favour of increased melt rates, and the effects of surface albedo lowering (accumulation of dust) and mass balance vs. altitude (fast vertical ice thickness loss) act as self-reinforcing mechanisms (Haeberli et al., 2007; Oerlemans et al., 2009). North-east-exposed glaciers lost most mass, southwest-exposed glaciers lost least (Fig. 11d). This is likely due to the different sensitivity to changes in air temperature increase for the respective classes of mean aspect (Evans and Cox, 2005; Evans, 2006). Because of the stronger influence of the shortwave radiation component and the fact that they are located at higher elevation, south-exposed glaciers generally react less sensitively to air temperature changes than north-exposed glaciers. Also, south-exposed glaciers are often smaller and thinner and therefore generally have a shorter response time and thus less negative mass balance.

6 Comparison to other studies for the European Alps

For the 1420 glaciers included in the SGI2010, the average mass balance of -0.62 ± 0.07 m w.e. yr^{-1} calculated over the reference period 1980–2010 is comparable to observed mass changes reported for other glacierized regions of the European Alps for recent years. Carturan et al. (2013a) computed an average mass budget of -0.69 ± 0.12 m w.e. yr^{-1} between the early 1980s and the mid-2000s for glaciers of the Ortler Alps group on the southwestern border of South Tyrol, Italy. Applying our method for temporal homogenization of mass changes derived from DEM differencing (cf. Sect. 3.1, Fig. 3), we calculate -0.65 m w.e. yr^{-1} for the Swiss Alps when averaged over the same time period as analyzed by Carturan et al. (2013a). From area and volume changes reported by Abermann et al. (2009) for the Austrian Ötztal Alps between 1969 and 2006, we calculate a mean mass balance of -0.40 m w.e. yr^{-1} averaged for 81 glaciers. The average mass balance for all Swiss glaciers is -0.39 m w.e. yr^{-1} from 1969 to 2006.

To derive surface elevation and mass changes for the entire Swiss Alps, Paul and Haeberli (2008) compared the

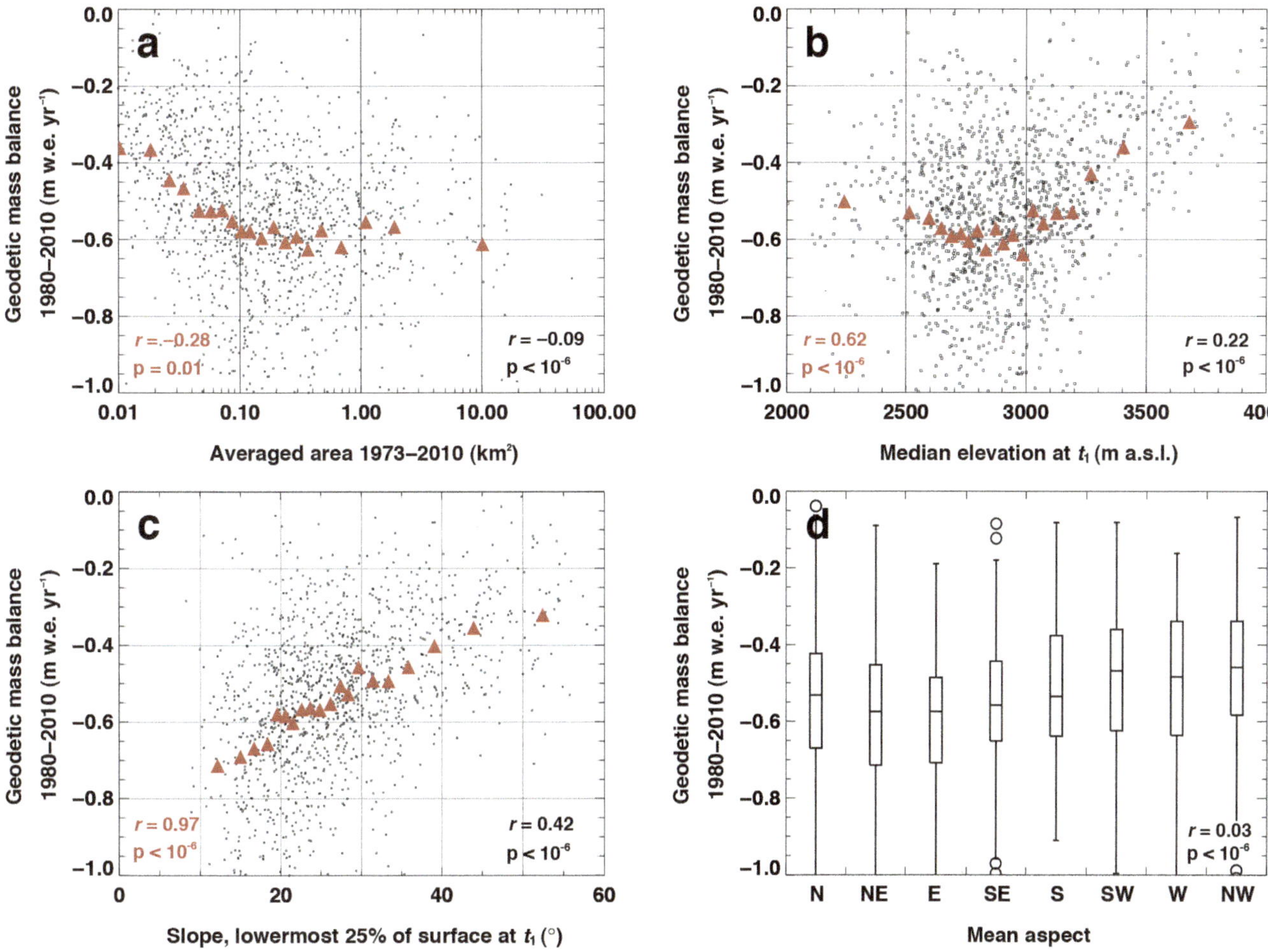

Figure 11. Correlation of temporally homogenized geodetic mass balance 1980–2010 and several geometrical indices. **(a)** Average area 1973–2010, **(b)** median elevation at t_1, **(c)** surface slope averaged over the lowermost 25 % of the glacier and **(d)** mean aspect, where black boxes enclose the interquartile range (iqr) of the data, defined at iqr75–iqr25, and whiskers extend out to the maximum or minimum value of the data, or to 1.5 times either the iqr75 or iqr25 when there is data beyond this range. Outliers are identified with small circles. Red triangles show mean values for 5 % quantiles of the data. r is the linear correlation coefficient and p the significance level of the respective relation (black always refers to the whole data set, while red refers to the mean values for the 5 % quantiles only).

DHM25 Level 1 DEMs to the SRTM DEM from February 2000 and combined the former with the SGI1973 and the latter with the SGI2000 created from medium-resolution (30 m) satellite imagery. They assumed $t_1 = 1985$ as constant and report an average mass balance of -0.78 m w.e. yr^{-1} between 1985 and 1999. Over the same reference period, we find a cumulative average mass balance for all Swiss glaciers of -0.60 m w.e. yr^{-1} by temporally homogenizing mass changes derived from DEM differencing (cf. Eq. 3). Both the quality of the different source data used and methodologies applied contribute to the differences in derived average mass balance. According to Jarvis et al. (2008), the vertical accuracy of the 90 m resolution SRTM DEM is ± 30 m. Over glacierized areas in Switzerland, however, it is probably considerably more accurate (Paul, 2008). Nevertheless, the quality of the SRTM DEM is not comparable to the recently compiled 2 m swissALTI^{3D} DEMs. If the SRTM DEM is used, the impacts of the penetration of radar waves into snow and ice need to be considered (Berthier et al., 2006; Gardelle et al., 2012b) because penetration can reach as much as 10 m for the SRTM C-band radar (Dall et al., 2001). This could explain the more negative surface elevation changes over accumulation areas observed by Paul and Haeberli (2008) who compared the SRTM DEM to photogrammetrically derived DEMs by Bauder et al. (2007). If all of the difference between the average 1985–1999 mass balance from Paul and Haeberli (2008) and our approach was attributed to the penetration of the radar signal into snow and ice, one could infer a mean penetration of the SRTM C-band of about 8 m beneath the February 2000 surface (calculated as 0.18 m w.e. yr^{-1} $\times$ 14 years, divided by an assumed mean firn density of 650 kg m^{-3} and an accumulation area ratio of 0.5). This would be somewhat higher than values reported for the Karakoram (Gardelle et al., 2012b) or southeastern

Alaska (Melkonian et al., 2014). In addition, preprocessing of the DEM prior to differencing may cause uncertainty if the source DEMs have to be downscaled to the same grid cell resolution (Carturan et al., 2013a). For our study, the 2 m resolution swissALTI^{3D} DEMs were upscaled to 25 m, corresponding to the cell size of the DHM25 Level 1 DEMs. The quality of both the source DEMs and the glacier outlines used is important, especially if changes of small and very small glaciers have to be assessed. Fischer et al. (2014) showed that the accuracy of glacier outlines derived from semi-automated satellite remote sensing approaches using medium-resolution satellite imagery does not satisfy the requirements for change assessments of glaciers $< 1.0\,\text{km}^2$.

Related to the methodological approaches used, the following issues likely also contribute to the differences in calculated average mass balances between Paul and Haeberli (2008) and this study: assuming a density of the volumetric change of $900\,\text{kg m}^{-3}$ implies that neither the mean firn density nor the firn thickness and area changed over the observation period. For glaciers in the European Alps, however, significant changes in both firn extent and density have been reported (e.g. Carturan et al., 2013a; Helfricht et al., 2014). Following Huss (2013), we apply a mean conversion factor of $850\,\text{kg m}^{-3}$ here. Furthermore, the average mass balance of individual glaciers (Eq. 2) can be significantly biased if differences in the reference years of the source data are not considered. For Grosser Aletschgletscher, assuming t_1 as 1985 instead of 1980 changes the result by $0.14\,\text{m w.e. yr}^{-1}$ (-0.87 instead of $-0.73\,\text{m w.e. yr}^{-1}$). Defining t_1 as 1985 for all glaciers in Switzerland would change the mean geodetic mass balance by $0.03\,\text{m w.e. yr}^{-1}$. Hence, differences in the reference years of the source data explain 17 %, differences in the assumptions made regarding the density of volume change 24 %, and differences in the quality of the source data used 59 % of the disagreement between the calculated average mass balances reported here and those reported by Paul and Haeberli (2008).

7 Conclusions

We presented glacier-specific changes in surface elevation, volume and geodetic mass balance for all glaciers in the Swiss Alps between 1980 and 2010. The DHM25 Level 1 DEMs acquired from 1961 to 1991 over glacierized areas were combined with the 1973 glacier inventory and compared to the swissALTI^{3D} DEMs from 2008 to 2011 and glacier outlines from the new Swiss Glacier Inventory SGI2010. By making use of these recently compiled new source data of improved quality, we were able to recompute and extend the data set of surface elevation changes in the Swiss Alps by Paul and Haeberli (2008). From the DEM differencing we derive an overall volume loss of $-22.51 \pm 1.76\,\text{km}^3$ for the entire Swiss Alps for the observation period. Glaciers still present in 2010 account for $-22.37\,\text{km}^3$ of this change. Observed glacier volume changes were greatest between 2700 and 2800 m a.s.l. and still noticeable above 3500 m a.s.l. No elevation bands with positive elevation changes were identified. Results were validated against elevation changes from independent photogrammetrically derived DEMs. An in-depth accuracy assessment of the computed volume and mass changes of the glaciers investigated was performed. Elevation-, slope- and aspect-dependent errors were investigated by comparison of both source DEMs over stable terrain outside the glaciers. Because of significant differences in the acquisition dates of individual DEMs, annual mountain-range mass balance variability (Huss, 2012) was used to temporally homogenize the geodetic mass changes to the reference time interval 1980–2010. Over this period, the area-weighted mean geodetic mass balance for the entire Swiss Alps was $-0.62 \pm 0.07\,\text{m w.e. yr}^{-1}$. For the main hydrological catchments of Switzerland, mean balances ranged from -0.52 to $-1.07\,\text{m w.e. yr}^{-1}$.

To better understand the spatial variability of the observed surface elevation and mass changes, we investigated the relationship between observed mass changes and topographic factors. Overall, as shown by several previous studies, differences in glacier hypsometry can partly explain the general pattern of glacier responses to changes in climatic forcing. We found strongest correlations for the geometrical indices terminus slope (i.e. mean over the lowermost 25 %) and median elevation.

The data set presented in this article is useful for various future studies and applications. For instance, both mass-balance-driven glacier evolution models for the entire Swiss Alps and extrapolations of the measured mass balances of single glaciers to entire catchments can be validated and improved. This would also imply a reduction in the uncertainty of future runoff projections from glacierized basins in Switzerland. Moreover, it is a valuable starting point for testing more sophisticated methods to explain the spatial variability in long-term mass balance.

Acknowledgements. This study was supported by the Swiss National Science Foundation (SNSF), grant 200021_137586. The SWISSIMAGE and swissALTI^{3D} data sets were purchased from the Swiss Federal Office of Topography, license no. 5704003629. Special thanks to the Laboratory of Hydraulics, Hydrology and Glaciology (VAW) at the ETH Zurich for providing us with photogrammetrically derived DEMs available for selected Swiss glaciers over our period of interest, to M. Rickenbacher, Y. Weidmann, and to the Department of Geography at the University of Zurich for providing us with the 1973 outlines in digital format. Finally, thanks to the scientific editor M. Sharp and to three anonymous reviewers, whose comments were helpful for finalizing the paper.

Edited by: M. Sharp

References

Abermann, J., Lambrecht, A., Fischer, A., and Kuhn, M.: Quantifying changes and trends in glacier area and volume in the Austrian Ötztal Alps (1969-1997-2006), The Cryosphere, 3, 205–215, doi:10.5194/tc-3-205-2009, 2009.

Abermann, J., Fischer, A., Lambrecht, A., and Geist, T.: On the potential of very high-resolution repeat DEMs in glacial and periglacial environments, The Cryosphere, 4, 53–65, doi:10.5194/tc-4-53-2010, 2010.

Abermann, J., Kuhn, M., and Fischer, A.: Climatic controls of glacier distribution and glacier changes in Austria, Ann. Glaciol., 52, 83–90, 2011.

Auer, I., Böhm, R., Jurkovic, A., Lipa, W., Orlik, A., Potzmann, R., Schöner, W., Ungersböck, M., Matulla, C., Briffa, K., Jones, P., Efthymiadis, D., Brunetti, M., Nanni, T., Maugeri, M., Mercalli, L., Mestre, O., Moisselin, J., Begert, M., Müller-Westermeier, G., Kveton, V., Bochnicek, O., Stastny, P., Lapin, M., Szalai, S., Szentimrey, T., Cegnar, T., Dolinar, M., Gajic-Capka, M., Zaninovic, K., Majstorovic, Z., and Nieplova, E.: HISTALP – historical instrumental climatological surface time series of the Greater Alpine Region, Int. J. Climatol., 27, 17–46, 2007.

Bauder, A., Funk, M., and Huss, M.: Ice-volume changes of selected glaciers in the Swiss Alps since the end of the 19th century, Ann. Glaciol., 46, 145–149, 2007.

Benn, D. I. and Evans, D. J. A.: Glaciers and Glaciation, Hodder Education, London, 802 pp., 2010.

Benn, D. I. and Lehmkuhl, F.: Mass balance and equilibrium-line altitudes of glaciers in high-mountain environments, Quat. Int., 65–66, 15–29, 2000.

Berthier, E., Arnaud, Y., Vincent, C., and Rémy, F.: Biases of SRTM in high-mountain areas: implications for the monitoring of glacier volume changes, Geophys. Res. Lett., 33, L08502, doi:10.1029/2006GL025862, 2006.

Berthier, E., Schiefer, E., Clarke, G. K. C., Menounos, B., and Rémy, F.: Contribution of Alaskan glaciers to sea-level rise derived from satellite imagery, Nat. Geosci., 3, 92–95, doi:10.1038/ngeo737, 2010.

Bolch, T., Buchroithner, M., Pieczonka, T., and Kunert, A.: Planimetric and volumetric glacier changes in the Khumbu Himal, Nepal, since 1962 using Corona, Landsat TM and ASTER data, J. Glaciol., 54, 592–600, 2008.

Braithwaite, R. J., Raper, S. C. B., and Candela, R.: Recent changes (1991–2010) in glacier mass balance and air temperature in the European Alps, Ann. Glaciol., 54, 139–146, doi:10.3189/2013AoG63A285, 2013.

Carturan, L., Filippi, R., Seppi, R., Gabrielli, P., Notarnicola, C., Bertoldi, L., Paul, F., Rastner, P., Cazorzi, F., Dinale, R., and Dalla Fontana, G.: Area and volume loss of the glaciers in the Ortles-Cevedale group (Eastern Italian Alps): controls and imbalance of the remaining glaciers, The Cryosphere, 7, 1339–1359, doi:10.5194/tc-7-1339-2013, 2013a.

Carturan, L., Baldassini, G. A., Bondesan, A., Calligaro, S., Carton, A., Cazorzi, F., Dalla Fontana, G., Francese, R., Guarnieri, A., Milan, N., Moro, D., and Tarolli, P.: Current behaviour and dynamics of the lowermost Italian glacier (Montasio Occidentale, Julian Alps), Geogr. Ann. A, 95, 79–96, doi:10.1111/geoa.12002, 2013b.

Cox, L. H. and March, R. S.: Comparison of geodetic and glaciological mass-balance techniques, Gulkana Glacier, Alaska, U.S.A., J. Glaciol., 50, 363–370, 2004.

Dall, J., Madsen, S. N., Keller, K., and Forsberg, R.: Topography and Penetration of the Greenland Ice Sheet Measured with Airborne SAR Interferometry, Geophys. Res. Lett., 28, 1703–1706, 2001.

DeBeer, C. and Sharp, M.: Topographic influences on recent changes of very small glaciers in the Monashee Mountains, British Columbia, Canada, J. Glaciol., 55, 691–700, 2009.

Diolaiuti, G. A., Bocchiola, D., Vagliasindi, M., D'Agata, C., and Smiraglia, C.: The 1975–2005 glacier changes in Aosta Valley (Italy) and the relations with climate evolution, Progr. Phys. Geogr., 36, 764–785, 2012.

Evans, I. S.: Glacier distribution in the Alps: statistical modelling of altitude and aspect, Geogr. Ann. A, 88, 115–133, 2006.

Evans, I. S. and Cox, N. J.: Global variations of local asymmetry in glacier altitude: separation of north-south and east-west components, J. Glaciol., 51, 469–482, 2005.

Fischer, A.: Comparison of direct and geodetic mass balances on a multi-annual time scale, The Cryosphere, 5, 107–124, doi:10.5194/tc-5-107-2011, 2011.

Fischer, M., Huss, M., Barboux, C., and Hoelzle, M.: The new Swiss Glacier Inventory SGI2010: Relevance of using high-resolution source data in areas dominated by very small glaciers, Arct., Antarct. Alp. Res., 46, 935–947, doi:10.1657/1938-4246-46.4.933, 2014.

Frei, C. and Schär, C.: A precipitation climatology of the Alps from high-resolution rain-gauge observations, Int. J. Climatol., 18, 873–900, 1998.

Furbish, D. J. and Andrews, J. T.: The use of hypsometry to indicate long-term stability and response of valley glaciers to changes in mass transfer, J. Glaciol., 30, 199–211, 1984.

Gardelle, J., Berthier, E., and Arnaud, Y.: Slight mass gain of Karakoram glaciers in the early twenty-first century, Nat. Geosci., 5, 322–325, doi:10.1038/ngeo1450, 2012a.

Gardelle, J., Berthier, E., and Arnaud, Y.: Impact of resolution and radar penetration on glacier elevation changes computed from DEM differencing, J. Glaciol., 58, 419–422, 2012b.

Gardent, M., Rabatel, A., Dedieu, J.-P., and Deline, P.: Multitemporal glacier inventory of the French Alps fromthe late 1960s to the late 2000s, Global Planet. Change, 120, 24–37, 2014.

Gardner, A. S., Moholdt, G., Cogley, J. G., Wouters, B., Arendt, A. A., Wahr, J., Berthier, E., Hock, R., Pfeffer, W. T., Kaser, G., Ligtenberg, S. R. M., Bolch, T., Sharp, M. J., Hagen, J. O., van den Broecke, M. R., and Paul, F.: A reconciled estimate of glacier contributions to sea level rise: 2003 to 2009, Science, 340, 852–857, 2013.

Glaciological Reports: The Swiss Glaciers, 1958/1959–2010/11, No. 80–132, Yearbooks of the Cryospheric Commission of the Swiss Academy of Sciences (SCNAT), Laboratory of Hydraulics, Hydrology and Glaciology (VAW) of ETH Zurich, Zurich, 1960–2015.

Haeberli, W., Alean, J.-C., Müller, P., and Funk, M.: Assessing risks from glacier hazards in high mountain regions: some experiences in the Swiss Alps, Ann. Glaciol., 13, 77–101, 1989.

Haeberli, W., Hoelzle, M., Paul, F., and Zemp, M.: Integrated monitoring of mountain glaciers as key indicators of global climate change: the European Alps, Ann. Glaciol., 46, 150–160, 2007.

Helfricht, K., Kuhn, M., Keuschnig, M., and Heilig, A.: Lidar snow cover studies on glaciers in the Ötztal Alps (Austria): comparison with snow depths calculated from GPR measurements, The Cryosphere, 8, 41–57, doi:10.5194/tc-8-41-2014, 2014.

Hoelzle, M., Haeberli, W., Dischl, M., and Peschke, W.: Secular glacier mass balances derived from cumulative glacier length changes, Global Planet. Change, 36, 295–306, 2003.

Hughes, P. D.: Twenty-first century glaciers and climate in the Prokletije Mountains, Albania, Arct., Antarct. Alp. Res., 41, 455–459, 2009.

Huss, M.: Present and future contribution of glaciers to runoff from macroscale drainage basins in Europe, Water Resour. Res., 47, W07511, doi:10.1029/2010WR010299, 2011.

Huss, M.: Extrapolating glacier mass balance to the mountain-range scale: the European Alps 1900–2100, The Cryosphere, 6, 713–727, doi:10.5194/tc-6-713-2012, 2012.

Huss, M.: Density assumptions for converting geodetic glacier volume change to mass change, The Cryosphere, 7, 877–887, 2013, http://www.the-cryosphere-discuss.net/7/877/2013/.

Huss, M. and Farinotti, D.: Distributed ice thickness and volume of all glaciers around the globe, J. Geophys. Res., 117, F04010, doi:10.1029/2012JF002523, 2012.

Huss, M., Bauder, A., and Funk, M.: Homogenization of long-term mass balance time series, Ann. Glaciol., 50, 198–206, 2009.

Huss, M., Hock, R., Bauder, A., and Funk, M.: 100-year mass changes in the Swiss Alps linked to the Atlantic Multidecadal Oscillation, Geophys. Res. Lett., 37, L10501, doi:10.1029/2010GL042616, 2010a.

Huss, M., Usselmann, S., Farinotti, D., and Bauder, A: Glacier mass balance in the south-eastern Swiss Alps since 1900 and perspectives for the future, Erdkunde, 64, 119–140, doi:10.3112/erdkunde.2010.02.02, 2010b.

Huss, M., Hock, R., Bauder, A., and Funk., M.: Conventional versus reference-surface mass balance, J. Glaciol., 58, 278–286, 2012.

IPCC: Climate Change 2013, The Physical Science Basis, Working Group I Contribution to the Fifth Assessment Report of the Intergovernmental Panel on Climate Change, WMO / UNEP, Cambridge University Press, Geneva, 2013.

Jarvis, A., Reuter, H. I., Nelson, A., and Guevara, E.: Hole-filled SRTM for the globe Version 4, available from the CGIAR-CSI SRTM 90 m Database, 2008.

Jóhannesson, T., Raymond, C., and Waddington, E.: Time-scale for adjustment of glaciers to changes in mass balance, J. Glaciol., 35, 355–369, 1989.

Kaser, G., Grosshauser, M., and Marzeion, B.: Contribution potential of glaciers to water availability in different climate regimes, Proc. Natl. Acad. Sci. USA, 107, 20223–20227, 2010.

Kuhn, M.: The mass balance of very small glaciers, Z. Gletscherkd. Glazialgeol., 31, 171–179, 1995.

Kuhn, M., Markl, G., Kaser, G., Nickus, U., Obleitner, F., and Schneider, H.: Fluctuations of climate and mass balances: different responses of two adjacent glaciers, Z. Gletscherkd. Glazialgeol., 21, 409–416, 1985.

Lambrecht, A. and Kuhn, M.: Glacier changes in the Austrian Alps during the last three decades, derived from the Austrian glacier inventory, Ann. Glaciol., 46, 177–184, 2007.

Larsen, C. F., Motyka, R. J., Arendt, A. A., Echelmeyer, K. A., and Geissler, P. E.: Glacier changes in southeast Alaska and northwest British Columbia and contribution to sea level rise, J. Geophys. Res., 112, F01007, 10.1029/2006JF000586, 2007.

Maisch, M., Wipf, A., Denneler, B., Battaglia, J., and Benz, C.: Die Gletscher der Schweizer Alpen: Gletscherhochstand 1850, Aktuelle Vergletscherung, Gletscherschwundszenarien, vdf Hochschulverlag, Zürich, 2nd Edn., 2000.

Marzeion, B., Jarosch, A. H., and Hofer, M.: Past and future sea-level change from the surface mass balance of glaciers, The Cryosphere, 6, 1295–1322, doi:10.5194/tc-6-1295-2012, 2012.

Melkonian, A. K., Willis, M. J., and Pritchard, M. E.: Satellite-derived volume loss rates and glacier speeds for the Juneau Icefield, Alaska, J. Glaciol., 60, 743–760, doi:10.3189/2014JoG13J181, 2014.

Motyka, R. J., Fahnestock, M., and Truffer, M.: Volume change of Jakobshavn Isbrae, West Greenland: 1985–1997–2007, J. Glaciol., 56, 635–646, 2010.

Müller, F., Caflish, T., and Müller, G.: Firn und Eis der Schweizer Alpen: Gletscherinventar, No. 57, Geographisches Institut der ETH Zürich, Zürich, 1976.

Nuth, C. and Kääb, A.: Co-registration and bias corrections of satellite elevation data sets for quantifying glacier thickness change, The Cryosphere, 5, 271–290, doi:10.5194/tc-5-271-2011, 2011.

Nuth, C., Moholdt, G., Kohler, J., Hagen, J. O., and Kääb, A.: Svalbard glacier elevation changes and contribution to sea level rise, J. Geophys. Res., 115, F01008, doi:10.1029/2008JF001223, 2010.

Oerlemans, J.: The microclimate of valley glaciers, Igitur, Utrecht University, Utrecht, 2010.

Oerlemans, J. and Reichert, B. K.: Relating glacier mass balance to meteorological data using a Seasonal Sensitivity Characteristic (SSC), J. Glaciol., 46, 1–6, 2000.

Oerlemans, J., Giesen, R. H., and van den Broeke, M. R.: Retreating alpine glaciers: increased melt rates due to accumulation of dust (Vadret da Morteratsch, Switzerland), J. Glaciol., 55, 729–736, 2009.

Paul, F.: The new Swiss glacier inventory 2000: Application of remote sensing and GIS, PhD thesis, Department of Geography, University of Zurich, Zurich, 2004.

Paul, F.: Calculation of glacier elevation changes with SRTM: is there an elevation dependent bias? J. Glaciol., 54, 945–946, 2008.

Paul, F. and Haeberli, W.: Spatial variability of glacier elevation changes in the Swiss Alps obtained from two digital elevation models, Geophys. Res. Lett., 35, L21502, doi:10.1029/2008GL034718, 2008.

Paul, F., Kääb, A., Maisch, M., Kellenberger, T., and Haeberli, W.: Rapid disintegration of Alpine glaciers observed with satellite data, Geophys. Res. Lett., 31, L21402, doi:10.1029/2004GL020816, 2004.

Paul, F., Frey, H., and LeBris, R.: A new glacier inventory for the European Alps from Landsat TM scenes of 2003: challenges and results, Ann. Glaciol., 52, 144–152, 2011.

Radić, V. and Hock, R.: Glaciers in the Earth's hydrological cycle: assessments of glacier mass and runoff changes on global and regional scales, Surv. Geophys., 35, 813–837, doi:10.1007/s10712-013-9262-y, 2014.

Rickenbacher, M.: Die digitale Modellierung des Hochgebirges im DHM25 des Bundesamtes für Landestopographie, Wiener Schr. Geogr. Kartogr., 11, 49–55, 1999.

Rignot, E., Rivera, A., and Casassa, G.: Contribution of the Patagonia Icefields of South America to sea level rise, Science, 302, 434–437, doi:10.1126/science.1087393, 2003.

Rolstad, C., Haug, T., and Denby, B.: Spatially integrated geodetic glacier mass balance and its uncertainty based on geostatistical analysis: application to the western Svartisen ice cap, Norway, J. Glaciol., 55, 666–680, 2009.

Sapiano, J. J., Harrison, W. D., and Echelmeyer, K. A.: Elevation, volume and terminus changes of nine glaciers in North America, J. Glaciol., 44, 119–135, 1998.

Schwitter, M. P. and Raymond, C. F.: Changes in the longitudinal profiles of glaciers during advance and retreat, J. Glaciol., 39, 582–590, 1993.

Shumsky, P. A.: Principles of structural glaciology, Translated from the Russian by D. Kraus, Dover Publications, Inc., New York, 497 pp., 1964.

Sorg, A., Bolch, T., Stoffel, M., Solomina, O., and Beniston, M.: Climate change impacts on glaciers and runoff in Tien Shan (Central Asia), Nat. Clim. Change, 2, 725–731, 2012.

Swiss Federal Office of Topography, swisstopo: DHM25 – Das digitale Höhenmodell der Schweiz, Bundesamt für Landestopographie swisstopo, Wabern, 2000.

Swiss Federal Office of Topography, swisstop: swissALTI^{3D}, Ausgabebericht 2013, Bundesamt für Landestopographie swisstopo, Wabern, 2013.

Thibert, E., Blanc, R., Vincent, C., and Eckert, N.: Glaciological and volumetric mass-balance measurements: error analysis over 51 years for Glacier de Sarennes, French Alps, J. Glaciol, 54, 522–532, 2008.

Trüssel, B. L., Motyka, R. J., Truffer, M., and Larsen, C. F.: Rapid thinning of lake-calving Yakutat Glacier and the collapse of the Yakutat Icefield, southeast Alaska, USA, J. Glaciol., 59, 149–161, 2013.

VanLooy, J. A. and Forster, R. R.: Use of historical elevation data to calculate surface-elevation and volume changes of Ha-Iltzuk Icefield, southwest British Columbia, Canada, 1970 to mid-1980s, Ann. Glaciol., 52, 109–115, 2011.

Vincent, C.: Influence of climate change over the 20th century on four French glacier mass balances, J. Geophys. Res., 107, 4375, doi:10.1029/2001JD000832, 2002.

WGMS (World Glacier Monitoring Service): Fluctuations of Glaciers 2005–2010 (Vol. X), edited by: Zemp, M., Frey, H., Gärtner-Roer, I., Nussbaumer, S. U., Hoelzle, M., Paul, F., and Haeberli, W., ICSU (WDS) / IUGG (IACS) / UNEP / UNESCO / WMO, World Glacier Monitoring Service, Zurich, Switzerland, 336 pp., publication based on database version: doi:10.5904/wgms-fog-2012-11, 2012.

Zemp, M., Paul, F., Hoelzle, M., and Haeberli, W.: Glacier fluctuations in the European Alps 1850–2000: an overview and spatiotemporal analysis of available data, in: The darkening peaks: Glacial retreat in scientific and social context, edited by: Orlove, B., Wiegandt, E., and Luckman, B. H., University of California Press, 152–167, 2008.

Zemp, M., Hoelzle, M., and Haeberli, W.: Six decades of glacier mass-balance observations: a review of the worldwide monitoring network, Ann. Glaciol., 50, 101–111, 2009.

Zemp, M., Thibert, E., Huss, M., Stumm, D., Rolstad Denby, C., Nuth, C., Nussbaumer, S. U., Moholdt, G., Mercer, A., Mayer, C., Joerg, P. C., Jansson, P., Hynek, B., Fischer, A., Escher-Vetter, H., Elvehøy, H., and Andreassen, L. M.: Reanalysing glacier mass balance measurement series, The Cryosphere, 7, 1227–1245, doi:10.5194/tc-7-1227-2013, 2013.

6

Updated cloud physics in a regional atmospheric climate model improves the modelled surface energy balance of Antarctica

J. M. van Wessem[1], C. H. Reijmer[1], J. T. M. Lenaerts[1], W. J. van de Berg[1], M. R. van den Broeke[1], and E. van Meijgaard[2]

[1]Institute for Marine and Atmospheric Research Utrecht, Utrecht University, Utrecht, the Netherlands
[2]Royal Netherlands Meteorological Institute, De Bilt, the Netherlands

Correspondence to: J. M. van Wessem (j.m.vanwessem@uu.nl)

Abstract. In this study the effects of changes in the physics package of the regional atmospheric climate model RACMO2 on the modelled surface energy balance, near-surface temperature and wind speed of Antarctica are presented. The physics package update primarily consists of an improved turbulent and radiative flux scheme and a revised cloud scheme that includes a parameterisation for ice cloud super-saturation. The ice cloud super-saturation has led to more moisture being transported onto the continent, resulting in more and optically thicker clouds and more downward long-wave radiation. Overall, the updated model better represents the surface energy balance, based on a comparison with > 750 months of data from nine automatic weather stations located in East Antarctica. Especially the representation of the turbulent sensible heat flux and net long-wave radiative flux has improved with a decrease in biases of up to 40 %. As a result, modelled surface temperatures have increased and the bias, when compared to 10 m snow temperatures from 64 ice-core observations, has decreased from −2.3 K to −1.3 K. The weaker surface temperature inversion consequently improves the representation of the sensible heat flux, whereas wind speed biases remain unchanged. However, significant model biases remain, partly because RACMO2 at a resolution of 27 km is unable to resolve steep topography.

1 Introduction

Understanding the climate of the polar regions and changes therein requires a thorough understanding of the surface energy balance (SEB), which describes the exchange of heat and moisture at the earth's surface. Regional atmospheric climate models (RCMs) are important tools to improve our understanding of polar SEB. They provide a physically consistent representation of the climate in areas with a low spatial and temporal coverage of observations. RCMs are also capable of resolving detailed features that are not captured by global circulation models. Specifically, RCMs have been successfully applied to remote areas such as Antarctica (e.g. Van Lipzig et al., 2002) and Greenland (Fettweis, 2007; Ettema et al., 2010a) to assess the SEB and surface mass balance (SMB) of these ice sheets (e.g. Van Lipzig et al., 1999; Van de Berg et al., 2005; Lenaerts et al., 2012a). Moreover, RCM output can be used to enhance the interpretation of remote sensing data such as GRACE (Chen et al., 2006), InSAR (Rignot et al., 2008) and radar/laser altimetry (Ligtenberg et al., 2012).

The Regional Atmospheric Climate Model RACMO2, which has been adapted for specific use over the polar regions, has recently undergone a major update of its physics package. In the present study we examine whether this update from version RACMO2.1 to RACMO2.3 has improved the representation of the Antarctic near-surface climate, with its extreme temperatures and winds, and surface energy balance. Even though RACMO2.1 has proved to realistically simulate the Antarctic near-surface climate (Van de Berg et al., 2005; Lenaerts et al., 2012a), previous model evaluations showed that downward long-wave radiation is generally underestimated by the model (Van de Berg et al., 2007) resulting in a significant cold surface bias (Van den Broeke, 2008). To see whether this has improved in RACMO2.3, we will assess the changes for Antarctica in SEB, near-surface wind speed and surface temperature and compare these to available observations. The Antarctic SEB historically has been the focus of many studies, based both on observations (Schlatter, 1970; Carroll, 1982; Reijmer and Oerlemans, 2002) and modelling (Van de Berg et al., 2007), including evaluations of models like RACMO (Van Lipzig et al., 1999; Reijmer et al., 2005). Here we use new data collected at Antarctic automatic weather stations (AWSs) that have been specifically designed to close and quantify the SEB (Reijmer and Oerlemans, 2002).

Section 2 discusses the model, the changes in model formulation and the observational data used for evaluation. In Sect. 3 the effects of the model changes on clouds, the SEB and the near-surface temperature and wind are presented and a comparison is made to observational data (Sects. 3.3–3.5), followed by conclusions in Sect. 4.

2 Data and methods

2.1 RACMO2 physics update

RACMO2 combines the dynamical processes of the High Resolution Limited Area Model (HIRLAM) (Undén et al., 2002) with the physics package of the European Centre for Medium-range Weather Forecasts (ECMWF) Integrated Forecast System (IFS). RACMO2 has been specifically adapted for use over the large ice sheets of Greenland and Antarctica (e.g. Reijmer et al., 2005). It is interactively coupled to a multilayer snow model that calculates melt, percolation, refreezing and runoff of meltwater (Ettema et al., 2010b; Greuell and Konzelmann, 1994). Surface albedo is based on a prognostic scheme for snow grain size (Kuipers Munneke et al., 2011) and a drifting snow routine simulates the interactions of drifting snow with the surface and the lower atmosphere (Lenaerts et al., 2012a). A horizontal resolution of ~ 27 km and a vertical resolution of 40 levels is used. The model is forced by ERA-Interim re-analysis data (January 1979–December 2011, Dee et al., 2011) at the ocean and lateral boundaries, while the domain interior is allowed to evolve freely.

Here we analyse changes in the modelled Antarctic near-surface climate after the ECMWF IFS physics package cycle CY23r4 in RACMO2.1 (White, 2001) has been updated to cycle CY33r1 in RACMO2.3 (ECWMF-IFS, 2008). The updates that have the most impact on Antarctic applications are the changes in the cloud scheme, the cloud microphysics and the radiation and turbulence schemes. All changes will be described below and are discussed in more detail in relation with the results in Sects. 3.3–3.5.

An important change in the cloud scheme is the inclusion of a parameterization for ice super-saturation as described by Tompkins and Gierens (2007). As a result, the specific humidity of cold air parcels (at temperatures where the difference between liquid water and ice saturation pressure is large) has to reach a higher value in order for condensation to occur. This leads to an improved representation of clouds and moisture concentrations in the (upper) troposphere (Tompkins and Gierens, 2007). Aircraft observations with the Microwave Limb Sounder (MLS) have shown that super-saturation frequently occurs over the steep coastal regions of Antarctica (Spichtinger et al., 2003). Simulations with the ECMWF IFS have already shown that the new parameterisation leads to a better global distribution of supersaturated atmospheres, albeit with a slight underestimation for Antarctica (Tompkins and Gierens, 2007).

Another change in the physics is the introduction of the McRad radiation scheme (Morcrette et al., 2008). It describes shortwave and long-wave radiation transfer through clouds, based on the Monte Carlo Independent Column Approximation (McICA, Barker et al., 2008), and a revision of cloud optical properties making the parameterisations that use these properties more accurate. This improves the interaction of multi-layer cloud cover with shortwave and long-wave radiation, but is believed to be of minor importance for Antarctica, considering the low occurrence frequency of these cloud types in this region. In the shortwave radiation scheme (SRTM, Mlawer and Clough, 1997) the Fouqart–Bonnel scheme is replaced by a scheme that is based on the correlated *k* method (Lacis, 1991). The latter is shown to lead to an overall improved accuracy in calculated fluxes and heating rates (ECWMF-IFS, 2008).

The last relevant physics change is the newly implemented Eddy-Diffusivity Mass Flux (EDMF, Siebesma et al., 2007) scheme for boundary-layer turbulence/shallow convection. This scheme distinguishes between large-scale (updraughts) and small-scale (turbulence) mixing processes in the surface and boundary layer by describing them with either mass fluxes or diffusion. The surface flux relies on Monin–Obukhov similarity theory but takes into account form drag (Beljaars et al., 2004) that is dependent on subscale orography. For topographically rough areas like the Antarctic Peninsula these changes are expected to be especially important.

Furthermore, the RACMO2 model update incorporates changes in the HIRLAM dynamical core. These are mostly of numerical nature and are not addressed here. For a more detailed and complete description of the entire RACMO2 update the reader is referred to Van Meijgaard et al. (2012) and ECWMF-IFS (2008) and references therein. The changes in the cloud scheme influence precipitation patterns over Antarctica. Whether these changes represent an improvement, and how these changes effect the mass budget of the ice sheet will be addressed in a forthcoming study. The impact of precipitation changes on the SEB is small, as will be shown in Sect. 3.3.

Fig. 1. Map of Antarctica with locations of the AWS (red diamonds) the 64 coring sites (black dots) and the position of the latitudinal cross-section used in Fig. 3. Also shown are the ice-shelf edge and grounding line (solid lines) and height intervals every 500 m (dashed lines) based on a digital elevation model from Liu et al. (2001).

2.2 Observational data

The near-surface wind, temperature and SEB are evaluated using observational data from nine automatic weather stations, maintained by the Institute for Marine and Atmospheric research of Utrecht University (UU/IMAU), see the IMAU AWS website http://www.projects.science.uu.nl/iceclimate/aws. Importantly, these AWSs measure all four radiation components as well as humidity and snow temperature, and therefore enable a full closure of the SEB. All AWSs are of similar design: single level measurements of wind speed/direction, temperature and relative humidity are performed at a height of approximately 3 m. The individual radiation components (SW↓, SW↑, LW↓, LW↑) are measured with a single sensor. Treatment of the radiation fluxes follows Van den Broeke (2004). Figure 1 shows the locations of the AWSs. They are located in Dronning Maud Land (DML) in different climate regimes: from relatively mild and wet coastal sites (AWS 4 and 11) to the steep escarpment region of DML (AWS 5, 6 and 16), the south dome of Berkner Island (AWS 10) and the high and cold East Antarctic plateau (AWS 8, 9 and 12). Observation lengths range from 4 to 15 yr of data. Due to instrumental problems and icing of the sensors some months of the data record are of lower quality. A summary of the location and data records of the AWSs is provided in Table 1. For more details see Van den Broeke et al. (2005a, b) and Reijmer and Oerlemans (2002).

The SEB can be written as

$$M = \mathrm{SW_{net}} + \mathrm{LW_{net}} + \mathrm{SHF} + \mathrm{LHF} + G, \tag{1}$$

where fluxes directed towards the surface are defined positive with units of W m^{-2}, M is melt energy ($M = 0$ if the surface temperature $T_\mathrm{s} < 273.15$ K), $\mathrm{SW_{net}}$ and $\mathrm{LW_{net}}$ are the net shortwave and long-wave radiative fluxes, SHF and LHF are the sensible and latent heat fluxes and G is the subsurface conductive heat flux. The sensible- and latent-turbulent heat fluxes are calculated using Monin–Obukhov similarity theory using the bulk method (Van den Broeke et al., 2005b). This implies that the turbulent fluxes and surface temperatures are calculated values and not direct measurements. All data from the nine AWSs are monthly averaged (resulting in 770 station-months) and compared with data from the same months of the two RACMO cycles. An assessment of the quality of the observational data can be found in Reijmer and Oerlemans (2002); Van den Broeke et al. (2004).

As the AWSs only cover a limited part of East Antarctica, 64 snow temperature observations (Fig. 1) are additionally used to evaluate the spatial performance of RACMO2 for T_s. The 10 m snow temperature is assumed to represent the annual mean surface temperature, a reasonable assumption in areas without melt. Note that at a model resolution of 27 km, the observational data are compared with data from the nearest model grid point. For the 10 m snow temperatures, this causes four locations to fall outside of the ice mask. For these points the nearest grid point that does fall within the ice mask is used.

3 Results

3.1 General climate characteristics

In Antarctica a negative to zero net radiation budget prevails during most months. In summer the radiation budget regularly becomes positive due to absorption of shortwave radiation at the surface. In winter, the radiation budget is balanced mainly by a positive (downward directed) SHF, as LHF is generally small due to the low humidity and thus small near-surface moisture gradients exist over the Antarctic ice sheet. The negative radiation budget prevails and cools the surface, resulting in a quasi-permanent surface-based temperature inversion. In combination with a sloping surface, this leads to the characteristic persistent katabatic winds over the Antarctic ice sheet. As the cooling is stronger in winter, the katabatic winds increase in strength in winter. Stronger katabatic winds enhance downward sensible heat transport, which counteracts the strength of the surface temperature inversion by increasing the surface temperature. This results in a weaker seasonality of (near-)surface temperature in high wind speed areas.

To illustrate these interactive processes, Fig. 2 shows the monthly mean values of 10 m wind speed ($V_{10\,\mathrm{m}}$), surface

Table 1. The AWS topographic characteristics and period of operation (until December 2011). No. months represents the number of available months over total months (725/770) of the model period (January 1979–December 2011). If no end time is indicated, the AWS is still operational end of 2011. Observed surface slope ($m\,km^{-1}$) is based on a 1 km × 1 km digital elevation model (Liu et al., 2001), modelled surface slope is based on the 27 km interpolated grid.

AWS	4	5	6	8	9	10	11	12	16
Latitude	72°45′ S	73°06′ S	74°28′ S	76°00′ S	75°00′ S	79°34′ S	71°09′ S	78°39′ S	71°57′ S
Longitude	15°29.9′ W	13°09.9′ W	11°31.0′ W	08°03′ W	00°00′ E/W	45°47′ W	06°42′ W	35°38′ E	23°20′ E
Elevation (obs)	34 m	363 m	1160 m	2400 m	2892 m	890 m	700 m	3620 m	1300 m
Elevation (mod)	23 m	332 m	1219 m	2405 m	2856 m	789 m	224 m	3621 m	1130 m
Slope (obs)	1.0	23.1	38.4	2.0	1.5	1.1	15.5	2.0	15.6
Slope (mod)	3.4	7.8	28.9	3.5	2.0	5.1	10.7	2.2	16.2
Start	Dec 1997	Feb 1998	Jan 1998	Jan 1998	Dec 1997	Jan 2001	Jan 2007	Dec 2007	Feb 2009
End	Dec 2002	–	Jan 2009	Jan 2003	–	Jan 2006	–	–	–
no. months	60/60	167/167	134/134	19/44	162/168	48/54	57/59	49/49	29/35

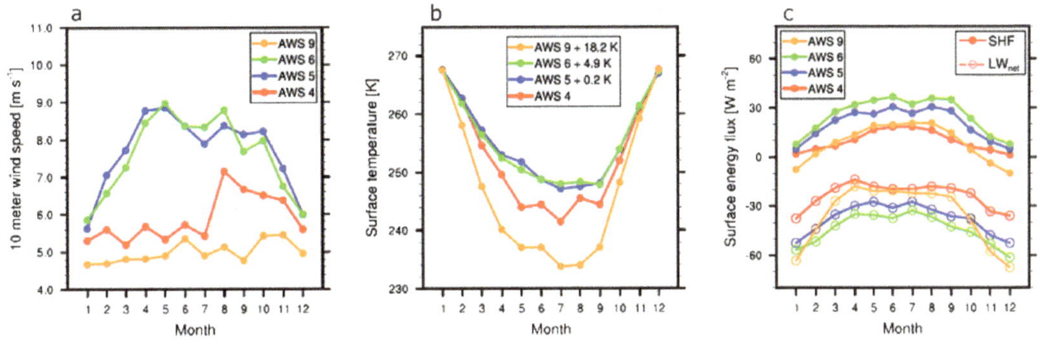

Fig. 2. Monthly mean **(a)** 10 m wind speed ($V_{10\,m}$), **(b)** surface temperature (T_s) and **(c)** sensible heat flux (SHF) and net long-wave radiation (LW_{net}) for AWS 4, AWS 5, AWS 6 and AWS 9. The temperature curves for AWS 5, 6 and 9 (Fig. 2b) are shifted upward by 0.2, 4.9 and 18.2 K, respectively, to correct for elevation and continentality differences (Van den Broeke, 2004).

temperature (T_s), net long-wave radiation (LW_{net}) and sensible heat flux (SHF) for four AWSs in different climate zones of the Antarctic ice sheet (AWS 4, 5, 6 and 9). Figure 2a shows that monthly mean wind speeds for AWS 5 and 6, in the steep escarpment region, exhibit a strong seasonal cycle due to stronger katabatic forcing in winter, with monthly mean wind speeds up to $9\,m\,s^{-1}$. These katabatic winds mix warm air downward to the surface (large SHF), increasing surface temperature (Fig. 2b), and hence upward long-wave radiation (Fig. 2c). For AWS 4 and AWS 9, located on the relatively flat coastal ice shelf and interior ice sheet, respectively, wind speeds are lower and show no seasonal cycle. At these sites the seasonal amplitude in temperature is larger than at the sites dominated by katabatic winds (AWS 5 and 6) mainly because the wintertime surface temperature inversion is stronger.

3.2 Changes in cloud properties and impact on simulated near-surface variables

The new parameterisation for cloud ice super-saturation changes the total amount of modelled clouds over Antarctica, most notably over the East Antarctic plateau. To illustrate this effect, Fig. 3 shows a latitudinal cross-section of the vertical distribution of total cloud water/ice content, averaged over the period 1979–2011. A significant increase of modelled cloud content is found over the East Antarctic plateau, while cloud content has decreased along the coastline and over the ocean. With the new parameterisation, moist air that reaches the continent has to exceed 100 % relative humidity by up to 50 % in order to form clouds. As a result, clouds form farther inland and higher up in the troposphere, resulting in more clouds simulated by RACMO2.3 in the interior.

The increase in clouds has caused more downward long-wave radiation to be emitted as seen in Fig. 4, where the difference fields (RACMO2.3–RACMO2.1) for LW↓ (a), T_s (b), $V_{10\,m}$ (c) and SHF (d) are shown. The increase in LW↓ is found on most of the Antarctic ice sheet, but is strongest on the East Antarctic plateau, and has led to higher surface temperatures (Fig. 4b), reducing the temperature gradients in the surface layer and resulting in lower SHF values (Fig. 4d). A related pattern in near-surface wind speed is not seen (Fig. 4c), with changes being smaller than 5 %.

The increase in clouds over the continent also results in an increase of precipitation. This is illustrated in Fig. 5, which shows the changes in total precipitation (mostly snowfall) for DML, where the AWSs are located. The generally increased

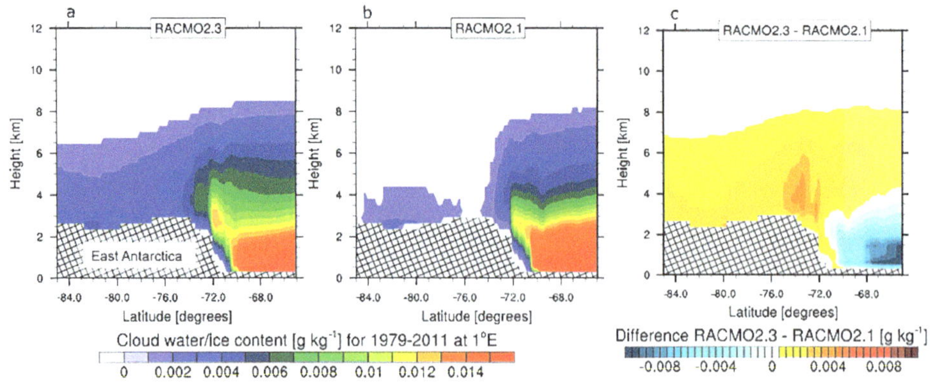

Fig. 3. Latitudinal cross-section of yearly (1979–2011) averaged total cloud water/ice content for **(a)** RACMO2.3, **(b)** RACMO2.1 and **(c)** the difference (RACMO2.3–RACMO2.1) at 1° E. Location of cross-section is indicated in Fig. 1.

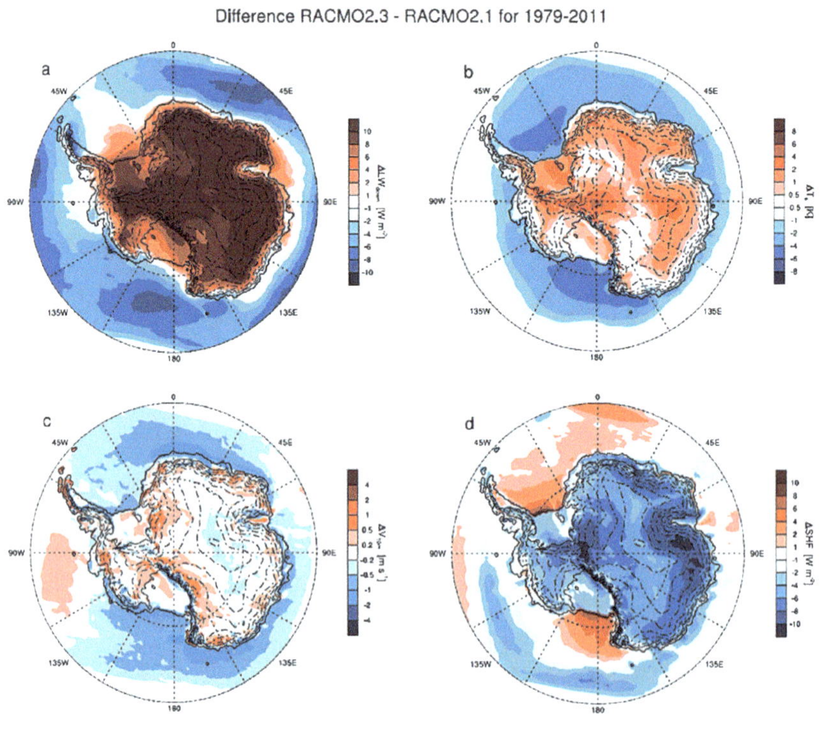

Fig. 4. Spatial distribution of difference (RACMO2.3–RACMO2.1) for LW_{down} **(a)**, T_s **(b)**, $V_{10\,m}$ **(c)** and SHF **(d)**.

precipitation will have an effect on the SEB through increased albedo (impacting the shortwave radiation) and sublimation (impacting latent heat flux). As will be shown in the next section, the effects on SW_{net} are small and limited to the short austral summer, while effects on LHF are generally small as well.

3.3 Impact on simulation of the surface energy balance

Figure 6 shows the difference between the monthly averaged modelled and observed SEB fluxes for all nine AWSs (>750 months) for RACMO2.3 and RACMO2.1. The average bias and correlation coefficients are summarized in Table 2, as well as the bias standard deviation σ_{bias} and root mean square deviation (RMSD). Most biases (and RMSD) are reduced in RACMO2.3: for SHF from $10.5\,W\,m^{-2}$ to $7.1\,W\,m^{-2}$, for LW_{net} from $-10.4\,W\,m^{-2}$ to $-6.3\,W\,m^{-2}$. The changes in LHF and SW_{net} are small. For SW_{net} the bias increased from $-1.3\,W\,m^{-2}$ to $-2.0\,W\,m^{-2}$ but the correlation remains high ($r^2 \simeq 0.93$). For LHF the slight improvement from $0.6\,W\,m^{-2}$ to $0.4\,W\,m^{-2}$ is of little significance due to its small magnitude and the uncertainty of the observed fluxes. The standard deviation σ_{bias} has the same

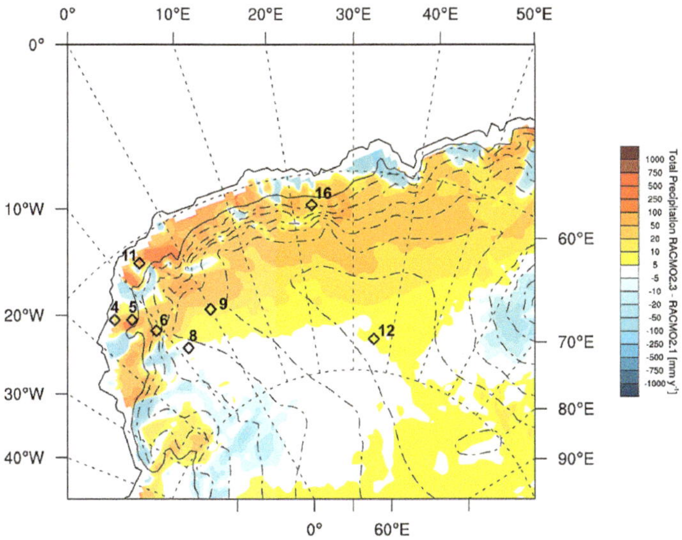

Fig. 5. Spatial distribution of difference (RACMO2.3–RACMO2.1) for total precipitation (rain + snowfall) in mm water equivalent per year over Dronning Maud Land, for 1979–2011. Shown are all the AWS, except AWS 10 that falls just outside of the boundary.

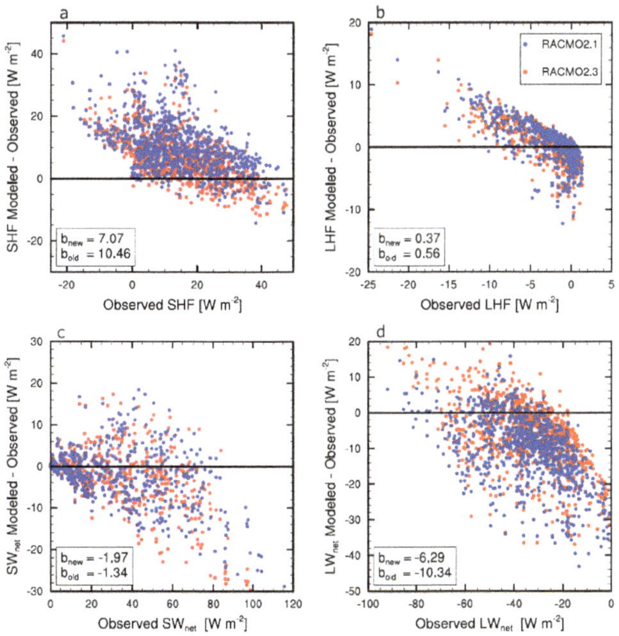

Fig. 6. Difference (modelled−observed of **(a)** sensible heat flux (SHF), **(b)** latent heat flux (LHF), **(c)** net shortwave radiation (SW_{net}) and **(d)** net long-wave radiation (LW_{net}) as a function of the AWS observations. Shown are RACMO2.3 (red) with bias b_{new} and RACMO2.1 (blue) with bias b_{old}. Biases (in [$W\,m^{-2}$]) are averages over all monthly average weather station data (> 750 data points).

order of magnitude as the bias, suggesting that the improvements are not statistically significant because of the significant noise.

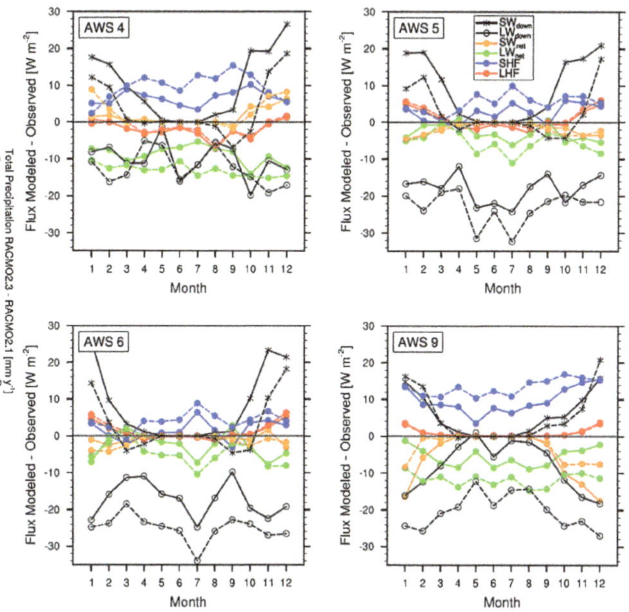

Fig. 7. Annual cycle of monthly mean difference (modelled−observed) of the SEB components (SHF (blue), LHF (red), SW_{net} (orange), LW_{net} (green)), LW↓ (black circles) and SW↓ (black asterisks) for **(a)** AWS 4, **(b)** AWS 5, **(c)** AWS 6 and **(d)** AWS 9. Shown are RACMO2.3 (solid line) and RACMO2.1 (dashed line).

To investigate the seasonal effects, Fig. 7 shows the monthly mean difference (modelled−observed) of SEB fluxes, LW↓ and SW↓ for AWS 4, 5, 6 and 9. LW_{net} is underestimated at the four AWS locations, and most significantly at AWS 4 and 9. The overestimated surface slope at these two sites is responsible for activating a katabatic feedback in winter, resulting in overestimated SHF. In RACMO2.3, the improved LW↓ reduces this problem for the stations located more inland, as can be seen in Fig. 7, most significantly for AWS 9. For AWS 4 the improvement is not related to the increased LW↓, but to a decreased LW↑ (not shown). For this coastal site, $V_{10\,m}$ that was overestimated has slightly decreased (the performance of the near-surface wind speed will be discussed in Sect. 3.4), resulting in a lower T_s and less long-wave cooling. For all AWS except AWS 9 the difference in SW↓ has increased, most notably for the coastal AWSs. Here, changes in the radiation scheme have led to increased atmospheric transmissivity. For the more inland AWSs the effect of increased transmissivity is increasingly balanced by increased snowfall (see Fig. 5), raising the surface albedo and SW↑.

The underestimated surface slope results in too low wind speeds at AWS 5 and 6, resulting in an overestimation of the surface temperature inversion (Fig. 9d). As a result, SHF is reasonably well represented, because stability effects remain small even at the underestimated wind speeds (Van den Broeke et al., 2005b). The problem is further reduced

Table 2. Mean, bias (model-observation), standard deviation of the bias σ, root mean square deviation and correlation coefficient (r^2 with significance level $p < 0.0001$) for sensible heat flux (SHF), latent heat flux (LHF), net shortwave radiation (SW_{net}), net long-wave radiation (LW_{net}), 10 m wind speed ($V_{10\,m}$) and surface temperature T_s as modelled by RACMO2.1 and RACMO2.3 and as observed with the 9 AWSs (770 months). Also shown are the values for annual averaged T_s in comparison with 10 m snow temperatures from 64 ice-core measurements. For calculation of the statistics all 770 months are used.

		Obs	RACMO2.1				RACMO2.3			
		mean	bias	σ_{bias}	RMSD	r^2	bias	σ_{bias}	RMSD	r^2
SHF	[Wm^{-2}]	14.24	10.46	8.53	13.58	0.57	7.07	8.12	10.78	0.6
LHF	[Wm^{-2}]	−2.11	0.56	2.95	3.0	0.31	0.36	2.77	2.79	0.38
SW_{net}	[Wm^{-2}]	22.73	−1.33	6.23	6.41	0.94	−1.96	7.60	7.86	0.92
LW_{net}	[Wm^{-2}]	−35.03	−10.36	9.83	14.22	0.66	−6.31	9.82	11.65	0.66
$V_{10\,m}$	[ms^{-1}]	6.38	−0.48	1.67	1.74	0.28	−0.51	1.64	1.72	0.27
T_s(AWS)	[K]	244.28	−3.24	4.5	5.57	0.91	−1.91	4.36	4.77	0.91
T_s (ice cores)	[K]	240.6	−2.32	2.29	3.23	0.96	−1.28	2.13	2.45	0.98

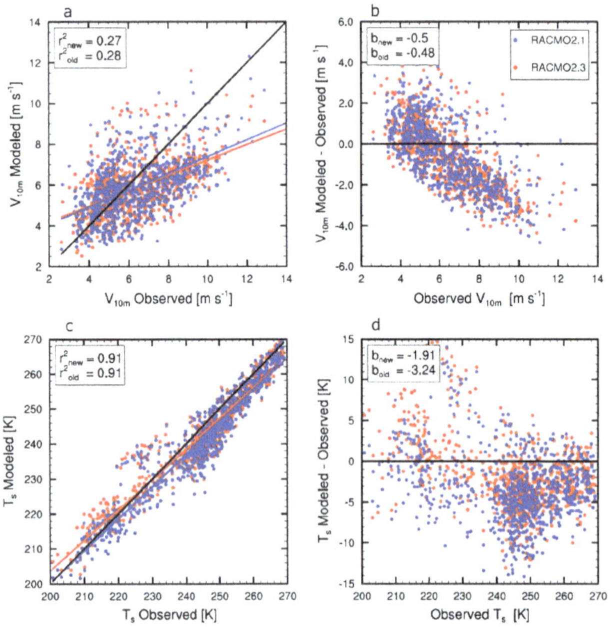

Fig. 8. Modelled and difference (modelled−observed) as a function of the AWS observations of (**a**, **b**) monthly averaged 10 m wind speed ($V_{10\,m}$) and (**c**, **d**) surface temperature (T_s). Shown are RACMO2.3 (red) with correlation r^2_{new} and bias b_{new} and RACMO2.1 (blue) with correlation r^2_{old} and bias b_{old}. Biases are averages over all data with units [m s^{-1}] for $V_{10\,m}$ and [K] for T_s.

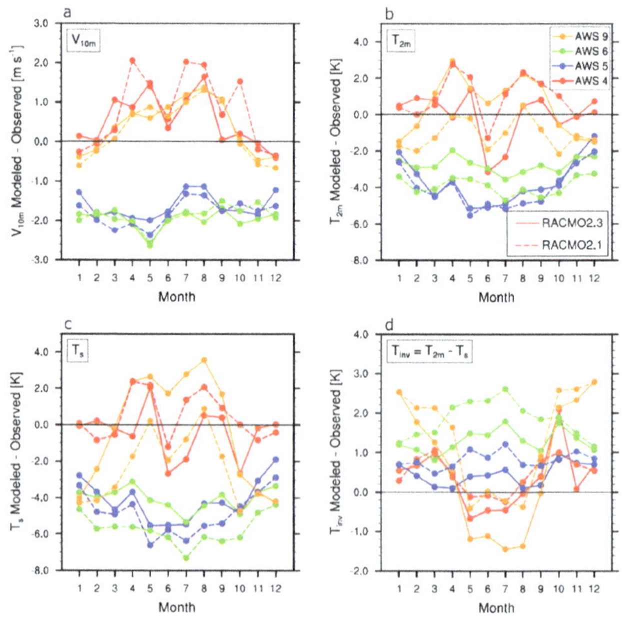

Fig. 9. Annual cycle of monthly mean difference (modelled−observed) of **(a)** 10 m wind speed ($V_{10\,m}$), **(b)** 2 m temperature ($T_{2\,m}$), **(c)** surface temperature (T_s) and **(d)** the surface temperature inversion ($T_{inv} = T_{2\,m} - T_s$) for AWS 4, AWS 5, AWS 6 and AWS 9. Shown are RACMO2.3 (solid line) and RACMO2.1 (dashed line).

in RACMO2.3, in which the SEB at AWS 5 and 6 is well represented.

In summer the SHF bias at AWS 4 is smaller because SW_{net} and LHF help to balance the excess long-wave cooling. For AWS 9, however, there is an increased negative bias in SW_{net} in RACMO2.3 due to an overestimated albedo (Van de Berg et al., 2007). To conclude, for most of the climate zones of the Antarctic ice sheet the improved representation of SEB components mainly results from a better representation of LW↓, but also changes in the turbulence scheme (Sect. 3.5) contribute to the improvement.

Figure 6a confirms that high wintertime SHF values in the escarpment zone are well represented, and that RACMO2 generally overestimates SHF in flatter areas. This leads to overestimated T_s and hence too negative LW_{net} (Fig. 6d). Both biases are significantly reduced in RACMO2.3, by 33 % (SHF) and 39 % (LW_{net}). In summer, RACMO2 underestimates SW_{net} in the high interior (Figs. 6c and 7d). As a result of the underestimation of SW_{net}, T_s is underestimated (Figs. 8c, d and 9c), while the surface temperature inversion and the SHF are overestimated (Figs. 6a, 7d). As a result of too low T_s, sublimation (negative LHF) is underestimated and summertime convection (upward SHF) is not modelled on the ice sheet (see Sect. 3.5).

3.4 Impact on simulation of temperature and near-surface wind speed

Figure 8 shows modelled values (Fig. 8a, c) and difference (model-observation) (Fig. 8b, d) of the monthly averaged wind speed and surface temperature of all AWSs as a function of the observed value. The figure also shows correlation coefficient (r^2) and average bias (b), also denoted in Table 2. Figure 8a, b shows that the wind speed representation has not improved in RACMO2.3 when compared to RACMO2.1 (for both data sets: $b \simeq 0.5$, $r^2 \simeq 0.27$). Both model cycles generally underestimate high wind speeds and overestimate low wind speeds. Since near-surface winds over the Antarctic ice sheet are dominated by katabatic forcing, this is caused by an overestimation of surface slopes in relatively flat areas (AWS 4, 8, 9, 10, 12) and an underestimation of surface slopes in steep areas (AWS 5, 6, 11) in the escarpment region of DML (and elsewhere in Antarctica), as a result of the limited horizontal resolution of the model (Reijmer et al., 2005; Lenaerts et al., 2012b). The small differences in $V_{10\,m}$ are due to the combined effect of model changes but the errors in wind speed with respect to the observations are dominated by the smoothed model topography (note that the model topography is the same in both model cycles).

Figure 9a shows the monthly mean difference (model-observation) of $V_{10\,m}$ for AWS 4, 5, 6 and 9. For AWS 5 and AWS 6 (slope > $10\,m\,km^{-1}$) wind speed is underestimated year round. For AWS 4 and AWS 9 the wind speed is overestimated in winter, when katabatic forcing is overestimated, and therefore shows a seasonality that is too pronounced. This has a strong effect on surface and 2 m temperature, as shown in Fig. 9b, c, where temperature is underestimated at AWS 5 and 6. The surface temperature inversion, defined here as $T_{inv} = T_{2\,m} - T_s$, is underestimated when wind speed is overestimated (AWS 4 and 9) (Fig. 9d), which is intuitively expected.

In contrast to wind speed, a clear improvement in surface temperature T_s (Fig. 8c, d) in RACMO2.3 over RACMO2.1 is seen, due to the increased LW↓. The cold bias has been reduced from 3.2 K to 1.9 K while the correlation has not changed ($r^2 = 0.91$). This improvement occurs year round for all the AWSs except for the coastal AWSs (see AWS 4), where the representation was already good due in part to the overestimated wind speed. A comparison of monthly averaged $V_{10\,m}$ and $T_{2\,m}$ from the Reference Antarctic Data for Environmental Research (READER, Turner et al., 2004) AWS and surface station data gave the same results (not shown).

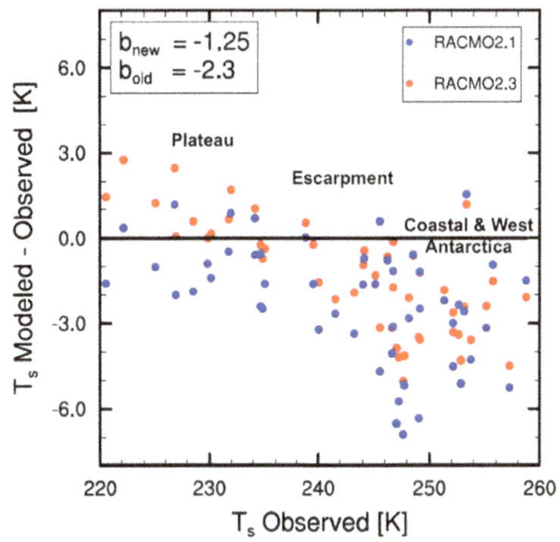

Fig. 10. Difference (modelled−observed) of the annual period averaged (1979–2011) surface temperature (T_s) as a function of 10 m ice-core observations. Shown are RACMO2.3 (red) with bias b_{new} and RACMO2.1 (blue) with bias b_{old} (in [K]).

Since the AWS are located mainly in DML, which has a topography that is not typical for the entire ice sheet, we use 10 m-snow temperature data to obtain a better spatial coverage for the T_s evaluation (Fig. 1). Figure 10 shows the difference between modelled and observed T_s as a function of the latter, averaged over the model time span. Average values, bias, σ_{bias}, RMSD and correlation are given in Table 2. Surface temperatures in RACMO2.1 are underestimated at almost all locations and on average are too low by 2.3 K. RACMO2.3 reduces this bias to −1.3 K. Overall the spatial variability of surface temperatures is well represented by both model versions ($r^2 = 0.96$ for RACMO2.1 and $r^2 = 0.98$ for RACMO2.3) although there seems to be a tendency towards larger underestimations at higher temperature locations. Figure 10 shows that the best agreement is found on the cold East Antarctic plateau, where the overestimated wind speeds compensate for the bias in temperature caused by the underestimated LW↓. Because of the improved LW↓, T_s on the plateau has changed from being slightly underestimated in RACMO2.1 to being slightly overestimated in RACMO2.3. In coastal and West Antarctica, T_s is underestimated the most, due to both $V_{10\,m}$ and LW_{net} being underestimated, and it is here that RACMO2.3 produces the largest improvement.

3.5 Impact on SHF regimes

To assess the impact of changes in the atmospheric surface layer scheme, Fig. 11 shows monthly averaged SHF for all AWSs (a), RACMO2.3 (b) and RACMO2.1 (c) as a function of the surface temperature inversion and wind speed (colour scheme). The filled circles represent winter conditions (April–September), open circles conditions for October–March. Figure 11a clearly shows four regimes.

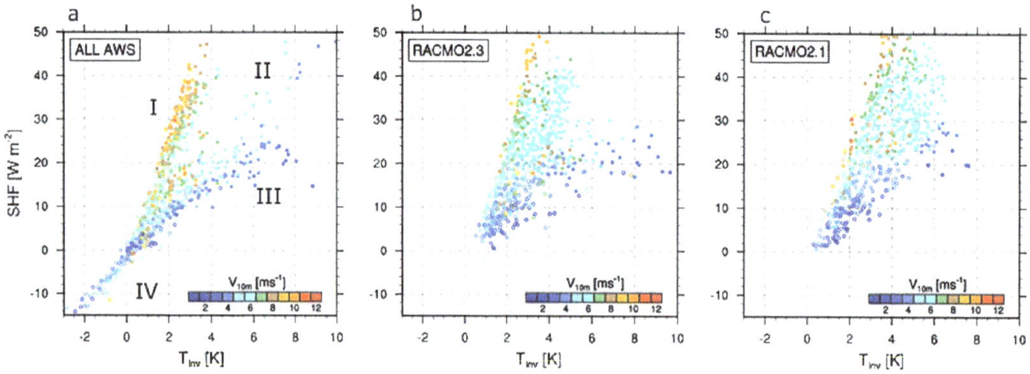

Fig. 11. Monthly averaged SHF as a function of the surface temperature inversion ($T_{\text{inv}} = T_{2\,\text{m}} - T_{\text{s}}$) for **(a)** all AWSs, **(b)** RACMO2.3 and **(c)** RACMO2.1. The colour scheme represents wind speed $V_{10\,\text{m}}$ and model data is from the same months and locations as the observational data. Filled circles are winter values (months 4–9) and open circles represent months (10–3). Four climate regimes are denoted, and explained in Sect. 3.5.

Regime I represents the katabatic wind zone where SHF increases quadratically with the katabatic wind forcing (inversion strength). Regime II represents the exceptional conditions at AWS 16 (Thiery et al., 2012; Gorodetskaya et al., 2013), where despite stable conditions and low wind speeds, SHF values are high probably due to large-scale circulation effects. The model (Fig. 11b, c) does not simulate this regime accurately due to the limited spatial resolution as the station is positioned in a topographically complex region. Regime III represents the stable conditions of the AWSs in flat areas where static stability effects become important at high values of the temperature inversion, suppressing SHF.

For lower stabilities and towards summer conditions the branches join and SHF shows a linear dependence on T_{inv}, the negative values indicating the convective summertime conditions at plateau stations AWS 9 and 12 (Regime IV). This regime is exclusively found on the plateau, where the low summertime temperatures prevent sublimation (LHF) to act as a surface energy sink (King et al., 2006). Because RACMO2 overestimates albedo and underestimates atmospheric transmissivity, a positive radiation balance is not simulated and convection does not occur (at least not in the monthly mean sense).

Figure 11b, c shows the inability of RACMO2 to simulate regimes II and IV. The behaviour in the katabatic wind zone is represented well, although the branch is less pronounced due to the underestimation of the slope and hence wind speeds. RACMO2.3 simulates an improved separation of the most important regimes II and III compared to RACMO2.1, because of the improved surface layer turbulence scheme and general changes in the simulated results.

4 Conclusions

The physics package of the regional atmospheric climate model RACMO2 adopted from the ECMWF-IFS has been upgraded from cycle CY23r4 (RACMO2.1) to CY33r1 (RACMO2.3). This study evaluates the effects of this change on the Antarctic surface energy balance (SEB), 10 m wind speed $V_{10\,\text{m}}$ and surface temperature T_{s} by comparing both cycles with observational SEB data gathered from nine automatic weather stations in East Antarctica and 64 deep snow temperature sites. The model has improved in several aspects. Due to the inclusion of a parameterization for cloud ice super-saturation, more clouds and increased moisture content are simulated in the upper troposphere. As a result, more clouds and an increased cloud optical thickness in the interior have resulted in more downward long-wave radiation. Consequently, in RACMO2.3 the biases in the sensible heat flux (SHF) and the net long-wave radiation (LW_{net}) have decreased from 10.5 to 7.1 W m^{-2} and -10.4 to -6.3 W m^{-2}, respectively. The change in long-wave radiation has improved the bias in the SHF through its tight coupling with T_{s} and wind speed: the bias in T_{s}, based on the deep snow temperature observations, has decreased from -2.3 K to -1.3 K. Near-surface air temperatures have also increased but less so than T_{s}, decreasing the surface-based temperature inversion. The bias in $V_{10\,\text{m}}$, which is mainly due to the flattened ice sheet topography at the limited model resolution, remains unchanged.

Acknowledgements. We are grateful for the financial support of NWO/ALW, Netherlands Polar Programme. The installation and maintenance of AWS 16 was financed by the Belgian Science Policy Office under grant number EN/01/4B supervised by N. van Lipzig KU Leuven. We thank I. Gorodetskaya for providing AWS 16 data.

Edited by: E. Hanna

References

Barker, H. W., Cole, J. N. S., Morcrette, J., and Pincus, R.: The Monte Carlo Independent Column Approximation : An assessment using several global atmospheric models, Q. J. Roy. Meteor. Soc., 134, 1463–1478, doi:10.1002/qj.303, 2008.

Beljaars, A. C., Brown, A. R., and Wood, N.: A new parametrization of turbulent orographic form drag, Q. J. Roy. Meteor. Soc., 130, 1327–1347, doi:10.1256/qj.03.73, 2004.

Carroll, J. J.: Long-Term Means and Short-term Variablity of the Surface Energy Balance Components at the South Pole, J. Geophys. Res., 87, 4277–4286, 1982.

Chen, J. L., Wilson, C. R., Blankenship, D. D., and Tapley, B. D.: Antarctic mass rates from GRACE, Geophys. Res. Lett., 33, L11502, doi:10.1029/2006GL026369, 2006.

Dee, D. P., Uppala, S. M., Simmons, A. J., Berrisford, P., Poli, P., Kobayashi, S., Andrae, U., Balmaseda, M. A., Balsamo, G., Bauer, P., Bechtold, P., Beljaars, A. C. M., Van de Berg, L., Bidlot, J., Bormann, N., Delsol, C., Dragani, R., Fuentes, M., Geer, A. J., Haimberger, L., Healy, S. B., Hersbach, H., Hólm, E. V., Isaksen, L., Kå llberg, P., Köhler, M., Matricardi, M., McNally, A. P., Monge-Sanz, B. M., Morcrette, J. J., Park, B. K., Peubey, C., de Rosnay, P., Tavolato, C., Thépaut, J. N., and Vitart, F.: The ERA-Interim reanalysis: configuration and performance of the data assimilation system, Q. J. Roy. Meteor. Soc., 137, 553–597, doi:10.1002/qj.828, 2011.

ECWMF-IFS: Part IV : Physical Processes (CY33R1), Tech. Rep. June, http://www.ecmwf.int/research/ifsdocs/CY33r1/PHYSICS/IFSPart4.pdf (last access: 29 October 2012), 2008.

Ettema, J., van den Broeke, M. R., van Meijgaard, E., and van de Berg, W. J.: Climate of the Greenland ice sheet using a high-resolution climate model – Part 2: Near-surface climate and energy balance, The Cryosphere, 4, 529–544, doi:10.5194/tc-4-529-2010, 2010a.

Ettema, J., van den Broeke, M. R., van Meijgaard, E., van de Berg, W. J., Box, J. E., and Steffen, K.: Climate of the Greenland ice sheet using a high-resolution climate model – Part 1: Evaluation, The Cryosphere, 4, 511–527, doi:10.5194/tc-4-511-2010, 2010b.

Fettweis, X.: Reconstruction of the 1979–2006 Greenland ice sheet surface mass balance using the regional climate model MAR, The Cryosphere, 1, 21–40, doi:10.5194/tc-1-21-2007, 2007.

Gorodetskaya, I. V., Van Lipzig, N. P. M., Van den Broeke, M. R., Mangold, A., Boot, W., and Reijmer, C. H.: Meteorological regimes and accumulation patterns at Utsteinen, Dronning Maud Land, East Antarctica: Analysis of two contrasting years, J. Geophys. Res., 118, 1700–1715, doi:10.1002/jgrd.50177, 2013.

Greuell, W. and Konzelmann, T.: Numerical modelling of the energy balance and the englacial temperature of the Greenland Ice Sheet. Calculations for the ETH-Camp location (West Greenland, 1155 m a.s.l.), Global Planet. Change, 9, 91–114, doi:10.1016/0921-8181(94)90010-8, 1994.

King, J. C., Argentini, S. A., and Anderson, P. S.: Contrasts between the summertime surface energy balance and boundary layer structure at Dome C and Halley stations, Antarctica, J. Geophys. Res., 111, D02105, doi:10.1029/2005JD006130, 2006.

Kuipers Munneke, P., Van den Broeke, M. R., Lenaerts, J. T. M., Flanner, M. G., Gardner, A. S., and Van de Berg, W. J.: A new albedo parameterization for use in climate models over the Antarctic ice sheet, J. Geophys. Res., 116, D05114, doi:10.1029/2010JD015113, 2011.

Lacis, A.: A Description of the Correlated k Distribution Method for Modeling Nongray Gaseous Absorption , Thermal Emission , and Multiple Scattering in Vertically Inhomogeneous Atmospheres to N2). Comparison cooling, J. Geophys. Res., 96, 9027–9063, 1991.

Lenaerts, J. T. M., Van den Broeke, M. R., Déry, S. J., Van Meijgaard, E., Van de Berg, W. J., Palm, S. P., and Sanz Rodrigo, J.: Modeling drifting snow in Antarctica with a regional climate model: 1. Methods and model evaluation, J. Geophys. Res., 117, D05108, doi:10.1029/2011JD016145, 2012a.

Lenaerts, J. T. M., Van Den Broeke, M. R., Scarchilli, C., and Agosta, C.: Impact of model resolution on simulated wind, drifting snow and surface mass balance in Terre Adélie, East Antarctica, J. Glaciol., 58, 821–829, doi:10.3189/2012JoG12J020, 2012b.

Ligtenberg, S. R. M., Horwath, M., Van den Broeke, M. R., and Legrésy, B.: Quantifying the seasonal breathing of the Antarctic ice sheet, Geophys. Res. Lett., 39, L23501, doi:10.1029/2012GL053628, 2012.

Liu, H., Jezek, K., Li, B., and Zhao, Z.: Radarsat Antarctic Mapping Project Digital Elevation Model Version 2., Tech. rep., 2001.

Mlawer, E. J. and Clough, S. A.: Shortwave and Longwave Enhancements in the Rapid Radiative Transfer Model, in: Proc. 7th Atmospheric Radiation Measurement (ARM) Science Team Meeting, 1992, 409–413, 1997.

Morcrette, J.-J., Barker, H. W., Cole, J. N. S., Iacono, M. J., and Pincus, R.: Impact of a New Radiation Package, McRad, in the ECMWF Integrated Forecasting System, Mon. Weather Rev., 136, 4773–4798, doi:10.1175/2008MWR2363.1, 2008.

Reijmer, C. H. and Oerlemans, J.: Temporal and spatial variability of the surface energy balance in Dronning Maud Land, East Antarctica, J. Geophys. Res., 107, 4759, doi:10.1029/2000JD000110, 2002.

Reijmer, C. H., Van Meijgaard, E., and Van den Broeke, M. R.: Evaluation of temperature and wind over Antarctica in a Regional Atmospheric Climate Model using 1 year of automatic weather station data and upper air observations, J. Geophys. Res., 110, D04103, doi:10.1029/2004JD005234, 2005.

Rignot, E., Bamber, J. L., Van den Broeke, M. R., Davis, C., Li, Y., Van de Berg, W. J., and Van Meijgaard, E.: Recent Antarctic ice mass loss from radar interferometry and regional climate modelling, Nat. Geosci., 1, 106–110, doi:10.1038/ngeo102, 2008.

Schlatter, T. W.: The Local Surface Energy Balance and Subsurface Temperature Regime in Antarctica, J. Appl. Meteorol., 11, 1048–1062, 1970.

Siebesma, A. P., Soares, P. M. M., and Teixeira, J.: A Combined Eddy-Diffusivity Mass-Flux Approach for the Convective Boundary Layer, J. Atmos. Sci., 64, 1230–1248, doi:10.1175/JAS3888.1, 2007.

Spichtinger, P., Gierens, K., and Read, W.: The global distribution of ice-supersaturated regions as seen by the Microwave Limb Sounder, Q. J. Roy. Meteor. Soc., 129, 3391–3410, doi:10.1256/qj.02.141, 2003.

Thiery, W., Gorodetskaya, I. V., Bintanja, R., Van Lipzig, N. P. M., Van den Broeke, M. R., Reijmer, C. H., and Kuipers Munneke, P.: Surface and snowdrift sublimation at Princess Elisabeth station, East Antarctica, The Cryosphere, 6, 841–857, doi:10.5194/tc-6-841-2012, 2012.

Tompkins, A. M. and Gierens, K.: Ice supersaturation in the ECMWF integrated forecast, Q. J. Roy. Meteor. Soc., 133, 53–63, doi:10.1002/qj.14, 2007.

Turner, J., Colwell, S. R., Marshall, G. J., Lachlan-Cope, T. A., Carleton, A. M., Jones, P. D., and Reid, P. A.: The SCAR READER Project : Toward a High-Quality Database of Mean Antarctic Meteorological Observations, J. Climate, 17, 2890–2898, 2004.

Undén, P., Rontu, L., Jarvinen, H., Lynch, P., Calvo, J., Cats, G., Cuxart, J., Eerola, K., Fortelius, C., Garcia-moya, J. A., Jones, C., Lenderlink, G., Mcdonald, A., Mcgrath, R., and Navascues, B.: HIRLAM-5 Scientific Documentation, Tech. Rep. December, Swedish Meteorology and Hydrology Institute, 2002.

Van de Berg, W. J., Van den Broeke, M. R., Reijmer, C. H., and Van Meijgaard, E.: Characteristics of the Antarctic surface mass balance, 1958–2002, using a regional atmospheric climate model, Ann. Glaciol., 41, 97–104, doi:10.3189/172756405781813302, 2005.

Van de Berg, W. J., Van den Broeke, M. R., and Van Meijgaard, E.: Heat budget of the East Antarctic lower atmosphere derived from a regional atmospheric climate model, J. Geophys. Res., 112, D23101, doi:10.1029/2007JD008613, 2007.

Van den Broeke, M. R.: Surface radiation balance in Antarctica as measured with automatic weather stations, J. Geophys. Res., 109, D04103, doi:10.1029/2003JD004394, 2004.

Van den Broeke, M. R.: Depth and density of the Antarctic firn layer, Arct. Antarct. Alp. Res., 40, 432–438, doi:10.1657/1523-0430(07-021)[BROEKE]2.0.CO;2, 2008.

Van den Broeke, M. R., Van As, D., Reijmer, C. H., and Van de Wal, R. S. W.: Assessing and Improving the Quality of Unattended Radiation Observations in Antarctica, J. Atmos. Ocean. Tech., 21, 1417–1431, doi:10.1175/1520-0426(2004)021<1417:AAITQO>2.0.CO;2, 2004.

Van den Broeke, M. R., Reijmer, C. H., and Van As, D.: Seasonal cycles of Antarctic surface energy balance from automatic weather stations, Ann. Glaciol., 41, 131–139, 2005a.

Van den Broeke, M. R., Van As, D., Reijmer, C. H., and Van de Wal, R. S. W.: Sensible heat exchange at the Antarctic snow surface: a study with automatic weather stations, Int. J. Climatol., 25, 1081–1101, doi:10.1002/joc.1152, 2005b.

Van Lipzig, N. P. M., Van Meijgaard, E., and Oerlemans, J.: Evaluation of a Regional Atmospheric Model Using Measurements of Surface Heat Exchange Processes from a Site in Antarctica, Mon. Weather Rev., 127, 1994–2011, 1999.

Van Lipzig, N. P. M., Van Meijgaard, E., and Oerlemans, J.: The spatial and temporal variability of the surface mass balance in Antarctica: results from a regional atmospheric climate model, Int. J. Climatol., 22, 1197–1217, doi:10.1002/joc.798, 2002.

Van Meijgaard, E., Van Ulft, L. H., Lenderink, G., De Roode, S. R., Wipfler, L., Boers, R., and Timmermans, A.: Refinement and application of a regional atmospheric model for climate scenario calculations of Western Europe, Climate changes Spatial Planning publication: KvR 054/12, 2012.

White, P. W.: Part IV : PHYSICAL PROCESSES (CY23R4), Tech. Rep. September, 2001.

7

A satellite-based snow cover climatology (1985–2011) for the European Alps derived from AVHRR data

F. Hüsler[1], **T. Jonas**[2], **M. Riffler**[1], **J. P. Musial**[1], **and S. Wunderle**[1]

[1]Institute of Geography and Oeschger Center for Climate Change Research, University of Bern, Bern, Switzerland
[2]WSL Institute for Snow and Avalanche Research SLF, Davos Dorf, Switzerland

Correspondence to: F. Hüsler (fabia.huesler@giub.unibe.ch)

Abstract. Seasonal snow cover is of great environmental and socio-economic importance for the European Alps. Therefore a high priority has been assigned to quantifying its temporal and spatial variability. Complementary to land-based monitoring networks, optical satellite observations can be used to derive spatially comprehensive information on snow cover extent. For understanding long-term changes in alpine snow cover extent, the data acquired by the Advanced Very High Resolution Radiometer (AVHRR) sensors mounted onboard the National Oceanic and Atmospheric Association (NOAA) and Meteorological Operational satellite (MetOp) platforms offer a unique source of information.

In this paper, we present the first space-borne 1 km snow extent climatology for the Alpine region derived from AVHRR data over the period 1985–2011. The objective of this study is twofold: first, to generate a new set of cloud-free satellite snow products using a specific cloud gap-filling technique and second, to examine the spatiotemporal distribution of snow cover in the European Alps over the last 27 yr from the satellite perspective. For this purpose, snow parameters such as snow onset day, snow cover duration (SCD), melt-out date and the snow cover area percentage (SCA) were employed to analyze spatiotemporal variability of snow cover over the course of three decades. On the regional scale, significant trends were found toward a shorter SCD at lower elevations in the south-east and south-west. However, our results do not show any significant trends in the monthly mean SCA over the last 27 yr. This is in agreement with other research findings and may indicate a deceleration of the decreasing snow trend in the Alpine region. Furthermore, such data may provide spatially and temporally homogeneous snow information for comprehensive use in related research fields (i.e., hydrologic and economic applications) or can serve as a reference for climate models.

1 Introduction

Alpine snow cover not only represents a significant geophysical variable in the climate system (IPCC, 2007) but also plays a major role in regulating mountain ecosystems (Jones, 2001; Jonas et al., 2008) and influences hydrological regimes (EEA, 2009). Furthermore, the alpine snow pack is considered an important economic factor, which has a large impacts on tourism (Elsasser and Bürki, 2002; Agrawala, 2007) and hydropower production (Hänggi and Weingartner, 2012). Likewise, snow is a critical variable in flood forecasting (Jasper et al., 2002) and reservoir management but also for avalanche warning and public safety services (Fuchs and Brandl, 2005). Therefore, it is of paramount interest to determine and quantify its variability on various spatial and temporal scales. In this context, "snow cover" has been officially declared an Essential Climate Variable (ECV) by the Global Climate Observing System (GCOS) and high priority is assigned to enhancing and maintaining snow cover observations (WMO, 2010, 2011).

Snow parameters and their spatiotemporal distribution are commonly monitored using three methodologies (or combinations of them), each one having its strengths and limitations. Ground-based monitoring networks provide valuable long time series data to investigate temporal variability and long-term trends (Beniston, 1997; Scherrer et al., 2004; Scherrer and Appenzeller, 2006; Laternser and Schneebeli, 2003; Marty, 2008;

Hantel and Hirtl-Wielke, 2007; Marty and Meister, 2012). However spatial analyses are constrained to dense station networks. Modeling approaches provide another possibility to increase our knowledge of the cryospheric processes. For snow cover, models range from considering simple relations between precipitation and air temperature (Rango and Martinec, 1995) through to complex physical interactions at different spatial scales (Marks et al., 1999; Lehning et al., 2006). However, while conceptual models require regional calibration, complex process-oriented models are often challenged by the enormous requirements as to input data (Magnusson et al., 2011). In addition, they may still rely on a set of assumptions and auxiliary data, which by themselves include uncertainties. Complementary rather than substitutionally, satellite remote sensing (RS) has become an attractive option to study snow cover variability on a wide range of spatial and temporal scales (König et al., 2001; Dietz et al., 2012; Metsämäki et al., 2012). The major advantage of visible RS is its ability to provide area-wide and spatially comprehensive surface information with a regular repeatability of measurements, even in remote areas. However, limitations include cloud cover impeding surface view, the obstruction of snow by dense vegetation as well as the surface heterogeneity in mountain areas, which complicates the interpretation of RS data. Nevertheless, climatological snow cover studies based on RS data have been successfully compiled on regional to hemispherical scales (Robinson and Frei, 2000; Armstrong and Brodzik, 2001; Zhao and Fernandes, 2009; Foppa and Seiz, 2012).

Starting in 1960, RS technology has undergone rapid changes. With advanced sensor technologies, new dimensions of monitoring snow properties have been achieved and the measurement accuracy has increased (Rees, 2005; Nolin, 2011; Dietz et al., 2012). Indeed, the exceptional value of older sensors lies in their unique long-term data set. Among these, the AVHRR instrument provides the longest record of visible satellite imagery on a daily basis. The University of Bern has archived daily full-resolution AVHRR data over Europe for almost 30 yr (1984–2012). Using AVHRR for long-term climate studies, however, means having to properly deal with the many limitations of this sensor, such as calibration uncertainties and changing channel configuration, which have been addressed in various studies (James and Kalluri, 1994; Cihlar et al., 2004; Heidinger et al., 2010; Gutman and Masek, 2012). The main difficulty therefore lies in the appropriate raw data processing, which is a key requirement to access the full potential of the archived data.

This study presents and analyzes the first consistent space-borne 1 km snow coverage climatology derived from AVHRR instruments over the European Alps for the period 1985–2011. A robust and validated snow retrieval algorithm (Hüsler et al., 2012) was applied to derive real surface variability with consistent results. Hence, the objectives of this study are (1) to compile a set of cloud-free satellite snow composites using a spatial and temporal gap-filling method and (2) to employ these composites to analyze the spatiotemporal snow cover variability in the Alpine region over the last three decades. This is done using snow parameters such as snow onset day, snow cover duration, snow melt-out date, and the snow cover area percentage (SCA).

Given the need for spatial representations of the alpine snow cover and its variations for many research fields, including hydrological and ecological applications, socio-economic questions or albedo parameterization in climate models, this data set is considered a powerful contribution to a regional climate observing system.

2 Study area

The topography and delineation of the Alpine region as defined by the Alpine Convention (http://www.alpconv.org/index_en, accessed December 2012) are shown in Fig. 1. The complex topography of the European Alps, coupled with the varying altitudinal and climatic conditions, essentially influences the spatial and temporal occurrence and duration of the snow cover. Consequently, it allows us to investigate highly variable climatic conditions on comparatively short horizontal and vertical distances. For our investigations, the study area was additionally subdivided into four distinct climatic regions which are assumed to reflect different snow cover regimes. These subregions are defined according to the statistical regionalization of different climate elements such as precipitation and temperature (Auer et al., 2007). The most distinct climate border is found along the central Alpine crest. It represents the continental transition zone from the temperate westerly to the Mediterranean subtropical climate. A second continental scale climate border is identified between the (western) oceanic influenced and the (eastern) continental part of Eurasia. This rather uniform border roughly follows the 12th degree eastern longitude and was not found to be caused by any topographic feature (Auer et al., 2007). The size, altitudinal distribution and climatic classification of the study area and each subregion are listed in Table 1.

3 Data

3.1 AVHRR data set

The data set consists of daily 1 km gridded binary snow cover maps generated from a full-resolution AVHRR data archive, which is unique in its form as it represents the longest and most comprehensive full-resolution RS data set for the European Alps (Hüsler et al., 2011). In this study, the data is restricted to the five-channel instruments AVHRR/2 and AVHRR/3 to ease the discrimination of thin cirrus clouds. The snow detection relies on a threshold approach that capitalizes on the spectral properties of snow in the visible and the near infrared spectrum and it is applicable to any kind of AVHRR sensor generation with consistent results. Originally

Table 1. Distinct properties of the Alps and its subregions in terms of altitude distribution, extent and climatic influence. Color encoding corresponds to Fig. 1.

Region	Altitude distribution	Extent (km^2)	Climatic features (Auer et al., 2007)
Alpine region	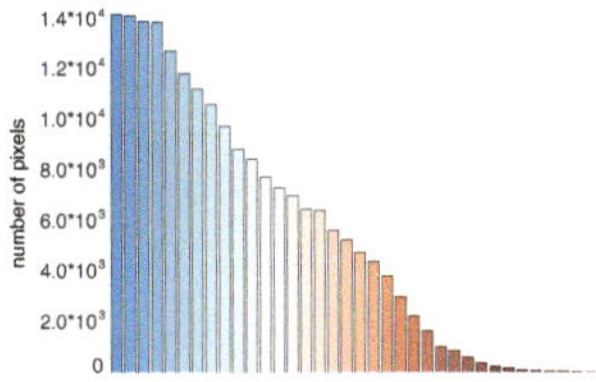	226 000	
North-west (NW)	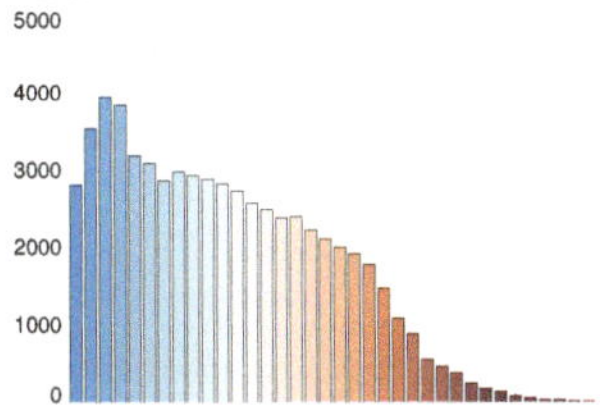	67 000	temperate westerly, oceanic features
North-east (NE)	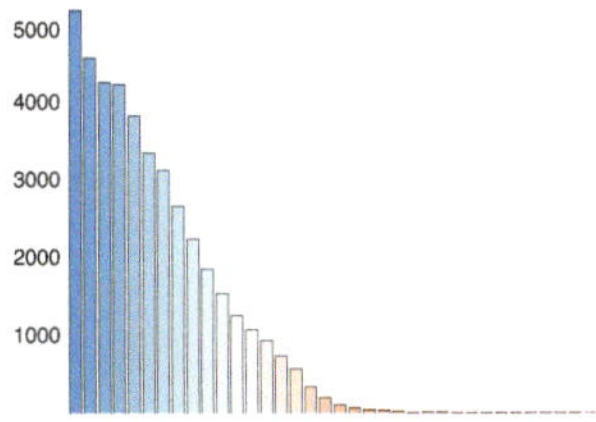	50 000	temperate westerly, continental features
South-east (SE)	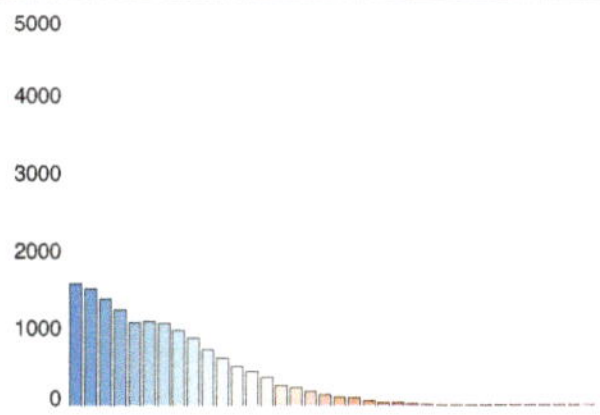	17 000	mediterranean subtropical, contintental features
South-west (SW)	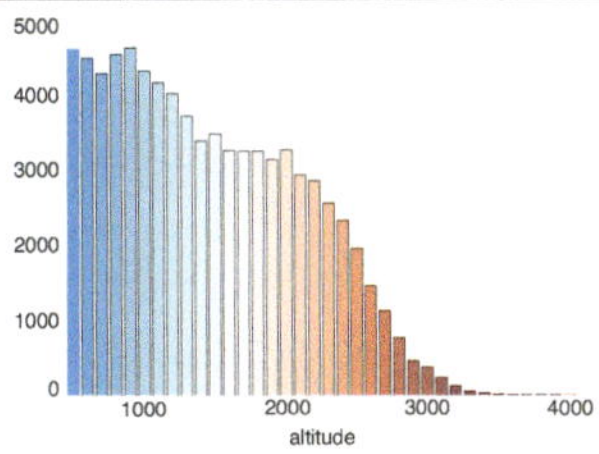	92 000	mediterranean subtropical, oceanic features

developed for rather flat areas in Canada (Khlopenkov and Trishchenko, 2007), the algorithm was adapted and optimized for use in complex terrain. The modifications of the original algorithm basically include changes in the thresholds, taking into account land cover types and topographic features such as terrain shadow. In addition, a distinct snow probability map comes with each snow map, providing a pixel-wise probability index of snow occurrence based on logistic regression (Hosmer and Lemeshow, 2000). A detailed description of the algorithm itself and an extensive validation of the product can be found in Hüsler et al. (2012). The overall classification accuracy of snow detection was found to be good, with an average probability of detection of up to 90 % depending on land cover class, acquisition geometry, and aspect.

The number of available scenes per season is displayed in Fig. 2. Included platforms are NOAA-09, 11, and 14 equipped with AVHRR/2 and NOAA-16, 17, 18, 19, and MetOp-A carrying an AVHRR/3 instrument onboard. To constrain the influence of the snow's distinct bidirectional distribution function effects (Wiscombe and Warren, 1980) and strong shadowing due to low sun angles, we only use midday and early-afternoon overpasses. This, however, comes at the cost of two small data gaps in 1994 as well as in 2000, when only early-morning platforms were in orbit. Additionally, the maximum sensor zenith angle was constrained

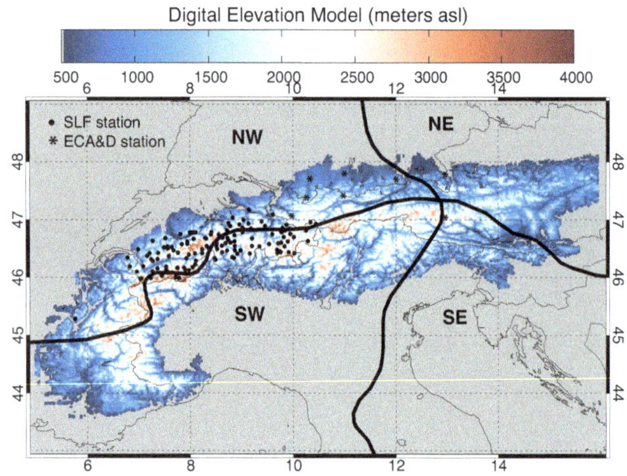

Fig. 1. Digital elevation model of the Alpine region. Black dots represent station locations used for the validation of the product. Bold lines indicate the subdivision into the four subregions (north-west, north-east, south-east and south-west) according to Auer et al. (2007). Data for the extra-alpine areas were not displayed.

to 35° since the snow mask accuracy was found to decrease with the distance to nadir.

An irregular distribution over such a long period is inevitable, mainly due to a varying number of platforms simultaneously in orbit. This needs to be taken into account in time series applications. Generally, there is an increasing trend in the data availability. While the number of scenes from 1985 to 1989 can primarily be used for temporal applications relying on monthly mean data, the record provides an almost complete daily coverage from 1989–2002 and reaches a maximum of 2–3 overpasses per day from 2003 to 2011. Therefore, depending on the type of application, two different time periods are considered in the study at hand. First, 1985 to 2011 for the monthly mean snow cover time series and second, 1991–2011 for spatial applications relying on daily data coverage.

3.2 Gridded station data set and snow depth measurements

As a reference data set, a gridded snow depth climatology based on in situ measurements (cf. Rhyner et al., 2002; Jonas et al., 2009) available for Switzerland was employed. This data set consists of daily snow depth measurements from 133 stations for the period of 1989–2009 and is therefore considered stable over time (see also Hüsler et al., 2012). To avoid the problem of comparing point scale data (stations) to area scale data (satellite), the snow depth data were mapped to the AVHRR grid using a snow data assimilation scheme based on Auer et al. (2004) and available from December to April.

In addition, in situ snow depth information from 48 automatic stations over the time period 2003–2010 were used to validate the gap-filling procedure to include longer time periods, to assess the seasonal performance, and to ensure that the gap-filling does not significantly lower the initial accuracy. The stations are distributed all over Switzerland covering different elevation ranges and the data are provided by the WSL Institute for Snow and Avalanche Research SLF. Snow-depth data is recorded automatically and available daily. To expand the validation beyond Switzerland, the 8 European Climate Assessment & Data stations (Haylock et al., 2008) lying within the Alpine Region were additionally included in the reference data set. The locations of all stations are illustrated in Fig. 1 and they are mainly positioned at elevations between 480 and 3100 m a.s.l.

4 Methods

4.1 Snow mask gap-filling procedure

One major drawback of optical sensors such as AVHRR is their inability to provide surface information under cloudy conditions. This strongly limits the number of useful snow observations as abundant cloud cover is present in the Alpine region, particularly during the winter season.

Various techniques have been developed to reduce cloud-covered pixels from optical imagery. They mostly include spatial and temporal filtering techniques (i.e., Voigt, 2000; Gafurov and Bárdossy, 2009; Parajka et al., 2010), multi-day maximum snow cover composites, or a combination of acquisitions from similar sensors (i.e., Parajka and Blöschl, 2008; Wang et al., 2008; Hall et al., 2010).

In this study, we use a combination of spatial and temporal gap-filling techniques to mitigate the influence of clouds. In the first step we applied straightforward spatial filtering using the regional snow line estimation (SNOWL) method as suggested by Parajka et al. (2010). This procedure reclassifies cloudy pixels to snow or snow-free land, according to their position, relative to the current snow line elevation. In their publication, the accuracy of this method was assessed using 754 climate stations and it was found that the SNOWL method provides robust snow cover mapping over Austria, even under abundant cloud cover of 90 % (Parajka et al., 2010). To avoid blurring regional differences by assuming one snow line elevation for the whole Alpine region, the reclassification was done separately for each of the four distinct climatic regions (cf. Sect. 2). In the second step, a forward (pixel value of closest clear-sky observation in the forward time direction) and a backward (pixel value of closest clear-sky observation in the backward time direction) gap-filling procedure (Foppa and Seiz, 2012) was carried out over a period of 10 days in both directions. Generally, the time period for filtering needs to be as short as possible but long enough to get cloud-free composites, especially during winter when cloud cover is frequent. Hence, it is a trade-off between remaining clouds and the blurring of snow information. Analyses of the amount of cloud and no data value

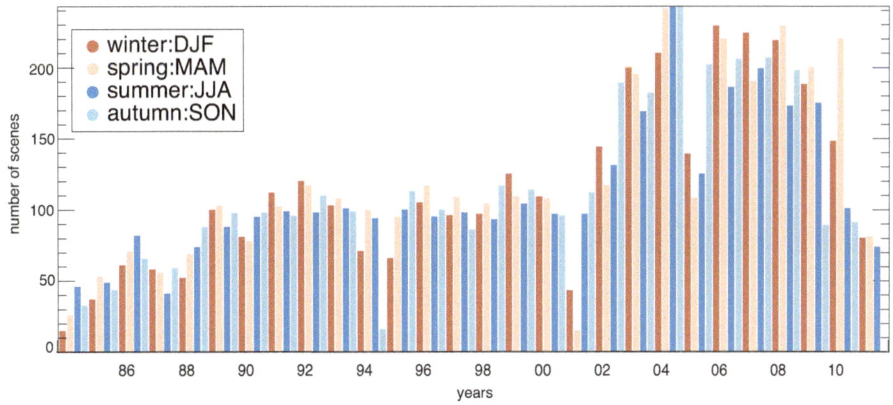

Fig. 2. Archived AVHRR overpasses illustrated by season and number of scenes for the period 1984–2011.

reduction showed that a minimum of 10 days is required to render cloud-free images. Even though this period might be slightly shorter for later years with abundant data, the number of days for the filtering is determined by the results for earlier years with limited data availability. Furthermore, it needs to be kept in mind that the closest clear-sky observation is used for the composite, which means that 10 days is the maximum amount of days to search for clear-sky observations and only applies to a small amount of pixels.

For the final product, the mean SCA value between the two values (the one from the forward and the one from the backward filtered product for a specific day) was taken. This is based on the assumption that the under- or overestimation in one time direction is minimized by averaging the results from the forward and the backward gap-filling procedures (Foppa and Seiz, 2012).

4.2 Snow gap-filling accuracy assessment

To assess the accuracy and consistency of the gap-filled product, two types of reference data sets were used: gridded snow depth data and point snow depth data. For the comparison of the gridded data set to the satellite gap-filled product, the SCA over Switzerland was calculated for both products for each day as described in Sect. 4.3. Different thresholds are suggested in literature to infer snow coverage from measured snow depth, which is assumed to approximately represent a 50 % fractional snow cover in complex alpine terrain at 1 km resolution. These commonly range from 1 cm (Parajka and Bloeschl, 2006; Foppa and Seiz, 2012) over 4 cm (Wang et al., 2009) to 5 cm (Romanov et al., 2002). Here, a threshold of 5 cm was used to calculate snow cover percentage to avoid shallow snow depth or patchy snow cover not detectable by satellite. Slopes steeper than 20° were excluded from the analyses as they are not considered being well represented in the two data sets.

For the point-wise validation the automated snow stations were considered as ground truth and were compared to the pixel closest to each station. The snow depth is measured from above with an ultrasonic snow depth sensor (Lehning et al., 1999). During the snow-free season, the same sensor measures vegetation height. Hence, a pre-processing of the station data was necessary. For this purpose, a robust automated correction algorithm was implemented to find the first and the last occurrence of a location specific minimum value to determine the snow melt out and onset timing. Values in between were set to zero as these are assumed to correspond to vegetation height rather than to snow depth. Quality checking indicated a good performance of the correction although short-term snow events after complete melt out and before permanent snow onset might not be represented in the reference data. Methodologically, the comparison was carried out by means of investigating the accuracy index (ACC), defined as the sum of correctly classified pixels divided by the total number of compared (i.e., cloud-free) pixels. As above, the station was classified to be snow covered when the measured snow depth was equal or higher to 5 cm. Furthermore, the misclassified pixels were evaluated in terms of snow over- (SO) and underestimation (SU), both relating the sum of incorrectly classified pixels to the total number of compared (i.e., cloud-free) pixels as suggested by Parajka et al. (2010). For this type of validation it must be kept in mind, that, in mountainous regions, steep elevation gradients are common and one single AVHRR pixel may include both valleys and mountain peaks with elevation differences up to 1200 m at alpine sites. As a consequence, differences may be observed when a grid cell value is compared to a in situ point observation in complex terrain.

4.3 Snow cover parameters

To estimate the temporal distribution, the snow cover area percentage (SCA) is a frequently used measure (Armstrong and Brodzik, 2001; Gao et al., 2012). It is defined as the percentage area of a certain reference area (i.e., the Alpine region) covered by snow. More specifically, it represents the ratio of snow covered pixels in the reference area to all pixels in the reference area at a specific point in time. For time series analysis, standardized anomalies with respect to month are computed using:

$$A = (X - \mu)/\sigma, \tag{1}$$

where X is the single monthly mean value, calculated as the mean of all daily values available for a specific month. The μ value is the long-term mean and σ the standard deviation of the respective month.

The snow cover duration (SCD) is another parameter of broad interest, that is, for winter tourism, water resource management, agricultural and disaster/flood forecasting. To make our results comparable in a broader context, we adapted the measures suggested by Wang and Xie (2009) and Gao et al. (2012). Using the gap-filled data product, we calculated the annual SCD using all available scenes within a hydrological year starting from 1 October to 30 September as follows:

$$\mathrm{SCD} = \sum_{i=0}^{N} D_i, \tag{2}$$

where N is the total number of scenes within the hydrological year; D_i is the actual pixel value for a one day composite within the period (0 for snow free, 1 for snow). This parameter basically counts pixel-wise the number of snow values and represents the total number of snow days for one hydrological year. The SCD anomalies provide a measure for the difference between the SCD value of a specific year compared to the long-term mean SCD. Likewise, monthly anomalies represent the difference between the actual monthly value and the long-term mean value for the respective month.

In the context of an anticipated climate change, the snow cover onset, as well as melt-out timing, represent valuable indicators. Following the explanations of Wang and Xie (2009), these measures are captured in the Snow Cover Onset Day (SCOD), which represents the first day of snow occurrence for a certain pixel,

$$\mathrm{SCOD} = D - \mathrm{SCD}', \tag{3}$$

where SCD$'$ is the snow cover duration within the period from 1 October (DOY = 274 or 275) to 31 December (DOY = 365 or 366) and D is the Day Of Year (DOY) of 31 December (DOY = 365 or 366), when the area is supposed to be covered by snow. The Snow Cover Melting Day (SCMD), that is, the last day with snow occurrence within a pixel, is defined as

$$\mathrm{SCMD} = D + \mathrm{SCD}'', \tag{4}$$

where SCD$''$ is the snow cover duration within the period from 1 January (DOY = 1) to 31 May (DOY = 151 or 152) and D is January 1 (DOY = 1) of the respective year.

Since these measures rely on the continuous availability of data, errors might arise from single missing scenes. Therefore, only the period of 1991–2011 (excluding the years 1995 and 2001) was considered when the amount of adjacent missing values is reasonably small (< 10). The SCOD and SCMD parameters are most meaningful in areas with stable snow cover (i.e., with a continuous snow cover between the onset and the melt out). For regions with intermittent snow cover, SCOD and SCMD count the snow days within the indicated period but are not affected by repeated short-term snow events that often occur at lower elevations. A slightly modified approach was tested, setting the SCOD (SCMD) date when a pixel was snow-covered for a certain number of consecutive days. This, however, resulted in single missing or gap-filled scenes which completely distorted the parameters. Therefore, the measures described here were concluded to be useful bearing in mind that the results are of most interest in those regions with continuous seasonal snow cover.

5 Results

5.1 Gap-filled product

The results of the spatial and temporal filtering process for all available scenes for the period January to April 2006 are illustrated in Fig. 3. This analysis was restricted to Switzerland due to the limited extent of the station reference data set (see Sect. 3.2). The values of the relative area percentage of the classes snow, snow free, cloud and invalid pixels per scene change considerably from the unfiltered scenes (a) to the final filtered scenes (c). As an intermediate step, panel (b) shows the relative redistribution of the classes after the spatial gap-filling but before the temporal gap-filling. Merely minor changes become apparent as this step affects only a small number of the pixels, which are primarily located at high altitudes. Panel (d) shows the difference between the forward and the backward gap-filling direction and the averaged value as compared to the gridded station data set. The average value is calculated as the mean between the SCA value derived from the forward filtering product and the SCA value of the backward filtering product for a specific day. Only minor differences in the range of 0–10% between the averaged value and the reference data set are found. Therefore the gap-filling procedure is considered appropriate. The quantitative improvement in terms of cloud reduction after each filtering step is listed in Table 2. While the change from the unfiltered data to the spatially filtered data is relatively small (change in median from 37.2 % to 35.7 %), the temporal filtering step strongly impacts on the number of cloudy pixels. The amount of clouds and non-valid pixels is, on average, reduced by 35 % (clouds) and

3 % (non-valid). The maximum percentage for the remaining clouds after temporal filtering is only 0.9 %. Vice versa, the number of clear-sky pixels (snow covered and snow free) is significantly increased. The snow percentage raises from 26.3 % up to 76.7 % and finally corresponds very well to the same measure derived from the gridded station data set with a median of snow percentage of 73 %. Likewise, the number of clear-sky snow-free cases shows an increase of 14 %.

The results of the station validation to assess the seasonal and filtered accuracy of the snow product are shown in Fig. 4 and consist of three box plots. The topmost plot (a) illustrates the ACC for the four products (unfiltered, spatially filtered, forward and backward filtered) for all months while (b) and (c) show the SU and SO errors, respectively. The mean annual accuracy ranges from 90 % (for unfiltered and spatially filtered) up to 91 % (for the temporally filtered products). As expected, the ACC is high for the full snow season (December–March) and the summer season with values constantly over 90 %, whereas it decreases to 82–89 % in the transition season. The mixed pixels including patchy snow cover, different land cover types such as forest, combined with strong elevation differences are suggested to be the reason for this. Also the IQR increases for the transition season, however the general pattern is quite consistent throughout all the years expressed in the comparatively small IQR values.

Except for the month of June and October, the temporally filtered product does not decrease in ACC through the filtering process. On the contrary, it even slightly increases the performance of the snow mapping. This is assumed to be caused by the reduction of misclassified cloudy as snowy pixels (or vice versa). Concerning the errors, they show constantly very low values close to 0 %. The reduction of ACC in the transition season is attributed to a slight underestimation of the snow cover in spring (SU error of 6 % (May) and 7.5 % (June) for the unfiltered product). As stated above, this is most probably caused by mixed pixels (i.e., snow lying under trees or snow patches and therefore not being detectable by means of remote sensing). The temporally filtered products show a comparatively higher SU error (up to 12–15% in June). As daily values are compared, this kind of validation is particularly sensitive to the uncertainty of ±10 days inherent in the composite product. During the snow onset season (October, November), however, there is a tendency towards snow overestimation by the temporally filtered products. This is more pronounced for the backward filtered product (SO error of 11 %) and is probably explained by the filling of pixels with clear-sky information from a couple of days (maximum 10) ahead of the actual day. Especially at lower elevations, the snow-onset process is characterized by several snow-onset events interrupted by melting events. Likewise, it needs to be considered that the correction of the station data described in Sect. 4.2 also leads to some uncertainties in the snow melt-out and onset timing as derived from the station. Particularly short-term spring or summer snowfall events after complete melt-out and minor snowfall events before permanent snow onset are not represented in the station measurements.

In summary, the influence of the filtering procedure on the accuracy of the product is small. Hence, such spatial and temporal interpolation techniques are considered powerful tools to retrieve more comprehensive surface information. However, they also introduce uncertainties into the data set. Designed to make assumptions about conditions beneath the cloud cover, confidence in a surface value decreases proportionally to the length of the cloud cover persistence. As introduced by Hall et al. (2010), a piece of associated cloud-persistence information for each pixel is provided as an indicator of confidence in the actual pixel value. The cloud persistence values are low during clear conditions indicating high confidence and they increase with persisting cloudy conditions, resulting in lower confidence. Furthermore, for interpretation, one should consider that products based on gap filling methods only provide estimates of cloud-free observations of multiple days (±10 days) and pixel values remain subject to this uncertainty. Likewise, initial uncertainties arise from erroneous classification of clouds as snow and vice versa since the two features exhibit very similar spectral properties and remain difficult to discriminate. Based on this satellite-based snow cover climatology, the spatiotemporal distribution of snow in the Alps over the last two decades is analyzed with respect to spatial distribution, annual and inter-annual variations in snow cover. The time period considered for each application is indicated.

5.2 Spatial distribution of snow cover (1991–2011)

The long-term annual means (1991–2011) for the parameters SCD, SCOD and SCMD for the Alpine region are displayed in Fig. 5 for each AVHRR pixel. The overall uncertainty lies at ±10 days as reported in Sect. 4.1. Generally, altitude is the main influencing factor due to the temperature gradient decreasing with height. All parameters clearly follow the distinct topography. Focussing on SCD (Fig. 5, left panel), the highest areas are shown in dark red indicating year-round (365 days) snow cover above approximately 2900 m a.s.l. The SCD gradually decreases along the altitudinal steps and reaches a minimum of < 25 days in the lowest regions in the south-west as well as in the south towards the Po Valley in Northern Italy. Similarly, regional differences become apparent. For example, the SCD in the south-western parts is much lower (i.e., 40 days on average) than in the north-eastern part (i.e., 75 days on average), even though lying at about the same altitude range (i.e., between 600 and 700 m a.s.l.). This occurs presumably due to different climatic influences (see Sect. 2). Yet, these regional differences decrease toward higher altitudes and the SCD becomes similar in magnitude in all regions.

Since the SCD is closely linked – at least in altitudes with stable snow cover – to the SCOD, when snow starts to accumulate, and to the SCMD, when snow disappears, a similar

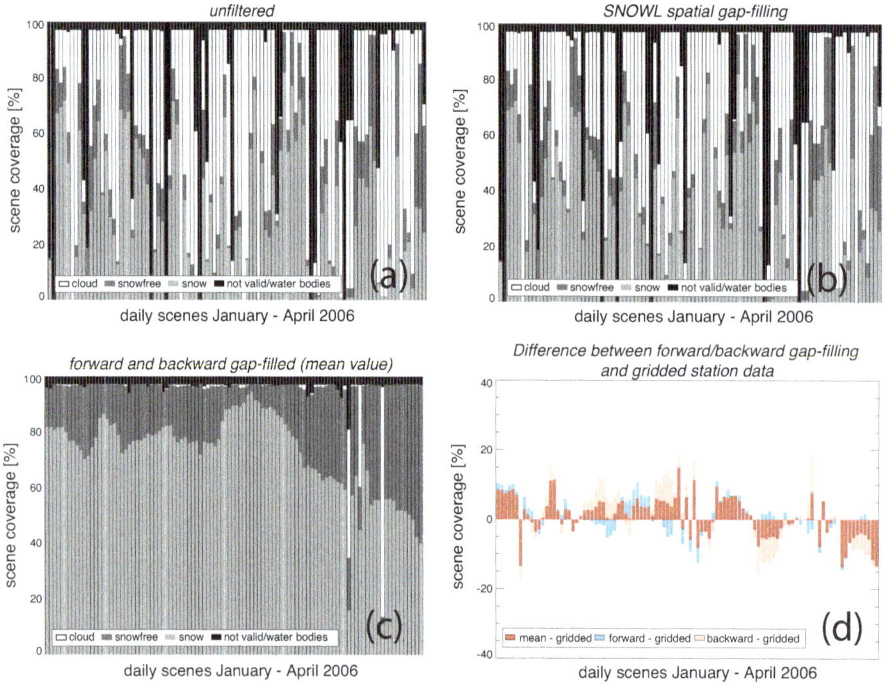

Fig. 3. Stepwise results from the spatial and temporal cloud removal process for the period January–April 2006: respective area percentage of Switzerland for the classes snow, snow free, cloud and not valid for **(a)** unfiltered scenes, **(b)** after SNOWL spatial gap-filling filter, **(c)** after temporal backward and forward gap-filling (mean value), **(d)** differences between forward and backward direction and averaged value thereof (calculated as the mean between the SCA value from the forward gap-filled product and the SCA value from the backward gap-filled product for the same day) and gridded station data.

Table 2. Quantitative assessment of classification redistribution after each type of filtering according to Fig. 3.

		Min (%)	Lower quartile (%)	Median (%)	Upper quartile (%)	Max (%)
Cloud	no filter	0.0	12.6	37.2	65.2	97.0
	spatial filter	0.0	14.0	35.7	61.0	97.0
	temporal filter	0.0	0.0	0.1	0.4	0.9
No data	no filter	0.0	3.0	3.0	6.4	10.3
	spatial filter	2.9	3.0	3.0	7.0	10.3
	temporal filter	2.9	2.9	2.9	3.0	3.0
Snow	no filter	0.0	6.2	26.3	51.2	81.9
	spatial filter	0.0	11.3	33.9	53.3	82.0
	temporal filter	40.0	62.9	76.7	82.2	94.6
	snow gridded	53.9	68.2	73.0	78.1	92.6
Snow free	no filter	0.0	1.5	5.5	14.1	29.8
	spatial filter	0.0	1.5	5.2	13.6	28.2
	temporal filter	2.4	14.2	19.3	31.8	56.7

pattern can be found in the center and right panels of Fig. 5. Even though the topography clearly emerges in both maps, regional differences such as the west–east as well as south–north gradient are present. A comparatively more blurred pattern is observed in the SCOD map as this measure depends not only on temperature but also on availability of precipitable water in atmosphere. The SCMD, in contrast, shows very subtle differences indicating a slow and gradual decrease of snow between January (with values between 25 and 59 at low altitudes) and the end of May (at altitudes of around 2300 m a.s.l.; values of 225–244) caused by the spring temperature increase. These results are perfectly in line with observations made by Jonas et al. (2008) illustrating that SCOD

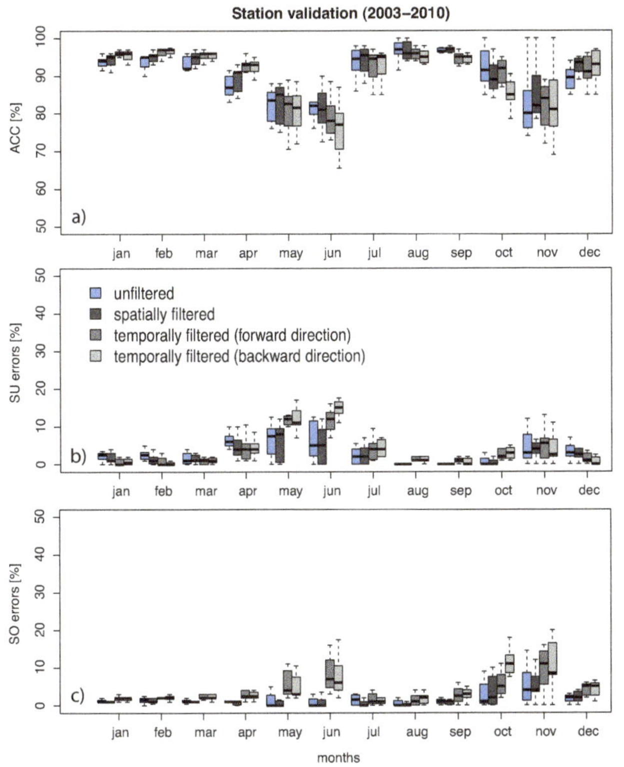

Fig. 4. Monthly median and Inter Quartile Range (IQR) of **(a)** ACC, **(b)** SU error, and **(c)** SO error from station validation for the period 2003–2010. Different colors refer to different types of filtering procedures (see legend).

(SCMD) shows little (high) correlation with elevation over time.

Figure 6 displays the SCD as a function of altitude and was calculated from all the pixel information available per 100 m altitude in the respective region. As expected, a clear positive relation between altitude and SCD is found with an increase of ca. 10 days per 100 m a.s.l. with a standard deviation of 25 days. This relationship, however, seems to depend on the climatic region. While in the southern regions exhibit constantly lower SCD at elevations up to 2200 m, the northern regions show above-average SCD at the same altitudes. It is also remarkable that the regional differences are more pronounced at lower elevations and tend to disappear towards the high elevations. The same behavior is also found in the standard deviations. This may imply a higher sensitivity of SCD on temperature at lower regions and a stronger dependency of SCD on precipitation at the higher elevations.

5.3 Annual variation of snow cover (1985–2011)

The amount of snow present in the Alps varies considerably over the course of one year. The seasonal snow cover generally starts to increase in November, reaches a maximum between January and February and starts to melt out in mid-March. By the end of May, almost all snow has disappeared except for the very high areas lying above 2500 m a.s.l.

Figure 7 quantitatively illustrates the monthly statistics, derived as median value of all the available daily scenes of a respective month, averaged over the period 1985–2011 retrieved from the satellite record. Focussing on the entire Alpine region, the snow cover peaks in January and February with a monthly median of 68 %, that corresponds to an area of approximately 155 000 km^2. During the melting season, it gradually decreases, reaching a minimum extent during August and September, when only the perennial snow and ice remain on the ground (approximately 2600 $km^2 \pm 25$ % in the Alps according to Fontana et al., 2010). The (IQR) is largest in February, March and November, when the highest inter-annual variability is observed due to variations of SCA in the transition zone (approximately 700–2100 m a.s.l.). As expected, the smallest variation is found during summer when a minimum snow cover is present, which is reasonably stable over the whole record's period.

Even though the subregions are not comparable in terms of absolute numbers as the altitude distribution is quite different (see Table 1), the qualitative annual snow cycle for each region can be seen from Fig. 7. Besides the general annual behavior, the subregions exhibit some interesting features: the north-western part of the Alps shows constantly higher values of SCA due to higher elevations while the south-western part (comparable in terms of altitudinal distribution) shows lower values in winter and relatively high values starting in April. The SCA decrease in the south-west and north-west, hosting the highest elevations, however, seems to be slightly slower than in the eastern regions by virtue of the melting process being decelerated as elevation increases. Obviously, the snow persistence varies in different elevations as onset of snow melt is naturally postponed with increases in the terrain elevation. Generally, the spread is highest from November through March but it varies among the four regions: the inter-annual variability in the north-west and south-west seems to be least pronounced (e.g., 10 % in the north-west in January) resulting in a lower IQR than in the eastern regions (e.g., 20 % in the north-east in January).

5.4 Inter-annual variation of snow cover (1991–2011)

In this section, we focus on the year-to-year variability of the snow cover, as it has repeatedly been shown to be considerable (Scherrer et al., 2004). The annual pixel-wise departures from the long-term mean (1991–2011) are displayed in Fig. 8 for the entire Alpine region. The highest regions (> 2900 m a.s.l.) in the Central Alps exhibit stable snow cover and the anomalies constantly lie within the uncertainty range of ±10 days. It can be further assumed that the SCD in the years 1998, 2002 and 2003 was below average, with 2007 being the poorest winter on record. In contrast, comprehensive above-average snow cover conditions are found in 1992, 1993, 2009 and 2010.

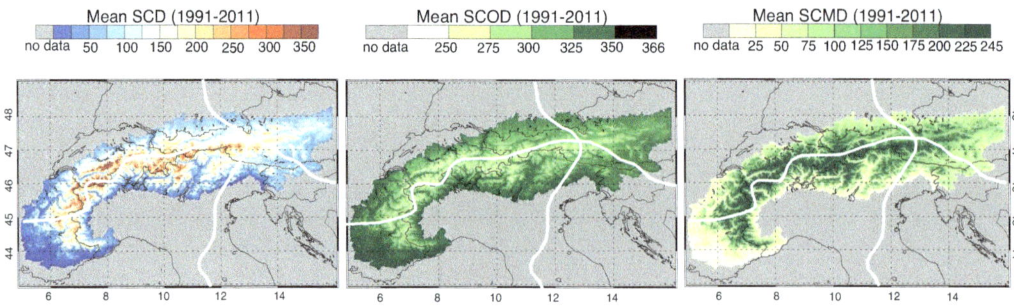

Fig. 5. Spatial representation of the long-term annual mean of snow cover parameters for the Alpine region (1991–2011): SCD (left) in numbers of snow days, SCOD (center) and SCMD (right), given as DOY for each pixel. Overall, a very strong correlation to the topography is observed but regional differences also become apparent. White lines indicate the climatic regions described in Sect. 2.

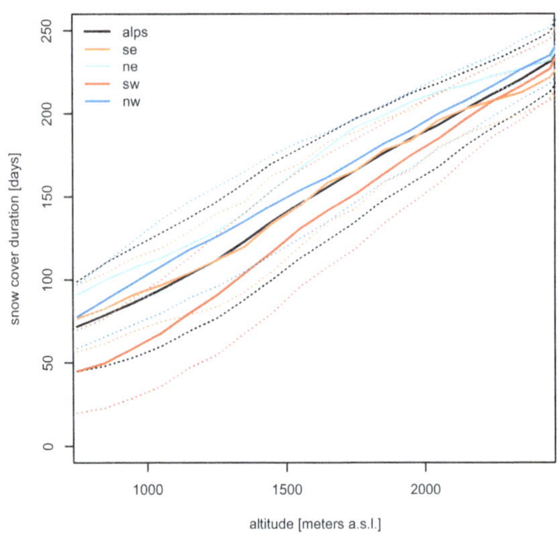

Fig. 6. SCD as a function of altitude for the Alps and the four regions separately. Bold lines represent the altitude mean SCD while the dotted lines indicate the standard deviation. The analysis was done at elevation steps of 100m.

It is striking that some anomaly patterns seem to occur systematically: firstly, a uniform pattern with above- (below-) average snow conditions, which prevails over the entire Alps (e.g., 1992, 1998, 2002 and 2007). Secondly, a pronounced north–south pattern can be observed in 1991, 1994 and 2005 when the Alpine ridge roughly divides the anomalies in a positive and a negative region. A third pattern follows an altitudinal gradient, for example, in 2003, 2006, 2008 and 2009. Lower levels clearly show below- (above-)average conditions while the higher altitudinal levels approach average conditions. Fourth, an east–west pattern, characterized by a meridional anomaly gradient, is presumed in 1996, 2000, 2004 and 2010.

Hence, the regions and altitudes were investigated in more detail as the varying climatic features seem to have a major influence on the snow cover distribution. Figure 9 illustrates the annual departures from the 1991–2011 median value (as it does not follow a normal distribution) for the three parameters SCD, SCOD and SCMD for each region with an altitudinal resolution of 100 m (from 700 to 2500 m, which corresponds to the altitude levels experiencing seasonal snow cover). Generally it can be observed, that the altitudinal variation is more pronounced in SCMD while the variability in the snow onset is lower between different altitudes. The year-to-year variability though, was found to be higher in the SCOD for all altitudes as concluded from the IQR plot (figure not shown) where the spread for SCOD was much larger than for SCMD.

A closer look at the altitudinal distribution of variation reveals that the variability at lower regions (displayed in bluish colors in Fig. 9) is increased in comparison to the values of higher altitudes (red colors). This again implies the strong dependency of snow cover variation on the vertical temperature gradient as the likelihood of frequent transitions over freezing point is highest at lower elevations.

Regarding the inter-regional differences, some interesting features become visible. Besides the lower altitudes, the inter-annual spatial variations are more apparent in the southern regions of the Alps, where the SCD and SCMD parameters especially show a higher spread in the higher regions (i.e., south-west and south-east in 2002 and 2010). Furthermore, the pronounced anomalies in the lower regions (+40 days SCD) in 1996 only affected the eastern part of the Alps. The pattern of comparatively (short) long SCD at (higher) lower altitudes in 2006 is assumed to be caused by a late snow onset (at all altitudes) and repeated snowfall events in February and March (Stucki et al., 2011). These led to record snow amounts also at lower altitudes which lasted longer and delayed the melt-out day in spring 2006. In the case of 1998, all regions show increased negative low-altitude anomalies and positive high-altitude anomalies, except for the south-eastern parts.

As these annual values are not affected by a cyclic (seasonal) pattern and no significant autocorrelation has been found, simple linear trends were estimated and tested for significance for each altitudinal range and each region. Most

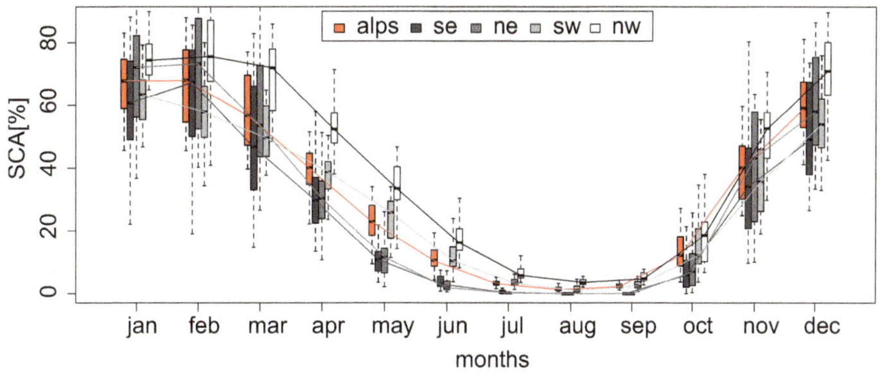

Fig. 7. Monthly median and IQR of SCA averaged over the period 1985–2011 for the whole Alpine region (red) and all four subregions (gray tones) over the course of the year. SCA values are given in numbers relative to the size of the respective region (cf. Table 1). To facilitate interpretation, lines connect median values from each month.

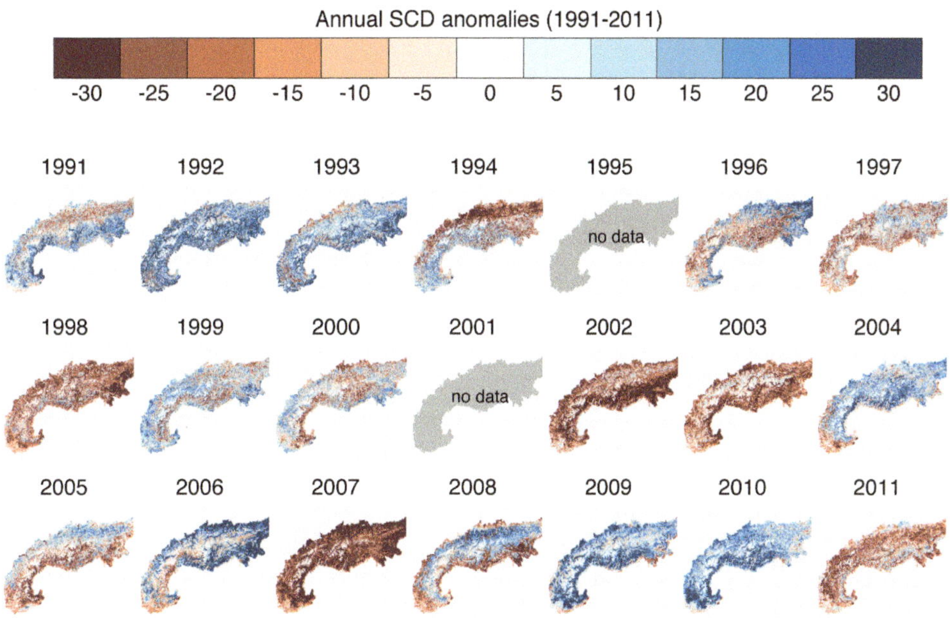

Fig. 8. Annual departures from the long-term mean (1991–2011) given in days. Years refer to 1 January. The data coverage in 1995 and 2001 was insufficient for the application.

of them showed a decreasing tendency but were found to be insignificant, including the trends for the Alpine region as a whole. For some particular altitudes in single regions, however, evidence for decreasing (or increasing) values was found. These mostly affected the lowest (and partially also the highest) altitudes in the southern regions: for SCD, a slight evidence against the null hypothesis ($0.05 < p < 0.1$) was found in SW at altitudes 700–900 m (trend = -2.1 days yr^{-1}) and 2100–2300 (trend = -1.2 days yr^{-1}) as well as in SE at altitudes 700 and 800 (trend = $-1.7/-1.8$ days yr^{-1}). Concerning SCOD, slight evidence for a later snow onset was also found in the lowest (900–1100) and highest altitude (2100–2300) levels in SW and SE. For altitudes 700–900, even moderate evidence ($p < 0.05$) was found, implying a significant trend of $+0.9$ days yr^{-1}. Also for the SE, significant trends ($p < 0.05$) toward a later snow onset in the lower altitudes (700–1200) were found (with trends around $+1$ day yr^{-1}). In summary and even though the time series is comparatively short, the lowest regions of the southern Alps seem to be the most affected by climate change over the last 20 yr while the changes in northern and more elevated parts do not show any significant tendency.

5.5 Time series of SCA (1985–2011)

Finally, an SCA time series analysis based on monthly data was carried out. To demonstrate the temporal stability of the

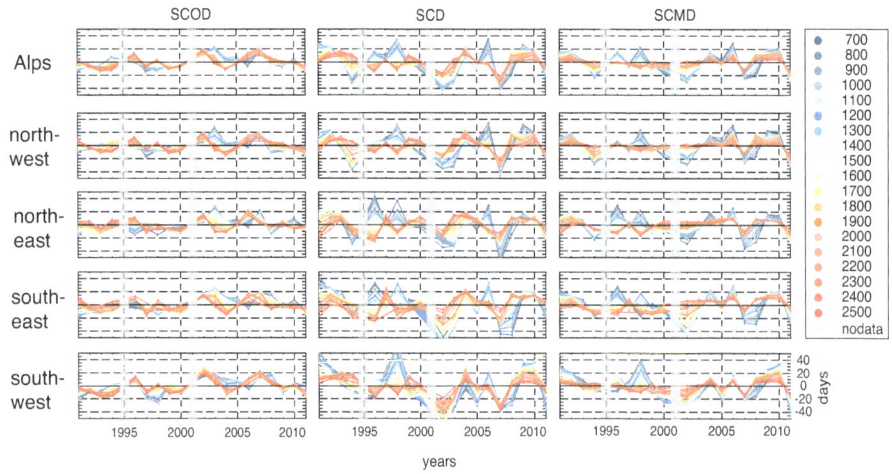

Fig. 9. Annual departures (1991–2011) of SCOD, SCD and SCMD from the long-term median for the Alpine region and all subregions given in days for different altitude levels. Years refer to 1 January. Connecting lines are displayed for easier readability of the figure only. The data coverage in 1995 and 2001 was insufficient for the application.

SCA data set derived from several generations of AVHRR sensors, monthly mean values from the satellite-derived SCA were compared to the same variable derived from the gridded station data set (available December–May, 1989–2009, for Switzerland). The results are displayed in Fig. 10. A high consistency is found, not only for the relative dynamic of the SCA, but also for the absolute numbers of monthly mean values, especially in the years after 2001. The satellite record even captures small features such as the two SCA peaks in the winters 1991, 2007 or 2008. A slight overestimation of satellite-derived SCA compared to ground SCA is found in 1992 and 1993, however, and in 1995 and 1999 the satellite SCA seems not to resolve the highest peaks in January and February. In 1990, the real snow cover onset is found to be somewhat shifted to earlier dates in the satellite data but the absolute numbers of the winter maximum, lying at 50 %, are well represented.

The irregular distribution of AVHRR data (cf. Fig. 2) needs special attention before time series analysis. Therefore, we tested whether the increasing availability of scenes influences the accuracy of the monthly values. This was done by artificially reducing the number of scenes in 2002–2011 to the level of 1990–2000 when only one satellite was in orbit at a time. The obtained monthly mean values are over-plotted in red in the upper graph and the absolute differences between the two data sets are shown in the lower graph in Fig. 10. Except for three outliers, 95 % of the values lie within ± 6 %. Overall, the monthly mean values do not significantly differ (t tested, $p = 0.8$). The mean difference lies at 1.1 % with a standard deviation of 3.6 %. Therefore, it is concluded that the data distribution does not have an essential influence on the derived snow climatology. In addition, we hypothesize that the slightly advanced sensor technology of AVHRR/3, operating on N16 to N19 from 2002 onwards, rather than the data disposability, is responsible for the high agreement with the ground data in Fig. 10 after 2002.

Having assessed the accuracy and consistency of the satellite record, a 27 yr time series of SCA was compiled and analyzed for the whole extent of the Alpine region. As missing values were inevitable due to satellite data unavailability, data gaps (namely three months in 1995 and four months in 2001) were filled by the long-term mean value of the respective months. The pronounced seasonal cycle of snow cover (see also Fig. 10) was removed by using the monthly standardized anomalies.

It has been shown that the precision of trend estimates in environmental data is critically influenced by the variability and autocorrelation of the underlying noise process (Tiao et al., 1990). In particular, autocorrelation tends to interact with the linear trend and therefore increases the length of the time series required to detect a certain trend (Weatherhead et al., 1998). The necessary length of record (monthly data) can therefore be assessed by investigating the standard deviation and the autocorrelation of the residuals after the seasonal component has been removed (i.e., the standardized anomalies).

As expected, the autocorrelation of the time series (residuals) is significant (> 0.5 lag-1) and the standard deviation of the residuals is $\sigma = 7.9$. Applying the Eq. (3) in Weatherhead et al. (1998), we calculated the number of years of data required to detect a real trend of a certain magnitude. Using the autocorrelation and variation of the time series residuals, this results in a minimum length of time period of 21 yr to detect a significant trend of the magnitude of 5 % per decade. A time period of approximately 61 yr would be required to detect a less strong but significant trend of 1 % per decade (always at a 90 % confidence level) for the SCA parameter. Therefore, we conclude, that (depending on the magnitude

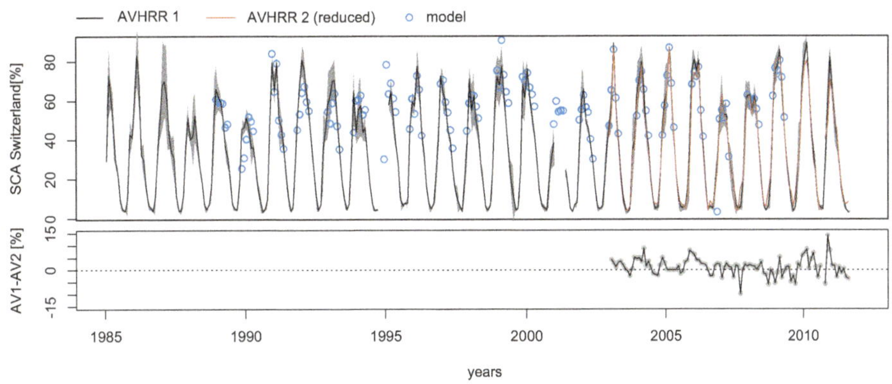

Fig. 10. Time series (1985–2011) of monthly values of SCA for Switzerland (AVHRR 1; black) including monthly standard deviation (gray shaded area) compared to the same measures derived from gridded station data (blue dots). AVHRR 2 (red) displays the snow cover data set with reduced number of scenes. Differences between the full data set (AVHRR 1) and the artificially reduced data set (AVHRR 2) are illustrated in the lower graph.

Table 3. Regression parameters trend ω per decade, p value, autocorrelation φ at lag-1 and standard deviation of the residuals σ for the SCA standardized anomaly time series illustrated in Figs. 11 and 12.

Region	ω/decade [SD]	p value	φ	σ
Alps	0.02	0.77	0.55	7.9
NE	0.08	0.21	0.36	11.13
NW	0.08	0.23	0.45	8.26
SE	0.03	0.62	0.49	11.17
SW	−0.04	0.55	0.50	7.77

of trend) the time series should be analyzed for trends with care.

Figure 11 illustrates the monthly SCA standardized anomalies derived from the satellite record over the European Alps (1985–2011). The respective regression parameters and the autocorrelation value are provided in Table 3. As has been observed in Fig. 8, the years 1988–1990, 2002 and 2007 are extreme with a relatively small SCA. On the contrary, the years 1991, 1992, 2005, 2009 and 2010 represent years with relatively high SCA values. Concerning the Alpine region as a whole, no statistically significant trend ($p = 0.7$) in SCA over the last 27 yr was found in annual SCA in the Alpine region.

To revisit the regional issue, the same analysis was carried out for all subregions (the absolute SCA time series for each region are plotted in Fig. A1). The standardized monthly anomalies are illustrated in Fig. 12. No significant trends were found here either, as can be seen from Table 3, which summarizes the regression parameters and the autocorrelation values for each region separately. Slight differences in the anomalies are still apparent between the subregions (e.g., in 1994 and 1996) and although not significant, the linear trend-lines tend towards increasing SCA in the northern parts and towards decreasing SCA in the southwestern region with a decadal trend of −0.04 standard deviations (cf. Table 3).

6 General discussion

The above findings generally indicate that alpine snow cover is highly variable in time and space. Due to its medium resolution of 1 km, this record is not able to investigate small scale features such as snow line dependence on aspect as a pixel in complex terrain typically includes various aspects. Rather, it is intended to capture the relative large scale snow dynamics that seem to be well reproduced. Four patterns relating to spatial anomaly occurrence were suggested: uniform, altitudinal, meridional and zonal. On the regional scale, these patterns generally agree with the findings by Laternser and Schneebeli (2003) who presented three priorities on which the degree of snowiness in Switzerland is based: first (time) with periods of abundant or poor snow conditions spanning large areas. Due to the relative brevity of the record, this pattern is not as obvious here but presumably resembles the first pattern. Second (space) with distinct regions showing more or less pronounced anomalies such as the north-south pattern found here. And third (altitude) with relative differences in anomalies along altitudinal steps. Analogous patterns were also found by Scherrer and Appenzeller (2006). In their investigations, they identified three major patterns of large-scale snow variability for Switzerland (for the parameters December-January-February period new snow sum, snow depth and snow days): uniform, north–south and low–high altitude. However, both studies were restricted to the area of Switzerland, that is, they were located (as defined here) mostly in the north-west and minor parts in the south-west. Upon extending the analysis to the entire Alpine region, an east–west gradient also becomes apparent in this study.

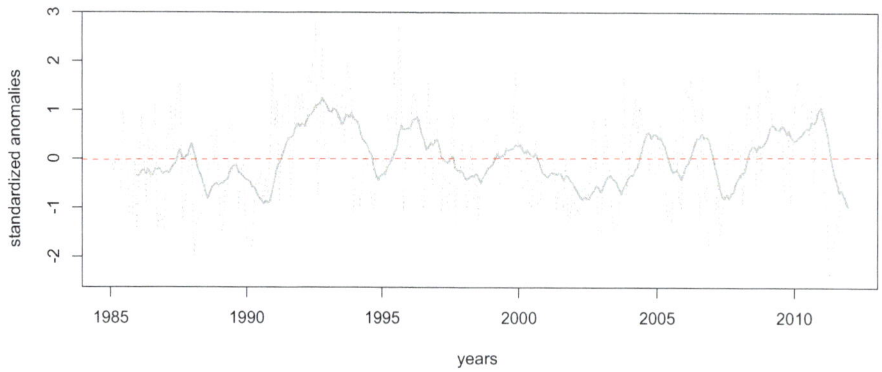

Fig. 11. Time series of Alpine region SCA monthly standardized anomalies derived from AVHRR data 1985–2011. The dotted line indicates the monthly values while the grey solid line depicts the 12-month standard moving average. The red line represents the respective linear trend.

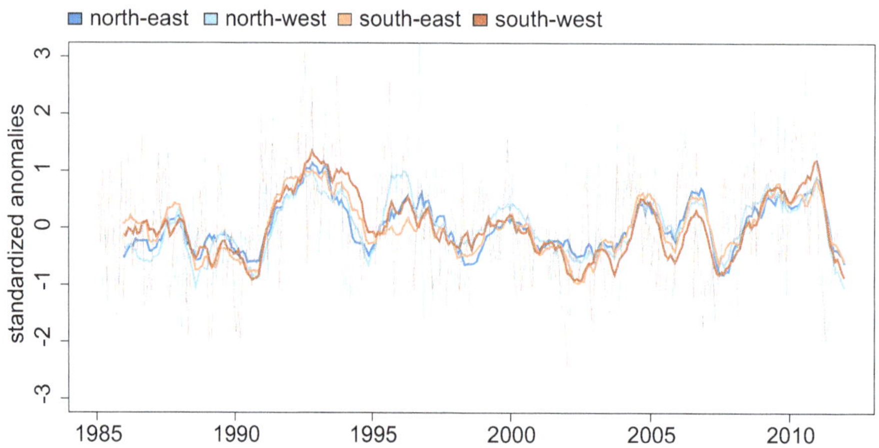

Fig. 12. Same as Fig. 11 but resolved into the four climatic regions.

The results obtained from the novel satellite-based SCA time series reveal interesting details on the temporal distribution of the snow cover extent in the Alpine region over the last 27 yr. Indeed, many long-term in situ snow climatologies in different regions have been analyzed investigating different snow parameters, generally showing a significant decrease of snow abundance (Marty, 2008; Beniston, 1997; EEA, 2009; Durand et al., 2009) particularly at lower altitudes. Most of them, however, analyze earlier periods or smaller areas than the record presented here. The 1990s were previously found to be the years with the lowest new snow sums and snow days on record (Laternser and Schneebeli, 2003), while the newest results show that the snow conditions seem to have partly recovered since 2000, especially at low elevations (Scherrer et al., 2010; Valt and Cianfarra, 2010). Recently, Scherrer et al. (2013) even showed that the strongly decreasing trends found in the late 1980s and 1990s are followed by a trend reversal in the last 10 yr (2000–2009), which was attributed to a recent "plateauing" or even slightly decreasing temperatures in Switzerland. Apparently, this behavior could be appropriately reproduced in the satellite record and seems to apply to the whole Alpine region. Furthermore, this may explain the absence of a significant trend in the standardized SCA anomalies. Therefore, our results do not conflict with previous findings but may also indicate – confirming the findings by Scherrer et al. (2013) – a deceleration of the decreasing snow trend in the Alpine region, even though the time period is comparatively short. Yet, indications towards a shorter SCD in lower regions in the southern parts were found.

Finally, on the hemispheric scale, the AVHRR snow time series is in good agreement with the satellite-based snow cover climatology of the Northern Hemisphere (NH) weekly updated at the Rutgers Snow Lab (http://climate.rutgers.edu/snowcover/; accessed March 2013). This climatology also suggests a weakening (even a reversal in the DJF) in the negative trend found for the period under investigation here. Unfortunately, the prominent abrupt shift in NH snow cover extent in the 1980s, described in Robinson and Frei (2000), lies beyond the length of this record. Nevertheless, the major advantage of this record, compared to similar products such as Moderate Resolution Imaging Spectroradiometer (MODIS) or (Advanced) Along-Track Scanning Radiometer ((A)ATSR), is its unique length. Likewise, the relatively high spatial resolution compared to existing long-term satellite records such as the Rutgers snow lab product (24 km) enables snow cover investigations even in complex terrain.

7 Conclusions

In this study, we presented a novel 1 km satellite-based snow cover data set for the Alpine region. The first objective was to generate a new set of cloud-free satellite snow products for use in climate research. This was done using a cloud gap-filling technique based on a combination of spatial and temporal compositing. Thereby, the amount of cloud could be substantially reduced and a comparison of the composite product with gridded station data suggests good accuracy. Furthermore, critical issues such as data record length or irregular data availability, potentially affecting the consistency and significance of the data set, were addressed. Based on the relation between the autocorrelation and the standard deviation of the residuals (see Weatherhead et al., 1998), the length of the recorded period is concluded to be sufficient for limited magnitudes of the trend for the SCA parameter. By artificially reducing the data availability to the level of one satellite in orbit at a time, it was shown that the values almost exactly coincide with the SCA derived from the full data set. Consequently, we assume data availability as having no essential influence on the SCA climatology. Yet, small data gaps, pixel misclassifications or cloud persistence longer than the filling period (10 days here) inevitably introduce minor errors in parameters such as SCOD, SCD or SCMD.

Based on this data set, the second objective was to investigate spatiotemporal distribution of snow cover in the European Alps over the last 27 yr (21 yr for spatial applications) from the satellite perspective. The long-term mean spatial representation of SCD, SCOD and SCMD showed a strong correlation of these parameters with altitude due to the vertical temperature gradient. However, regional differences became apparent, especially between the northern and the southern parts of the Alpine main crest, but also along the east-west direction. Similar patterns, namely uniform, altitudinal, meridional and zonal, were found in the annual departures of SCD from the long-term mean for the period 1991–2011. Regional analyses indicated significant trends towards shorter SCD at low elevations (700–900 m a.s.l.) in the south-west and south-east.

Overall, the analysis of the 27 yr SCA monthly mean time series (1985–2011) indicated no significant linear trend for the entire Alpine region. The same is true for regional analyses even though some tendencies toward increasing (decreasing) SCD in the northern (southern) parts were found. Although this result is somewhat surprising in the context of an anticipated climate change, it is in accordance with other research findings.

As this study presents a broad overview on the spatiotemporal snow cover variability, it also implies the need for more detailed investigations. For example, future work could be directed towards seasonal trend analysis. It is also recommended to focus on large-scale atmospheric modes of climate variability (North Atlantic Oscillation or Arctic Oscillation), that were found to have a certain influence on European snow cover (Bartolini et al., 2010; Kim et al., 2012). In addition, the use of a simple degree-day approach or the zero-degree isotherm might be included in the analysis to discern the influence of temperature and precipitation on snow cover variability. Furthermore, the SCMD data may feature an interesting bridge to the analysis of future alpine plant growth as this parameter was found to be highly correlated to the timing of snow melt out (Rammig et al., 2010).

Following the above considerations, these results establish the confidence in RS products for the purpose of long-term snow applications as we succeeded in (a) suppressing the influence of changing sensors and increasing data availability and (b) the gap filling methods seem to work well. Both, (a) and (b) raise confidence in the data set and demonstrate the valuable role of a long-term satellite-based snow climatology for complementing in situ time series. This study is therefore considered an important contribution to a regional climate observing system. Basically, the methodology used to compile this data set could be applied to other regions of the world with sparser ground data than the Alpine region. Nevertheless, this critically depends on the availability of long-term AVHRR archives at the sensor's full spatial resolution. An extension of the data set to cover whole Europe is planned within the scope of the Globsnow-2 project funded by the European Space Agency (ESA). Finally, this novel data record bears the potential to serve as a reference for climate models, in which snow cover is still not adequately represented, and to provide spatially and temporally comprehensive information for related research fields.

Appendix A

Fig. A1 Time series (1985–2011) of absolute monthly values of SCA for each region including monthly standard deviation (gray shaded area).

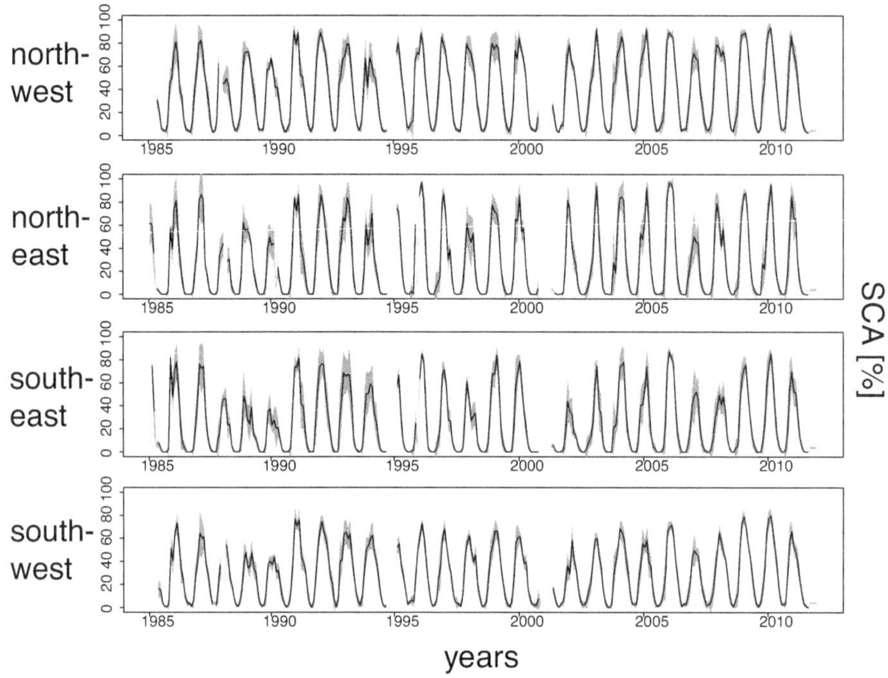

Acknowledgements. The AVHRR 1 km data were partially acquired from the NOAA Comprehensive Large Array-data Stewardship System (CLASS; http://www.nsof.class.noaa.gov), the German Aerospace Center (DLR) and the Freie Universität Berlin. Special thanks go to D. Koslowsky. We acknowledge the E-OBS data set from the EU-FP6 project ENSEMBLES and the data providers in the ECA & D project (http://www.ecad.eu). The authors wish to express their gratitude to R. Armstrong and N. Foppa for their valuable inputs on the manuscript. This work was funded by the Hyper-Swiss-Net and the ESA Globsnow-2 project.

Edited by: D. Hall

References

Agrawala, S.: Climate change in the European Alps: adapting winter tourism and natural hazards management, Organisation for Economic Co-operation and Development, 2007.

Armstrong, R. L. and Brodzik, M. J.: Recent Northern Hemisphere snow extent: A comparison of data derived from visible and microwave satellite sensors, Geophys. Res. Lett., 28, 3673–3676, 2001.

Auer, I., Böhm, R., Jurkovic, A., Lipa, W., Orlik, A., Potzmann, R., Schöner, W., Ungersböck, M., Matulla, C., Briffa, K., Jones, P., Efthymiadis, D., Brunetti, M., Nanni, T., Maugeri, M., Mercalli, L., Mestre, O., Moisselin, J.-M., Begert, M., Müller-Westermeier, G., Kveton, V., Bochnicek, O., Stastny, P., Lapin, M., Szalai, S., Szentimrey, T., Cegnar, T., Dolinar, M., Gajic-Capka, M., Zaninovic, K., Majstorovic, Z., and Nieplova, E.: HISTALP-historical instrumental climatological surface time series of the Greater Alpine Region, Int. J. Climatol., 27, 17–46, doi:10.1002/joc.1377, 2007.

Auer, M., Meister, R., Stoffel, A., and Weingartner, R.: Analyse und Darstellung der mittleren monatlichen Schneehöhen in der Schweiz, Wasser Energ. Luft, 96 (7/8), 173–178, 2004.

Bartolini, E., Claps, P., and D'Odorico, P.: Connecting European snow cover variability with large scale atmospheric patterns, Adv. Geosci., 26, 93–97, doi:10.5194/adgeo-26-93-2010, 2010.

Beniston, M.: Variations of snow depth and duration in the Swiss Alps over the last 50 years: links to changes in large-scale climatic forcings, Clim. Change, 36, 281–300, 1997.

Cihlar, J., Latifovic, R., Chena, J., Trishchenko, A., Duc, Y., Fedosejevsa, G., and Guindona, B.: Systematic corrections of AVHRR image composites for temporal studies, Remote Sens. Environ., 89, 217–233, doi:10.1016/j.rse.2002.06.007, 2004.

Dietz, A. J., Kuenzer, C., Gessner, U., and Dech, S.: Remote sensing of snow – a review of available methods, Int. J. Remote Sens., 33, 4094–4134, doi:10.1080/01431161.2011.640964, 2012.

Durand, Y., Giraud, G., Laternser, M., Etchevers, P., Mérindol, L., and Lesaffre, B.: Reanalysis of 47 Years of Climate in the French Alps (1958–2005): Climatology and Trends for Snow Cover, J. Appl. Meteorol. Clim., 48, 2487–2512, doi:10.1175/2009JAMC1810.1, 2009.

EEA: Regional climate change and adaptation, Tech. rep., European Environmental Agency, doi:10.2800/12552, Copenhagen, 2009.

Elsasser, H. and Bürki, R.: Climate change as a threat to tourism in the Alps, Clim. Res., 20, 253–257, doi:10.3354/cr020253, 2002.

Fontana, F. M. A., Trishchenko, A. P., Luo, Y., Khlopenkov, K. V., Nussbaumer, S. U., and Wunderle, S.: Perennial snow and ice variations (2000–2008) in the Arctic circumpolar land area from satellite observations, J. Geophys. Res., 115, F04020, doi:10.1029/2010JF001664, 2010.

Foppa, N. and Seiz, G.: Inter-annual variations of snow days over Switzerland from 2000–2010 derived from MODIS satellite data, The Cryosphere, 6, 331–342, doi:10.5194/tc-6-331-2012, 2012.

Fuchs, S. and Brandl, M.: Damage Potential and Losses Resulting from Snow Avalanches in Settlements of the Canton of Grisons, Switzerland, Nat. Hazards, 34, 53–69, doi:10.1007/s11069-004-0784-y, 2005.

Gafurov, A. and Bárdossy, A.: Cloud removal methodology from MODIS snow cover product, Hydrol. Earth Syst. Sci., 13, 1361–1373, doi:10.5194/hess-13-1361-2009, 2009.

Gao, J., Williams, M. W., Fu, X., and Gong, T.: Spatiotemporal distribution of snow in eastern Tibet and the response to climate change, Remote Sens. Environ., 121, 1–9, doi:10.1016/j.rse.2012.01.006, 2012.

Gutman, G. and Masek, J. G.: Long-term time series of the Earth's land-surface observations from space, Int. J. Remote Sens., 33, 4700–4719, doi:10.1080/01431161.2011.638341, 2012.

Hall, D. K., Riggs, G. A., Foster, J. L., and Kumar, S. V.: Development and evaluation of a cloud-gap-filled MODIS daily snow-cover product, Remote Sens. Environ., 114, 496–503, doi:10.1016/j.rse.2009.10.007, 2010.

Hänggi, P. and Weingartner, R.: Variations in Discharge Volumes for Hydropower Generation in Switzerland, Water Resour. Manag., 26, 1231–1252, doi:10.1007/s11269-011-9956-1, 2012.

Hantel, M. and Hirtl-Wielke, L. M.: Sensitivity of Alpine snow cover to European temperature, Int. J. Climatol., 27, 1265–1275, doi:10.1002/joc.1472, 2007.

Haylock, M. R., Hofstra, N., Klein Tank, A. M. G., Klok, E. J., Jones, P. D., and New, M.: A European daily high-resolution gridded data set of surface temperature and precipitation for 1950-2006, J. Geophys. Res., 113, D20119, doi:10.1029/2008JD010201, 2008.

Heidinger, A. K., Straka, W. C., Molling, C. C., Sullivan, J. T., and Wu, X. Q.: Deriving an inter-sensor consistent calibration for the AVHRR solar reflectance data record, Int. J. Remote Sens., 31, 6493–6517, doi:10.1080/01431161.2010.496472, 2010.

Hosmer, D. W. and Lemeshow, S.: Applied logistic regression, Wiley series in probability and statistics: Texts and references section, Wiley, 2000.

Hüsler, F., Fontana, F., Riffler, M., Neuhaus, C., Musial, J., and Wunderle, S.: AVHRR Archive and Processing Facility at the University of Bern: A Comprehensive 1-km Satellite Data Set for Climate Change Studies, EARSeL eProc, 10, 83–101, 2011.

Hüsler, F., Jonas, T., Wunderle, S., and Albrecht, S.: Validation of a modified snow cover retrieval algorithm from historical 1-km AVHRR data over the European Alps, Remote Sens. Environ., 121, 497–515, doi:10.1016/j.rse.2012.02.018, 2012.

IPCC: Climate Change 2007: The Physical Science Basis, Tech. rep., Intergovernmental Panel on Climate Change (IPCC), Geneva, Switzerland, 2007.

James, M. E. and Kalluri, S. N.: The Pathfinder AVHRR land data set: An improved coarse resolution data set for terrestrial monitoring, Int. J. Remote Sens., 15, 3347–3363, 1994.

Jasper, K., Gurtz, J., and Lang, H.: Advanced flood forecasting in Alpine watersheds by coupling meteorological observations and forecasts with a distributed hydrological model, J. Hydrol., 267, 40–52, doi:10.1016/S0022-1694(02)00138-5, 2002.

Jonas, T., Rixen, C., Sturm, M., and Stöckli, V.: How alpine plant growth is linked to snow cover and climate variability, J. Geophys. Res.-Biogeo., 113, G03013, doi:10.1029/2007JG000680, 2008.

Jonas, T., Marty, C., and Magnusson, J.: Estimating the snow water equivalent from snow depth measurements in the Swiss Alps, J. Hydrol., 378, 161–167, doi:10.1016/j.jhydrol.2009.09.021, 2009.

Jones, J.: Habitat selection studies in avian ecology: a critical review, The Auk, 118, 557–562, 2001.

Khlopenkov, K. V. and Trishchenko, A. P.: SPARC: New cloud, snow, and cloud shadow detection scheme for historical 1-km AVHRR data over Canada, J. Atmos. Ocean. Tech., 24, 322–343, 2007.

Kim, Y., Kim, K.-Y., and Kim, B.-M.: Physical mechanisms of European winter snow cover variability and its relationship to the NAO, Clim. Dynam., 40, 1657–1669, doi:10.1007/s00382-012-1365-5, 2012.

König, M., Winther, J.-G., and Isaksson, E.: Measuring snow and glacier ice properties from satellite, Rev. Geophys., 39, 1–27, doi:10.1029/1999RG000076, 2001.

Laternser, M. and Schneebeli, M.: Long-term snow climate trends of the Swiss Alps (1931-99), Int. J. Climatol., 23, 733–750, 2003.

Lehning, M., Bartelt, P., Brown, B., Russi, T., Stockli, U., and Zimmerli, M.: SNOWPACK model calculations for avalanche warning based upon a new network of weather and snow stations, Cold Reg. Sci. Technol., 30, 145–157, 1999.

Lehning, M., Völksch, I., Gustafsson, D., Nguyen, T. A., Stähli, M., and Zappa, M.: ALPINE3D : a detailed model of mountain surface processes and its application to snow hydrology, Hydrol. Process., 20, 2111–2128, doi:10.1002/hyp.6204, 2006.

Magnusson, J., Farinotti, D., Jonas, T., and Bavay, M.: Quantitative evaluation of different hydrological modelling approaches in a partly glacierized Swiss watershed, Hydrol. Process., 25, 2071–2084, doi:10.1002/hyp.7958, 2011.

Marks, D., Domingo, J., Susong, D., Link, T., and Garen, D.: A spatially distributed energy balance snowmelt model for application in mountain basins, Hydrol. Process., 13, 1935–1959, doi:10.1002/(SICI)1099-1085(199909)13:12/13<1935::AID-HYP868>3.0.CO;2-C, 1999.

Marty, C.: Regime shift of snow days in Switzerland, Geophys. Res. Lett., 35, L12501, doi:10.1029/2008GL033998, 2008.

Marty, C. and Meister, R.: Long-term snow and weather observations at Weissfluhjoch and its relation to other high-altitude observatories in the Alps, Theor. Appl. Climatol., 110, 573–583, doi:10.1007/s00704-012-0584-3, 2012.

Metsämäki, S., Mattila, O.-P., Pulliainen, J., Niemi, K., Luojus, K., and Böttcher, K.: An optical reflectance model-based method for fractional snow cover mapping applicable to continental scale, Remote Sens. Environ., 123, 508–521, doi:10.1016/j.rse.2012.04.010, 2012.

Nolin, A. W.: Recent advances in remote sensing of seasonal snow, J. Glaciol., 56, 1141-1150, doi:10.3189/002214311796406077, 2011.

Parajka, J. and Blöschl, G.: Validation of MODIS snow cover images over Austria, Hydrol. Earth Syst. Sci., 10, 679–689, doi:10.5194/hess-10-679-2006, 2006.

Parajka, J. and Blöschl, G.: Spatio-temporal combination of MODIS images-potential for snow cover mapping, Water Resour. Res., 44, W03406, doi:10.1029/2007WR006204, 2008.

Parajka, J., Pepe, M., Rampini, A., Rossi, S., and Blöschl, G.: A regional snow-line method for estimating snow cover from MODIS during cloud cover, J. Hydrol., 381, 203–212, doi:10.1016/j.jhydrol.2009.11.042, 2010.

Rammig, A., Jonas, T., Zimmermann, N. E., and Rixen, C.: Changes in alpine plant growth under future climate conditions, Biogeosciences, 7, 2013–2024, doi:10.5194/bg-7-2013-2010, 2010.

Rango, A. and Martinec, J.: Revisiting the degree-day method for snowmelt computations, Water Resour. Bull., 31, 657–669, 1995.

Rees, W. G.: Remote Sensing of Snow and Ice, CRC Press, Boca Raton (FL), 2005.

Rhyner, J., Bründl, M., Etter, H. J., Steiniger, M., Stöckli, U., Stucki, T., Zimmerli, M., and Amman, W.: Avalanche warning Switzerland – consequences of the avalanche winter 1999, in: Proceedings of the 13th Int. Snow Science Workshop, 30 September–3 October, Penticton, B.C., Canada, 2002.

Robinson, D. A. and Frei, A.: Seasonal Variability of Northern Hemisphere Snow Extent Using Visible Satellite Data, The Professional Geographer, 582, 307–314, 2000.

Romanov, P., Gutman, G., and Csiszar, I.: Satellite-derived snow cover maps for north America: Accuracy Assesement, Adv. Space Res., 30, 2455–2460, 2002.

Scherrer, S. C. and Appenzeller, C.: Swiss Alpine snow pack variability: major patterns and links to local climate and large-scale flow, Clim. Res., 32, 187–199, doi:10.3354/cr032187, 2006.

Scherrer, S. C., Appenzeller, C., and Laternser, M.: Trends in Swiss Alpine snow days: The role of local- and large-scale climate variability, Geophys. Res. Lett., 31, L13215, doi:10.1029/2004GL020255, 2004.

Scherrer, S. C., Wüthrich, C., Croci-Maspoli, M., Weingartner, R., and Appenzeller, C.: Analyses of newly digitized and reconstructed snow series over the last 100+ years in Switzerland, EMS Annual Meeting Abstracts Vol. 7, EMS2010-514, 13–17 September , Zürich, 2010.

Scherrer, S. C., Wüthrich, C., Croci-Maspoli, M., Weingartner, R., and Appenzeller, C.: Snow variability in the Swiss Alps 1864-2009, Int. J. Climatol., 33, 3162–3173, doi:10.1002/joc.3653, 2013.

Stucki, T., Duerr, L., and Etter, H.: Schnee und Lawinen in den Schweizer Alpen. Winter 2005/06. Wetter, Schneedecke und Lawinengefahr. Winterbericht SLF, Tech. rep., Davos, WSL-Institut für Schnee- und Lawinenforschung SLF, 2011.

Tiao, G. C., Reinsel, G. C., Xu, D., Pedrick, J. H., Zhu, X., Miller, A. J., DeLuisi, J. J., Mateer, C. L., and Wuebbles, D. J.: Effects of autocorrelation and temporal sampling schemes on estimates of trend and spatial correlation, J. Geophys. Res., 95, 20507–20517, 1990.

Valt, M. and Cianfarra, P.: Recent snow cover variability in the Italian Alps, Cold Reg. Sci. Technol., 64, 146–157, doi:10.1016/j.coldregions.2010.08.008, 2010.

Voigt, S.: Advanced methods for operational mapping of Alpine snow cover using medium resolution optical satellite data, Inauguraldissertation, University of Bern, Switzerland, 2000.

Wang, X. and Xie, H.: New methods for studying the spatiotemporal variation of snow cover based on combination products of MODIS Terra and Aqua, J. Hydrol., 371, 192–200, doi:10.1016/j.jhydrol.2009.03.028, 2009.

Wang, X., Xie, H., and Liang, T.: Evaluation of MODIS snow cover and cloud mask and its application in Northern Xinjiang, China, Remote Sens. Environ., 112, 1497–1513, 2008.

Wang, X., Xie, H., Liang, T., and Huang, X.: Comparison and validation of MODIS standard and new combination of Terra and Aqua snow cover products in northern Xinjiang , China, Hydrol. Process., 429, 419–429, doi:10.1002/hyp.7151, 2009.

Weatherhead, E. C., Reinsel, G. C., Tiao, G. C., Meng, X.-L., Choi, D., Cheang, W.-K., Keller, T., DeLuisi, J., Wuebbles, D. J., Kerr, J. B., Miller, A. J., Oltmans, S. J., and Frederick, J. E.: Factors affecting the detection of trends: Statistical considerations and applications to environmental data, J. Geophys. Res., 103, 17149–17161, doi:10.1029/98JD00995, 1998.

Wiscombe, W. J. and Warren, S. G.: Model for the Spectral Albedo of Snow. I : Pure Snow, Engineering, 37, 2712–2733, 1980.

WMO: Implementation Plan for the Global Observing System for Climate in Support of the UNFCCC (2010 Update), Tech. Rep. 1523, Global Climate Observing System, http://eprints.soton.ac.uk/162953/ (last access: 18 June 2012), 2010.

WMO: Systematic Observation Requirement for Satellite-based Data Products for Climate (Update 2011), GCOS-154, Tech. rep., World Meteorological Organization, Geneva, 2011.

Zhao, H. and Fernandes, R.: Daily snow cover estimation from Advanced Very High Resolution Radiometer Polar Pathfinder data over Northern Hemisphere land surfaces during 1982–2004, J. Geophys. Res., 114, D05113, doi:10.1029/2008JD011272, 2009.

Stable climate and surface mass balance in Svalbard over 1979–2013 despite the Arctic warming

C. Lang, X. Fettweis, and M. Erpicum

Département de Géographie, Université de Liège, Liège, Belgium

Correspondence to: C. Lang (charlotte.lang@ulg.ac.be)

Abstract. With the help of the regional climate model MAR (Modèle Atmosphérique Régional) forced by the ERA-Interim reanalysis (MAR_{ERA}) and the MIROC5 (Model for Interdisciplinary Research on Climate) global model (MAR_{MIROC5}) from the CMIP5 (Coupled Model Intercomparison Project) database, we have modelled the climate and surface mass balance of Svalbard at a 10 km resolution over 1979–2013. The integrated total surface mass balance (SMB) over Svalbard modelled by MAR_{ERA} is negative ($-1.6\,Gt\,yr^{-1}$) with a large interannual variability (7.1 Gt) but, unlike over Greenland, there has been no acceleration of the surface melt over the past 35 years because of the recent change in atmospheric circulation bringing northwesterly flows in summer over Svalbard, contrasting the recent observed Arctic warming. However, in 2013, the atmospheric circulation changed to a south–southwesterly flow over Svalbard causing record melt, SMB ($-20.4\,Gt\,yr^{-1}$) and summer temperature. MIROC5 is significantly colder than ERA-Interim over 1980–2005 but MAR_{MIROC5} is able to improve the near-surface MIROC5 results by simulating not significant SMB differences with MAR_{ERA} over 1980–2005. On the other hand, MIROC5 does not represent the recent atmospheric circulation shift in summer and induces in MAR_{MIROC5} a significant trend of decreasing SMB ($-0.6\,Gt\,yr^{-2}$) over 1980–2005.

1 Introduction

In the context of global warming, it is important to evaluate the impact of climate change on high latitude zones, that are known to be very sensitive to a rise in temperature (IPCC AR5, 2013). Over 1961–2004, the Arctic has been the second largest contribution (excluding the Antarctic and Greenland ice sheets) to sea level rise (Kaser et al., 2006). According to Gardner et al. (2013) and Shepherd et al. (2012), between 2003 and 2009, glaciers and ice caps (including peripheral glaciers in Greenland and Antarctica) have contributed to sea level rise as much as the Greenland and Antarctic ice sheets, contributing together to 61 % of the total sea level rise. Arctic ice loss represents almost 50 % of the total glacier and ice caps loss but Svalbard contributed to only 4 % of the total Arctic contribution (Gardner et al., 2013).

Svalbard seems to be the ice cap least sensitive to the recent Arctic warming in summer (Serreze et al., 2009) and while melt records have been broken several times in Greenland in the second half of the 2000s (Fettweis et al., 2013a), the surface mass balance (SMB) of Svalbard has been closer to balance after 2004 (Moholdt et al., 2010). Fettweis et al. (2013a) attributed it to atmospheric circulation changes in summer damping the warming over Svalbard, as we will discuss in Section 4. However, this recent stabilisation of the Svalbard SMB needs to be seen from a larger perspective, which has been missed until now.

The Svalbard (surface) mass balance has already been intensively studied but previously published studies involved either long time series but on only a few glaciers or over extended areas but on a shorter timescale and mostly used statistically or empirically based mass balance models. On a local scale (i.e. on one or a few glaciers), various types of mass balance reconstructions using different tools have been performed over Svalbard: Lefauconnier and Hagen (1990) used correlations between the measured mass balance and climatic parameters measured at Ny-Ålesund weather station (northwestern coast of Spitsbergen), Rasmussen and Kohler (2007) used a model running with meteorological data from

NCEP/NCAR reanalysis, de Woul and Hock (2005) used a Positive Degree-Day (PDD) model and temperature and precipitation data from Ny-Ålesund. Energy balance models have also been used: van Pelt et al. (2012) used an energy balance model coupled to a snow model forced by the regional climate model (RCM) RACMO on Nordenskiöldbreen and Rye et al. (2010, 2012) used a coupled surface-subsurface energy balance model forced by the ERA-40 reanalysis to reconstruct the SMB of Midre Lovénbreen. On a larger scale, Schuler et al. (2007) have modelled the surface mass balance of the Austfonna ice cap with a model based on weather data and SMB measurements. Førland et al. (2011) used the 25 km outputs of the NorACIA-RCM (Førland et al., 2009) to statistically downscale temperature and precipitation at the location of several weather stations and Benestad et al. (2002) empirically downscaled temperature using principal component analysis. Day et al. (2012) compared precipitation from the HadRM3 RCM (25 km) to the SMB measurements from Pinglot et al. (1999) and performed future projections of the Svalbard climate. Finally, Bamber et al. (2004, 2005) estimated elevation changes of glaciers and ice caps between 1996 and 2002 using airborne lasers and, more recently, Moholdt et al. (2010) computed elevation changes from 2003 to 2008 based on measurements made by the ICESat satellite.

Knowing that the SMB (precipitation contributing to the accumulation and melt caused by positive temperature) is the component of mass balance most sensitive to climate change, this necessitates the study of the SMB over the entirety of Svalbard on a longer time period, in view of the current apparent stabilisation of the Svalbard SMB in full opposition to the other Arctic ice caps. To do so, regional climate models (RCMs), fully coupled with energy balance models are ideal tools. Indeed, they allow high spatial resolution simulations with models that are calibrated for specific regions. In addition, the RCMs also give us the opportunity to perform more reliable future projections compared to forced energy balance models, which do not simulate the surface albedo positive feedback for example.

Among the available RCMs, the regional climate model MAR (Modèle Atmosphérique Régional) has been extensively evaluated over Greenland and is able to accurately simulate the Greenland ice sheet SMB (e.g. Fettweis et al. (2013b) and Franco et al. (2013) and references therein). That is why we propose here to study the current (1979–2013) SMB evolution over the entirety of Svalbard with MAR at a resolution of 10 km. Both the ERA-Interim reanalysis (Dee et al., 2011) and the global model MIROC5 (Model for Interdisciplinary Research on Climate) (Watanabe et al., 2010; Sakamoto et al., 2012) from the CMIP5 (Coupled Model Intercomparison Project) database were used to force MAR over the current climate. The MIROC5 model has been chosen as it is one of the best CMIP5 models for simulating the current climate and atmospheric circulation over Greenland and surrounding areas with respect to ERA-Interim (Fettweis et al., 2013b).

Section 2 of this article gives a description of MAR and the forcings of our simulations. Section 3 evaluates MAR over Svalbard by comparing its results to near-surface measurements from weather stations and SMB measurements. We discuss its results over the current climate in Sect. 4. In Sect. 5, we compare ERA-Interim with MIROC5 over Svalbard as well as MAR forced by both of those in Sect. 6, with the aim of performing future projections in a companion paper (Lang et al., 2014).

2 Model and forcings

2.1 The MAR model

MAR (Modèle Atmosphérique Régional) is a regional atmospheric climate model specifically developed for the study of polar regions (Gallée and Schayes, 1994) and consists of an atmospheric model coupled with the surface model SISVAT (Soil Ice Snow Vegetation Atmosphere Transfer) (de Ridder and Gallée, 1998; Gallée et al., 2001) through the exchange of energy fluxes, momentum, precipitation, etc.

The SISVAT model is a vertical 1-D multi-layered model consisting of a soil/vegetation module and a snow/ice energy balance module resolving most of the processes occurring at the surface of the snow/ice pack and is based on the CROCUS model (Brun, 1989) from the CEN (Centre d'Études de la Neige).

The soil/vegetation module simulates the exchanges of heat and moisture with the atmosphere above the land without snow or ice while the snow/ice module deals with the exchanges between the atmosphere and the sea ice, the glaciers and ice sheets and the snow covering the land. Snow layers are characterised by their temperature, density, height, age, liquid water content, dendricity and sphericity of the crystals and grain size. These snow properties evolve with time in the snow metamorphism module according to Brun (1989). The energy balance between the soil and the snow is computed in the thermodynamic module through the absorbed shortwave flux, the longwave fluxes (upward and downward), the sensible and latent heat fluxes at the surface, the melting and sublimation of snow/ice and evaporation and refreezing of the meltwater heat fluxes, the heat fluxes due to precipitation (solid and liquid), condensation and deposition and the heat flux from the ground.

The model is not coupled with a 3-D ice sheet model, which prevents us from modelling dynamical processes and constrains us to use a fixed topography and ice extent throughout the entire simulation. That is why we focus only on the surface mass balance. The MAR version used here is 3.3 and the configuration is the one used in Fettweis et al. (2013b).

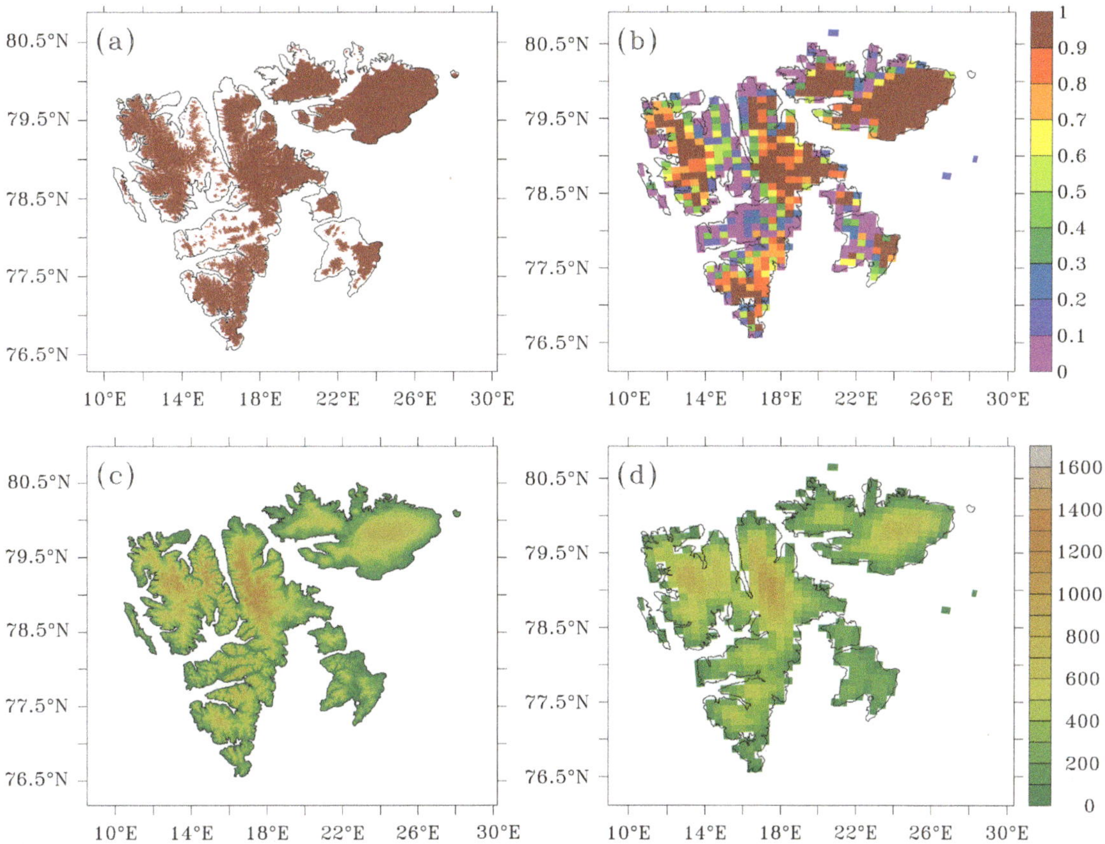

Figure 1. (a) Permanent ice mask from Nuth et al. (2013) interpolated on a 250 m grid. **(b)** Permanent 10 km fractional ice mask as used in MAR. **(c)** Svalbard topography (m a.s.l.) from the Norsk Polarinstitutt interpolated on a 250 m grid. **(d)** 10 km topography as used in MAR (m a.s.l.).

Table 1. Stations used for validation. Period over which data are available, coordinates and elevation of the stations (m), elevation of the corresponding model pixel used in the validation (m) and distance between the station and the pixel (km).

Station	Period observation		Coordinates		Elev	Pixel elev (m)			Dist station-pixel (km)		
	Temperature	Precipitation			(m)	MAR	ERA	ASR	MAR	ERA	ASR
Hornsund	2005–2013	1996–2013	77.00° N	15.50° E	10	178	22	227	6.0	57.0	36.8
Kapp Heuglin	2006–2013	–	78.25° N	22.82° E	14	54	87	166	6.7	28.7	21.9
Ny-Ålesund	1979–2013	1979–2013	78.92° N	11.93° E	8	137	215	361	3.3	64.5	52.3
Svalbard Lufthavn	1979–2013	1979–2013	78.25° N	14.47° E	28	188	214	242	4.2	30.3	13.4
Sveagruva	1979–2013	1979–2002	77.88° N	16.72° E	9	284	234	247	1.1	54.1	57.9

2.2 Model forcings

We have run MAR over the period 1979–2013 at a spatial resolution of 10 km. The lateral and oceanic boundaries were forced every 6 h (temperature, wind and humidity at each vertical level as well as sea surface temperature and sea-ice cover over the ocean) by the ERA-Interim (MAR_{ERA}) reanalysis (0.75 ° resolution) and the MIROC5 (MAR_{MIROC5}) general circulation model (1.4 ° resolution). The reanalysis and the GCM fields were also used to initialise the simulations at the beginning (1 September 1974). Five years of spin-up are required to reduce the impact of the snowpack initialisation (in particular the snow density) on our results.

2.3 Errors in the 10 km interpolated topography

The fractional permanent ice mask and topography used in MAR over Svalbard (Fig. 1b and d) have been interpolated at 10 km from the glacier inventory of Nuth et al. (2013) (Fig. 1a) and from the topography from the Norsk Polarinstitutt (NPI) respectively (Fig. 1c). Both shapefile data sets had previously been interpolated on a 250 m grid to produce

Table 2. 1979–2013 mean annual and summer correlation (R^2), RMSE and bias (°C) between MAR_{ERA} and the observed daily temperature and percentage of missing observations (% MO).

Station	Annual				Summer			
	R^2	RMSE (°C)	Bias (°C)	% MO	R^2	RMSE (°C)	Bias (°C)	% MO
Hornsund	0.94	3.79	−3.25	21	0.48	2.81	−2.45	21
Kapp Heuglin	0.93	3.66	−2.48	32	0.78	1.35	−0.68	38
Ny-Ålesund	0.94	2.49	−1.31	3	0.74	2.60	−2.27	1
Svalbard Lufthavn	0.93	3.66	−2.77	1	0.72	3.30	−2.91	0
Sveagruva	0.92	4.92	−4.00	3	0.65	4.39	−4.08	5

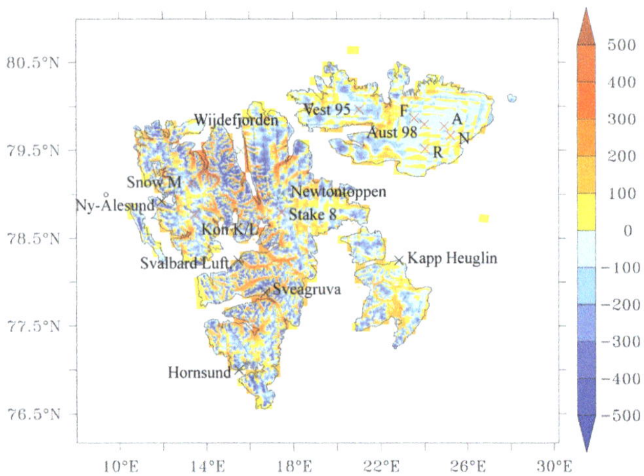

Figure 2. Elevation difference (m) between the 10 km MAR topography (based on the topography of the Norsk Polarinstitutt – NPI) and the topography of the NPI interpolated on a 250 m grid. The black X show the location of the weather stations used in the validation and the red ones show the location of the stakes from Pinglot et al. (1999, 2001).

files that can be read by MAR. The total area (Kvitøya excluded) of the ice mask is 33 264 km^2, which is about 55 % of the total land surface. With the 10 km interpolated ice mask, the total permanent ice area is 30 042 km^2 and corresponds to 49 % of the 10 km interpolated land surface.

In version 3 of MAR, the ice mask is fractional, i.e. each pixel is associated with a proportion of its area that is covered with permanent ice (Fig. 1b). In our analysis, the ice sheet area corresponds to the pixels covered with at least 50 % of permanent ice. If not, we consider these pixels to be the tundra zone. In all calculations showing integrated values over the entire ice sheet, we have computed weighted averages according to the percentage of ice covering each ice pixel (i.e. having a permanent ice area higher than 50 %).

Figure 2 shows that using a resolution of 10 km underestimates the elevation, especially on both sides of Wijdefjorden (northern Spitsbergen) where the difference is greater than 500 m due to a very steep topography.

We have divided the elevation range into 16 classes (Supplement Table S1) and computed the mean elevation error between the interpolated 10 km elevation (as used in MAR) and the 250 m topography, as well as the mean absolute error. The mean error gives an indication of whether a certain elevation range is underestimated (negative value) or overestimated (positive value) in the MAR topography, and the absolute error gives the mean value of the elevation bias, regardless of its sign. Due to the smoothing of the topography at a resolution of 10 km, the topography used in our MAR simulations underestimates all the elevations above 1000 m (classes 11 to 16) whereas for classes 3 to 10 (200–1000 m), the elevation is mostly underestimated but there are also areas where it is overestimated in the MAR topography. Finally, close to the coastline, where the elevation is lower than 200 m, the MAR topography slightly overestimates the elevation for most of the pixels.

As a consequence of the elevation underestimation at a resolution of 10 km, a lot of glaciers are too low in altitude in MAR than in reality, which could impact their simulated surface mass balance as shown by Lenaerts et al. (2013), who found that the bias increases when the SMB is increasingly negative. In extreme cases, some glaciers at a resolution of 10 km could be at such low elevations that they should not even exist under the present climate in the 10 km grid. As a result, the accumulation zone is missing and the melt is overestimated and could introduce a bias when considering the integrated surface mass balance of Svalbard. However, the interannual variability of the surface mass balance should not significantly be affected by the smoothing of the topography at a resolution of 10 km. Based on the histogram (Supplement Fig. S1), corrections of the SMB integrated over the entirety of Svalbard will be given in Section 4. Finally, as we consider only the pixels covered with more than 50 % of ice to be ice pixels, a lot of small glaciers (corresponding to 10 % of the permanent ice area) are left out of our analysis.

3 Evaluation of MAR forced by ERA-Interim

In order to evaluate our model over the present climate, we have compared the MAR results (called MAR_{ERA} hereafter)

Table 3. 1979–2013 mean annual and summer correlation (R^2), RMSE and bias (°C) between the ERA-Interim reanalysis and the observed temperature.

Station	Annual			Summer		
	R^2	RMSE (°C)	Bias (°C)	R^2	RMSE (°C)	Bias (°C)
Hornsund	0.94	3.01	2.24	0.57	1.31	0.63
Kapp Heuglin	0.95	2.18	1.14	0.66	1.54	0.55
Ny-Ålesund	0.93	3.03	−1.95	0.78	2.33	−2.00
Svalbard Lufthavn	0.96	2.31	−1.40	0.79	2.12	−1.66
Sveagruva	0.93	2.75	−1.07	0.72	2.40	−1.98

Table 4. 1979–2013 mean annual measured precipitation ($mm\,yr^{-1}$), proportion of that precipitation that is simulated by MAR_{ERA} and percentage of missing observations (% MO).

Station	Pobs ($mm\,yr^{-1}$)	Pmod/Pobs	% MO
Hornsund	378	1.16	50
Kapp Heuglin	–	–	100
Ny-Ålesund	409	0.77	47
Svalbard Lufthavn	187	1.52	39
Sveagruva	252	1.52	42

forced by ERA-Interim from ECMWF (Dee et al., 2011) over 1979–2013 to near-surface measurements of temperature and precipitation from weather stations as well as surface mass balance measurements. For comparison, these weather measurements have also been compared to the ASR (Wilson et al., 2011) and the ERA-Interim reanalysis, knowing that no observation is assimilated into the MAR model.

3.1 Comparison of MAR with weather stations

First, we have compared the daily near-surface temperature (TAS) of the stations listed in Table 1 and shown in Fig. 2 to the daily MAR_{ERA} TAS (corresponding to the 2–3 m level) of the pixel that is the closest to each station. As 1979 is the first year of ERA-Interim reanalysis, the comparison has been made over 1979–2013 when data were available.

While MAR is too cold compared to the observations, the daily variability of the temperature is very well simulated by MAR (Table 2). Part of these biases are however caused by the overestimation of the stations' elevation in MAR induced by the used 10 km resolution and the very steep topography near the coast. However, given the values of both temperature and elevation biases, MAR is anyway too cold even if no elevation bias was present as the temperature vertical gradient is 1 °C $100\,m^{-1}$ at its maximum. As summer temperature has more impact on the surface mass balance (through the melt) than the annual mean temperature, we have separately evaluated the JJA (June, July and August) temperatures. In summer, the MAR cold bias is smaller than the annual bias at Hornsund and Kapp Heuglin and comparable at Svalbard Lufthavn and Sveagruva and the RMSE is smaller at every station except Ny-Ålesund. The daily observed variability is however less well reproduced by MAR in summer.

The main effect of this MAR cold bias on the modelled SMB is a likely underestimation of the amount of melt in summer. Colder air can also contain less moisture and therefore a cold bias should imply an underestimation of snowfall. However, it is the temperature in altitude that influences the moisture content rather than TAS, and a negative TAS bias does not necessarily mean that the free atmosphere temperature bias will also be cold. It is therefore difficult to interpret the TAS cold bias in terms of snowfall underestimation. Moreover, the weather stations used in the validation are all located at the coast and most of them in fjords. As the resolution of ERA-Interim is too low to represent those fjords, we interpolated the ERA-Interim sea-ice cover (SIC) and sea surface temperature (SST) on the 10 km grid, then extrapolated SIC and SST in the fjords using the values of the nearest pixels. But this extrapolated SIC/SST may not be representative of the SIC/SST of the fjords and therefore could cause a temperature bias that would not be present further inland, as the ocean conditions influences the coastal regions a lot. However, we have no observations far inland and in the free atmosphere to confirm this.

It is well known that the outputs of regional climate models are strongly dependent on the reanalysis or global model used to force their boundaries. However, as the MAR and ERA-Interim reanalysis biases (Table 3) are different and of opposite signs for some of the stations (MAR is too cold while ERA-Interim is too warm), this shows well that MAR is totally free in the boundary layer and that the MAR cold bias does not come from the lateral boundaries. Moreover, in Ny-Ålesund, the MAR bias averaged over 2000–2010 (Supplement Table S2) is smaller than the ASR bias (Supplement Table S3), both on the annual timescale and during summer. This suggests that MAR compares well with other model outputs using data assimilation. On the annual timescale, MAR is better than ASR at reproducing the daily variability of the temperature for every station and comparable to ERA whereas in summer, the three products are comparable.

MAR underestimates the annual mean amount of precipitation at Ny-Ålesund and overestimates it at the other three

Table 5. Sites of Pinglot et al. (1999, 2001) used in the comparison.

Stake	Coordinates		Elevation stake (m)	Elevation MAR (m)	Distance pixel-stake (km)	Period
Stake 8	78°48′ N	17°28′ E	1173	895	3.66	1986–1996
Kon K	78°47′ N	13°17′ E	639	586	4.77	1986–1988
Kon L	78°46′ N	13°27′ E	726	586	4.63	1986–1991
Snow M	79°08′ N	13°18′ E	1170	849	4.62	1986–1991
Vest 95	79°58′ N	21°01′ E	600	478	4.78	1986–1994
F	79°52′ N	23°32′ E	727	651	3.09	1986–1999
Aust 98	79°48′ N	24°00′ E	740	710	2.26	1986–1997
A	79°50′ N	24°56′ E	729	623	3.72	1986–1998
N	79°40′ N	25°14′ E	491	518	3.76	1986–1999
R	79°31′ N	24°02′ E	511	469	1.82	1986–1999

Table 6. Annual measured SMB (m w.e. yr^{-1}) from Pinglot et al. (1999, 2001) and simulated by MAR_{ERA}, SMB difference (% and m w.e. yr^{-1}) between the MAR outputs and the measurements, MAR interannual variability of the SMB (m w.e. yr^{-1}). MAE = Mean absolute error.

Stake	Mean annual SMB (m w.e. yr^{-1})		SMB difference		MAR interannual variability (m w.e. yr^{-1})
	Pinglot	MAR	(%)	(m w.e. yr^{-1})	
Stake 8	0.75	0.74	−1.3	−0.01	0.18
Kon K	0.48	0.28	−41.7	−0.20	0.28
Kon L	0.62	0.31	−50.0	−0.31	0.19
Snow M	0.57	0.67	17.5	0.10	0.18
Vest 95	0.41	0.30	−26.8	−0.11	0.18
F	0.37	0.37	0.0	0.00	0.26
Aust 98	0.52	0.46	−11.5	−0.06	0.17
A	0.42	0.43	2.4	0.01	0.24
N	0.20	0.34	70.0	0.14	0.26
R	0.23	0.29	26.1	0.06	0.27
MAE			24.7	0.10	

Table 7. 2003–2008 mean elevation changes (dh/dt, m yr^{-1}) from MAR_{ERA} and Moholdt et al. (2010) for different regions.

Region	dh/dt (m yr^{-1})	
	MAR_{ERA}	Moholdt et al. (2010)
Austfonna	0.00	0.11 ± 0.04
Vestfonna	−0.14	−0.16 ± 0.08
Northwest Spitsbergen	−0.22	−0.54 ± 0.10
Northeast Spitsbergen	0.02	0.06 ± 0.06
South Spitsbergen	−0.45	−0.15 ± 0.16
Barentsøya/Edgeøya	−0.38	−0.17 ± 0.11
Total	−0.13	−0.12 ± 0.04

stations (Table 4). It is obvious that we can not resolve the complex spatial variability of precipitation along the coast at a resolution of 10 km but it is also difficult to gauge the snowfall amount in this windy region. Moreover, a lot of data are missing for all the stations. Therefore, we can not draw any conclusion about a likely overestimation of the MAR precipitation by using only precipitation measurements from coastal weather stations.

3.2 Comparison of MAR with SMB measurements

As validation of the SMB, we have compared MAR to SMB measurements from Pinglot et al. (1999) and Pinglot et al. (2001), as indicated in Tables 5 and 6 and Fig. 2.

The MAR model underestimates the SMB for 5 of the 10 sites and overestimates it for the remaining 5 (Table 6) so there is no systematic bias. The mean error is −0.20 m w.e. yr^{-1}, corresponding to −2 %, and the absolute error is 0.10 m w.e. yr^{-1} (i.e. 25 %), but none of the differ-

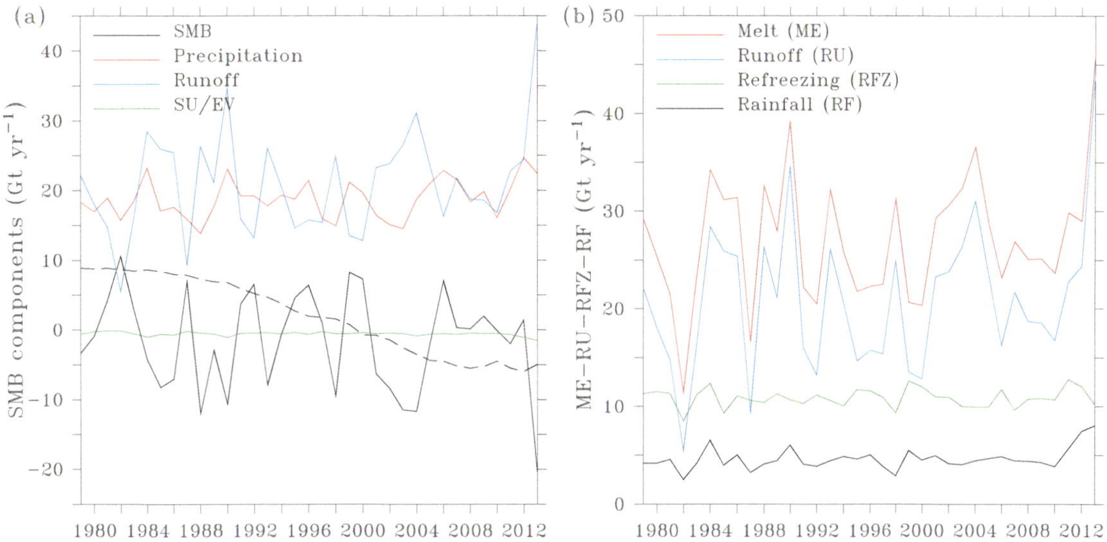

Figure 3. **(a)** Evolution of the MAR_{ERA}-based SMB integrated over the permanent ice mask (Gt yr^{-1}) and its components: (solid plus liquid) precipitation, meltwater runoff and sublimation and evaporation (SU/EV) over 1979–2013. The dashed black curve represents the MAR_{MIROC5}-based SMB (after having applied a 10 yr running mean). **(b)** Same as **(a)** but for the liquid water (melt, runoff and liquid precipitation) and refreezing (Gt yr^{-1}).

Table 8. Comparison between SMB simulated by MAR (mm w.e. yr^{-1}) and different studies. (1) The SMB estimate has been calculated as the net mass balance (km^3 yr^{-1}) minus the estimated calving flux from Błaszczyk et al. (2009) (6.75 km^3 yr^{-1}), then converted in mm w.e. yr^{-1} by dividing it by the surface of the glaciated area. (2) The MAR_{ERA} SMB has been estimated over 1979–2013.

Reference	Time period	Mass balance (km^3 yr^{-1})	SMB estimate (mm w.e. yr^{-1})$^{(1)}$	MAR_{ERA} estimate (mm w.e. yr^{-1})
Mémin et al. (2011)	2003–2008	−9.1	−65	−98
Mémin et al. (2011)	2003–2008	−15.5	−243	−98
Mémin et al. (2011)	1998–2007	−25.0	−508	−88
Wouters et al. (2008)	2003–2007	−8.8	−49	−118
Nuth et al. (2010)	65/90–03/07$^{(2)}$	−9.7	−106	−75

ences are significant with respect to the MAR interannual variability (the difference is significant at the 95 % confidence level if it is higher than twice the interannual variability of the MAR SMB).

On the Austfonna and Vestfonna ice caps, where the slopes are gentle and a resolution of 10 km is enough to represent the main variations of the topography, the SMB is generally well modelled, except for stake N where the difference is a bit larger (0.14 m w.e. yr^{-1}, corresponding to a difference of +70 %). On Spitsbergen, on the contrary, the topography is so steep that a 10 km resolution is not enough to represent it and elevation biases are huge. The precipitation pattern is more complex than on the ice caps because of the "barrier effect" induced by the topography and, therefore, as a result of the elevation underestimation discussed in the previous section, there could be local precipitation biases influencing the modelled surface mass balance. For example, between Kongsvegen (Kon K and Kon L) and the ocean lies an area where the elevation is highly underestimated. At Ny-Ålesund weather station, located in that area, the modelled precipitation is underestimated by 25 %. In our topography, the "barrier effect" of the elevated topography is not present and orographic precipitation may occur in another region. As a consequence, the SMB modelled at Kongsvegen is underestimated quite a lot (−0.2 and −0.31 m w.e. yr^{-1}, corresponding to −42 and −50 %). Using the WRF model (Weather Research and Forecasting), Claremar et al. (2012) investigated the effect of model resolution on wind speed in Svalbard, which strongly depends on the topography as does precipitation. They also conclude that, over very hilly topography, wind speed biases are large and a very high resolution is needed. From this comparison, we can conclude that MAR simulates well the surface mass balance but a resolution of 10 km is likely too coarse to model correctly the SMB and its components over Spitsbergen, where the topography is more complex.

Day et al. (2012) have compared some of the measurements from Pinglot et al. (1999) to the precipitation from the

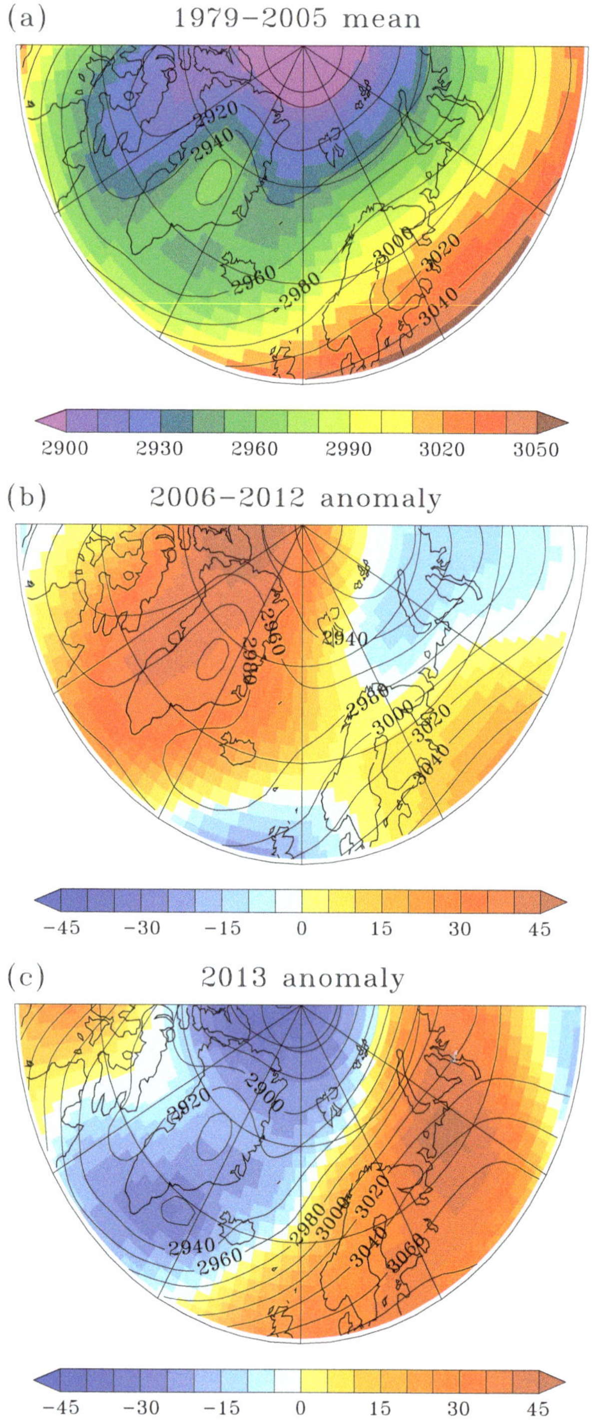

Figure 4. **(a)** Mean 1979–2005 JJA geopotential height at 700hPa ($Z700_{JJA}$) (m) from ERA-Interim. **(b)** 2006–2012 mean $z700_{JJA}$ anomaly (m) with respect to the 1979–2005 mean. **(c)** Same as **(b)** but for 2013. The black lines represent the mean $Z700_{JJA}$ for each period.

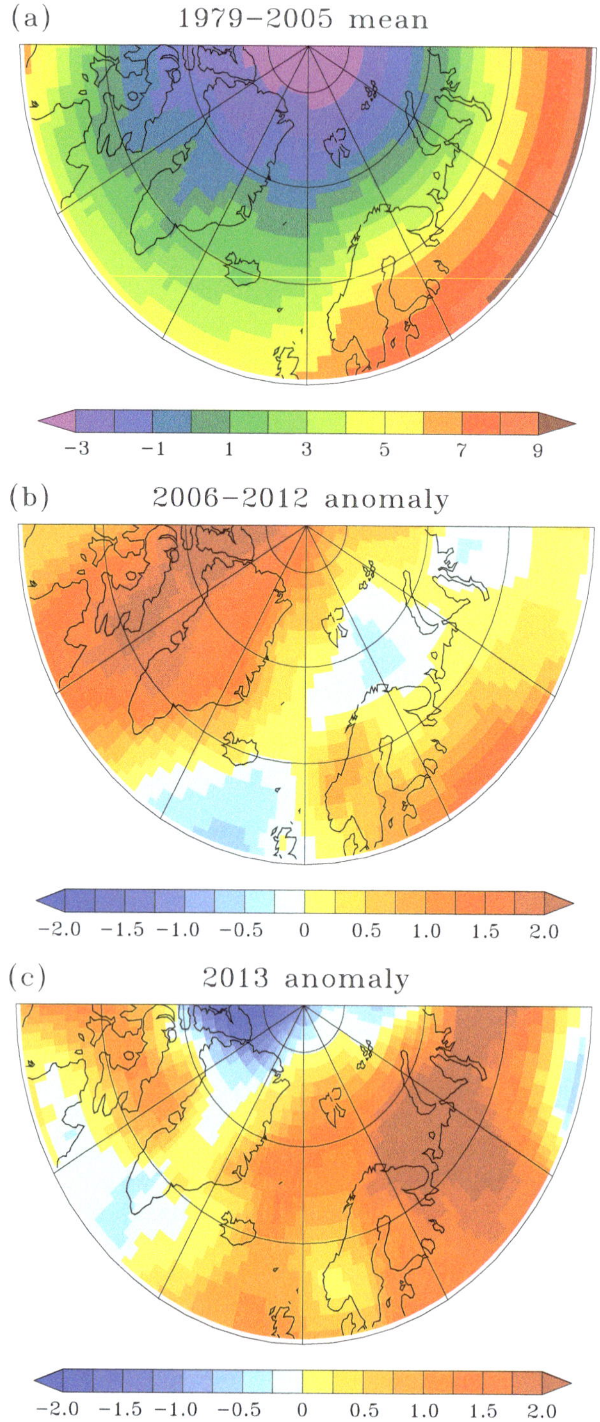

Figure 5. **(a)** 1979–2005 $T850_{JJA}$ mean (°C) from ERA-Interim. **(b)** 2006–2012 mean $T850_{JJA}$ anomaly (°C) with respect to the 1979–2005 mean. **(c)** Same as **(b)** but for 2013.

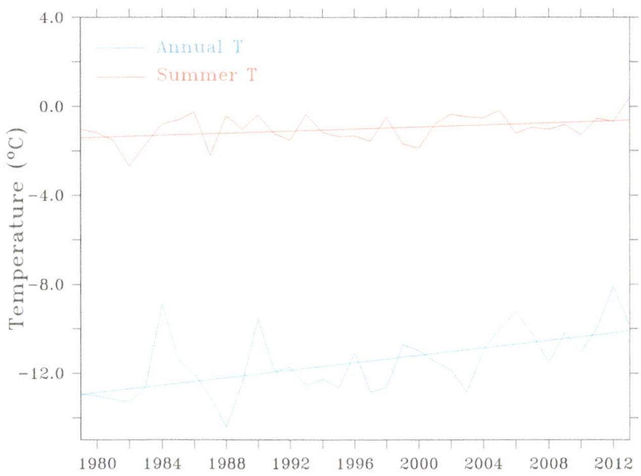

Figure 6. Evolution of the mean annual (blue) and summer (red) MAR_{ERA} TAS (°C) integrated over the permanent ice area between 1979 and 2013 with their linear trend.

HadRM3 RCM (as the ice cores of Pinglot et al. (1999) were retrieved in the accumulation zone). Their biases are similar for the Aust 98 stake, but we have better results at Vest 95 (-0.11 m w.e. yr^{-1} vs. -0.24). However, Day et al. (2012) results do not show a huge bias on Kongsvegen like ours do.

Finally, we can compare the MAR_{ERA} mean elevation change rate (dh/dt in m yr^{-1}) over 2003–2008 to Moholdt et al. (2010) (Supplement Fig. S2, to be compared to Fig. 1 of Moholdt et al. (2010)). In northwestern Spitsbergen, MAR_{ERA} simulates a thickening of the interior whereas the entirety of northwestern Spitsbergen is thinning in Moholdt et al. (2010) and in south Spitsbergen, MAR_{ERA} barely simulates the thickening on the east coast. As a result, the regionaly averaged thinning of northwestern Spitsbergen is underestimated and the averaged thinning of south Spitsbergen is overestimated (Table 7). On Austfonna and in northeastern Spitsbergen, on the other hand, the pattern of the mean elevation change rate compares well with Moholdt et al. (2010). MAR_{ERA} represents well the thickening of the interior of Austfonna and northeastern Spitsbergen but slightly overestimates the margins thinning in northeastern Spitsbergen and slightly overestimates it on Austfonna. Regionally averaged, dh/dt of northeastern Spitsbergen and Austfonna also compares well with Moholdt et al. (2010). Integrated over Svalbard, the MAR_{ERA} mean elevation change is the same as Moholdt et al. (2010) (-0.13 m yr^{-1} vs. -0.12 ± 0.04). MAR_{ERA} therefore compares well with Moholdt et al. (2010), considering the biases associated to the 10km topography and the fact that only the first ~ 10 m of ice and snow are modelled in MAR and that the compaction of the deep snow/ice layers is therefore not taken into account.

4 Results of MAR over the present climate

The mean annual total SMB integrated over Svalbard (Fig. 3a, black curve) simulated by MAR_{ERA} between 1979 and 2013 is -1.6 ± 7.1 Gt yr^{-1}, corresponding to -54 ± 236 mm w.e. yr^{-1} with our ice sheet mask. The bias resulting from the use of a 10 km topography can be corrected based on the elevation classes histogram in Supplement Fig. S1 and gives an estimated mean annual value of 0.4 Gt yr^{-1} (corresponding to 12 mm w.e. yr^{-1}), fully included in the uncertainty given here by the 1979–2013 interannual variability (standard deviation of 7.1 Gt). The high interannual variability is mainly a result of the variability of the meltwater runoff ($R^2 = 0.85$ between the SMB and runoff), itself mainly due to the interannual variability of the JJA mean TAS. Based on measurements made in the 1960s–1990s, Hagen et al. (2003) estimated the SMB integrated over Svalbard to be -14 ± 3 mm w.e. yr^{-1} or -0.5 ± 0.1 Gt yr^{-1}. Our mean value of -54 mm w.e. yr^{-1} over 1979–2013 therefore aligns with the values of Hagen et al. (2003), considering the large interannual variability of our SMB and the fact that the time period over which the simulations were performed are not the same (e.g. our mean value would have been -36 mm w.e. yr^{-1} if we had not considered the year 2013). Calving has been estimated by Hagen et al. (2003) to be 4.5 Gt yr^{-1} (~ 110 mm w.e. yr^{-1}) and is therefore a very important component of the net mass balance compared to their estimation of SMB. However, it is small compared to the contribution of surface runoff to the total mass loss (680 mm w.e. yr^{-1} in Hagen et al. (2003) and 695 mm w.e. yr^{-1} simulated by MAR_{ERA}). Błaszczyk et al. (2009) estimated a calving flux of 6.75 ± 1.75 km^3 yr^{-1} over 2000–2006 from ASTER imagery and we used this value to estimate SMB values from different mass balance (MB) estimates (Table 8). Considering again the large interannual variability of the SMB, our MAR_{ERA} estimates compares well with Wouters et al. (2008) and the low value of Mémin et al. (2011) (corresponding to the MB of -9.1 km^3 yr^{-1}), both obtained from GRACE measurements. It also compares well with Nuth et al. (2010), knowing that the time period of their estimate is different from ours and different from the time period over which the calving flux was estimated. The high GRACE value of Mémin et al. (2011) (MB of -15.5 km^3 yr^{-1}) and the value obtained by ground gravity observations (MB of -25 km^3 yr^{-1}) give a surface loss much larger than ours but those values are also quite large compared to the other studies. To sum up, MAR_{ERA} compares well with studies for which the SMB has been estimated and also gives satisfying results compared to other studies for which we had to estimate the SMB contribution using a calving flux value estimated over the same period.

SMB measurements starting in the 1960s on individual glaciers show a stability of the SMB until the late 1990s (Hagen et al. (2003) and references therein). The SMB of these glaciers located near the coast was negative, mean-

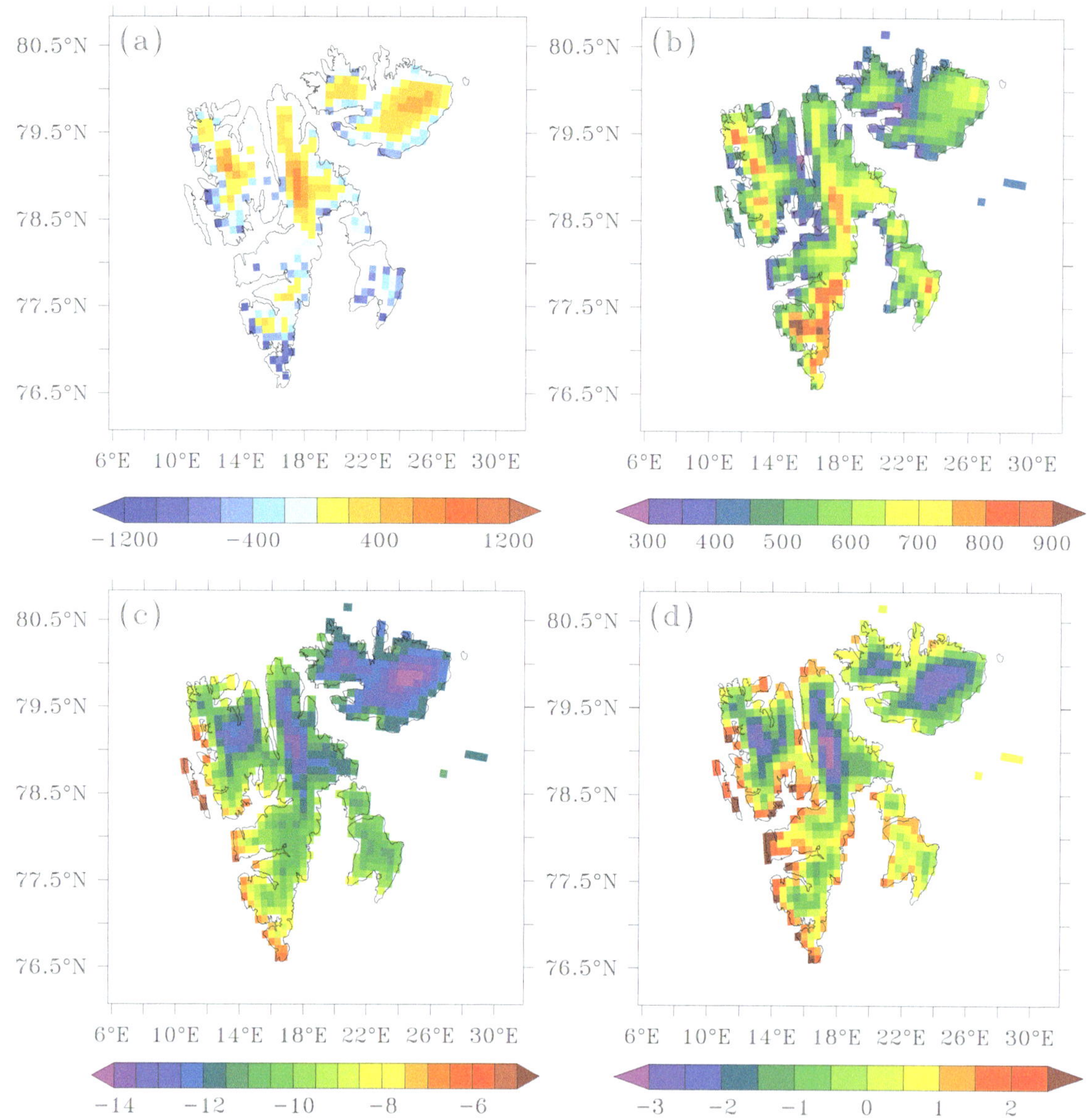

Figure 7. **(a)** Mean annual SMB (mm w.e. yr^{-1}) averaged over 1979–2013. **(b)** Same as **(a)** for the annual precipitation (mm w.e. yr^{-1}). **(c)** Same as **(a)** for the mean annual near-surface temperature (TAS) (°C). **(d)** Same as **(a)** for the mean summer near-surface temperature (TAS_{JJA}) (°C).

ing that the glaciers are losing mass but without any acceleration nor deceleration of the surface mass loss. However, some glaciers like Kongsvegen and Kronebreen experienced increased melting in the late 1990s (Nuth et al., 2012) but their SMB stabilised in the second half of the 2000s. Integrated over the entirety of Svalbard, the 1979–2013 linear temporal trend of the MAR_{ERA} SMB (−0.1 Gt yr^{-2}) is not statistically significant and therefore suggests stability. Contrary to individual SMB measurements, we can not affirm that the integrated SMB is really negative as the averaged MAR_{ERA} SMB is close to zero and given the biases associated to the used 10 km resolution. Moholdt et al. (2010) highlighted a very negative SMB in 2003–2004 followed by a series of more balanced values between 2004 and 2007. MAR_{ERA} also suggests very low values of SMB in 2003 and 2004 (~ -12 Gt) and more balanced values over 2005–2012. We can therefore conclude that the SMB has been stable (yet negative) over the past 35 years when integrated over the entirety of Svalbard. The recent trend is however opposite to what has been occurring over the Greenland ice sheet, where the SMB has been stable until the end of the 1990s and record melts have been observed since 2006 and can be explained by the recent change in atmospheric flow frequencies in summer, causing more frequent southerly flows over Greenland but rather northerly flows over Svalbard in summer (Fettweis et al., 2013a).

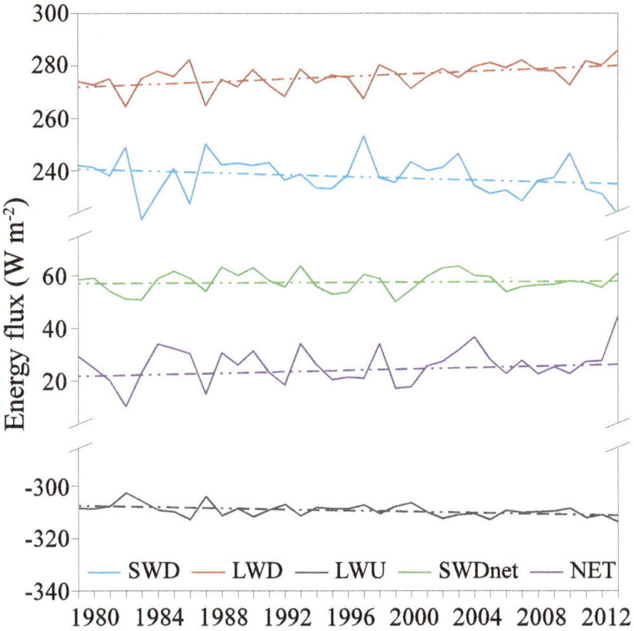

Figure 8. Evolution of the JJA energy balance components fluxes ($W\,m^{-2}$) simulated by MAR_{ERA} over the permanent ice area with their linear trends in dashed lines.

Over Svalbard, the mean 1979–2005 summer 700 hPa ERA-Interim atmospheric circulation was a westerly or west–southwesterly flow (Fig. 4a). After 2005, however, the circulation changed as a result of more frequent North Atlantic Oscillation (NAO) negative phases in summer. $Z700_{JJA}$ (summer 700 hPa geopotential height, representing the general circulation) increased more over Greenland than it did over Svalbard (Fig. 4b) resulting in northwesterly flows over Svalbard and more anticyclonic conditions over Greenland (Fettweis et al., 2013a). Consequently, a summer temperature increase breaking melt records has been observed over Greenland since 2006 (Fettweis et al., 2013a). Over Svalbard, on the contrary, the northerly flow brings colder air (Fig. 5b, showing temperature at 850 hPa (T850) as it drives the melt variability, according to Fettweis et al. (2013b)), and the surface mass balance has remained stable over the period 1979–2012 despite the recent observed Arctic warming (Anisimov et al., 2007). In summer 2013, however, the 700 hPa summer atmospheric circulation was again a westerly flow and could not oppose Arctic warming anymore. As a result, the 2013 JJA ERA-Interim T850 anomaly with respect to the 1979–2005 mean (Fig. 5c) was positive, contrary to the 2006–2012 period, and MAR_{ERA} simulated the highest TAS_{JJA} of the last 35 years (Fig. 6), causing the meltwater runoff to break records ($44.1\,Gt\,yr^{-1}$). The 2013 MAR_{ERA} SMB was also the lowest of the last 35 years ($-20.4\,Gt\,yr^{-1}$), whereas the precipitation was higher than average but not significantly different (Fig. 3a).

The recent change in the 700 hPa summer atmospheric circulation, damping the effect of the observed Arctic warming over Svalbard is also responsible for the stabilisation of the summer TAS (TAS_{JJA} simulated by MAR_{ERA} in Fig. 6) in the second half of the 2000s, as opposed to Greenland. Over the past 35 years, the annual Svalbard temperature has risen by 2.8 °C as shown in Fig. 6 whereas the mean summer temperature increase is more moderate (0.79 °C for the past 35 years). Both linear trends are statistically significant but, for the summer temperature, the very high value of 2013 has a large influence on the value of the trend given its position at the end of the time series. If we exclude summer 2013, the JJA temperature trend is not statistically significant.

While the mean annual values of precipitation and run-off are quite similar (18.7 and $20.9\,Gt\,yr^{-1}$, Fig. 3a), the precipitation amount has been stable over 1979–2013 (interannual variability of $2.7\,Gt\,yr^{-1}$), whereas the interannual variability of run-off is high ($7.4\,Gt\,yr^{-1}$). Sublimation and evaporation, for their part, are quite constant and contribute very little to the SMB variability. Their negative values for every year indicate that MAR simulates a greater deposition than sublimation and evaporation. About $64 \pm 9\,\%$ of the total liquid water (melt plus liquid precipitation) runs off and the remaining 36 % refreezes (Fig. 3b). Contrary to runoff and melt, the amount of water that refreezes is constant from year to year (standard deviation of $0.95\,Gt\,yr^{-1}$ for the refreezing vs. $6.6\,Gt\,yr^{-1}$ for the melt). Here again, none of the linear trends over 1979–2013 are significant.

The surface mass balance (Fig. 7a) is positive only on the ice caps on Nordaustlandet and in high elevation zones in north Spitsbergen where temperatures are low or precipitation high or both. The mean annual TAS (Fig. 7c) goes from −5 °C on the west coast of Spitsbergen to almost −15 °C in the centre of the ice caps and in Newtontoppen region (highest elevations, Fig. 2). Moreover, there is a west-to-east temperature gradient showing the effect of the North Atlantic Drift bringing oceanic heat to the west coast of the archipelago. This temperature gradient is enhanced by the larger SIC along the east coast that further cools it and increase the contrast with the west coast. The mean TAS_{JJA} (Fig. 7d) is positive along the coasts, except in northeastern Spitsbergen and Austfonna where sea ice is still present in summer. The west-to-east gradient is less pronounced in summer than on the annual timescale as the sea ice strongly decreases on the east coast of Spitsbergen in summer. Precipitation (Fig. 7b) is lower on the west coast of Spitsbergen than on the east coast (e.g. on Austfonna) because of frequent depressions in the Barents Sea bringing humid air on the eastern coast of Svalbard (Winther et al., 1998; Hisdal, 1976). Due to the underestimation of the elevation, we can expect precipitation to be lower than observed as MAR likely underestimates the amount of humidity/clouds blocked by the mountains. The impact of the humidity underestimation on the longwave radiation also explains in part the MAR cold bias. According to Liestøl (1993), the maximum amount of

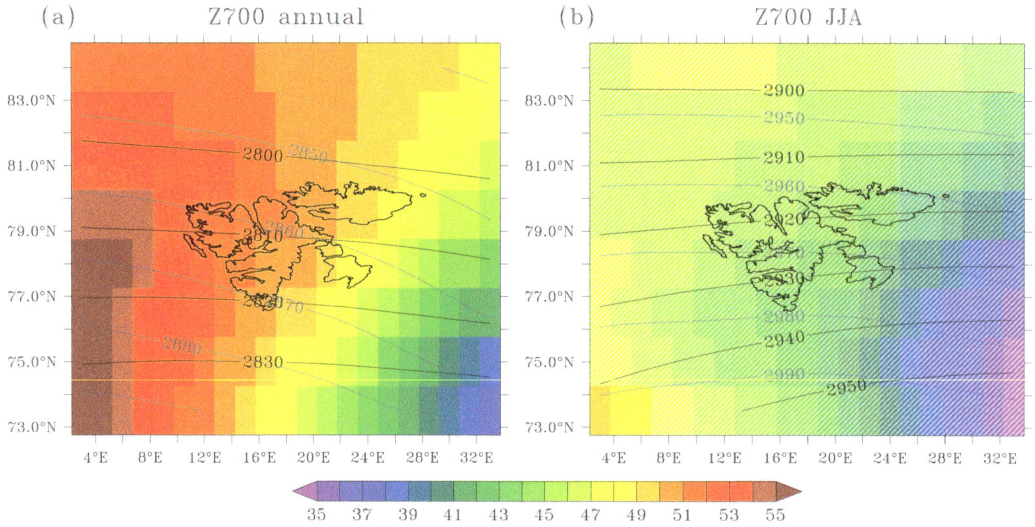

Figure 9. (a) Difference of mean annual geopotential height (m) at 700 hPa (Z700) between MIROC5 and ERA-Interim over 1980–2005. The black lines show Z700 for ERA-Interim and the blue lines are for MIROC5. **(b)** Same as **(a)** but in summer (JJA). The non-hatched areas correspond to the areas for which the difference is significant at the 95 % confidence level (with respect to the ERA-Interim-based 1980–2005 interannual variability), whereas the hatched areas corresponds to non-significant differences.

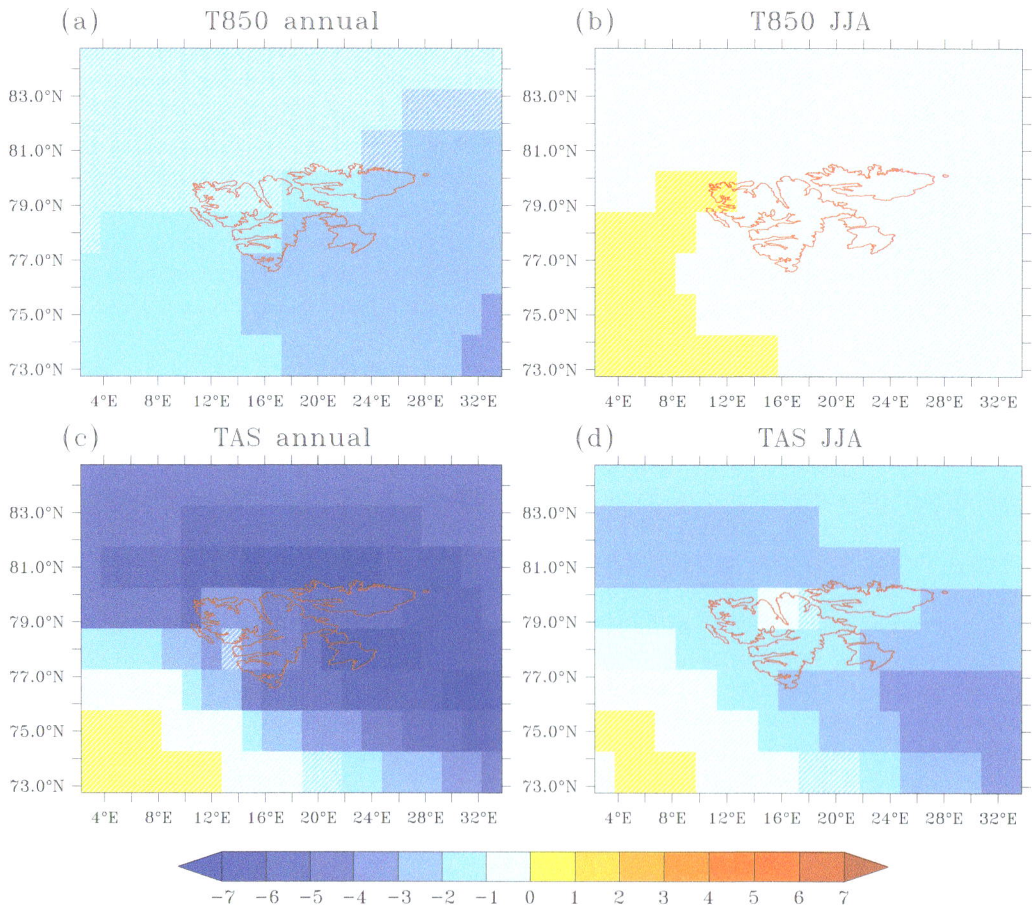

Figure 10. (a) Annual 850 hPa temperature (T850) difference (°C) between MIROC5 and ERA-Interim over 1980–2005. **(b)** Same as **(a)** but in summer (JJA). **(c)** Same as **(a)** but for the near-surface temperature (TAS, °C). **(d)** Same as **(c)** but in summer (JJA). The non-hatched areas correspond to the areas for which the difference is significant at the 95 % confidence level.

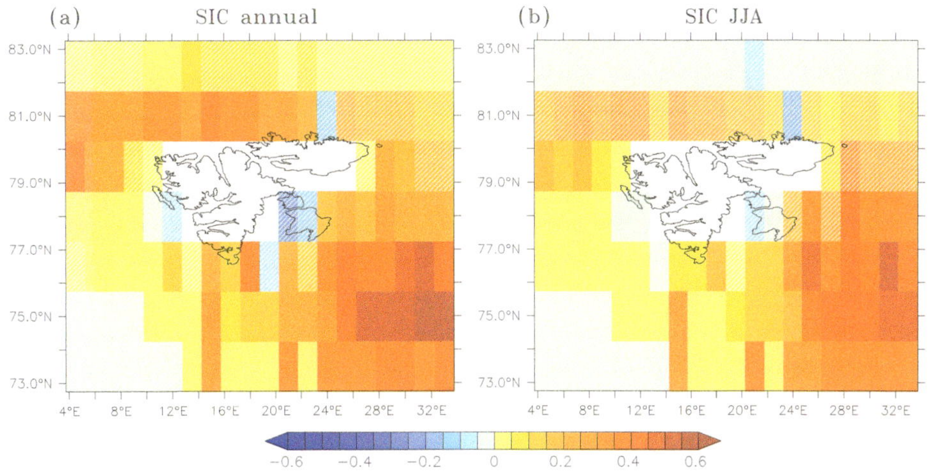

Figure 11. **(a)** Annual sea-ice cover (SIC) difference between MIROC5 and ERA over 1980–2005. The SIC go from zero to one according to the portion of the oceanic pixel covered in sea ice (0 = ice-free pixel, 1 = pixel completely covered with sea ice). **(b)** Same as **(a)** but in summer (JJA). The non-hatched areas correspond to the areas for which the difference is significant at the 95 % confidence level.

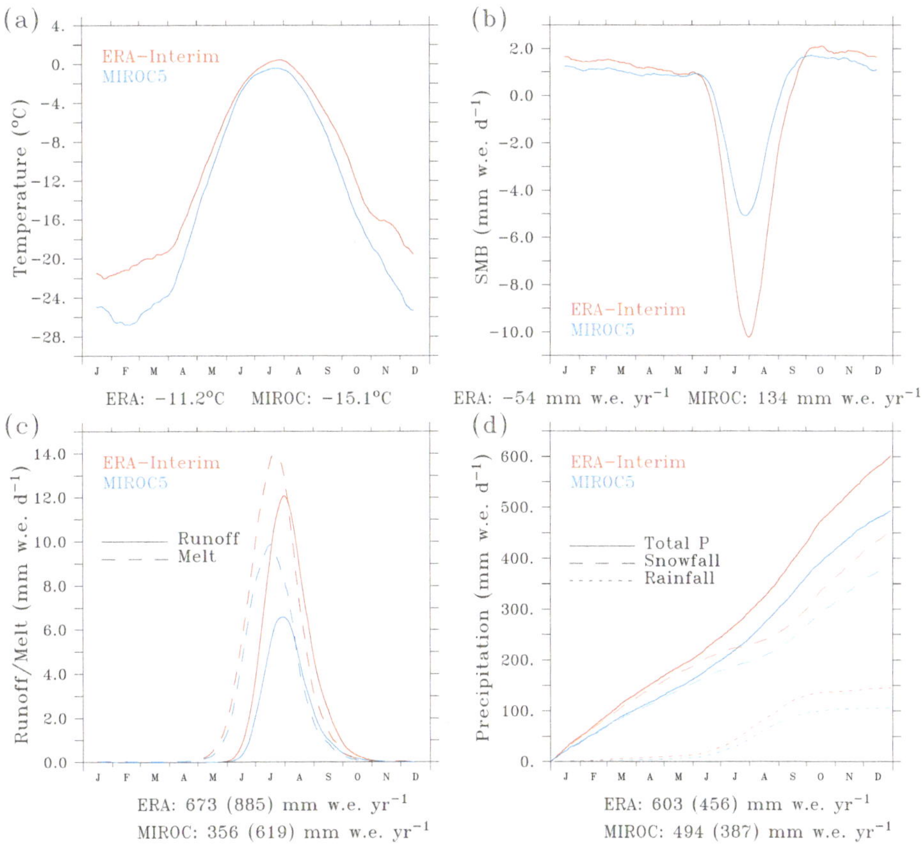

Figure 12. **(a)** 1980–2005 mean annual near-surface temperature cycle (°C) for MAR_{ERA} (red) and MAR_{MIROC5} (blue). A 30-day running mean has been applied to smooth the curves. The numbers give the mean annual temperature integrated over the permanent ice area for MAR_{ERA} and MAR_{MIROC5}. **(b)** Same as **(a)** but for the surface mass balance (mm w.e. day^{-1}). The numbers give the annual SMB (mm w.e. yr^{-1}) integrated over the permanent ice area for MAR_{ERA} and MAR_{MIROC5}. **(c)** Same as **(a)** but for the runoff (solid line) and melt (dashed line) (mm w.e. day^{-1}). The listed numbers give the mean annual runoff, with the melt over the permanent ice area in brackets (mm w.e. yr^{-1}). **(d)** Same as **(a)** but for the cumulated total precipitation (solid line), snowfall (large dashes) and rainfall (small dashes) (mm w.e. day^{-1}). The numbers give the mean annual precipitation and snowfall (shown in brackets) over the permanent ice area (mm w.e. yr^{-1}).

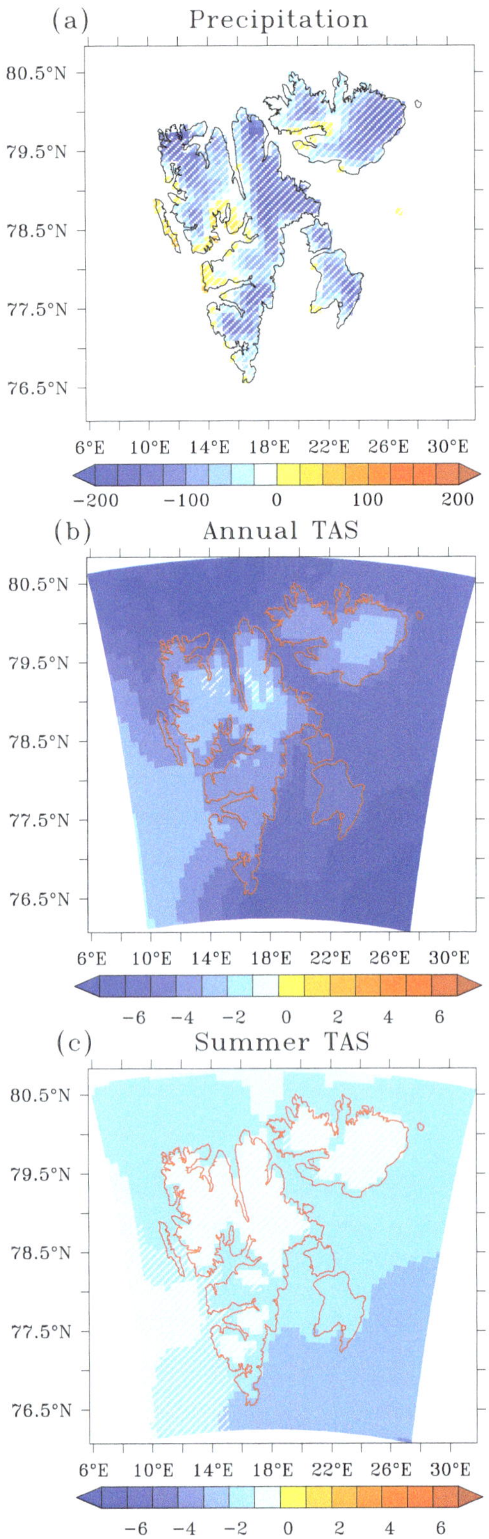

Figure 13. **(a)** Annual precipitation difference (mm yr^{-1}) between MAR_{MIROC5} and MAR_{ERA} averaged over 1980–2005. **(b)** Same as **(a)** but for the annual near-surface temperature (°C). **(c)** Same as **(b)** but for summer (JJA). The non-hatched areas correspond to the areas where the difference is significant (i.e. higher than twice the interannual variability of MAR_{ERA}).

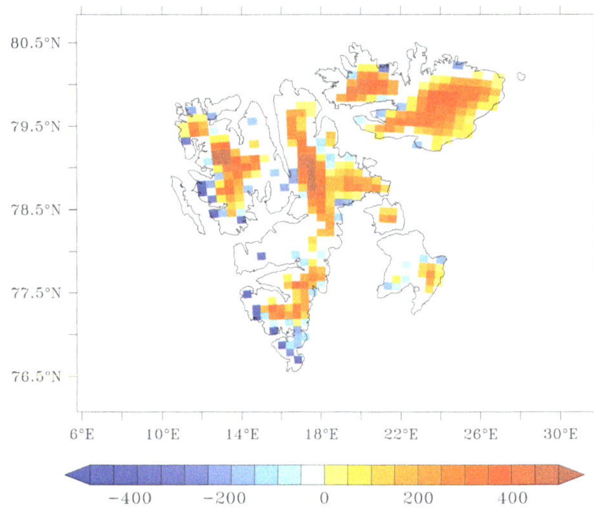

Figure 14. 1980–2005 mean annual SMB (mm w.e. yr^{-1}) modelled by MAR_{MIROC5}.

precipitation (more than 1000 mm) is found in the southeastern part of Spitsbergen, where the wind transports humid air onto the mountain slopes. In our case, the maximum is also located in the south of Spitsbergen but it is underestimated (900–950 mm).

The net energy flux (NET) available at the surface for the melt can be written

$$NET = SWDnet + LWnet + SHF + LHF \quad (W\,m^{-2}),$$

where

- SWDnet = SWD × (1 − a) is the net downward shortwave radiation, i.e. the amount of the downward shortwave (= solar radiation) energy flux (SWD) that is absorbed by the surface following its albedo (a).
- LWnet = LWD-LWU is the net longwave radiation, i.e. the difference between the downward longwave radiation coming from the atmosphere and the upward longwave radiation emitted by the surface.
- SHF and LHF are the sensible and latent heat fluxes. These fluxes are negligible with respect to the solar and infrared fluxes and are therefore not shown in Fig. 8.

Over 1979–2013, the net energy flux at the surface in summer has increased (Fig. 8), as a result of increases in both net downward shortwave (SWDnet) and longwave (LWDnet) summer energy fluxes, giving more energy for the melt. However, only the longwave energy fluxes linear trends are significant. As the summer SWD has been decreasing, the dominant factor causing the increase of SWDnet over 1979–2013 is the decrease of the surface albedo. The increase in LWD is a consequence of the increasing greenhouse effect-induced atmosphere warming as well as the significant increase of summer cloud cover (5 % in 35 years with a 1979–2013 mean cloud cover of 73±3 %). The cloudiness increase is also responsible for the SWD decrease through the greater reflection of the sunlight by the clouds.

5 Comparison of ERA-Interim and MIROC5 over Svalbard

With the aim of performing future projections with MAR in a companion paper (Lang et al., 2014), we need a global model (GCM) to force its lateral boundaries, and we need to be sure that MAR forced by the GCM is able to correctly simulate the current climate over Svalbard. Indeed, the future projections will be more uncertain if MAR forced by the GCM shows huge biases over the present. Given the dependency of RCM outputs to the forcings (Fettweis et al., 2013b), it is first necessary to evaluate the GCM itself over Svalbard to be able to explain the possible biases when MAR is forced by the GCM with respect to MAR forced by ERA-Interim (chosen here as reference over current climate as in Fettweis et al., 2013b). Suitable GCMs are those that are capable of modelling the free atmosphere as MAR is not able to correct possible biases in the free atmosphere in view of the dimension of our integration domain. To achieve this, we have compared Z700, representing the atmospheric circulation and T850, as well as TAS and SIC, as MAR is forced by SST and SIC over ocean. Among the CMIP5 models evaluated in Fettweis et al. (2013b), MIROC5 (Model for Interdisciplinary Research On Climate, Watanabe et al., 2010; Sakamoto et al., 2012) is one of the best GCMs over Greenland with respect to ERA-Interim. The MIROC5 global model works also well over Svalbard as we will show hereafter. As the historical run ends in 2005, the comparison period extends here from 1980 to 2005.

According to Fig. 9, the annual mean MIROC5-based Z700 is higher than ERA's by about 40 to 55 m. The annual positive anomaly is significant at the 95 % confidence level (for clarity, the significant anomalies correspond to the non-hatched areas) whereas the summer difference barely is. We consider the difference significant at the 95 % confidence level if it is higher than twice the standard deviation of the ERA-Interim-based Z700 (representing the interannual variability of Z700). The solid lines, showing the mean Z700, suggests that the circulation is slightly diverted (clockwise for the annual circulation and anticlockwise for the JJA circulation), from ERA-Interim to MIROC5 going from a westerly flow in ERA to a west–northwesterly flow in MIROC5. At the annual timescale (1980–2005), MIROC5 is colder at 850 hPa than ERA-Interim by 2 to 4 °C, and this difference is significant over the south and east of Svalbard (Fig. 10a, annual mean). In summer, this T850 anomaly is not significant (Fig. 10b). At the surface (TAS, Fig. 10c and d), the cold bias is even larger, especially at the annual timescale, except in the southwestern corner, where we have a non-significant positive bias. Contrary to the JJA T850 anomaly, the TAS_{JJA} cold bias of MIROC5 with respect to ERA-Interim is significant.

The temperature difference is linked in part to the larger SIC in MIROC5 with respect to ERA-Interim (Fig. 11). The only zone where the temperature anomaly is positive corresponds to the zone where the MIROC5 SIC is lower than the ERA-Interim SIC. Conversely, the area in the southeastern corner where MIROC5 prescribes more than 50 % of ice coverage, whereas the ocean is mostly ice free with ERA, corresponds to the zone where MIROC5 is the coldest compared to ERA-Interim.

Contrary to air temperatures that are forced only at the lateral boundaries of our integration domain, SIC and sea surface temperatures (SST) are 6 hourly forced over the entire MAR domain as MAR is not coupled with an oceanic model. Their biases therefore impact on the climate modelled by MAR, especially near the coast, where most of the weather stations are located.

6 Comparison of MAR forced by ERA-Interim and MIROC5

According to Fig. 12a showing the annual cycle of TAS, MAR forced by MIROC5 (hereafter MAR_{MIROC5}) is colder than MAR_{ERA} through the whole year: during summer, the difference is close to zero, but it is larger than 5 °C in winter. Integrated over the entirety of Svalbard, the annual SMB is positive with MAR_{MIROC5} (+4.0 Gt yr^{-1} on average over 1980–2005, corresponding to 134 mm w.e. yr^{-1}), whereas Svalbard loses mass on average with MAR_{ERA} (−1.6 Gt yr^{-1} or −54 mm w.e. yr^{-1}) over 1980–2005. The SMB differences occur mainly in summer through meltwater runoff (Fig. 12c) as the precipitation difference (Fig. 12d) between MAR_{ERA} and MAR_{MIROC5} is much smaller than the runoff difference (only 58 % of the MAR_{ERA} runoff is modelled by MAR_{MIROC5}, whereas 82 % of the total amount of MAR_{ERA} snowfall is modelled). The melt season is shorter for MAR_{MIROC5} than for MAR_{ERA} ($\sim$ 145 day yr^{-1} vs. $\sim$ 155 day yr^{-1}) and the magnitude of surface melt is also smaller (68 % of the amount of melt in MAR_{ERA}) with MAR_{MIROC5} (Fig. 12c).

The amount of precipitation in MAR_{MIROC5} is lower than in MAR_{ERA} (Figs. 12d and 13a). The difference is caused by (i) the cold bias of MIROC5 (the atmosphere can contain less moisture) and (ii) the difference in SIC between ERA-Interim and MIROC5. When the ocean is covered with ice, the exchange of moisture between the ocean and the atmosphere is strongly reduced and so is the amount of water available for precipitation (Noël et al., 2014). An overestimation of SIC also results in a decrease of inland TAS (Noël et al., 2014). The larger SIC in MIROC5 therefore also causes MAR_{MIROC5} to be colder than MAR_{ERA}, especially in the northwestern and southeastern corners of our integration domain according to Figs. 13b and 11. The bias, still

Table 9. 1980–2005 mean SMB, its components ($Gt\,yr^{-1}$) and annual and JJA TAS (°C) with their linear trends ($Gt\,yr^{-2}$ and $°C\,yr^{-1}$) from MAR_{ERA} and MAR_{MIROC5}. The statistically significant (i.e. superior to twice the interannual variability of the variable in MAR_{ERA} or MAR_{MIROC5}) linear trends are in bold.

	Mean ($Gt\,yr^{-1}$ – °C)		Trend ($Gt\,yr^{-2}$ – $°C\,yr^{-1}$)	
	MAR_{ERA}	MAR_{MIROC5}	MAR_{ERA}	MAR_{MIROC5}
SMB	−1.6	4.0	−0.2	**−0.6**
Runoff	20.2	10.7	0.2	**0.7**
Precipitation	18.1	14.8	−0.004	0.1
Sublimation/evaporation	−0.57	0.15	−0.007	−0.02
Temperature (annual)	−11.2	−15.1	0.05	0.1
Temperature (summer)	−0.9	−1.9	0.03	**0.07**

greater than 1.0 °C in the high elevation central regions in the north of Spitsbergen and in the interior of Austfonna, indicates that Svalbard is extremely impacted by the sea surface conditions, even far inland. However, this TAS negative anomaly is also induced by the MIROC5-based free atmosphere, which is too cold at the MAR lateral boundaries. Due to positive feedbacks, it is likely that the SIC overestimation and the too-cold free atmosphere are linked in MIROC5.

Finally, the TAS bias is reduced over land (about 1.5 °C for the annual bias and 1 °C in summer, Fig. 13b and c) in MAR_{MIROC5} compared to the MIROC5 bias and becomes significant in MAR_{MIROC5} in summer only near the coast where the sea-ice bias has the greatest influence (Noël et al., 2014), suggesting that MAR is really able to improve the MIROC5 inputs (showing a significant cold bias over most of the land area).

Over 1980–2005, the SMB simulated by MAR_{ERA} (Fig. 7 also holds for the 1980–2005 period, as the climate has been stable over the last 35 years) is positive only in the northwestern and northeastern central parts of Spitsbergen and on the ice caps whereas MAR_{MIROC5} predicts a mass gain over most of Svalbard (Fig. 14) due to the underestimation of the melt. The only locations where large amounts of mass are lost every year with MAR_{MIROC5} are the west coast and the very southern part of Spitsbergen. Over the reference period, MAR_{ERA} predicts that only 48 % of the area of Svalbard covered with permanent ice has a positive SMB while with MAR_{MIROC5}, 74 % of this area gains mass on average. Integrated over the entirety of Svalbard, the SMB modelled by MAR_{MIROC5} is therefore positive (Table 9).

Over most of Svalbard, the MAR_{MIROC5} SMB is larger than the MAR_{ERA} SMB because of the run-off deficit in MAR_{MIROC5} (Fig. 15a). However, the SMB bias is not significant, given its large interannual variability, except in some places in the very south of Spitsbergen and Edgeøya. The snowfall bias is almost never significant and the areas of significant runoff underestimation in MAR_{MIROC5} are mostly located where the JJA temperature bias is the largest (Fig. 15b and c). In the centre of northern Spitsbergen, the SMB modelled by MAR_{MIROC5} is lower than MAR_{ERA} SMB (coincidently where the later is positive) as it corresponds to areas where the precipitation bias is the largest, likely as a result of too-low temperature in MIROC5 disallowing significant precipitation. In south Spitsbergen and on Edgeøya on the other hand, the precipitation modelled by MAR_{MIROC5} is less underestimated whereas the runoff bias is very negative, hence causing a largely positive SMB bias.

Whereas we saw in Sect. 4 that there has been no SMB temporal trend over the past 35 years according to MAR_{ERA}, MAR_{MIROC5} shows a significant SMB trend ($-0.6\,Gt\,yr^{-2}$, Fig. 3). The runoff and summer temperature trends are also significant, contrary to MAR_{ERA} (Table 9). The negative SMB trend is due to the inability of MIROC5 to correctly represent the recent atmospheric circulation change damping the global warming impact over Svalbard, according to Fettweis et al. (2013a). As a result, the summer temperature rises significantly instead of remaining constant and the melt is overestimated in the second half of the 2000s.

7 Conclusions

In this study, MAR has been evaluated over Svalbard: although it is too cold, the modelled SMB is close to the measured one in areas where the 10 km resolution is enough to correctly represent the topography (i.e. on Austfonna). On Spitsbergen, on the other hand, the 10 km resolution is not able to resolve the complex topography and therefore the precipitation pattern. As a result, large biases in the SMB are present and higher resolutions are therefore needed to correctly simulate the SMB in this area.

What has been observed on several glaciers between the 1960s and the 1990s has been extended to the present day over the entirety of Svalbard in this study: there has been no significant temporal change of the surface mass balance over the last 35 years despite the global change-induced Arctic warming observed since the end of 1990s (Serreze et al., 2009). Because of the recent change in atmospheric circulation in summer (favouring northwesterly flow over Svalbard), there has not been any recent record surface melt in Svalbard

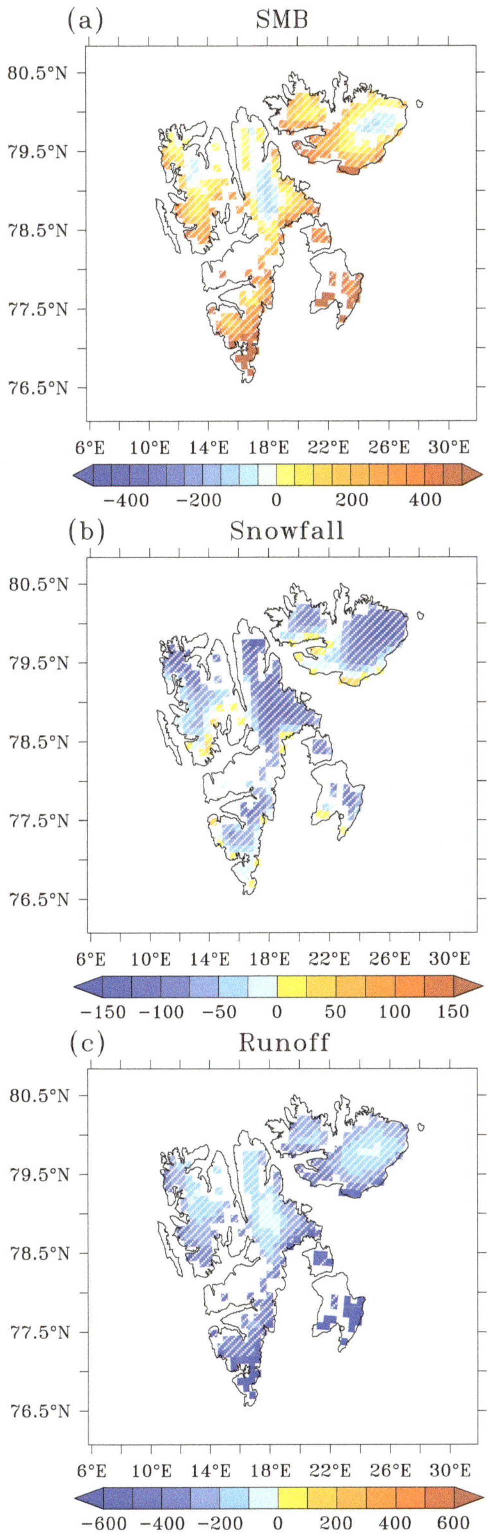

Figure 15. **(a)** Annual SMB difference (mm w.e. yr^{-1}) between MAR_{MIROC5} and MAR_{ERA} averaged over 1980–2005. **(b)** Same as **(a)** but for the snowfall (mm w.e. yr^{-1}). **(c)** Same as **(a)** but for the runoff (mm w.e. yr^{-1}). The non-hatched areas correspond to the areas where the difference is significant.

like in Greenland until 2013. In 2013 on the other hand, the atmospheric circulation was again a westerly flow over Svalbard, causing the SMB to break a record in Svalbard whereas the melt was much more moderate in Greenland. This shows the important role of general circulation anomalies in summer and the need to have time series long enough to know if these recent circulation changes are due to natural variability or not.

With the perspective of performing future simulations, we have compared MIROC5 and ERA-Interim over Svalbard as well as MAR forced by MIROC5 and ERA-Interim over 1980–2005. Averaged over 1980–2005, MIROC5 is significantly colder than ERA-Interim and prescribes more sea ice that impacts the temperature over land and the precipitation simulated by MAR. MAR_{MIROC5}, however, has proven able to improve the MIROC5 results, and the SMB, runoff and precipitation differences to the MAR_{ERA} simulations are barely significant. In summer, the near-surface temperature difference over the land is significant only close to the coastline.

If we look at the temporal evolution of the SMB, on the other hand, MAR_{MIROC5} SMB shows a significant negative trend, contrary to MAR_{ERA} SMB, because MIROC5 does not represent the recent atmospheric change that caused the SMB of Svalbard to remain constant on average. However, integrated over the entirety of Svalbard, the differences are not statistically significant, suggesting that the MIROC5-forced future projections should not be affected a lot by the biases over current climate with respect to the ERA-Interim-forced run.

Acknowledgements. C. Lang works under a PhD grant from the Fonds pour la formation à la Recherche dans l'Industrie et l'Agriculture (FRIA), Belgium. The meteorological data come from the Norwegian Meteorological Institute (www.eklima.no). The authors wish to thank J. Kohler and the Norsk Polarinstitutt for providing us the topography and C. Nuth for the ice mask. We also thank one anonymous reviewer, J. Lenaerts, M. Pelto and C. Rye, whose comments helped improve this manuscript a lot.

Edited by: J. O. Hagen

References

Anisimov, O. A., Vaughan, D. G., Callaghan, T. V., Furgal, C., Marchant, H., Prowse, T. D., Vilhjálmsson, H., Walsh, J. E., Parry, M. L., Canziani, O. F., Palutikof, J. P., van der Linden, P. J., and Hanson, C. E. (Eds.): Polar Regions (Arctic and Antarctic), Climate Change 2007: Impacts, Adaptation and Vulnerability. Contribution of Working Group II to the Fourth Assessment Report of the Intergovernmental Panel on Climate Change, Cambridge University Press, Cambridge, 653–685, 2007.

Bamber, J. L., Krabill, W., Raper, V., and Dowdeswell, J. A.: Anomalous recent growth of part of a large Arctic ice cap: Austfonna, Svalbard, Geophys. Res. Lett., 31, L12402, doi:10.1029/2004GL019667, 2004.

Bamber, J. L., Krabill, W., Raper, V., Dowdeswell, J. A., and Oerlemans, J.: Elevation changes measured on Svalbard glaciers and ice caps from airborne laser data, Ann. Glaciol., 42, 202–208, 2005.

Benestad, R. E., Førland, E. J., and Hanssen-Bauer, I.: Empirically downscaled temperature scenarios for Svalbard, Atmos. Sci. Lett., 3, 71–93, doi:10.1006/asle.2002.0051, 2002.

Błaszczyk, M., Jania, J. A., and Hagen J. O.: Tidewater glaciers of Svalbard: Recent changes and estimates of calving fluxes, Pol. Polar Res., 30, 85–142, 2009.

Brun, E.: Investigation on wet-snow metamorphism in respect of liquid-water content, Ann. Glaciol., 13, 22–26, 1989.

Claremar, B., Obleitner, F., Reijmer, C., Pohjola, V., Waxegård, A., Karner, F., and Rutgersson, A.: Applying a Mesoscale Atmospheric Model to Svalbard Glaciers, Adv. Meteorol., 2012, 321649, doi:10.1155/2012/321649, 2012.

Day, J. J., Bamber, J. L., Valdes, P. J., and Kohler, J.: The impact of a seasonally ice free Arctic Ocean on the temperature, precipitation and surface mass balance of Svalbard, The Cryosphere, 6, 35–50, doi:10.5194/tc-6-35-2012, 2012.

Dee, D. P., Uppala, S. M., Simmons, A. J., Berrisford, P., Poli, P., Kobayashi, S., Andrae, U., Balmaseda, M. A., Balsamo, G., Bauer, P., Bechtold, P., Beljaars, A. C. M., van de Berg, L., Bidlot, J., Bormann, N., Delsol, C., Dragani, R., Fuentes, M., Geer, A. J., Haimberger, L., Healy, S. B., Hersbach, H., Hólm, E. V., Isaksen, L., Kållberg, P., Köhler, M., Matricardi, M., McNally, A. P., Monge-Sanz, B. M., Morcrette, J.-J., Park, B.-K., Peubey, C., de Rosnay, P., Tavolato, C., Thépaut, J.-N., and Vitart, F.: The ERA-interim reanalysis: configuration and performance of the data assimilation system, Q. J. Roy. Meteor. Soc., 137, 553–597, doi:10.1002/qj.828, 2011.

de Ridder, K. and Gallée, H.: Land surface–induced regional climate change in Southern Israel, J. Appl. Meteorol., 37, 1470–1485, 1998.

de Woul, M. and Hock, R.: Static mass-balance sensitivity of Arctic glaciers and ice caps using a degree-day approach, Ann. Glaciol., 42, 217–224, 2005.

Fettweis, X., Hanna, E., Lang, C., Belleflamme, A., Erpicum, M., and Gallée, H.: *Brief communication* "Important role of the mid-tropospheric atmospheric circulation in the recent surface melt increase over the Greenland ice sheet", The Cryosphere, 7, 241–248, doi:10.5194/tc-7-241-2013, 2013a.

Fettweis, X., Franco, B., Tedesco, M., van Angelen, J. H., Lenaerts, J. T. M., van den Broeke, M. R., and Gallée, H.: Estimating the Greenland ice sheet surface mass balance contribution to future sea level rise using the regional atmospheric climate model MAR, The Cryosphere, 7, 469–489, doi:10.5194/tc-7-469-2013, 2013b.

Førland, E. J., Benestad, R. E., Flatøy, F., Hanssen-Bauer, I., Haugen, J. E., Isaksen, K., Sorteberg, A., and Ådlandsvik, B.: Climate development in North Norway and the Svalbard region during 1900–2100, Norwegian Polar Institute Report 128, Tromsø, Norway, 2009.

Førland, E. J., Benestad, R., Hanssen-Bauer, I., Haugen, J. E., and Skaugen, T. E.: Temperature and precipitation development at Svalbard 1900–2100, Adv. Meteorol., 2011, 893790, doi:10.1155/2011/893790, 2011.

Franco, B., Fettweis, X., and Erpicum, M.: Future projections of the Greenland ice sheet energy balance driving the surface melt, The Cryosphere, 7, 1–18, doi:10.5194/tc-7-1-2013, 2013.

Gallée, H. and Schayes, G.: Development of a three-dimensional meso-γ primitive equation model: katabatic winds simulation in the area of Terra Nova Bay, Antarctica, Mon. Weather Rev., 122, 671–685, 1994.

Gallée, H., Guyomarc'h, G., and Brun, E.: Impact of snow drift on the Antarctic ice sheet surface mass balance: possible sensitivity to snow-surface properties, Bound.-Lay. Meteorol., 99, 1–19, 2001.

Gardner, A. S., Moholdt, G., Cogley, J. G., Wouters, B., Arendt, A. A., Wahr, J., Berthier, E., Hock, R., Pfeffer, W. T., Kaser, G., Ligtenberg, S. R.M, Bolch, T., Sharp, M. J., Hagen, J. O., van den Broeke, M. R., and Paul, F.: A reconciled estimate of glacier contributions to sea level rise: 2003 to 2009, Science, 340, 852–857, 2013.

Hagen, J. O., Melvold, K., Pinglot, F., and Dowdeswell, J. A.: On the net mass balance of the glaciers and ice caps in Svalbard, Norwegian Arctic, Arct. Antarct. Alp. Res., 35, 264–270, 2003.

Hisdal, V.: Geography of Svalbard, 1 Edn., Norwegian Polar Research Institute, Oslo, Norway, 1976.

IPCC, Stocker, T. F., Qin, D., Plattner, G.-K., Tignor, M., Allen, S. K., Boschung, J., Nauels, A., Xia, Y., Bex, V., and Midgley, P. M. (Eds.): Climate Change 2013: The Physical Science Basis. Contribution of Working Group I to the Fifth Assessment Report of the Intergovernmental Panel on Climate Change, Cambridge University Press, Cambridge, UK and New York, NY, USA, 1535 pp., 2013.

Kaser, G., Cogley, J. G., Dyurgerov, M. B., Meier, M. F., and Ohmura, A.: Mass balance of glaciers and ice caps: consensus estimates for 1961–2004, Geophys. Res. Lett., 33, L19501, doi:10.1029/2006GL027511, 2006.

Lang, C., Fettweis, X., and Erpicum, M.: Future projections of the Svalbard climate and surface mass balance with the regional climate model MAR, The Cryosphere Discuss., 9, 115–140, doi:10.5194/tcd-9-115-2015, 2015.

Lefauconnier, B. and Hagen, J. O.: Glaciers and climate in Svalbard: Statistical analysis and reconstruction of the Brøggerbreen mass balance for the last 77 years, Ann. Glaciol., 14, 148–152, 1990.

Lenaerts, J. T. M., van Angelen, J. H., van den Broeke, M. R., Gardner, A. S., Wouters, B., and van Meijgaard, E.: Irreversible mass loss of Canadian Arctic Archipelago glaciers, Geophys. Res. Lett., 40, 870–874, doi:10.1002/grl.50214, 2013.

Liestøl, O.: Glaciers of Svalbard, Norway, US Geological Surv. Prof. Pap., 1386, E127–E151, 1993.

Mémin, A., Rogister, Y., Hinderer, J., Omang, O. C., and Luck, B.: Secular gravity variation at Svalbard (Norway) from ground observations and GRACE satellite data, Geophys. J. Int., 184, 1119–1130, doi:10.1111/j.1365-246X.2010.04922.x, 2011.

Moholdt, G., Nuth, C., Hagen, J. O., and Kohler, J.: Recent elevation changes of Svalbard glaciers derived from ICESat laser altimetry, Remote Sens. Environ., 114, 2756–2767, 2010.

Noël, B., Fettweis, X., van de Berg, W. J., van den Broeke, M. R., and Erpicum, M.: Sensitivity of Greenland Ice Sheet surface mass balance to perturbations in sea surface temperature and sea ice cover: a study with the regional climate model MAR, The Cryosphere, 8, 1871–1883, doi:10.5194/tc-8-1871-2014, 2014.

Nuth, C., Moholdt, G., Kohler, J., Hagen, J. O., and Kääb, A.: Svalbard glacier elevation changes and contribution to sea level rise, J. Geophys. Res., 115, F01008, doi:10.1029/2008JF001223, 2010.

Nuth, C., Schuler, T. V., Kohler, J., Altena, B., and Hagen, J. O.: Estimating the long-term calving flux of Kronebreen, Svalbard, from geodetic elevation changes and mass-balance modelling, J. Glaciol., 58, 119–133, doi:10.3189/2012JoG11J036, 2012.

Nuth, C., Kohler, J., König, M., von Deschwanden, A., Hagen, J. O., Kääb, A., Moholdt, G., and Pettersson, R.: Decadal changes from a multi-temporal glacier inventory of Svalbard, The Cryosphere, 7, 1603–1621, doi:10.5194/tc-7-1603-2013, 2013.

Pinglot, J. F., Pourchet, M., Lefauconnier, B., Hagen, J. O., Isaksson, E., Vaikmäe, R., and Kamiyama, K.: Accumulation in Svalbard glaciers deduced from ice cores with nuclear tests and Chernobyl reference layers, Polar Res., 18, 315–321, 1999.

Pinglot, J. F., Hagen, J. O., Melvold, K., and Eiken, T., and Vincent, C.: A mean net accumulation pattern derived from radioactive layers and radar soundings on Austfonna, Nordaustlandet, Svalbard, J. Glaciol., 47, 555–566, 2001.

Rasmussen, L. A. and Kohler, J.: Mass balance of three Svalbard glaciers reconstructed back to 1948, Polar Res., 26, 168–174, doi:10.1111/j.1751-8369.2007.00023.x, 2007.

Rye, C. J., Arnold, N. S., Willis, I. C., and Kohler, J.: Modeling the surface mass balance of a high Arctic glacier using the ERA-40 reanalysis, J. Geophys. Res.-Earth, 115, F02014, doi:10.1029/2009JF001364, 2010.

Rye, C. J., Willis, I. C., Arnold, N. S., and Kohler, J.: On the need for automated multiobjective optimization and uncertainty estimation of glacier mass balance models, J. Geophys. Res.-Earth, 117, F02005, doi:10.1029/2011JF002184, 2012.

Sakamoto, T. T., Komuro, Y., Nishimura, T., Ishii, M., Tatebe, H., Shiogama, H., Hasegawa, A., Toyoda, T., Mori, M., Suzuki, T., Imada, Y., Nozawa, T., Takata, K., Mochizuki, T., Ogochi, K., Emori, S., Hasumi, H., and Kimoto, M.: MIROC4h – A new high-resolution atmosphere-ocean coupled general circulation model, J. Meteorol. Soc. Jpn., 90, 325–359, doi:10.2151/jmsj.2012-301, 2012.

Schuler, T. V., Loe, E., Taurisiano, A., Eiken, T., Hagen, J. O., and Kohler, J.: Calibrating a surface mass-balance model for Austfonna ice cap, Svalbard, Ann. Glaciol., 46, 241–248, 2007.

Serreze, M. C., Barrett, A. P., Stroeve, J. C., Kindig, D. N., and Holland, M. M.: The emergence of surface-based Arctic amplification, The Cryosphere, 3, 11–19, doi:10.5194/tc-3-11-2009, 2009.

Shepherd, A., Ivins, E. R., A, G., Barletta, V. R., Bentley, M. J., Bettadpur, S., Briggs, K. H., Bromwich, D. H., Forsberg, R., Galin, N., Horwath, M., Jacobs, S., Joughin, I., King, M. A., Lenaerts, J. T. M., Li, J., Ligtenberg, S. R. M., Luckman, A., Luthcke, S. B., McMillan, M., Meister, R., Milne, G., Mouginot, J., Muir, A., Nicolas, J. P., Paden, J., Payne, A. J., Pritchard, H., Rignot, E., Rott, H., Sørensen, L. S., Scambos, T. A., Scheuchl, B., Schrama, E. J. O., Smith, B., Sundal, A. V., van Angelen, J. H., van de Berg, W. J., van den Broeke, M. R., Vaughan, D. G., Velicogna, I., Wahr, J., Whitehouse, P. L., Wingham, D. J.,Yi, D., Young, D., and Zwally, H. J.: A reconciled estimate of of ice-sheet mass balance, Science, 338, 1183–1189, 2012.

van Pelt, W. J. J., Oerlemans, J., Reijmer, C. H., Pohjola, V. A., Pettersson, R., and van Angelen, J. H.: Simulating melt, runoff and refreezing on Nordenskiöldbreen, Svalbard, using a coupled snow and energy balance model, The Cryosphere, 6, 641–659, doi:10.5194/tc-6-641-2012, 2012.

Watanabe, M., Suzuki, T., O'ishi, R., Komuro, Y., Watanabe, S., Emori, S., Takemura, T., Chikira, M., Ogura, T., Sekiguchi, M., Takata, K., Yamazaki, D., Yokohata, T., Nozawa, T., Hasumi, H., Tatebe, H., and Kimoto, M.: Improved climate simulation by MIROC5: mean states, variability, and climate sensitivity, J. Climate, 23, 6312–6335, 2010.

Wilson, A. B., Bromwich, D. H., and Hines, K. M.: Evaluation of Polar WRF forecasts on the Arctic System Reanalysis domain: Surface and upper air analysis, J. Geophys. Res., 116, D11112, doi:10.1029/2010JD015013, 2011.

Winther, J.-G., Bruland, O., Sand, K., Killingtveit, Å., and Marechal, D.: Snow accumulation distribution on Spitsbergen, Svalbard, in 1997, Polar Res., 17, 155–164, 1998.

Wouters, B., Chambers, D., and Schrama, E. J. O.: GRACE observes small-scale mass loss in Greenland, Geophys. Res. Lett., 35, L20501, doi:10.1029/2008GL034816, 2008.

9

Sea-ice extent and its trend provide limited metrics of model performance

D. Notz

Max Planck Institute for Meteorology, Hamburg, Germany

Correspondence to: D. Notz (dirk.notz@mpimet.mpg.de)

Abstract. We examine how the evaluation of modelled sea-ice coverage against reality is affected by uncertainties in the retrieval of sea-ice coverage from satellite, by the usage of sea-ice extent to overcome these uncertainties, and by internal variability. We find that for Arctic summer sea ice, model biases in sea-ice extent can be qualitatively different from biases in sea-ice area. This is because about half of the CMIP5 models and satellite retrievals based on the Bootstrap and the ASI algorithm show a compact ice cover in summer with large areas of high-concentration sea ice, while the other half of the CMIP5 models and satellite retrievals based on the NASA Team algorithm show a loose ice cover. For the Arctic winter sea-ice cover, differences in grid geometry can cause synthetic biases in sea-ice extent that are larger than the observational uncertainty. Comparing the uncertainty arising directly from the satellite retrievals with those that arise from internal variability, we find that the latter by far dominates the uncertainty estimate for trends in sea-ice extent and area: most of the differences between modelled and observed trends can simply be explained by internal variability. For absolute sea-ice area and sea-ice extent, however, internal variability cannot explain the difference between model and observations for about half the CMIP5 models that we analyse here. All models that we examined have regional biases, as expressed by the root-mean-square error in concentration, that are larger than the differences between individual satellite algorithms.

1 Introduction

The evaluation of climate-model simulations against reality is important both to build confidence in future projections from these models and to understand and improve their possible shortcomings. For a useful evaluation, two quantities must be known: first, the real evolution of the variable that is to be evaluated, and second, the degree to which one can expect agreement between simulation and reality in light of the internal variability of the climate system. In this contribution we examine how the evaluation of modelled sea-ice coverage is affected by the incomplete knowledge of both quantities and by the standard approach that is taken to overcome this incomplete knowledge.

The incomplete knowledge of the actual state of the sea-ice cover arises primarily from the difficulty in accurately determining sea-ice concentration from space, which is the only feasible method to obtain daily global data. Because of wide-spread cloud coverage, most often the passive microwave signature of the ocean surface as retrieved from satellites is used to derive the most likely sea-ice concentration in a specific area. This passive microwave signature is, however, strongly affected by meltwater at the ice surface and also by surface temperature, wind speed, humidity and other atmospheric properties. Because of these influencing factors, different retrieval algorithms result in different estimates of sea-ice concentration in a particular area (see, for example, Comiso et al., 1997; Kwok, 2002; Meier, 2005; Andersen et al., 2007).

These long-known discrepancies between sea-ice concentration estimates from different algorithms imply some uncertainty in our knowledge of the "true" sea-ice coverage. To still allow for the comparison of model simulations against a reliable "true" state of sea-ice coverage, most studies that aim at evaluating multiple models against reality have resorted to using a quantity called sea-ice extent that differs only minimally between the various algorithms. This quantity measures the total area of the ocean surface in which significant amounts of sea ice exist. To calculate sea-ice extent in gridded data, one usually adds the area of all grid cells with an ice concentration of more than 15 %. Hence, sea-ice extent in a certain area would be the same for an algorithm that sees a sea-ice concentration of, for example, 40 % and for an algorithm that sees a sea-ice concentration of 60 %. While sea-ice extent was initially only used to assess the observed long-term evolution of the sea-ice cover (e.g. Zwally et al., 1983; Parkinson et al., 1987), it has now become common practice to use sea-ice extent also as the primary (and often sole) variable to assess the quality of modelled sea-ice coverage (e.g. Stroeve et al., 2007, 2012; Massonnet et al., 2012).

Sea-ice extent is always larger than the more direct integrative measure sea-ice area, which is simply the total area of the sea-ice cover and as such a much more direct measure of ice coverage. Important physical quantities such as Arctic-wide average albedo, open-water fraction and thus ocean–atmosphere heat exchange depend therefore much more directly on sea-ice area than on the non-linear measure sea-ice extent. This was already acknowledged by early works on satellite remote sensing (cf. Zwally et al., 1983). The focus on sea-ice extent is, as described, nevertheless understandable since this parameter can be more reliably observed from ships, airplanes and satellites than sea-ice area. This then allows both for a better assessment of the long-term (including pre-satellite) evolution of the ice cover and reduces the uncertainty of the observational data against which model simulations are compared.

This reduction in uncertainty in the observational data comes, however, at a price, in that sea-ice extent can give misleading results regarding model quality. Consider the trivial, fictitious observed sea-ice cover in three grid cells shown in Fig. 1a. Compared to these observations, a model could simulate a smaller sea-ice area that nevertheless results in a larger sea-ice extent because of a slight shift in the location or the spatial distribution of the sea-ice cover (Fig. 1b). A model could also simulate a larger sea-ice area with a smaller sea-ice extent (Fig. 1c). Hence, small shifts in the location of the modelled sea-ice pack, in particular in the marginal ice zone with its strong gradients in sea-ice concentration, can result in misleading results regarding the actual bias in modelled sea-ice cover.

In addition to these grid-independent issues, there is also a grid-dependent issue related to the usage of sea-ice extent vs. sea-ice area. Generally, higher grid resolution causes a lower sea-ice extent. At very high resolution, sea-ice extent converges to the same value as sea-ice area, since then almost all grid cells will either be fully ice covered or fully ice free.

We became aware of these issues when we analysed results from the Max Planck Institute for Meteorology Earth System

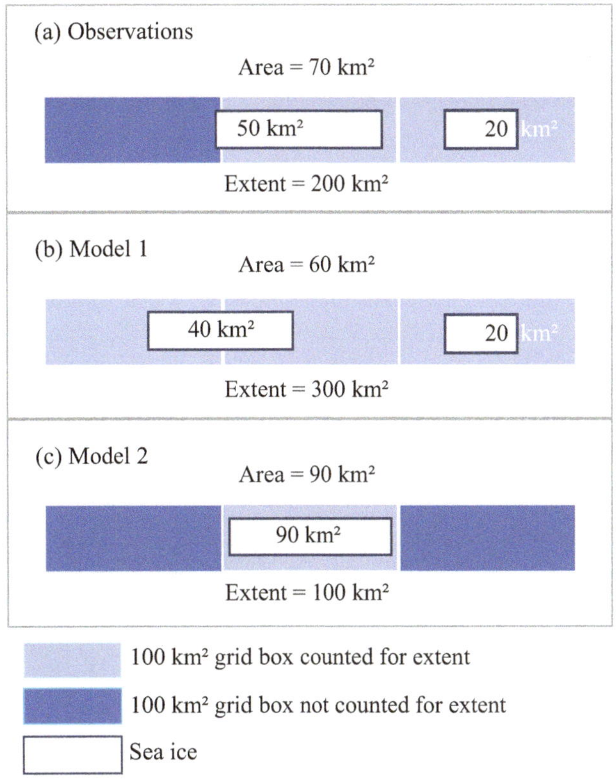

Fig. 1. A fictitious example to illustrate the possible non-intuitive relationship between sea-ice area and sea-ice extent. **(a)** In the observations, the ice pack is distributed such that two grid cells are covered by more than 15 % ice. **(b)** In a fictitious model simulation, less sea ice than in the observations is distributed such that three grid cells are covered by more than 15 % ice. **(c)** In a fictitious model simulation, more sea ice than in the observations is distributed such that only one grid cell is covered by more than 15 % ice.

Model MPI-ESM: compared to observations, this model has about 6 % too small a September Arctic sea-ice extent, but 20 % too small a sea-ice area (Notz et al., 2013). In contrast, this model's predecessor ECHAM5/MPIOM had about 20 % too large a September Arctic sea-ice extent, but only about 7 % too large a sea-ice area. This gave rise to the question of whether too strong a focus on sea-ice extent can give misleading results regarding the quality of modelled sea-ice coverage, and which implications this has for quantitative model evaluation.

In this contribution, we examine these questions by analysing output from models that have contributed to the Coupled Model Intercomparison Project, phase 5 (CMIP5). Our aim is to give the reader a quantitative assessment, and an explanation, for the different outcomes in model–data comparison studies based on sea-ice extent vs. sea-ice area and their trends. Because positive and negative regional biases cancel each other out in the calculation of either sea-ice extent or sea-ice area, we additionally analyse these measures' relationship to the mean absolute bias in sea-ice concentration, which avoids such cancellation of errors. We also touch upon the issue of local biases in sea-ice concentration, which are relevant for a more detailed analysis of model quality. Our aim is to allow the reader an informed assessment of which parameter to use for a specific purpose and how to handle the related observational uncertainty. In particular, we put our findings into the context of uncertainty that arises because of the internal variability of the Arctic climate system.

The satellite products and the model data that we use are introduced in Sect. 2. In Sect. 3.1, we analyse the compactness of the modelled and satellite-retrieved sea-ice cover, which is important to understand the analysis of the different biases in sea-ice extent, area, and in their trends, discussed in Sect. 3.2. In Sect. 3.3, we examine the impact of grid resolution, followed by an analysis of cancelling negative and positive biases in Sect. 3.4. Section 3.5 then contains an analysis of the impact of internal variability on the assessment of model quality. In Sect. 3.6 we briefly touch upon some issues related to the non-linearity of sea-ice extent. We discuss the implications of these findings for model-evaluation purposes in Sect. 4. Our main findings are then summarised in Sect. 5.

2 Models and data

For our analysis, we focus on the period 1979–2005, which is the overlapping period of the most-widely used satellite records of sea-ice coverage and the "historical" simulations of the CMIP5 protocol (Taylor et al., 2012). These historical simulations are forced by the observed evolution of greenhouse gases, solar radiation, etc. For all 117 historical simulations from 26 different models that we consider here, time series of monthly mean sea-ice extent and area are calculated from their monthly mean sea-ice concentration fields. The sea-ice extent is calculated as the total area of all grid cells with at least 15 % sea-ice concentration. For sea-ice area, the area of all grid cells is multiplied by their sea-ice concentration and then added. For sea-ice area and extent, linear trends are calculated as a least-squares fit to the time series. Ensemble-mean and multi-model mean time series of sea-ice extent and sea-ice area are calculated as the ensemble-mean and the multi-model mean of the individual simulations' time-series of these two parameters, and not from the ensemble-mean or multi-model mean concentration fields (compare Sect. 3.6).

The model results are compared against satellite retrievals of sea-ice concentration. As described in the introduction, different algorithms result in different estimates of sea-ice concentration because they are based on different transfer functions to derive sea-ice concentration from the measured passive-microwave signature. These differences are best documented (e.g. Comiso et al., 1997; Kwok, 2002) for the two satellite algorithms for sea-ice concentration that are most widely used for model-data intercomparison studies: the Bootstrap algorithm (Comiso, 1986) and the NASA Team algorithm (Cavalieri et al., 1984) that forms the basis for the NSIDC Sea-Ice Index (Fetterer et al., 2002, updated 2012). Both provide sea-ice concentration data from 1979 onwards, and will be used throughout this study to exemplify the shortcomings of a direct comparison of modelled sea-ice extent to one particular satellite algorithm. Additionally, we consider retrievals based on the ASI algorithm (Kaleschke et al., 2001; Spreen et al., 2008), which provides sea-ice concentration based on SSM/I data from 1991 onwards and based on AMSR-E data from 2002 onwards.

Sea-ice concentration retrieved through the Bootstrap algorithm is, especially in summertime, probably closer to the real sea-ice concentration than that from the NASA Team algorithm, because the latter has been found to be biased low compared to independent observations (e.g., Agnew and Howell, 2003; Partington et al., 2003). The Bootstrap algorithm, in contrast, results in estimates of sea-ice concentration that are very close to the "Climate Data Record of Passive Microwave Sea Ice Concentration" (CDR, Meier et al., 2011) that is a merged product of different algorithms with the aim to provide a consistent time series of sea-ice concentration. In summer, estimates of sea-ice area of the Bootstrap algorithm also agree favourably with estimates based on the ASI algorithm from SSMI satellite data and the higher resolved AMSR-E satellite data, while estimates of sea-ice area based on the NASA Team algorithm are significantly lower (Fig. 2). Since all passive-microwave algorithms will see surface melt ponds as open water, their estimates of sea-ice concentration in summer have been found to be lower than independent observations. Comiso and Nishio (2008) have therefore suggested to synthetically increase sea-ice extent by a 25 km-wide margin during the melt season. In line with existing model–satellite intercomparison studies we will not take such a measure for our model-satellite intercomparison in Sect. 3. We will, however, return to the issue of the low-bias in satellite retrievals in Sect. 4. There we will also discuss in more detail the greater uncertainty of the retrieved sea-ice area and the differences between the various algorithms shown in Fig. 2.

3 Results

3.1 The frequency distribution of sea-ice concentration

The Bootstrap and the NASA Team algorithms result in similar estimates of mean September Arctic sea-ice extent for the period 1979–2005, namely 7.3 million km^2 for the Bootstrap algorithm and 6.9 million km^2 for the NASA Team algorithm. It is usually assumed that this difference is small enough to allow for the meaningful, quantitative comparison of modelled sea-ice extent against the estimated extent from an individual satellite retrieval. The difference in September mean sea-ice area for the same period is much larger,

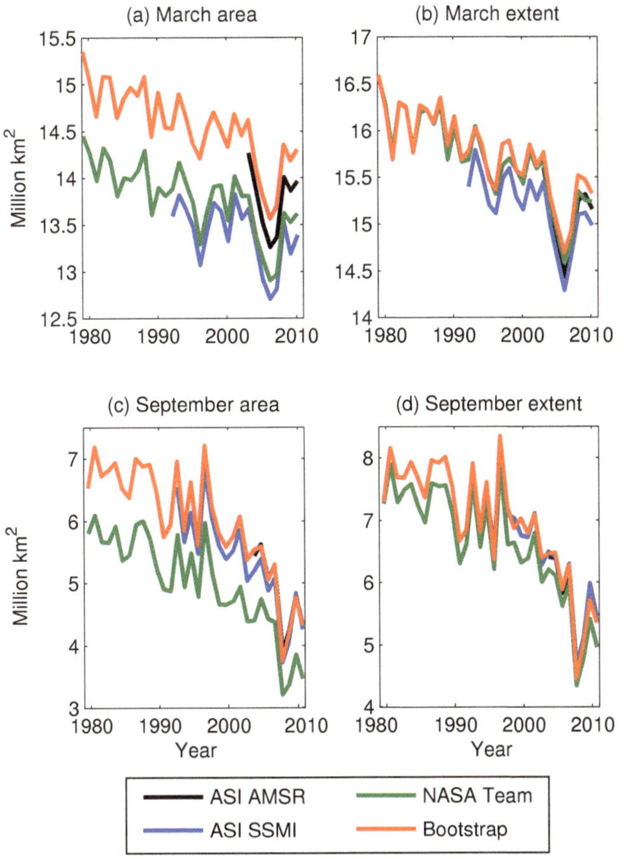

Fig. 2. September and March sea-ice area and sea-ice extent as retrieved from satellite for the period 1979–2010. Different colors denote different algorithms or satellites. Area and extent were calculated based on sea-ice concentration fields on EASE grids with 25 km resolution (NASA Team and Bootstrap, based on SMMR and SMM/I, 1979–2010), 12 km resolution (ASI SSM/I, 1992–2010) and 6 km resolution (ASI AMSR-E, 2002–2010).

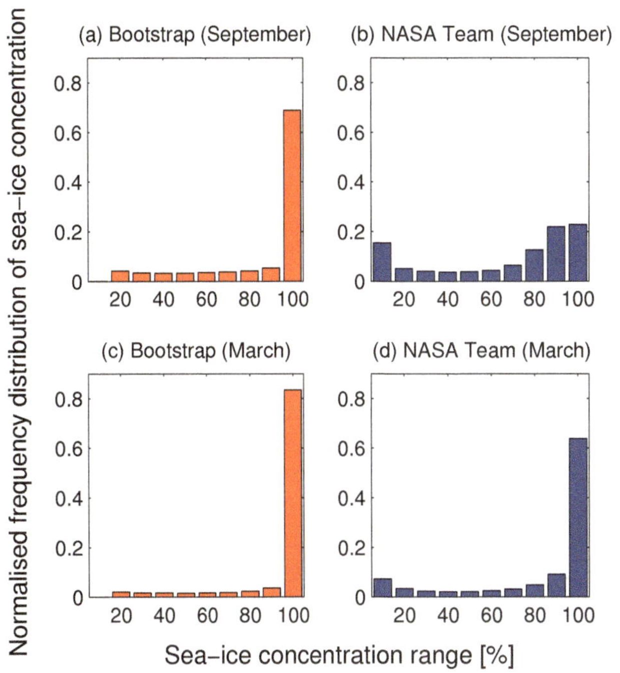

Fig. 3. Histogram of 1979–2005 sea-ice concentration in all areas with at least 0.1 % sea-ice concentration. **(a, c)** Satellite retrievals based on the Bootstrap algorithm and **(b, d)** satellite retrievals based on the NASA Team algorithm for **(a, b)** September and **(c, d)** March. The numbers on the x axis denote the upper limit of each bar: e.g. 20 denotes the concentration range 10 to 20 %.

with 6.3 million km^2 for the Bootstrap algorithm compared to only 5.2 million km^2 for the NASA Team algorithm. This much larger difference is the main reason why the sea-ice area estimate of an individual satellite retrieval is usually not used for model-evaluation purposes. Such large relative difference arises, however, only in summer: in March, both the estimates of sea-ice area and of sea-ice extent are similar between the two algorithms, as the mean 1979–2005 sea-ice extent is 15.9 million km^2 for Bootstrap and 15.8 million km^2 for NASA Team, while sea-ice area is 14.6 million km^2 for Bootstrap and 13.9 million km^2 for NASA Team.

Since our focus here is on sea-ice extent vs. sea-ice area, it is important to understand the cause for the different agreement between these two measures for the satellite algorithms. For this purpose, we consider the frequency distribution of sea-ice concentration that is displayed by the two algorithms. Of particular importance for the estimate of sea-ice area is the amount of ice-covered grid cells that have a very high ice concentration. According to the Bootstrap algorithm the ice cover is very compact in summer, with about 70 % of all ice-covered grid cells having more than 90 % ice concentration (Fig. 3a). In contrast, according to the NASA Team algorithm the ice cover is quite loose, with only about 20 % of all ice-covered grid cells having such high ice concentration in summer (Fig. 3b). This difference comes primarily about by the different treatment of sea ice that is covered by surface meltwater (Meier and Notz, 2010; L. T. Pedersen, personal communication, 2013): while both algorithms interpret the meltwater-covered sea ice as open water, the Bootstrap algorithm more strongly compensates for this well-known bias compared to the NASA Team algorithm. The two versions of the ASI algorithm that were analysed for the present study show a similarly compact ice cover as the Bootstrap algorithm.

The large difference between the NASA Team and the Bootstrap algorithms in the estimated frequency of high sea-ice concentration causes their large difference in estimated sea-ice area. In wintertime, the estimated frequency of high sea-ice concentration is much more similar for the two algorithms (Fig. 3c, d), which explains the smaller difference of estimated sea-ice area for that season. Differences in estimated sea-ice extent come about by different estimates of the frequency of low sea-ice concentration. Since at this end of the spectrum differences between the two algorithms are

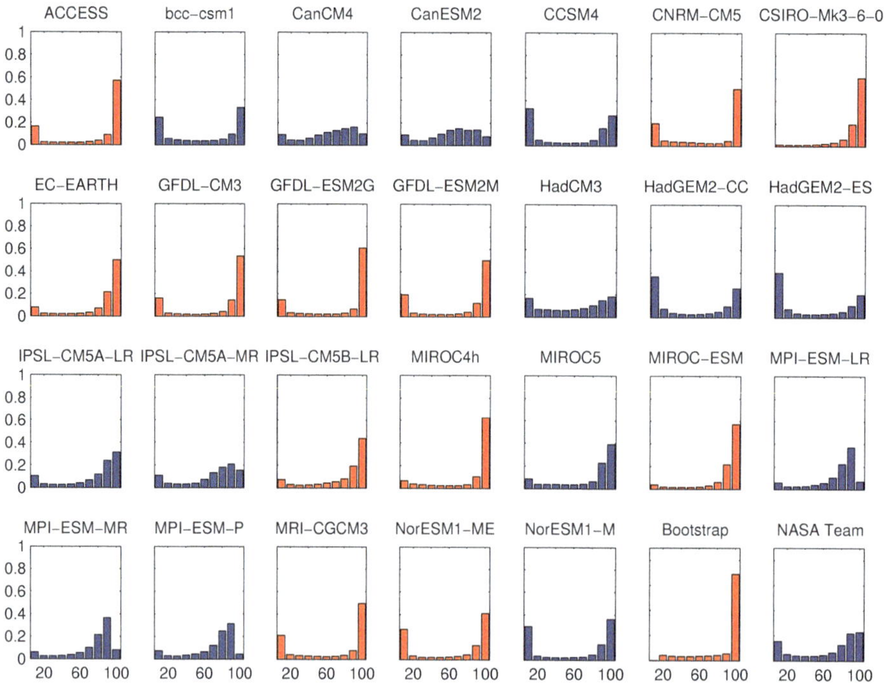

Fig. 4. Histogram of 1979–2005 September sea-ice concentration in all grid cells with at least 0.1 % sea-ice concentration in CMIP5 model simulations. The numbers on the x axis denote the upper limit of each bar: e.g. 20 denotes the concentration range 10 to 20 %. Red panels denote histograms with a compact sea-ice cover, while blue panels denote histograms with a loose sea-ice cover. For models with multiple simulations, the ensemble mean is shown.

small both in summer and winter, both algorithms result in similar estimates of sea-ice extent.

Examining the frequency distribution of summer sea-ice concentration in the CMIP5 model simulations, we find that these simulations can be divided into two groups. One group simulates a compact ice cover in summer (red panels in Fig. 4), while the other group simulates a loose ice cover (blue panels in Fig. 4). In winter, all models simulate a compact ice cover (not shown). Somewhat arbitrarily, we chose a normalised frequency of 0.4 for the 90...100 % concentration band as the dividing line between simulations with a compact ice cover and simulations with a loose ice cover. An alternative definition could be based on the ratio of the amount of sea ice in the highest concentration and that in the second-highest class. Depending on the demarcation line for this ratio, this would slightly modify the composition of the two classes without qualitatively affecting the results discussed in the following.

It would be interesting to examine why roughly half of the CMIP5 models produce a compact ice cover while the other half does not, in particular since this might allow further insights into the quality of the satellite retrievals. Some initial analyses point towards the relative distribution of melting between lateral melt and thinning in individual models to play some role, but a conclusive analysis of this question is beyond the scope of this paper.

What is important, however, is to reiterate the fact that the occurrence of a compact vs. a loose ice cover has different implications in the models compared to the satellite retrievals: in the models, this terminology does indeed refer to the actual simulated state of the ice cover. In the satellite retrievals, however, this differentiation is above all a reflection of the different treatment of surface meltwater by the different algorithms. An algorithm that interprets more of that meltwater as ice free will necessarily result in an "observed" loose ice cover, though this then has little to say about the real properties of the ice pack.

3.2 Extent vs. area

By nature of the definition of sea-ice extent, differences between sea-ice extent and sea-ice area are comparably small for compact sea ice, because of the large number of grid cells with a very high ice concentration. In contrast, the difference between extent and area is usually much larger for a loose ice cover (see Fig. 5a–c).

This has direct consequences for the analysis of model biases based on these two measures. We find for simulations with a compact sea-ice cover that biases relative to the Bootstrap retrieval are similar for sea-ice area and sea-ice extent (red dots are close to red line in Fig. 6a). In particular, all simulations with a compact sea-ice cover that are within ±10 % of the retrieved sea-ice extent are also within ±10 %

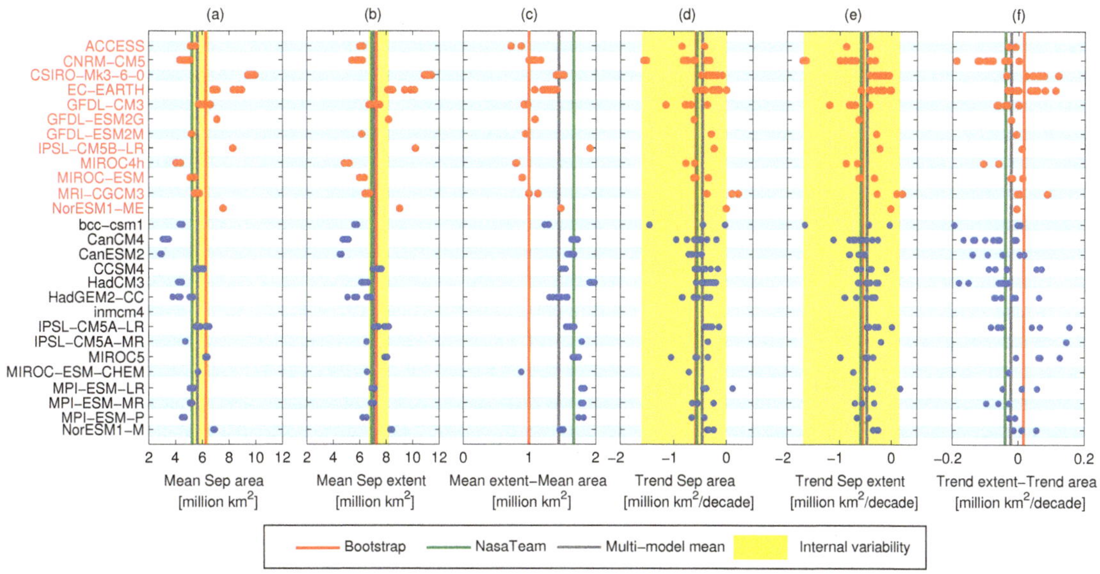

Fig. 5. Overview of the September sea-ice coverage in the 117 historical CMIP5 simulations analysed for this study. Each individual dot corresponds to a single simulation. The vertical lines show the values of the observational record and the mean of all simulations. The yellow shading indicates estimated internal variability. All data refer to the period 1979–2005. **(a)** Mean September sea-ice area, **(b)** mean September sea-ice extent, **(c)** difference between mean September sea-ice extent and mean September sea-ice area, **(d)** linear trend in September sea-ice area, **(e)** linear trend in September sea-ice extent, **(f)** difference between trend in September sea-ice extent and trend in September sea-ice area. Models with a compact ice cover are labelled in red.

of the retrieved sea-ice area. For the simulations with a loose ice cover, we find that those models that underestimate sea-ice extent relative to the Bootstrap retrieval have a stronger percentage bias in sea-ice area than they have in sea-ice extent, while those simulations that overestimate sea-ice extent have a smaller percentage bias in sea-ice area than in extent (Fig. 6a). A number of simulations with a loose ice cover that fall within $\pm 10\,\%$ of the retrieved sea-ice extent are clearly outside the $\pm 10\,\%$ range of the retrieved sea-ice area, and vice versa. Hence, a focus on sea-ice extent can give misleading results regarding model quality compared to a focus on sea-ice area (see also Fig. 5a, b, where a number of simulations with loose ice match the Bootstrap sea-ice extent well but are below Bootstrap sea-ice area).

Relative to the satellite-retrieved estimates based on the NASA Team algorithm, we find that biases for sea-ice extent are similar to biases for sea-ice area for simulations with a loose ice cover (blue dots close to green line in Fig. 6a). Simulations with a compact ice cover that overestimate the mean sea-ice extent compared to the NASA Team algorithm in contrast have a stronger percentage bias in sea-ice area, and vice versa.

For March, all simulations and both satellite retrievals have a compact ice cover. Hence, percentage biases in sea-ice area are for all simulations almost identical to the biases in sea-ice extent (Fig. 6b).

To understand this behaviour of simulations with a compact ice cover vs. those with a loose ice cover, we need to consider that the former have a small difference between sea-ice extent and sea-ice area, while the latter have a larger difference. Figure 7 illustrates how this explains the different behaviour of the two model families: if any of the loose-ice simulations with their comparably large difference between sea-ice extent and sea-ice area results in too small a mean sea-ice extent, this simulations' bias in sea-ice area will be comparably large. If, however, the simulation resulted in too large a sea-ice extent, its bias in sea-ice area would be comparably smaller – simply because the difference between extent and area is larger in the simulations than in the observations. For simulations with a compact ice cover, biases in extent and area relative to the Bootstrap algorithm are very similar, because these simulations' difference between sea-ice extent and sea-ice area is similar to that of the Bootstrap observations. Compared to observations based on the NASA Team algorithm, the simulations with a compact ice cover have generally a lower difference between extent and area, which explains their contrasting behaviour relative to the NASA Team algorithm.

In winter, all simulations result in a compact sea-ice cover. Therefore, in winter they have a difference between sea-ice extent and sea-ice area similar to the Bootstrap observations, which explains the consistent wintertime biases of all model simulations.

Examining trends in sea-ice area and sea-ice extent, we find that the Bootstrap retrieval gives almost the same number for both these measures, namely an average loss of

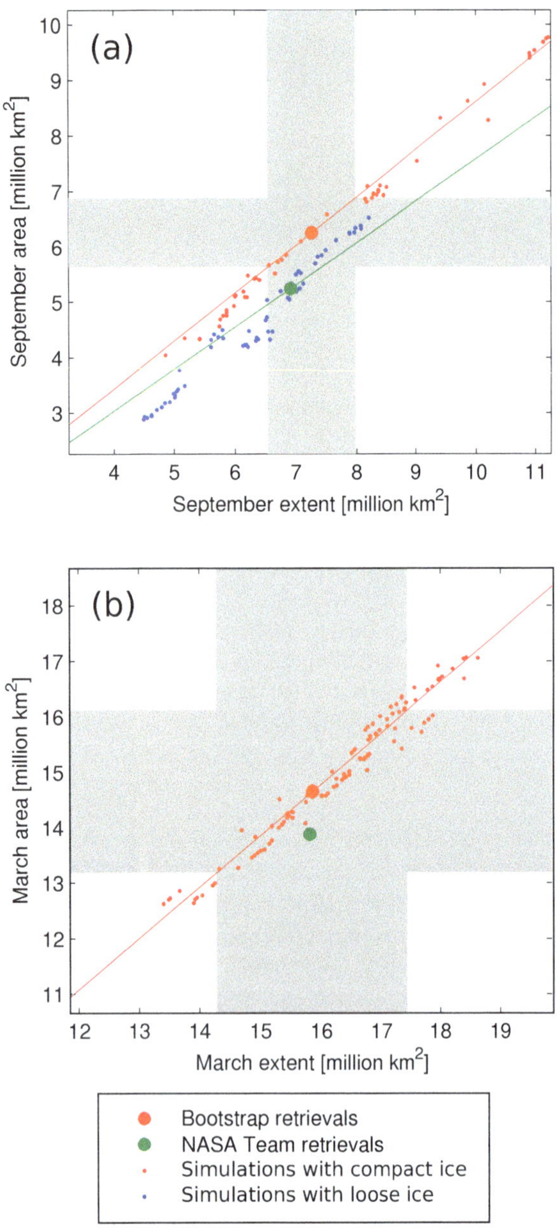

Fig. 6. **(a)** September and **(b)** March sea-ice area vs. sea-ice extent in models and satellite retrievals. The red line connects all value pairs that have the same percentage bias in sea-ice extent and in sea-ice area relative to the Bootstrap retrievals. The gray shading indicates a ±10 % range around the values obtained from the Bootstrap retrievals. Note that in March all simulations have compact ice, which is why there are no blue dots in **(b)**.

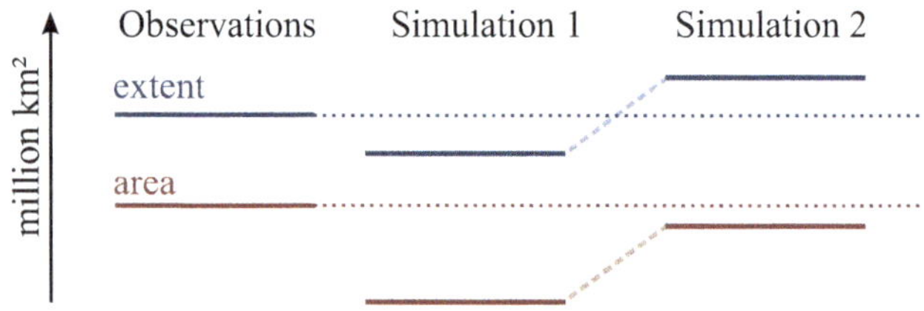

Fig. 7. Schematic to explain the findings in Fig. 6: because of the smaller difference between sea-ice extent and sea-ice area in observations with a compact ice cover than in simulations with a loose ice cover, models with a loose ice cover and slightly too large a simulated sea-ice extent result in a comparably small bias in simulated sea-ice area. The difference between simulated extent and simulated area is the same for both simulations.

0.56 million km^2 per decade in sea-ice extent and a loss of 0.58 million km^2 per decade in sea-ice area during the period 1979–2005. The models, in contrast, show inconsistent behaviour, with both smaller and larger trends in sea-ice area than in extent (Fig. 5d, f). The consistent trends in the satellite retrieval can be understood by analysing the individual trends for different ice-concentration ranges (Fig. 8). Almost all the ice loss in the Bootstrap retrievals happens within the ice-concentration range 90 to 100 %, with no compensating increase in lower ice-concentration ranges (second to last panel in Fig. 8). An ice loss at these high concentrations will have roughly the same impact on sea-ice area and on sea-ice extent. For most models, in contrast, the ice loss is spread over a wider range of sea-ice concentrations. In addition, the grid cells with high ice concentration often only lose some of their ice, which then causes an increase in the number of grid cells with intermediate ice concentration. This compensation then causes a smaller loss of sea-ice extent than of sea-ice area. Some models, however, also show a faster loss in sea-ice extent than in sea-ice area. This behaviour can be understood if a significant amount of grid cells with intermediate sea-ice concentration become ice free in a simulation. The entire area of these grid cells is then lost in terms of sea-ice extent, while only the fraction of these grid cells that was ice-covered is lost from sea-ice area.

The different biases in trends of area and extent in models vs. the satellite retrievals obviously have consequences for the assessment of model quality (Fig. 9a). A number of simulations result in trends that lie within ±20 % of the Bootstrap retrieved trends in sea-ice extent, while they lie outside the ±20 % range for the simulated trends in sea-ice area. In particular, models that have too fast a loss in sea-ice extent compared to Bootstrap retrievals sometimes have too slow a loss in sea-ice area compared to the Bootstrap retrievals. The same holds for the trends in winter sea-ice coverage (Fig. 9b). Hence, again, an assessment of model quality based on an analysis of trends in sea-ice extent can give misleading results.

3.3 Grid resolution

While the different histograms of sea-ice concentration explain most of the findings discussed so far, differences in model grids might also be relevant for different biases in sea-ice extent and in sea-ice area. As discussed in the introduction, one would generally expect a smaller difference

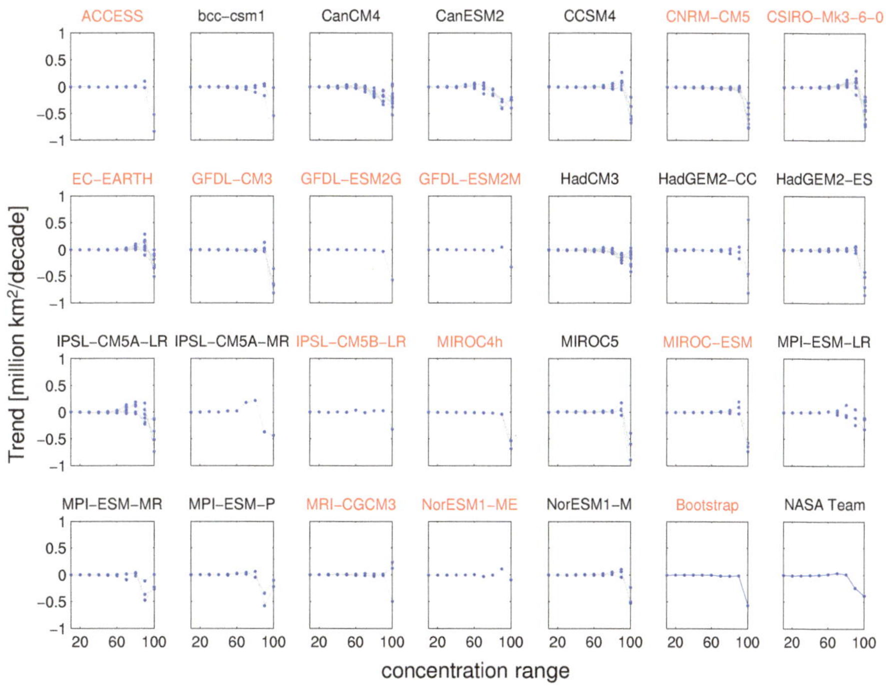

Fig. 8. Trends in September sea-ice area per ice-concentration category. The numbers on the x axis denote the upper limit of each bar: e.g. 20 denotes the concentration range 10 to 20.

between sea-ice extent and sea-ice area for higher grid resolution. The comparably high resolution of the satellite data set might therefore have contributed to the comparably small difference between sea-ice area and sea-ice extent for the Bootstrap algorithm (Fig. 5c).

To examine this possibility, we bilinearly interpolated the gridded Bootstrap-derived sea-ice concentration field for each month of the year 2007 from the original 25 km EASE grid to each individual model grid and then calculated area and extent on the model grids. We find that sea-ice area usually agrees well between the original grid and the individual model grids, with a multi-model mean difference of less than 50 000 km^2 all year round (blue line in Fig. 10). Individual models typically show a mean difference of less than 200 000 km^2 all year round, where the difference compared to the original EASE grid comes primarily about through roundoff errors, which is also exemplified by the fact that both positive and negative differences occur. Sea-ice extent as calculated from the interpolated sea-ice concentration on the lower-resolved model grids, however, is always larger than the one on the original 25 km EASE grid. In particular in winter, the multi-model mean difference reaches more than 800 000 km^2, decreasing to less than 200 000 km^2 around the summer minimum (red line in Fig. 10). For the calculation of sea-ice extent, grid resolution and grid geometry can hence strongly affect the comparison between model simulations and satellite retrievals for the large ice cover that is still typical for wintertime.

3.4 Cancelling biases

So far, we have examined possible misinterpretations that can arise when using sea-ice extent instead of sea-ice area for model-evaluation purposes. However, both measures allow for cancelling biases and hence render a regional assessment of model quality difficult: a model that has a large positive bias in sea-ice concentration in one region and a large negative bias in another region might simulate a better overall sea-ice area than a model that has weak negative biases in both regions. Therefore, an analysis of the mean absolute bias in sea-ice concentration gives a better indication of regional model performance compared to either sea-ice extent or sea-ice area.

We calculated for the period 1979 until 2005 the area-weighted, monthly mean bias and the area-weighted, monthly mean absolute bias in sea-ice concentration in the CMIP5 simulations relative to the Bootstrap retrievals. Doing so, we obviously find very good correlation between the mean percentage bias in the integrative measures extent or area and the mean bias in sea-ice concentration (compare Fig. 11a, b vs. c): for the mean bias in concentration, regional errors cancel in a similar way as they do for extent and area. Therefore, a linear regression of the biases in area vs. biases in mean concentration results in a high value of $R^2 = 0.93$. Because of the non-linearity of sea-ice extent, the linear regression of sea-ice extent on sea-ice concentration gives a slightly lower value of $R^2 = 0.85$. The fact that R^2 is not 1

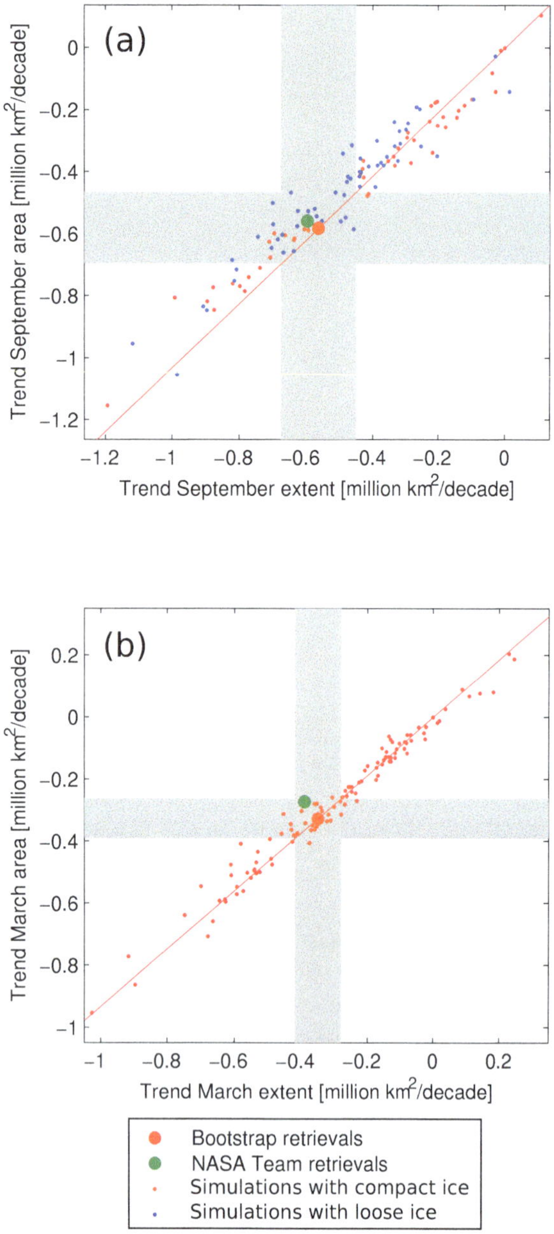

Fig. 9. **(a)** September and **(b)** March trends in sea-ice area vs. trends in sea-ice extent for the period 1979–2005. The red line connects all value pairs that have the same percentage bias in sea-ice extent and in sea-ice area relative to the Bootstrap retrievals. The gray shading indicates a ±20 % range around the trends obtained from the Bootstrap retrievals.

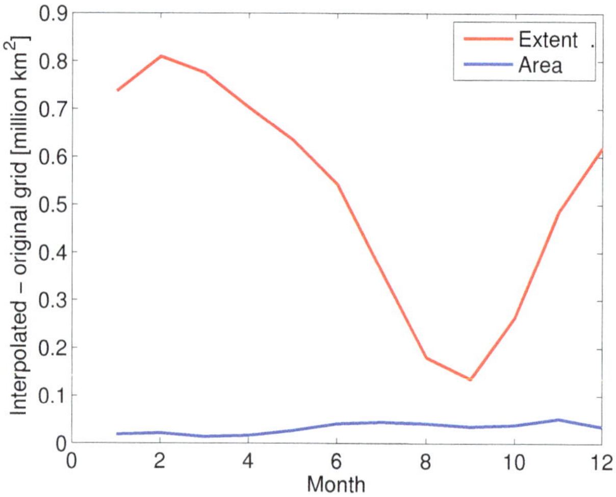

Fig. 10. Mean difference of sea-ice area and sea-ice extent between the original 25 km EASE-Grid and all CMIP5 model grids throughout the year 2007. For each month, the Bootstrap-derived sea-ice concentration was interpolated onto all individual model grids, from which then sea-ice extent and sea-ice area were calculated. The differences of all grids relative to the original EASE-Grid were averaged for this figure.

for the linear regression of area vs. mean concentration is primarily related to interpolation issues during the calculation of mean biases.

For the absolute biases in sea-ice concentration that prevent the cancellation of regional biases, however, correlation with the absolute percentage bias in the integrative measures sea-ice extent and sea-ice area is low, giving $R^2 \approx 0.5$ for both measures: some models with almost no bias in sea-ice extent or area still have comparably large mean absolute concentration biases.

For model-evaluation purposes this suggests that additional insights can be gained by considering not only sea-ice area, but also the root mean square bias of the sea-ice concentration fields. This allows for some estimate of the quality of modelled regional sea-ice distributions, while the integrated measure sea-ice area allows for an estimate of the quality of the overall sea-ice volume that is formed through the convergence and divergence of heat fluxes across the entire Arctic. Both measures would be particularly insightful if the magnitude and timing of their seasonal distribution were assessed.

3.5 Internal variability

In the previous sections, we have shown that the more reliably measurable sea-ice extent can give misleading results regarding model quality compared to the geophysically more meaningful sea-ice area. We will now examine how important these differences are in the light of internal variability. Such internal variability, which captures the chaotic nature of the climate system, often permits a broad range of possible responses in a specific variable to a specified evolution of the external forcing. For example, in a previous study examining the Earth-System Model MPI-ESM, we found in one of its CMIP5 historical simulations an increase in the sea-ice cover in the Arctic during the period 1979–2005, while another historical simulation showed a decrease almost as large as observed (Notz et al., 2013). Both simulations were driven by the same external forcing in greenhouse-gas concentration,

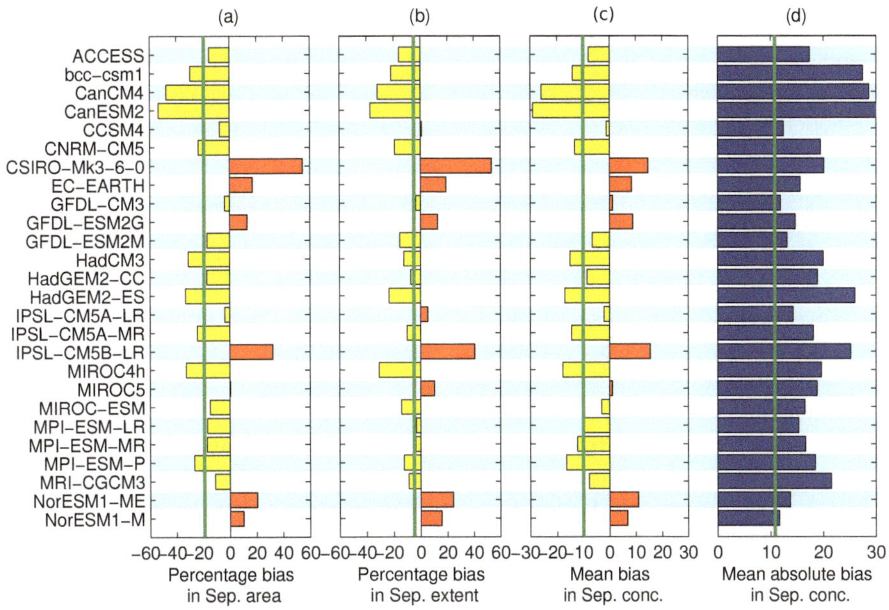

Fig. 11. Overview of biases in September sea-ice coverage in the 117 "historical" CMIP5 simulations analysed for this study relative to Bootstrap satellite retrievals. **(a)** Mean bias in September sea-ice area. **(b)** Mean bias in September sea-ice extent. **(c)** Mean bias in sea-ice concentration. **(d)** Mean absolute bias in sea-ice concentration. The vertical green lines denote the respective bias of the NASA Team retrieval relative to the Bootstrap retrieval.

solar activity, aerosol load etc. The only difference between the simulations were the initial weather conditions on the simulated 1 January 1850, which was the starting date for the simulations. This exemplifies the well-known fact that because of the chaotic nature of the climate system, the trend in the response of the climate system to a given trend in the external forcing can vary drastically on short timescales.

Because of this large internal variability of the Arctic climate system it is often impossible to judge whether a difference in sea-ice coverage between a model simulation and observations is simply random or caused by a model deficiency (cf. Winton, 2011). While there are observational estimates of decadal-scale internal variability of the Arctic sea-ice cover (cf. Notz and Marotzke, 2012), a reasonable range of sea-ice trends for previous decades can obviously not robustly be inferred from observations, since only one single trend has been realised by the real world. It is, however, possible to estimate from ensemble model simulations how much of a modelled trend is caused by the external forcing and how much of it is caused by internal variability, and to then translate these results to the real world (e.g. Kay et al., 2011; Day et al., 2012).

Here, we use a simple, straight-forward method to estimate a reasonable range of observed sea-ice area and sea-ice extent and of their trends from our CMIP5 simulation ensemble: we assume that this reasonable range is given by the maximum spread of ensemble simulations of all those models that encompass the observed evolution within their ensemble members. While this would ideally be done on a model-by-model basis, many models that we examine here do only provide a single ensemble member. We therefore generalise the spread from models that do provide multiple ensemble members to all simulations that we consider here.

Using this approach to examine mean September sea-ice area for the period 1979–2005 (yellow shading in Fig. 5a), we find an up to 1 million km^2 ensemble spread for those models that match the Bootstrap-observed value of 6.3 million km^2 in at least one of their ensemble members. This spread is comparable to the difference between the Bootstrap- and the NASA-Team-derived sea-ice area. Hence, for the period considered here the reasonable range of mean September sea-ice area as derived from model simulations is similar to the uncertainty range of the satellite observations. Simulations that fall outside of this range are most likely incompatible with the observed external forcing. Based on this reasoning, 14 of the 26 models that we analysed have too small a sea-ice area for that period in all their ensemble members, while 5 have too large a sea-ice area in all their ensemble members. The mean of all simulations, 5.6 million km^2, lies within the reasonable range.

For mean September sea-ice extent (Fig. 5b), the spread of individual ensemble members of those models whose simulations encompass the Bootstrap estimate of 7.3 million km^2 is 1.5 million km^2. Hence the model spread is about four times as large as the difference between the Bootstrap estimate and the 6.9 million km^2 estimate for the NASA Team algorithm. Of the 26 models that we analysed, 12 have too small a sea-ice extent for that period in all their ensemble members, while 4 have too large a sea-ice extent in all their ensemble members. The mean of all simulations, 7.1 million km^2, lies again within the reasonable range.

The large number of model simulations that fall outside the reasonable range for both sea-ice extent and sea-ice area indicates that the mean value of these two measures is in principle helpful for model-evaluation purposes, notwithstanding the differences that can arise for individual models for the two measures as discussed above. In contrast, the internal variability of the trends is so large that trends of individual simulations can hardly be used for model-evaluation purposes (Fig. 5d, e): for the period 1979–2005, many models which generated one simulation with a sea-ice trend similar to the observed one simulate for identical forcing and slightly different initial conditions trends that are twice as strongly negative, or trends that are even positive. Hence, any trend that falls within this range might be the consequence of internal variability affecting the modelled trend rather than a model deficiency. Using such criterion, almost all simulations that we consider here show a trend for the period 1979–2005 that is consistent with the observed increase in greenhouse-gas emissions. Since 2005, Arctic sea-ice coverage in summer has decreased rapidly. The trend in September sea-ice coverage for the extended period 1979–2012, however, remains below 1 million km^2 ice loss per decade both for extent and area. As such, the trend remains comfortably within our estimated range of modelled trends modified by internal variability. Hence, also for the extended temporal range until 2012 we cannot positively identify the modelled trends as inconsistent with the applied forcing.

In their evaluation of CMIP5 simulations, Stroeve et al. (2012) estimate a range for the trend in sea-ice extent that is consistent with the observed external forcing by calculating the standard deviation of the observations around the linear trend. Since the observed trend might be extraordinary for the observed forcing, we here instead assume that the reasonable range for the trend is given by the much larger ensemble spread of those models that encompass the observed trend within their ensemble spread. We then take this model ensemble spread to represent the range of possible trends that are consistent with an externally forced trend modified by internal variability over the previous decades. Since we find that almost all simulations that we consider here fall within this wider range of reasonable trends, we conclude that an assessment based on the difference between the observed trend and individual ensemble simulations only allows for very limited insights into model quality.

3.6 Non-linearity

For completing our discussion of the usage of sea-ice extent for model evaluation, we should finally note that for any comparison of modelled mean sea-ice extent with observations, the non-linearity of sea-ice extent must carefully be taken into account. Mean sea-ice extent should normally be calculated as the mean of the sea-ice extents of the individual simulations, and not as the sea-ice extent of the mean concentration of the simulations. Consider, for example, two simulations, one with 0 % ice concentration in a certain region and the other with 35 % ice concentration in that same region. The mean ice concentration of these simulations is larger than 15 %, and the sea-ice extent of the mean of the two simulations will be identical to the sea-ice extent of the simulation with the higher sea-ice concentration. The same issue arises when directly comparing sea-ice extent from daily observations with monthly mean fields of model output: the monthly-mean sea-ice extent as derived from a monthly-mean sea-ice concentration field will usually be larger than the monthly mean of daily estimates of sea-ice extent. Therefore for the purpose of this paper, all daily satellite data sets were averaged to monthly data before calculating sea-ice extent. Since sea-ice area scales linearly with ice coverage, these issues do not apply for any study using sea-ice area as a metric for model quality.

4 Discussion

In the previous section, we have shown that for a number of reasons the sole consideration of sea-ice extent for the evaluation of model quality can give misleading results. We therefore recommend that future studies that aim at evaluating the performance of sea ice move away from the sole consideration of sea-ice extent and also consider the model performance for the more meaningful integrative quantity sea-ice area.

In doing so, differences between different satellite algorithms will play a more prominent role than for sea-ice extent (see Fig. 2). Hence, such comparison will need to take more the form of a comparison of observational data with a specific uncertainty vs. model simulations with a specific internal variability. To quantify the uncertainty of the satellite data, we compared in more detail the four satellite algorithms shown in Fig. 2. We find that despite their large difference in retrieved sea-ice area, these algorithms have a similar year-to-year variability, which becomes apparent if anomalies of all satellite algorithms relative to the retrieved area in 2010 are plotted together (see Fig. 12a, b). Hence, the difference between the satellite products is largely caused by a constant offset and there is larger certainty in anomalies in sea-ice area than there is in its absolute value. This is important for any model simulation with assimilated sea-ice concentration fields: one should expect such a model to at least retrieve the anomaly structure of the satellite time series, which can be very reliably estimated.

To quantify the uncertainty of sea-ice area retrievals and of the retrieved trends, we calculated the mean seasonal cycle of sea-ice area and of trends in sea-ice area for the period 2003–2010, for which all four satellite products contain data (see Fig. 12c, d). We then for each month simply subtracted

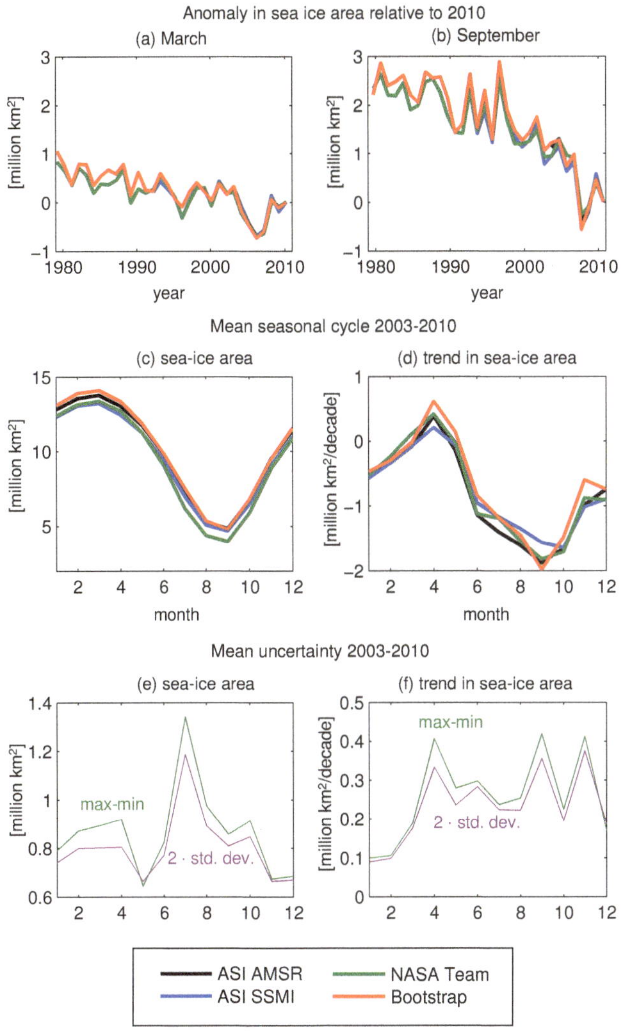

Fig. 12. **(a)** March and **(b)** September anomalies in sea-ice area as retrieved from satellites for the period 1979–2010. Different colours denote different algorithms or satellites. **(c)** Seasonal cycle in sea-ice area and **(d)** in sea-ice-area trend as retrieved from satellites for the period 2003–2010. **(e)** Uncertainty in retrieved sea-ice area and **(f)** in retrieved trend of sea-ice area.

the minimum value from the maximum value to obtain a time series of uncertainties based on passive microwave data. Doing so, we find that apart from July, differences in estimated sea-ice area are less than 1 million km^2 (green curve in Fig. 12e). The same is found for an estimate of twice the standard deviation (purple curve in Fig. 12e). Hence, a value of 1 million km^2 can be taken as a rough approximation of the uncertainty of retrieved sea-ice area throughout the year. This uncertainty is comparable to the one found by Comiso et al. (1997) in his comparison of estimated sea-ice area for the Bootstrap and the NASA Team algorithm. The true uncertainty is probably larger than this value, since we here only examine the differences between individual passive-microwave algorithm. Additional uncertainties that are common to all these algorithms are not reflected by this number. Such uncertainties include, for example, changes in snow-surface properties, seasonal changes in cloud cover, and the impact of thin ice.

Repeating a similar analysis for sea-ice trends, we find that uncertainties from passive-microwave products are less than 0.4 million km^2 $decade^{-1}$ throughout the year, with smaller values in wintertime (Fig. 12f). Hence, this value can be taken as an approximation of the uncertainty of retrieved trends in sea-ice area.

A number of models have smaller biases in sea-ice area than 1 million km^2 relative to satellite retrievals. For these models, biases in this integrative measure could therefore simply be explained by the uncertainty range of the satellite retrievals. For the absolute biases in mean concentration, however, all models show larger biases towards satellite retrievals than the retrievals do among each other. The integrated regional biases in the models are hence not explicable by measurement uncertainty.

For a more detailed analysis of modelled sea-ice coverage, the regional distribution of biases must be analysed. Therefore, the mapping of differences in modelled mean sea-ice concentration is a standard tool in examining model quality. However, again the interpretation of such an analysis hinges on the reliability of the underlying concentration field as obtained from satellite retrievals: in particular in summer, large differences arise between different algorithms (Fig. 13a). To allow for a rough quantification of the uncertainty of retrieved sea-ice concentration from satellite, we have calculated for each month the median of the gridded difference between sea-ice concentration obtained from the NASA Team algorithm and that obtained from the Bootstrap algorithm (Fig. 13b). This then allows one to estimate if a certain regional difference between model and satellite retrieval in a specific month still lies within the observational uncertainty. The figure confirms our analysis of the integrative measures discussed in the previous subsections: during wintertime, estimates of sea-ice concentration are very similar for different satellite products, while a median uncertainty of around 10 % is typical for summer and early autumn. Note that this assessment only gives a somewhat crude estimate of the reliability of retrieved sea-ice concentration from satellites: locally, differences between the two products considered here can exceed 50 % throughout the year.

Our analysis has also shown that internal variability gives rise to much larger uncertainty in the estimate of model quality than do the differences between individual satellite retrievals. This is particularly true for the assessment of modelled trends in sea-ice coverage, which usually vary rapidly in time (see also Notz et al., 2013). In the light of this finding, for model evaluation purposes an integrative assessment of the quality of modelled processes and statistical distributions is more insightful than a simple comparison of modelled time series. This includes, for example, an assessment of seasonal changes in the ice-thickness distribution, the response of the

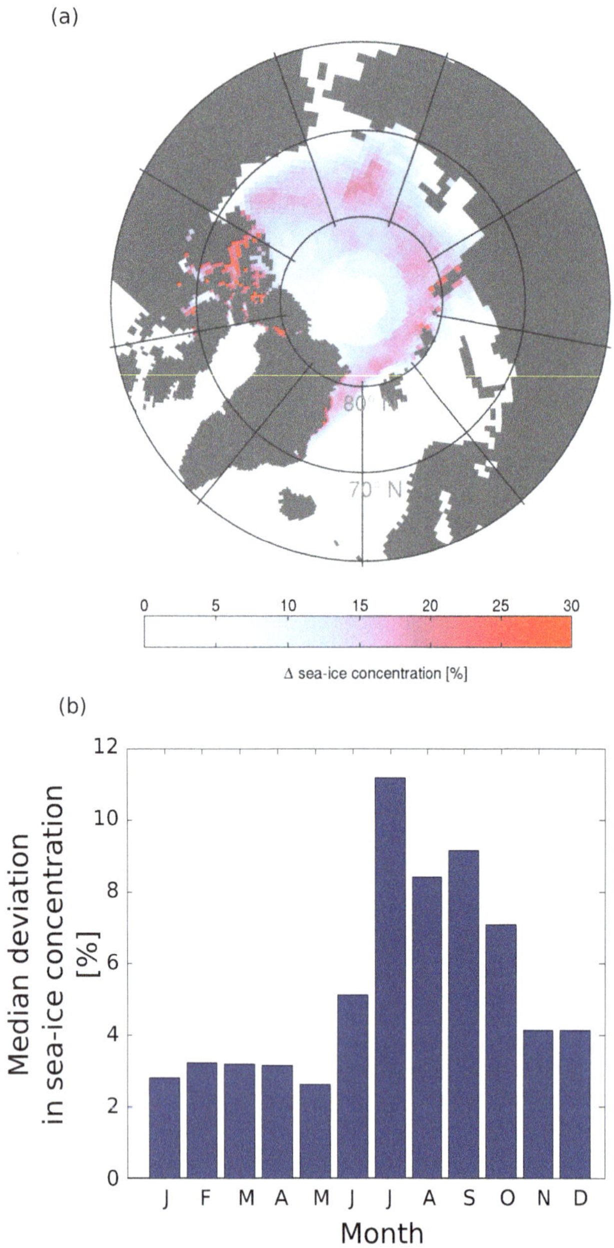

Fig. 13. **(a)** Mean difference in September sea-ice concentration between Bootstrap retrieval and NASA Team retrieval for the period 1979–2007 (Bootstrap minus NASA Team). **(b)** Monthly median deviation in sea-ice concentration between Bootstrap retrieval and NASA Team retrieval for the period 1979–2007.

ice cover to divergent wind fields, and an assessment of the statistical distribution of sea-ice concentration as carried out as part of the present study. Through such focused analysis, ideally across a number of satellite algorithms, we can identify shortcomings in these algorithms and at the same time work towards identifying deficits in our sea-ice model physics.

5 Conclusions

In this paper, we have discussed how the evaluation of modelled sea-ice coverage against observations is affected by the incomplete knowledge of the real evolution of the sea-ice cover, by internal variability, and by technical issues such as differences in model grids. For the quantitative assessment of model quality, all these factors need to be taken into account. Our results can be summarised as following:

5.1 Evaluation of sea-ice coverage

1. Summer biases between a particular model and a particular satellite retrieval can be different for sea-ice extent and for sea-ice area. This is because some models and some algorithms see the summer Arctic sea-ice cover as compact with a high fraction of high-concentration sea ice, while others do not. In winter, all algorithms and all models see the Arctic sea-ice cover as compact.

2. Simulations with a compact ice cover have a similar bias in sea-ice extent and in sea-ice area relative to satellite retrievals based on the Bootstrap algorithm or the ASI algorithm. Relative to these algorithms, simulations with a loose ice cover with a negative bias in sea-ice extent usually have an absolute larger bias in sea-ice area, while simulations with a positive bias in sea-ice extent usually have an absolute smaller bias in sea-ice area.

3. Internal variability of sea-ice area as estimated from CMIP5 simulations is comparable to the observational uncertainty as estimated from different passive-microwave algorithms, while internal variability of sea-ice extent from the simulations is about four times as large as the observational uncertainty.

4. For sea-ice area, 19 of the 26 models that we examined here and for sea-ice extent, 16 of the 26 models have all their ensemble members outside of the reasonable range that we estimated from the ensemble spread from those models that capture the observed value in at least one of their ensemble members.

5. The error that is introduced in the calculation of sea-ice extent by different grid geometries can be larger than the observational uncertainty in months with a large ice coverage.

6. Because biases in sea-ice extent can give misleading results regarding model quality, we recommend that biases in sea-ice area are also taken into account in the assessment of model quality. Based on differences between individual passive-microwave retrievals, we estimate the uncertainty in satellite-retrieved sea-ice area to be 1 million km^2 throughout

the year. The uncertainty in retrieved trends is less than 0.4 million km^2 $decade^{-1}$ throughout the year. The median uncertainties in retrieved sea-ice concentration range from below 5 % throughout winter and spring to about 10 % in summer. These numbers will have to be re-assessed (and probably increased) once reliable data sets of Arctic sea-ice coverage become available that are not based on passive microwave data.

7. There is little correlation between biases in the integrative measures sea-ice extent and sea-ice area compared to the mean absolute bias in sea-ice concentration. This is caused by the fact that for the integrative measures, regional positive and negative biases can cancel. The average absolute bias in sea-ice concentration relative to observations is therefore a useful additional estimator of model quality.

5.2 Evaluation of trends

1. Internal variability of sea-ice trends as estimated from the ensemble spread of CMIP5 model simulations is so large that almost all differences in trends between observations and simulations of CMIP5 models for the period 1979–2005 (and, indeed, until 2012, see Sect. 3.5) could be caused by internal variability. Many models show in one simulation a much stronger trend than has been observed, while a different simulation with the same model and the same forcing shows for slightly different initial conditions a much weaker trend than has been observed.

2. If despite the large internal variability differences between modelled and observed trends are of quantitative interest, one must note that model simulations with too fast a retreat of sea-ice extent have generally a smaller bias in simulated sea-ice-area trends relative to Bootstrap retrievals. Models that simulate too slow a retreat of sea-ice extent have generally a larger bias in sea-ice area trends. This is independent of the compactness of the ice cover.

Acknowledgements. I am very grateful to P. Griewank, A. Haumann, V. Kattsov, J. Marotzke, W. Meier, B. Merryfield and A. Stössel for insightful discussions and/or helpful comments on this manuscript. Comments by two reviewers were essential for improving this manuscript and are very gratefully acknowledged. I thank PCMDI for their management of CMIP5, the various modelling groups for carrying out the simulations used here, and the German Climate Computing Centre (DKRZ) whose data portal facilitated data access tremendously. NASA Team and Bootstrap algorithm sea-ice concentration data were obtained from the National Snow and Ice Data Center NSIDC, Boulder, Colorado, US. ASI Algorithm sea-ice concentration were obtained from the Integrated Climate Data Center (ICDC, http://icdc.zmaw.de/), University of Hamburg, Hamburg, Germany. This work has been funded through a Max Planck Research-Group Fellowship.

Edited by: D. Feltham

References

Agnew, T. and Howell, S.: The use of operational ice charts for evaluating passive microwave ice concentration data, Atmos. Ocean, 41, 317–331, doi:10.3137/ao.410405, 2003.

Andersen, S., Tonboe, R., Kaleschke, L., Heygster, G., and Pedersen, L. T.: Intercomparison of passive microwave sea ice concentration retrievals over the high-concentration Arctic sea ice, J. Geophys. Res., 112, 18 pp., doi:10.1029/2006JC003543, 2007.

Cavalieri, D. J., Gloersen, P., and Campbell, W. J.: Determination of Sea Ice Parameters With the NIMBUS 7 SMMR, J. Geophys. Res., 89, 5355–5369, doi:10.1029/JD089iD04p05355, 1984.

Comiso, J. C.: Characteristics of Arctic Winter Sea Ice From Satellite Multispectral Microwave Observations, J. Geophys. Res., 91, 975–994, doi:10.1029/JC091iC01p00975, 1986.

Comiso, J. C. and Nishio, F.: Trends in the sea ice cover using enhanced and compatible AMSR-E, SSM/I, and SMMR data, J. Geophys. Res., 113, C02S07, doi:10.1029/2007JC004257, 2008.

Comiso, J. C., Cavalieri, D. J., Parkinson, C. L., and Gloersen, P.: Passive microwave algorithms for sea ice concentration: A comparison of two techniques, Remote Sens. Environ., 60, 357–384, doi:10.1016/S0034-4257(96)00220-9, 1997.

Day, J. J., Hargreaves, J. C., Annan, J. D., and Abe-Ouchi, A.: Sources of multi-decadal variability in Arctic sea ice extent, Environ. Res. Lett., 7, 034011, 2012.

Fetterer, F., Knowles, K., Meier, W., and Savoie, M.: Sea Ice Index, Digital media, National Snow and Ice Data Center, Boulder, Colorado USA, 2002, updated, 2012.

Kaleschke, L., Heygster, G., Lüpkes, C., Bochert, A., Hartmann, J., Haarpaintner, J., and Vihma, T.: SSM/I sea ice remote sensing for mesoscale ocean-atmosphere interaction analysis: Ice and icebergs, Can. J. Remote Sens., 27, 526–537, 2001.

Kay, J. E., Holland, M. M., and Jahn, A.: Inter-annual to multi-decadal Arctic sea ice extent trends in a warming world, Geophys. Res. Lett., 38, L15708, doi:10.1029/2011GL048008, 2011.

Kwok, R.: Sea ice concentration estimates from satellite passive microwave radiometry and openings from SAR ice motion, Geophys. Res. Lett., 29, 25-1–25-4, doi:10.1029/2002GL014787, 2002.

Massonnet, F., Fichefet, T., Goosse, H., Bitz, C. M., Philippon-Berthier, G., Holland, M. M., and Barriat, P.-Y.: Constraining projections of summer Arctic sea ice, The Cryosphere, 6, 1383–1394, doi:10.5194/tc-6-1383-2012, 2012.

Meier, W. N.: Comparison of passive microwave ice concentration algorithm retrievals with AVHRR imagery, in Arctic peripheral seas, IEEE T. Geosci. Remote, 43, 1324–1337, doi:10.1109/TGRS.2005.846151, 2005.

Meier, W. and Notz, D.: A note on the accuracy and reliability of satellite-derived passive microwave estimates of sea-ice extent,

CliC Arctic sea ice working group consensus document, World Climate Research Program, 2010.

Meier, W., Fetterer, F., Savoie, M., Mallory, S., Duerr, R., and Stroeve, J.: NOAA/NSIDC Climate Data Record of Passive Microwave Sea Ice Concentration, Digital media, National Snow and Ice Data Center, Boulder, Colorado USA, 2011.

Notz, D. and Marotzke, J.: Observations reveal external driver for Arctic sea-ice retreat, Geophys. Res. Lett., 39, L051094, doi:10.1029/2012GL051094, 2012.

Notz, D., Haumann, A., Haak, H., Jungclaus, J., and Marotzke, J.: Arctic sea ice as simulated by MPI-ESM, J. Adv. Model. Earth Syst., 5, 173–194, doi:10.1002/jame.20016, 2013.

Parkinson, C. L., Comiso, J. C., Zwally, H. J., Cavalieri, D. J., Gloersen, P., and Campbell, W. J.: Arctic sea ice, 1973–1976: satellite passive-microwave observations, National Aeronautics and Space Administration, Washington, D.C., 1987.

Partington, K., Flynn, T., Lamb, D., Bertoia, C., and Dedrick, K.: Late twentieth century Northern Hemisphere sea-ice record from U.S. National Ice Center ice charts, J. Geophys. Res., 108, 3343, doi:10.1029/2002JC001623, 2003.

Spreen, G., Kaleschke, L., and Heygster, G.: Sea ice remote sensing using AMSR-E 89-GHz channels, J. Geophys. Res, 113, C02S03, doi:10.1029/2005JC003384, 2008.

Stroeve, J., Holland, M. M., Meier, W., Scambos, T., and Serreze, M.: Arctic sea ice decline: Faster than forecast, Geophys. Res. Lett., 34, L09501, doi:10.1029/2007GL029703, 2007.

Stroeve, J. C., Kattsov, V., Barrett, A., Serreze, M., Pavlova, T., Holland, M., and Meier, W. N.: Trends in Arctic sea ice extent from CMIP5, CMIP3 and observations, Geophys. Res. Lett., 39, L16502, doi:10.1029/2012GL052676, 2012

Taylor, K. E., Stouffer, R. J., and Meehl, G. A.: An Overview of CMIP5 and the experiment design, B. Am. Meteorol. Soc., 93, 485–498, doi:10.1175/BAMS-D-11-00094.1, 2012.

Winton, M.: Do Climate Models Underestimate the Sensitivity of Northern Hemisphere Sea Ice Cover?, J. Climate, 24, 3924–3934, doi:10.1175/2011JCLI4146.1, 2011.

Zwally, H. J., Parkinson, C. L., and Comiso, J. C.: Variability of Antarctic Sea Ice and Changes in Carbon Dioxide, Science, 220, 1005–1012, doi:10.1126/science.220.4601.1005, 1983.

10

Diffusive equilibration of N_2, O_2 and CO_2 mixing ratios in a 1.5-million-years-old ice core

B. Bereiter, H. Fischer, J. Schwander, and T. F. Stocker

Climate and Environmental Physics, Physics Institute, University of Bern, Sidlerstrasse 5, 3012 Bern, Switzerland

Oeschger Centre for Climate Change Research, University of Bern, 3012 Bern, Switzerland

Correspondence to: B. Bereiter (bereiter@climate.unibe.ch), H. Fischer (hfischer@climate.unibe.ch), J. Schwander (schwander@climate.unibe.ch), and T. F. Stocker (stocker@climate.unibe.ch)

Abstract. In the framework of the International Partnerships in Ice Core Sciences, one of the most important targets is to retrieve an Antarctic ice core that extends over the last 1.5 million years (i.e. an ice core that enters the climate era when glacial–interglacial cycles followed the obliquity cycles of the earth). In such an ice core the annual layers of the oldest ice would be thinned by a factor of about 100 and the climatic information of a 10 000 yr interval would be contained in less than 1 m of ice. The gas record in such an Antarctic ice core can potentially reveal the role of greenhouse gas forcing on these 40 000 yr cycles. However, besides the extreme thinning of the annual layers, also the long residence time of the trapped air in the ice and the relatively high ice temperatures near the bedrock favour diffusive exchanges. To investigate the changes in the O_2 / N_2 ratio, as well as the trapped CO_2 concentrations, we modelled the diffusive exchange of the trapped gases O_2, N_2 and CO_2 along the vertical axis. However, the boundary conditions of a potential drilling site are not yet well constrained and the uncertainties in the permeation coefficients of the air constituents in the ice are large. In our simulations, we have set the drill site ice thickness at 2700 m and the bedrock ice temperature at 5–10 K below the ice pressure melting point. Using these conditions and including all further uncertainties associated with the drill site and the permeation coefficients, the results suggest that in the oldest ice the precessional variations in the O_2 / N_2 ratio will be damped by 50–100 %, whereas CO_2 concentration changes associated with glacial–interglacial variations will likely be conserved (simulated damping 5 %). If the precessional O_2 / N_2 signal will have disappeared completely in this future ice core, orbital tuning of the ice-core age scale will be limited.

1 Introduction

In the framework of the International Partnerships in Ice Core Sciences (IPICS), one target is to retrieve an ice core reaching back in time as far as possible. The main objective of such a core would be to cover the Middle Pleistocene climate transition, the transition interval from the "40 kyr world" prior to about 900 thousand years before present (kyr BP) (Elderfield et al., 2012), where full glacial–interglacial cycles occurred in parallel to obliquity cycles, to the well-known "100 kyr world" thereafter, where rather short interglacials were observed every about 100 kyr. Accordingly, such an ice core should ideally disclose the last 1.5 million years (Myr) of climate history. The conditions, where such an ice core can be found, are only satisfied in central East Antarctica where it is cold and dry enough (see Fischer et al., 2013). The most important boundary condition for finding such old ice is that no melting occurs at the bottom of the ice sheet. Only in this case, the ancient ice may be preserved in the deepest part of the ice sheet. In the case of the EPICA Dome C (EDC) ice core, about 0.6 mm ice per year is melting at the bedrock (Parrenin et al., 2007b), which limits the age of the deepest (undisturbed) ice to about 800 kyr.

A consequence of a frozen bedrock and the preservation of ancient ice in the deep section of an ice core is that the annual layers get extremely thinned in the deepest part. Since, in theory, an annual layer could be conserved infinitely long at such a drill site, the annual layers would also be thinned infinitely. Therefore, the question arises by how much the annual layers can be thinned until gas diffusive processes start to deteriorate and eventually remove the signal stored in the trapped atmospheric air. In order to investigate this question we modified the diffusion model of Bereiter et al. (2009) and combined it with a one-dimensional ice flow and ice temperature model. The diffusive exchange of O_2, N_2 and CO_2 is calculated and the behaviour of an initially sinusoidal concentration sequences of these gases with different periods is simulated along the ice core axis, while the ice is compressed and moving down with the vertical ice flow (see results in Sect. 4).

To calculate the diffusive exchange of the air constituents in the ice, permeation coefficients (product of diffusion and solubility) are required for each constituent. These coefficients for O_2, N_2 and CO_2 are very small and a precise measurement is very challenging. Therefore, only a few estimates of these parameters, using different approaches, are found in the literature. For O_2 and N_2 we use the "set B" of Bereiter et al. (2009) and call it here "fast set" (FS), but we do not use the "set A" as it provided unrealistic results in that work. In addition, we also use the permeation parameters of Salamatin et al. (2001) (referred to as "slow set" (SS)) and a recent estimate of total air permeability of Uchida et al. (2011). For CO_2 we use only the permeation parameter of Ahn et al. (2008) due to the lack of other reliable estimates (see Fig. 2 and Sect. 2.3 for more details).

The two parameter sets FS and SS for O_2 and N_2 show differences of up to two orders of magnitudes. A possibility to further constrain these parameters based on ice core measurements of the O_2/N_2 ratio is presented in Sect. 3. Similar to the gas diffusion modelling work shown in Lüthi et al. (2010), which focused on the equilibration between layers of different CO_2 concentrations below the bubble to clathrate transition zone (BCTZ), the observed equilibration between layers of different O_2/N_2 ratios in the Vostok ice core is compared with a model using the two parameter sets. Such a layering is thought to be induced by a dynamic disequilibrium during clathrate formation (Lüthi et al., 2010). If the Vostok data indeed reflect the equilibration of layered O_2/N_2 ratios on a scale of centimetre, as we think, the experiments allow us to better constrain the effective permeation coefficients for the temperature regime below the BCTZ. However, since this temperature regime is very different from the regime in which a possible equilibration in the oldest ice core takes place, our simulations in Sect. 3 do not allow us to rule out any of the two permeation data sets used in our simulations.

2 Method

2.1 Model set-up

The gas diffusion part of the model is based on the model presented in Bereiter et al. (2009) with the difference that the diffusion is calculated vertically (along the ice core axis) and not in radial direction of the ice core. The diffusion equation is adjusted as follows:

$$\frac{\partial C_n}{\partial t} = \frac{\partial}{\partial z}\left(D_n \frac{\partial C_n}{\partial z}\right),$$

where C_n represents the concentration in the ice of the air constituent n (O_2, N_2 or CO_2), D_n the corresponding diffusion coefficient in ice, and z the vertical direction from $z = 0$ (bottom) to $z = H$ (surface of the ice sheet). The gas exchange of the air enclosures with the surrounding ice is calculated according to Bereiter et al. (2009). The diffusion model is run for discrete intervals of the ice core. D_n and the solubility are a function of the mean temperature of such an interval, which means that they are kept independent of z within each interval.

In order to calculate the annual layer thickness of the ice and the corresponding temperature at a certain depth, the following one-dimensional ice flow model is used (Parrenin et al., 2007b):

$$w(z) = -(A - m)\,u(z) - m, \qquad (1)$$

$$u(z) = 1 - \frac{p+2}{p+1}\left(1 - \frac{z}{H}\right) + \frac{1}{p+1}\left(1 - \frac{z}{H}\right)^{p+2}, \qquad (2)$$

where $w(z)$ denotes the vertical ice velocity at the height z over the bedrock, A the accumulation rate of ice at the surface, m the melt rate at the bedrock, H the vertical thickness of the ice sheet and p a tuning factor. A, m and H are expressed in ice equivalent of constant density (921 kg m^{-3}). The velocity profile is based on the work of Lliboutry (1979) and is also used for the latest ice-core age scales (Parrenin et al., 2007a).

For the temperature $T(z)$, besides H and A, also the surface temperature T_s and the geothermal heat flux Q_g are relevant. Based on the heat diffusion and advection equation

$$\frac{\partial T}{\partial t} = K\frac{\partial^2 T}{\partial z^2} - w\frac{\partial T}{\partial z}, \qquad (3)$$

the following equation for the stationary temperature profile $T(z)$ can be derived:

$$T(z) = T_s - C_0\left(\int e^{f(z')}\mathrm{d}z' - \int e^{f(z')}\mathrm{d}z'\right), \qquad (4)$$

$$f(z') = \frac{-1}{K}(A-m)\left(z' + p_1\left(1 - \frac{z'}{H}\right)^2 - p_2\left(1 - \frac{z'}{H}\right)^{p+3}\right) - \frac{1}{K}mz', \qquad (5)$$

Table 1. Drill site parameters for the different model experiments. The Vostok ice thickness (ice equivalent) is given by Salamatin et al. (1998). The other Vostok parameters are derived from the calibration procedure. For the EDC parameters the accumulation rate (temporal mean), the melt rate and the ice thickness (ice equivalent, temporal mean) of Parrenin et al. (2007a) are used. The EDC surface temperature (temporal mean) is from Jouzel et al. (2007). The EDC tuning parameter and the geothermal heat flux are derived from the calibration procedure. The oldest ice core parameters are artificial and set such that the age and temperature requirements (1.5 Myr 100 m above ground, bottom temperature below pressure melting point) are fulfilled.

Drill site parameters		Vostok	EDC	Oldest ice core
A	Accumulation rate [mm yr^{-1}]	19	18.2	15
m	Melt rate [mm yr^{-1}]	0.8	0.7	0
H	Ice thickness [m]	3690	3153	2700
p	Tuning parameter	20	3.8	7.3
T_s	Surface temperature [K]	214.5	212.2	213
Q_g	Geothermal heat flux [mW m^{-2}]	45	53.3	55

$$C_0 = \frac{-Q_g}{\lambda}\exp\left(\frac{1}{K}(A-m)(p_1-p_2)\right), \tag{6}$$

$$p_1 = \frac{p+2}{p+1}\frac{H}{2},$$

$$p_2 = \frac{1}{p+1}\frac{H}{p+3},$$

$$\lambda = 2.43\frac{W}{mK},$$

$$K = 1.42\cdot 10^{-6}\frac{m^2}{s},$$

where λ is the thermal conductivity, and K the thermal diffusivity of ice calculated for 243 K (Schwander et al., 1997). 243 K is about the mean ice temperature. Since we tune the ice temperature profile to measured borehole temperatures, our simulations are not sensitive to this value. Q_g is the part of the geothermal heat flux, which is not used for melting of ice at the bottom and is diffusing upwards by heat conduction. In case of no melting at the bottom (oldest ice) this is of course equal to the total geothermal heat flux.

Note that in reality λ and K are functions of temperature. Above, we assume λ and K to be constant and calculate the steady state, that is, we use average boundary conditions (surface temperature, accumulation rate and ice thickness). This allows us to derive the semi-analytical solution for $T(z)$ of Eq. (4). Therefore, the melt rates and geothermal heat fluxes derived in our model may differ quantitatively from the true values at Dome C. However, the temperature in the deeper part of the ice sheet is well represented as illustrated in Fig. 1. The temperature profile of a potential oldest ice core site is expected to be similar with the difference – as mentioned above – that the bedrock temperature must be lower to avoid ice melting.

In a first step, the depth–age distribution (age(z)) is calculated for the given input parameters A, H, m and p by integrating the inverse of the vertical velocity $w(z)$:

$$age(z) = \int \frac{1}{w(z')}\mathrm{d}z'.$$

Then, this scale is divided into intervals of equal temporal duration, meaning that an ice particle, which travels down in the ice sheet, would stay within each of these intervals for the same period of time. For each of these intervals an average annual layer thickness and an average ice temperature is calculated using the temperature profile described from solving Eqs. (4) to (6) (see Fig. 1).

The diffusion model is run for each of these intervals starting with the upper most one with a given trapped air composition (see Figs. 4 and 6 for corresponding starting distributions). The simulation time for each interval corresponds to the residence time in the intervals. After the gas composition changes have been calculated for the first interval, the resulting composition is used in the next step for the run in the next lower interval where the length of the trapped gas section is scaled according to the new annual layer thickness, and the permeation coefficients are changed according to the new temperature. This process is repeated downwards through all intervals until the lowest interval is reached.

2.2 Input parameters for the ice core model

In order to start the model, the boundary conditions for the ice flow and temperature model (ice parameters) need to be set as well as the diffusion and solubility coefficients of the gases in the ice (gas parameters). The ice parameters for the different experiments are given in Table 1. In the case of the Vostok and EDC parameters they have been tuned in order to fit the available age–depth relations (age(z)) and borehole temperatures ($T(z)$) as well as possible. However, the parameters were kept as close as possible to independently published values. Since the ice flow and temperature model used here assumes steady state conditions, the temporal average of the corresponding parameters are used.

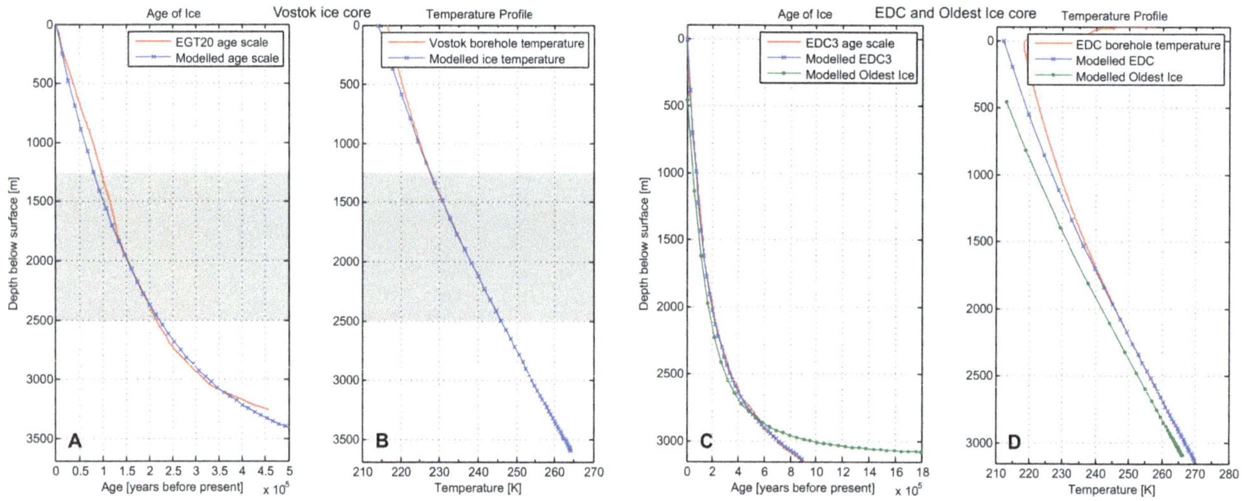

Fig. 1. Age of ice **(A, C)** and ice temperature **(B, D)** profiles of the Vostok **(A, B)** as well as of the EDC and oldest ice core **(C, D)**. Red lines: reference records for tuning the ice parameters (**A** EGT20 age scale (Petit et al., 1999); EDC3 age scale (Parrenin et al., 2007a); **(B)** Vostok bore hole temperature (Salamatin et al., 1994); **(D)** EDC bore hole temperature measured in December 2004). Blue lines: modelled ice age and ice temperature profiles using the Vostok **(A, B)** and EDC **(C, D)** ice parameters (Table 1). Green lines: modelled ice age and ice temperature profiles using the oldest ice core ice parameters. The grey shaded area marks the depth below the BCTZ for which the simulations of Sect. 3 have been performed. The depth level zero corresponds to the model ice sheet thickness at Vostok **(A, B)** and EDC **(C, D)** drill site in m ice equivalent. The depth of the oldest ice core is shifted by 453 m such that the bedrock is reached at the same depth as for the EDC core.

For most of the EDC ice parameters corresponding values can be found in the literature. The factors A, m, H and T_s are set such that they lie within their uncertainties or agree within 2 % with the values reported in Parrenin et al. (2007a) and Jouzel et al. (2007). The factor p is modified relative to the corresponding value given in Parrenin et al. (2007b) (1.97 ± 0.93) in order to improve the agreement of the modelled age(z) with the ED3 age scale (Parrenin et al., 2007a). The factor Q_g is set such that $T(0)$ fits the EDC bedrock temperature (see Fig. 1).

In the case of Vostok, corresponding values for all ice parameters are not found in the literature, and if so, they are mostly less precise compared to the EDC parameters and/or do not cover the full extent of the ice core. However, for the simulations performed under Vostok conditions (see Sect. 3), it is only necessary that the modelled age(z) and $T(z)$ roughly represent the corresponding reference records below the BCTZ, between 1250 and 2500 m depth (grey shaded area in Fig. 1), and not down to the bottom of the ice core (3379 m depth). We calibrate our models with the EGT20 age scale of Petit et al. (1999) and the borehole temperature of Salamatin et al. (1994). For the tuning of the Vostok ice parameters, the parameter H was kept fixed at the value presented in Salamatin et al. (1998) and the other parameters were varied.

The artificial parameters of the oldest ice core are chosen such that the ice 100 m above the bedrock reaches an age of roughly 1.5 million years (aim of IPICS oldest ice) and that the bottom ice temperature is 4–5 K below the pressure melting point of the ice. The reason for the 100 m criterion is that several ice cores show disturbed ice layering due to irregular flow below a certain depth (e.g. EPICA community members (2006), NEEM community members (2013)). In the EDC ice core this depth lies at about 100 m above the bedrock. The tuning parameter p is estimated using the theoretical calculation of Parrenin et al. (2007b). However, in the case of the EDC and Dome Fuji ice cores this theoretical value was found to be too high. Parrenin et al. (2007b) suggest that the bedrock topography might cause this discrepancy. Another reason might be that Eq. (2) has been derived for ice sheets far away from ice divides and ice domes (Lliboutry, 1979), which is not fulfilled for the two ice cores. Accordingly, also the p value for an oldest ice core might be closer to the EDC value, which results in less thinning of the deep ice and, hence, a weaker gas diffusion effect. The influence of this different thinning profile is discussed in Sect. 4 and in the Supplement.

As mentioned in Sect. 1, for finding such old ice it is of fundamental importance that the ice does not melt at the bedrock. Only in this case, old ice is preserved in the deep part of the ice sheet. In order to ensure a cold bedrock, the ice thickness is not allowed to be too large. Otherwise, the insulating effect of the ice enables the geothermal heat to bring the bottom ice to the pressure melting point. Observing this constraint we have arbitrarily chosen an ice thickness of

2700 m which permits a cold bedrock under the geothermal conditions we have set.

2.3 Input parameters for the gas diffusion model

The main reason for a limited predictability of the gas exchange effects in the oldest ice core is the uncertainty of the gas parameters (solubility and diffusion coefficients) of N_2, O_2 and CO_2. As mentioned in Sect. 1, these parameters are very difficult to assess since they are very small, making measurements under lab conditions very challenging. Therefore, the parameters have been estimated in an indirect way. The experimental approaches that were used do not permit to estimate solubility and diffusivity of the gases separately but only the product of both, the permeability. In fact, it is the permeability that determines the magnitude of the diffusive gas transport in the ice from one air inclusion to another. Thus, an increase of the diffusion coefficient can be compensated by a corresponding decrease of the solubility coefficient and vice versa.

For the CO_2 permeability only one reliable estimate of Ahn et al. (2008) is found which is used in all simulations. This estimate is based on analysis of the CO_2 concentration around a layer of refrozen melt water in the Siple Dome ice core. Older estimates for the permeability exist, but have shown unrealistic results in a previous diffusion study (Bereiter et al., 2009). Note, however, that also the estimate by Ahn et al. (2008) is missing an independent experimental proof. For the permeability of N_2 and O_2 we use two sets, which are based on different approaches and demonstrate the large spread of such estimates (see Fig. 2). The origin of the "slow set" (SS) is the work of Salamatin et al. (2001) in which the parameters are estimated by reconstructing the O_2 / N_2 ratio fractionation between bubbles and clathrates in the BCTZ. The "fast set" (FS) is based on the work of Ikeda-Fukazawa et al. (2004) and Ikeda-Fukazawa et al. (2005), who have calculated the parameters based on a computer model simulating the behaviour of the gas molecules in the ice lattice (molecular simulations). The second set corresponds also to the "set B" parameters used in Bereiter et al. (2009). Since the gas diffusion model requires a solubility and a diffusion coefficient for each gas species, but in most cases only the permeation coefficients are available, we used for all gas species and throughout our calculations diffusion coefficients derived by Ikeda-Fukazawa et al. (2004), which are also based on molecular simulations, and adjusted the solubility coefficients accordingly.

The FS parameters are entirely synthetic based only on molecular simulations (Ikeda-Fukazawa et al., 2004, 2005). The SS and the CO_2 parameters are based on a model that simulates observed air fractionation and diffusive smoothing, respectively, in an ice core (Salamatin et al., 2001). The temperature regimes at which the observations have been performed are indicated by the crosses in Fig. 2. Since they cover only a small temperature range of a few degrees, the

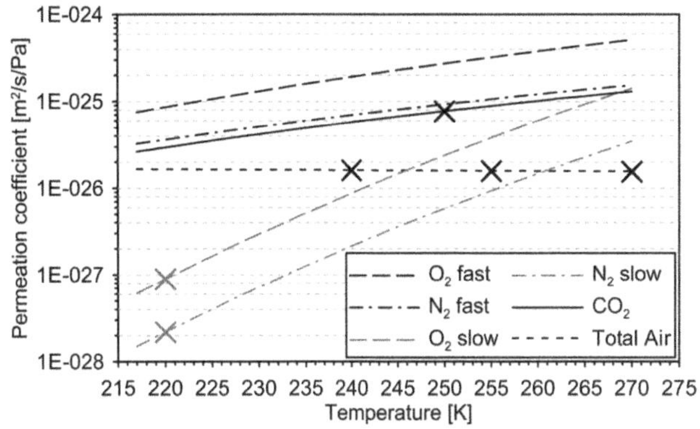

Fig. 2. Comparison of the different estimates of gas permeation coefficients in ice. The black dashed line denoted with "total air" represents the total air permeation value of Uchida et al. (2011). The black lines denoted with "fast" represent the "fast set" (FS) (Ikeda-Fukazawa et al., 2005) and the grey lines denoted with "slow" represent the "slow set" (SS) (Salamatin et al., 2001) used for the simulation presented here. The CO_2 permeation coefficient is the same for both sets (Ahn et al., 2008) (solid black line). Most permeation coefficients shown here consist of a solubility and a diffusion parameter. In all these cases the diffusion parameters of Ikeda-Fukazawa et al. (2004) are used whereas the solubility is adjusted accordingly. The crosses indicate for which ice temperature regime the corresponding parameters have an experimental support and, hence, show which parts of the parameters are extrapolated. In the case of the total air parameter the temperature regime covers a larger range as indicated by the crosses.

temperature sensitivities of the parameters are rather uncertain and the extrapolation of these parameters to temperatures far away from the observed regime may be inaccurate. In contrast to that, the molecular simulations used for the FS parameters have been performed over a wide range of ice temperatures (230–270 K) allowing to derive a theoretical temperature dependence of the parameters for the full temperature range of our study. These temperature sensitivities are much smaller than the ones of the SS parameters. However, since there is not enough independent proof for or against one of these parameter sets, both sets are used in our simulations.

A recently published estimate of total air permeability in ice (Uchida et al., 2011) is also shown in Fig. 2. This estimate is based on a model that reconstructs observed growth of clathrates in the Dome Fuji ice core due to air permeation in ice. The growth of the clathrates takes place over a large temperature range from about 240 K to 270 K (indicated by the crosses). Therefore, this estimate does not suffer from the extrapolation uncertainty of the other observationally based parameters. Provided that the reconstruction model is valid, this estimate suggests even weaker temperature sensitivities of the parameters compared to the other sets and shows the lowest permeabilities for temperatures above 263 K, which are most relevant for our oldest ice core experiments (Sect. 4), since the deep ice is exposed to such temperatures for several 100 kyr. However, we can not directly use this estimate for our simulations since separate parameters for the different air constituents are required in order to simulate the effect on the mixing ratios. Nevertheless, we estimate the effect of such a low air permeability on the O_2 / N_2 ratio in the oldest ice core simulations by running the model with the FS and SS parameters scaled down with a constant factor to match the total air permeability of Uchida et al. (2011) at about 265 K (see Sect. 4).

3 A constraint for the gas permeability in ice

In this section an experimental constraint for the permeation coefficients is described which is based on a natural phenomenon in deep ice cores. Based on the O_2 / N_2 ratio data from the Vostok ice core, the constraint is used to test the permeation coefficients of N_2 and O_2 of the two parameter sets used in this work.

Lüthi et al. (2010) showed that in the BCTZ of different Antarctic ice cores the O_2 / N_2 ratio and the CO_2 concentration of the trapped air vary between horizontal layers of a few centimetres vertical extent. This non-climatic signal is explained by a layer-wise transformation of bubbles to clathrates in the BCTZ and the preferential trapping of O_2 and CO_2 in the clathrates relative to N_2, leading to local fluctuations in the CO_2 concentration and O_2 / N_2 ratio between these layers. The amplitude of the CO_2 fluctuations declines exponentially below the BCTZ presumably due to gas diffusion through ice. The model presented in Sect. 2 allows us to simulate the decline of these fluctuations, which are compared with the decline in the signal amplitude of such layers found in ice core records. However, the available O_2 / N_2 records do not all resolve the decline. In order to do so, the record must cover the ice core section of approximately 1000 m just beneath the BCTZ with an adequate resolution and the vertical extent of the individual ice samples used for the record must not be much larger than the original layering. In the case of the O_2 / N_2 ratio records, there exists a record from the Vostok ice core (Bender, 2002) that fulfils these requirements. Note, however, that this record consists of discrete samples and no intervals of continuous O_2 / N_2 measurements exist for this core.

Figure 3 shows the Vostok O_2 / N_2 record in the top part (grey dots) and in the bottom part the absolute value of the residuals of the data points relative to the spline throughout the data after correcting for the measurement uncertainty (0.5 ‰) (grey crosses). The residuals represent the data variations on short time and length scales and, hence, contain also the variations on a centimetre scale due to the layering process in the BCTZ. The bottom of the BCTZ is found at 1250 m depth and an age of 83 kyr BP, respectively, where the residuals are approximately 5 ‰. From this point on the residuals decline with depth and time supposedly due to diffusive exchange in the ice. Therefore, the exponential fit through the residuals older than 83 kyr BP (black line) is a representation of the average diffusive decline of the layers with time.

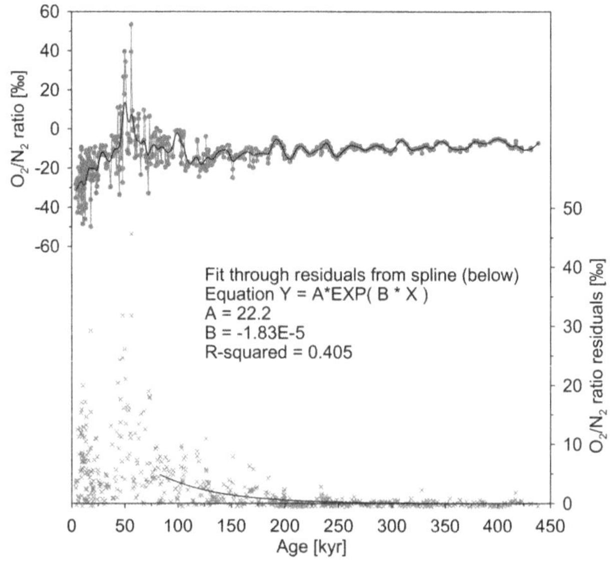

Fig. 3. Top: O_2 / N_2 ratio data from the Vostok ice core (Bender, 2002) (grey dots) and the data spline with a cut-off frequency of 11 kyr (Enting, 1987) (black line). Bottom: residuals of the O_2 / N_2 ratio values relative the the spline (grey crosses) and the exponential fit through the residuals older than 83 kyr (black line). The text in the middle of the figure shows the fit results.

In order to simulate this measured decline, the model of Sect. 2 is run with an O_2 / N_2 ratio distribution at the start as shown in the example of Fig. 4 (blue line). This distribution in Fig. 4 represents a layered O_2 / N_2 ratio with a layer thickness of 3 cm and an amplitude of 4.8 ‰. The amplitude corresponds to the exponential fit value at the bottom of the BCTZ, where we set the diffusion time to zero in Fig. 5. The range of layer thickness comes from the work of Lüthi et al. (2010) on EPICA Dronning Maud Land (EDML) and EDC ice; however, the effective thickness is quite uncertain. Therefore, simulations with different layer thicknesses were performed (see below). In the model experiments shown in this section, the simulation starts in the upper most interval of the grey shaded area in Fig. 1.

As seen in Fig. 4, the layers equilibrate with time due to the gas diffusion along the ice core axis. For the comparison of this simulated decline with the one derived from the data (Fig. 5), the averaged amplitude over one full layer is used. Four different experiments have been performed: (A) with 3 cm layer thickness and the FS parameters (Fig. 5 red line), (B) with a layer thickness of 3 cm and the SS parameters (Fig. 4 and Fig. 5 purple line), (C) with 2.5 cm layer thickness and the SS parameters (Fig. 5 left dashed magenta line), (D) with 5 cm layer thickness and the SS parameters (Fig. 5 right

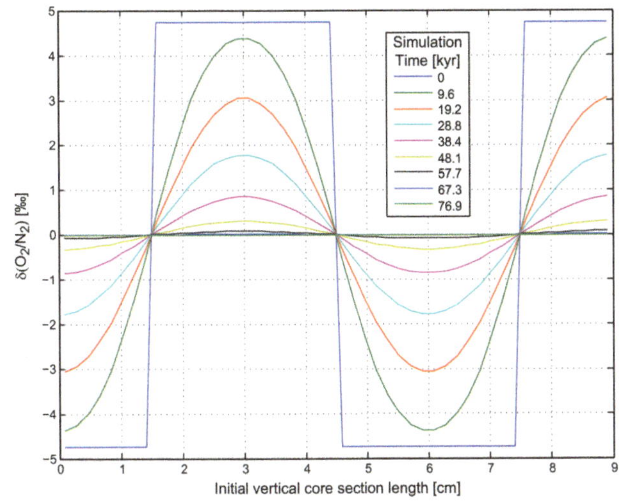

Fig. 4. Example of the model output of the O_2 / N_2 layering decline experiments using the SS parameters. The blue line shows the starting O_2 / N_2 distribution representing 3 layers (half a layer at each edge) with 3 cm thickness and an amplitude of 4.8 ‰. The different coloured lines represent simulated distributions after a certain duration in the ice as indicated by the legend (in 1000 yr [kyr]). From these results the corresponding decline in Fig. 5 is calculated (purple line).

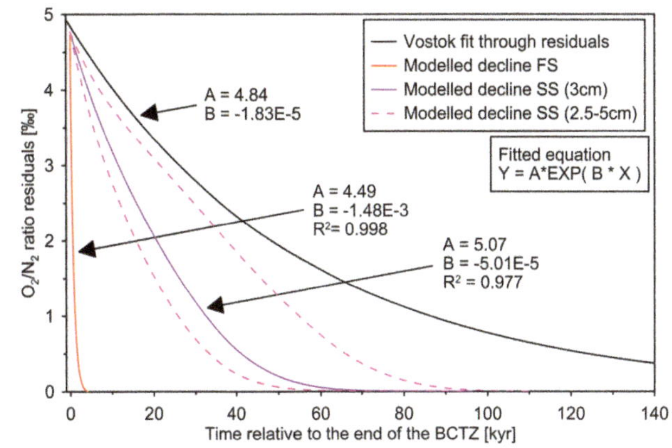

Fig. 5. Comparison of the amplitude decline of the O_2 / N_2 fluctuations below the BCTZ derived from the Vostok data (black line, see also Fig. 3) and the modelled decline using different parameter sets (FS or SS) and layer thicknesses. The red line shows the modelled decline using the FS and 3 cm-layer thickness, the purple line shows decline using the SS and 3 cm thickness and the magenta lines show the decline using the SS and 2.5 cm (left) and 5 cm (right) thicknesses. The results of the exponential fit through the Vostok data as shown in Fig. 3 as well as through two modelled declines are shown for comparison (arrows indicate the corresponding values).

dashed magenta line). The results from the FS parameters (red line) show clearly that this parameter set equilibrates the layered gas composition too fast. The associated lifetime of the variations is more than two orders of magnitudes shorter than what has been observed in the data. The results with the SS parameters also show a shorter lifetime of the variations compared to the data, but the difference is not more than a factor of 4.

One large uncertainty in these experiments is the actual thickness of the layers in the Vostok ice core as no continuous intervals of O_2 / N_2 measurements exist for this core that could resolve it. In the EDML ice core, where continuous measurements are available, a layer thickness of 2.5 to 3 cm was found (Lüthi et al., 2010). However, the boundary conditions (accumulation rate, number and size of individual precipitation events, mean surface temperature, etc.) of the EDML ice core are not the same as for the Vostok ice core. In this respect, the EDC ice core is much closer to the Vostok core. From the EDC core a short section of layer-resolving O_2 / N_2 measurements exist (Lüthi et al., 2010). Unfortunately, the vertical sample length used for this small data set is fairly close to the layer thickness itself and therefore is not clearly resolving the layers. A maximum layer thickness of 5 cm can be found in this set. Therefore, we also performed a corresponding simulation.

The results presented in this section suggest that the SS parameters are closer to the effective values compared to the FS parameters and the parameter of Uchida et al. (2011) (close to the FS parameters), at least for the temperature regime below the BCTZ at Vostok (about 233 K). In principle, an independent set of gas parameters could be derived from our approach, however, they would only be valid for the temperature regime below the BCTZ and not for the much warmer temperatures near the bedrock. Our approach does not provide a strong constraint on the temperature sensitivities and earlier estimates also show large discrepancies in this regard (see Sect. 2.3). For these reasons, an extrapolation of a set of gas parameters derived from our approach to higher temperatures found near the bedrock would be unreliable. Applying our approach to several O_2 / N_2 records from different ice cores might enable us to derive reliable temperature sensitivities of the parameters.

Probably the most delicate point in our approach used here is whether or not the data residuals from the Vostok data set (Fig. 3) reliably reflect the diffusive decline of the O_2 / N_2 fluctuations. They might be affected by artefacts due to, for example, bad core quality and gas loss in the section closer to the BCTZ. Therefore, this approach would benefit a lot from a denser set of O_2 / N_2 records. This could be obtained from a continuous flow analysis technique as shown in Lüthi et al. (2010), who used the measurement technique of Huber et al. (2003). Such a data set would enable to directly estimate the decline of the layer amplitudes below the BCTZ. Therefore, the approach presented here can provide more precise O_2 and N_2 permeation parameters when such continuous records from below the BCTZ and from different ice cores become available.

4 Gas diffusion in deep ice cores

In this section we present the results of the diffusive gas equilibration of trapped CO_2 and O_2 / N_2 signals in deep ice cores during the movement of the gas enclosures near the surface to the bedrock, as obtained by the model described in Sect. 2.1. In all experiments, the simulation starts with a sinusoidal CO_2 concentration and O_2 / N_2 ratio variation with an amplitude of 50 ppmv and 5 ‰, respectively, as shown in Fig. 6. The period of the variations are changed between 100, 40, 20, 5 and 1 kyr by adjusting the temporal length of the simulated windows accordingly. In the example of Fig. 6, a window length of 10 kyr has been used which corresponds to a period of 20 kyr. One simulation has been run for most of the possible combinations of the two different ice flow and temperature profiles of the EDC and oldest ice core (see Fig. 1), the two gas permeation parameter sets (SS and FS, see Fig. 2) and the different signal periods (100, 40, 20, 5 and 1 kyr) resulting in 15 runs in total. The resulting amplitude dampening factors derived from these simulations are summarised in Figs. 7 and 8. Note that the dampening factors are not only controlled by the duration of the diffusion process, but also by the ice temperature and layer thickness, which both change during the advection of the ice from the surface to the bottom of the ice sheet. Accordingly, the basic rule for diffusive processes, that diffusion time (and thus dampening) scales with the square of the diffusion length (thus with the length of some oscillation in metres of ice) is not strictly applicable here.

The five periods, which we have chosen for the simulations, roughly represent some typical oscillations of the CO_2 concentration or the O_2 / N_2 ratio found in ice cores: (A) 100 kyr represents CO_2 variations associated with the 100 kyr glacial–interglacial cycles (Lüthi et al., 2008), (B) 40 kyr represents CO_2 variations expected during the 40 kyr glacial–interglacial cycles, (C) 20 kyr represents precessional O_2 / N_2 variations (Kawamura et al., 2007), (D) 5 kyr represents CO_2 variations associated with carbon dioxide maxima/antarctic isotope maxima (CDM/AIM) events (Bereiter et al., 2012), and (E) 1 kyr represents about the minimum period that does get trapped in low accumulation ice cores without significant dampening by gas diffusion in the firn column (Spahni et al., 2003). The discussion below about the O_2 / N_2 ratio focuses on the orbital 20 kyr period since it is the only relevant feature found so far in this proxy. For the CO_2 concentration the 100 kyr, 40 kyr and 5 kyr periods are most important. However, in the figures the results for all periods are plotted for completeness.

The results of the EDC simulations (Fig. 7) show that neither the relevant periods of CO_2 (5 kyr and higher) nor of O_2 / N_2 (20 kyr) are strongly affected by the vertical gas exchange in this core, independent of the parameter sets used. For the 5 kyr period in CO_2 the dampening factor after 800 kyr (oldest part of the undisturbed EDC ice core) lies at 5 %, whereas larger periods are not affected at all. The amplitude dampening factor for the 20 kyr period in the O_2 / N_2 signal after 800 kyr is 7 % at maximum. Such small dampeners are currently not identifiable since no independent records from another ice core exist for comparison. As the O_2 / N_2 model results using the FS parameters show the maximum possible effect that is still in line with the EDC gas records, in which no signal equilibration has been identified so far, the O_2 / N_2 results for the oldest ice core simulations using the FS parameters can be interpreted as an upper limit of the equilibration effect that has to be expected. Since no other ice core reaches as far back in time as the EDC ice core and the accumulation rate and, hence, the annual layer thickness is relatively small in this ice core, we may safely conclude that vertical diffusion of gases in the ice does not substantially affect any ice core analysed to date.

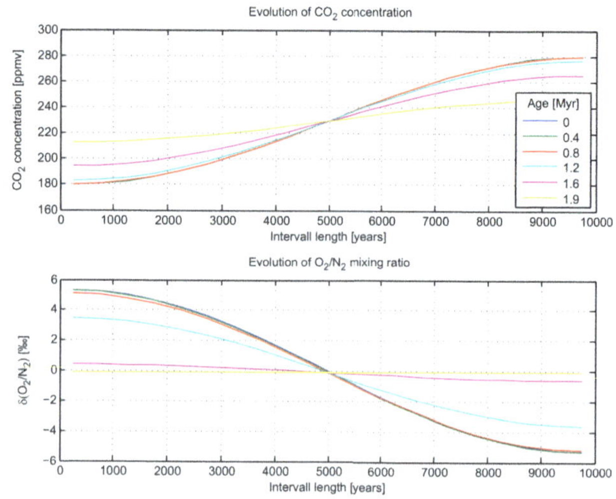

Fig. 6. Example of the model performance showing the simulated changes of the CO_2 concentration (top) and the O_2 / N_2 ratio (bottom) for the oldest ice core ice flow and temperature profile (Fig. 1) using the SS parameters. The starting gas distribution (dark blue line) is half of a sinusoidal oscillation (period 20 kyr) with an amplitude of 50 ppmv (CO_2) and 5 ‰ (O_2 / N_2) over the complete simulated time window of 10 kyr. The different coloured lines represent simulated distributions after a certain duration in the ice as indicated by the legend (in million years [Myr]). This simulation is used to calculate the amplitude dampening of the 20 kyr signal (see Fig. 8).

The simulations for the oldest ice core (Fig. 8), however, provide a different picture and show a clear equilibration effect for the relevant periods of the CO_2 concentration and the O_2 / N_2 ratio variations. This is due to the strong thinning of the ice by a factor of about 100 after 1.5 Myr and the long residence time of the gas in warm ice near the bedrock. The results imply that the CO_2 signals (Fig. 8) associated with CDM/AIM variations will virtually vanish after 1.5 Myr but that variations associated with glacial–interglacial changes will be dampened by not more than 5 %. This means that the CO_2 changes associated with glacial–interglacial variations are likely still resolvable in such old ice. Note, however, that

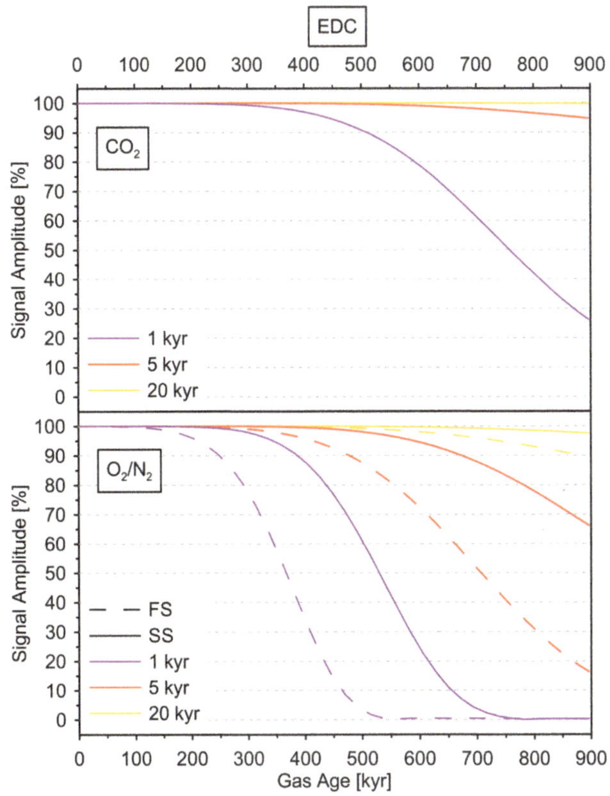

Fig. 7. Simulated amplitude dampening of CO_2 concentration (top) and O_2 / N_2 ratio (bottom) signals for the EDC ice flow and temperature profile (Fig. 1). The different coloured lines show the results for different periods of the signal as indicated by the legend.

this result depends critically on the used value for the CO_2 permeability in ice. Here we use the value derived by Ahn et al. (2008), which, however, has not been experimentally reproduced yet.

In the case of the trapped O_2 / N_2 ratio, the two parameter sets provide clearly different results (Fig. 8). While the SS simulates almost no influence on the 20 kyr period after 0.9 Myr, the FS simulations show already a significant damping of about 30 %. This is substantially more compared to what the model simulates for ice of similar age under EDC conditions using the same parameter set. The reason is the much stronger thinning of the ice older than about 600 kyr in the oldest ice core relative to the EDC ice (see Fig. 1), causing stronger diffusive equilibration in that ice. After 1.5 Myr both parameter sets show a strong equilibration of the O_2 / N_2 signal even for the 40 kyr period. For the 20 kyr period the FS simulation shows a nearly complete equilibration already after 1.3 Myr. In the SS simulations this signal is preserved with a dampening of the original amplitude by 80 % after 1.5 Myr.

In addition to the uncertainties in the gas parameters, also the oldest ice core parameters (Table 1) are not well constrained. We have tested the oldest ice core simulations for differences in the tuning parameter p and the bedrock temperature (see Supplement). As mentioned in Sect. 2.2, the tuning factor p of the oldest ice core simulations might be too high and thinning in the deep part might be overestimated. The test with a p value equal to the one of the EDC simulations shows a 15 % smaller dampening relative to the standard simulation with a total dampening of the original amplitude by 65 % after 1.5 Myr. The results with a 5 K lower bedrock temperature show a similar influence compared to the standard simulation. In regard to the large uncertainties of the gas parameters, the uncertainties in the p value and the bedrock temperatures are of secondary importance.

As mentioned in Sect. 3, an estimate of total air permeation in ice suggests lower permeability than the SS parameters at temperatures near the bedrock (Uchida et al., 2011). The same estimate suggests also a higher permeability for temperatures below the BCTZ which is not supported by the results in Sect. 3. Nevertheless, we tested the influence of lower permeabilities by running two oldest ice core simulations (20 kyr period) where in one case downscaled FS parameters, and in the other case downscaled SS parameters were used (see Supplement). The results suggest that also with such low air permeabilities the precessional O_2 / N_2 signal is dampened by 50 % in the 1.5 Myr old ice.

In summary, the simulations for the oldest ice core suggest that significant equilibration is likely to occur for O_2 / N_2 and CO_2 signals with periods of 40 kyr and 20 kyr, respectively, and periods below that. Slower variations should still be resolvable in 1.5 Myr old ice. Periods shorter than 5 kyr are likely to be smeared out completely. In the case of precessional O_2 / N_2 variations with a period of 20 kyr, a maximum of about 50 % is expected to be preserved of the original amplitude after 1.5 Myr. In the case of CO_2 CDM/AIM variations (approximately 5 kyr, Bereiter et al., 2012) no preservation is expected in such old ice. Due to the uncertainties in the gas parameters in ice, a wide range of amplitude dampening is possible. For example, for precessional O_2 / N_2 variations (20 kyr period) the reduction in amplitude varies between 50 % in 1.5 Myr old ice (parameters by Uchida et al., 2011) and complete vanishing (FS) already in 1.3 Myr old ice. However, as suggested by the simulations in Sect. 3, the FS parameters might be too high and, hence, the upper limit estimate is likely to be too pessimistic. Other uncertain parameters such as the ice temperature or the thinning in such an ice core also influence the results presented here, but compared to uncertainties in the gas parameters their influence is of secondary importance. Nevertheless, dating the deepest part of the oldest ice core based on orbital tuning of the O_2 / N_2 signal (Kawamura et al., 2007) could be problematic depending on the effective dampening that occurred. The O_2 / N_2 measurement uncertainty is in the range of 1 % of the precessional signal, so from this technical perspective a 50 % dampening might still be unproblematic. However, it is not clear whether the slow or fast parameters are valid, and also not whether the parameters change as homogeneously

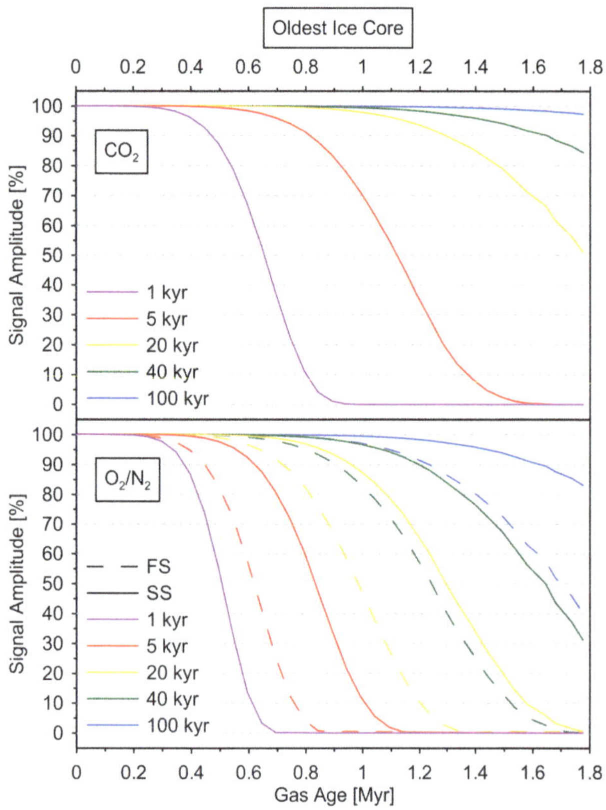

Fig. 8. Simulated amplitude dampening of CO_2 concentration (top) and O_2 / N_2 ratio (bottom) signals for the oldest ice core ice flow and temperature profile (Fig. 1). The different coloured lines show the results for different periods of the signal as indicated by the legend.

as assumed here or are strongly inhomogeneous on a scale of centimetres to metres. This clearly illustrates the need for more research on the permeation of gases in glacier ice to derive robust permeability estimates.

Changes of the ice properties are not included in our model. The size of the ice grains increases with depth (Durand et al., 2009). Larger ice grains might influence the gas permeability of the ice. With increasing depth and temperature water veins may be present along ice grain junctions in the deep part of the ice (Pol et al., 2010), which might increase the gas transport. However, a marginal fraction of clathrates is located along grain boundaries in deep ice (Uchida et al., 2011) and, hence, this effect is expected to be small. Finally, clathrates grow with increasing depth (Uchida et al., 2011), but how this influences the gas exchange between the clathrates is not clear.

Most gases trapped in ice have a lower permeability through the ice lattice than the water molecules of the ice itself since the gas molecules need to go through the step of dissolution in the ice before they can diffuse through the lattice. In addition, in deep ice warmer than 263 K, such as in the EDC core below 2900 m depth, diffusion along water veins at ice grain junctions is suggested to increase the diffusivity of water molecules causing a doubling of the diffusion length for stable water isotopes to 40 cm near the bedrock (Pol et al., 2010). In our artificial oldest ice cores, a water isotope signal associated with a glacial–interglacial cycle of 40 kyr duration will cover a vertical extent of only a few meters in the 1.5 Myr old part. Considering the roughly three times larger residence time of the ice in conditions above 263 K (roughly 700 kyr, see example Fig. 1) compared to the EDC core, the diffusion length for water isotopes is expected to increase to 1 m and more. In such a case the 40 kyr isotope signal could be completely lost and trapped gases as well as particulate mineral dust aerosol might then be the only proxy that allows reconstructions of the past glacial–interglacial cycles back to 1.5 Myr. In order to avoid a complete disappearance of the 40 kyr isotope signal, the bedrock ice temperature must be below 263 K.

5 Conclusions

The simulation of vertical diffusion of trapped gases in an ice sheet and the associated influence on the air composition in ice cores confirms that CO_2 and O_2 / N_2 records from existing ice cores do not suffer from a significant signal loss by gas diffusion. With regard to the IPICS target of retrieving a 1.5-million-years-old ice core, however, this signal loss becomes significant. The results presented here show that the vertical gas exchange will start to considerably influence the CO_2 variations associated with glacial–interglacial changes and precessional O_2 / N_2 changes in ice older than 1 Myr. For 1.5 Myr old ice the estimated amplitude damping of a 40 kyr glacial–interglacial CO_2 variation is on the order of 5 % suggesting that CO_2 variations associated with glacial–interglacial changes are likely to be preserved in the 1.5 Myr old ice. The damping of precessional O_2 / N_2 changes with a period of 20 kyr lies in the range of 50–80 % after 1.5 Myr. A worst case scenario is that after 1.3 Myr the signal has virtually equilibrated. This suggests that O_2 / N_2 variations associated with orbital changes will be significantly dampened or may even have completely disappeared in the deep part of the oldest ice core. Depending on the effective impact of this process, ice core dating by tuning the O_2 / N_2 ratio variations to orbital parameters (Kawamura et al., 2007) could be problematic.

The reason for the large uncertainty range of the simulated gas diffusion effect on the trapped O_2 / N_2 ratio is mainly due to the large uncertainties in the gas permeation coefficients in ice. A constraint for gas permeation coefficients presented in this work suggests that the FS parameters, which are used to derive the worst case scenario for the O_2 / N_2 ratio, are much too high for temperatures around 233 K. However, extrapolating this finding to ice temperatures of about 263 K, relevant for the effects in the oldest ice core, is critical as the temperature sensitivities of the parameters are weakly constrained. Nevertheless, the method presented here can provide further constraints on the permeation of gas in ice and might also provide better temperature sensitivities in future investigations by applying it to records of different ice cores. The reason for the too high FS values is not clear, however, it is likely that the air permeability of natural ice is also influenced by its crystallographic structure. The FS values have been derived by molecular simulations taking into account a monocrystalline ice structure (Ikeda-Fukazawa et al., 2004), whereas polar ice has a polycrystalline structure possibly leading to a reduction of the permeability. Once such an oldest ice core is drilled and the dampening profiles of the different gases are measured, an independent and quantitative estimate of the permeabilities can be deduced by inverting the model presented in this work.

Water isotopes are generally more mobile in the ice than the trapped gases. Therefore, the signal equilibration due to molecular diffusion is likely to be stronger for water isotope signals. Under conditions expected for the oldest ice core and used for the simulations here, it is expected that glacial–interglacial cycles recorded in the water isotopes will be damped substantially in 1.5 Myr old ice. It is possible that this signal will have completely vanished in such an ice core. In this case, trapped gases and mineral dust may be the only palaeo-climate indicators in this archive that will show glacial–interglacial cycles. Ice temperatures near the bedrock of below 263 K are required to preserve the isotope signal (Pol et al., 2010).

Acknowledgements. This work is funded by the Swiss National Science Foundation. We thank B. Stauffer for sharing his valuable experience and giving helpful inputs to this work.

Edited by: K. Kawamura

References

Ahn, J., Headly, M., Wahlen, M., Brook, E. J., Mayewski, P. A., and Taylor, K. C.: CO_2 diffusion in polar ice: observations from naturally formed CO_2 spikes in the Siple Dome (Antarctica) ice core, J. Glaciol., 54, 685–695, 2008.

Bender, M. L.: Orbital tuning chronology for the Vostok climate record supported by trapped gas composition, Earth Planet. Sci. Lett., 204, 275–289, 2002.

Bereiter, B., Schwander, J., Lüthi, D., and Stocker, T. F.: Change in CO_2 concentration and O_2/N_2 ratio in ice cores due to molecular diffusion, Geophys. Res. Lett., 36, L05703, doi:10.1029/2008GL036737, 2009.

Bereiter, B., Lüthi, D., Siegrist, M., Schüpbach, S., Stocker, T. F., and Fischer, H.: Mode change of millennial CO_2 variability during the last glacial cycle associated with a bipolar marine carbon seesaw, P. Natl. Acad. Sci., 109, 9755–9760, 2012.

Durand, G., Svensson, A., Persson, A., Gagliardini, O., Gillet-Chaulet, F., Sjolte, J., Montagnat, M., and Dahl-Jensen, D.: Evolution of the texture along the EPICA Dome C ice core, Physics of ice core records II, 68, 91–105, 2009.

Elderfield, H., Ferretti, P., Greaves, M., Crowhurst, S., McCave, I. N., Hodell, D., and Piotrowski, A. M.: Evolution of Ocean Temperature and Ice Volume Through the Mid-Pleistocene Climate Transition, Science, 337, 704–709, 2012.

Enting, I. G.: On the use of smoothing splines to filter CO_2 data, J. Geophys. Res., 92, 10977–10984, 1987.

EPICA community members: One-to-one coupling of glacial climate variability in Greenland and Antarctica, Nature, 444, 195–198, 2006.

Fischer, H., Severinghaus, J., Brook, E., Wolff, E., Albert, M., Alemany, O., Arthern, R., Bentley, C., Blankenship, D., Chappellaz, J., Creyts, T., Dahl-Jensen, D., Dinn, M., Frezzotti, M., Fujita, S., Gallee, H., Hindmarsh, R., Hudspeth, D., Jugie, G., Kawamura, K., Lipenkov, V., Miller, H., Mulvaney, R., Parrenin, F., Pattyn, F., Ritz, C., Schwander, J., Steinhage, D., van Ommen, T., and Wilhelms, F.: Where to find 1.5 million yr old ice for the IPICS "Oldest-Ice" ice core, Clim. Past, 9, 2489–2505, doi:10.5194/cp-9-2489-2013, 2013.

Huber, C., Leuenberger, M., and Zumbrunnen, O.: Continuous Extraction of Trapped Air from Bubble Ice or Water for On-Line Determination of Isotope Ratios, Anal. Chem., 75, 2324–2332, 2003.

Ikeda-Fukazawa, T., Kawamura, K., and Hondoh, T.: Mechanism of Molecular Diffusion in Ice Crystals, Molecular Simulation, 30, 973–979, 2004.

Ikeda-Fukazawa, T., Fukumizu, K., Kawamura, K., Aoki, S., Nakazawa, T., and Hondoh, T.: Effects of molecular diffusion on trapped gas composition in polar ice cores, EPSL, 229, 183–192, 2005.

Jouzel, J., Masson-Delmotte, V., Cattani, O., Dreyfus, G., Falourd, S., Hoffmann, G., Minster, B., Nouet, J., Barnola, J., Chappellaz, J., Fischer, H., Gallet, J. C., Johnsen, S., Leuenberger, M., Loulergue, L., Lüthi, D., Oerter, H., Parrenin, F., Raisbeck, G., Raynaud, D., Schilt, A., Schwander, J., Selmo, E., Souchez, R., Spahni, R., Stauffer, B., Steffensen, J. P., Stenni, B., Stocker, T. F., Tison, J., Werner, M., and Wolff, E. W.: Orbital and millennial Antarctic climate variability over the last 800,000 years, Science, 317, 793–796, 2007.

Kawamura, K., Parrenin, F., Lisiecki, L., Uemura, R., Vimeux, F., Severinghaus, J. P., Hutterli, M. A., Nakazawa, T., Aoki, S., Jouzel, J., Raymo, M. E., Matsumoto, K., Nakata, H., Motoyama, H., Fujita, S., Goto-Azuma, K., Fujii, Y., and Watanabe, O.: Northern Hemisphere forcing of climatic cycles in Antarctica over the past 360,000 years, Nature, 448, 912–917, 2007.

Lliboutry, L.: A critical review of analytical approximate solutions for steady state velocities and temperature in cold ice sheets, Z. Gletscherkd. Glacialgeol, 15, 135–148, 1979.

Lüthi, D., Le Floch, M., Bereiter, B., Blunier, T., Barnola, J., Siegenthaler, U., Raynaud, D., Jouzel, J., Fischer, H., Kawamura, K., and Stocker, T. F.: High-resolution carbon dioxide concentration record 650,000–800,000 years before present, Nature, 453, 379–382, 2008.

Lüthi, D., Bereiter, B., Stauffer, B., Winkler, R., Schwander, J., Kindler, P., Leuenberger, M., Kipfstuhl, S., Capron, E., Landais, A., Fischer, H., and Stocker, T. F.: CO_2 and O_2 / N_2 variations in and just below the bubble–clathrate transformation zone of Antarctic ice cores, Earth Planet. Sci. Lett., 297, 226–233, 2010.

NEEM community members: Eemian interglacial reconstructed from a Greenland folded ice core, Nature, 493, 489–494, 2013.

Parrenin, F., Barnola, J.-M., Beer, J., Blunier, T., Castellano, E., Chappellaz, J., Dreyfus, G., Fischer, H., Fujita, S., Jouzel, J., Kawamura, K., Lemieux-Dudon, B., Loulergue, L., Masson-Delmotte, V., Narcisi, B., Petit, J.-R., Raisbeck, G., Raynaud, D., Ruth, U., Schwander, J., Severi, M., Spahni, R., Steffensen, J. P., Svensson, A., Udisti, R., Waelbroeck, C., and Wolff, E.: The EDC3 chronology for the EPICA Dome C ice core, Clim. Past, 3, 485–497, doi:10.5194/cp-3-485-2007, 2007a.

Parrenin, F., Dreyfus, G., Durand, G., Fujita, S., Gagliardini, O., Gillet, F., Jouzel, J., Kawamura, K., Lhomme, N., Masson-Delmotte, V., Ritz, C., Schwander, J., Shoji, H., Uemura, R., Watanabe, O., and Yoshida, N.: 1-D-ice flow modelling at EPICA Dome C and Dome Fuji, East Antarctica, Clim. Past, 3, 243–259, doi:10.5194/cp-3-243-2007, 2007b.

Petit, J. R., Jouzel, J., Raynaud, D., Barkov, N. I., Barnola, J., Basile, I., Bender, M., Chappellaz, J., Davis, M., Delaygue, G., Delmotte, M., Kotlyakov, V. M., Legrand, M., Lipenkov, V. Y., Lorius, C., Pépin, L., Ritz, C., Saltzman, E., and Stievenard, M.: Climate and atmospheric history of the past 420,000 years from the Vostok ice core, Antarctica, Nature, 399, 429–436, 1999.

Pol, K., Masson-Delmotte, V., Johnsen, S., Bigler, M., Cattani, O., Durand, G., Falourd, S., Jouzel, J., Minster, B., Parrenin, F., Ritz, C., Steen-Larsen, H. C., and Stenni, B.: New MIS 19 EPICA Dome C high resolution deuterium data: Hints for a problematic preservation of climate variability at sub-millennial scale in the "oldest ice", Earth Planet. Sci. Lett., 298, 95–103, 2010.

Salamatin, A. N., Lipenkov, V. Y., and Blinov, K. V.: Vostok (Antarctica) climate record time-scale deduced from the analysis of a borehole-temperature profile, Ann. Glaciol., 20, 207–214, 1994.

Salamatin, A. N., Lipenkov, V. Y., Barkov, N., Jouzel, J., Petit, J. R., and Raynaud, D.: Ice core age dating and paleothermometer calibration based on isotope and temperature profiles from deep boreholes at Vostok Station (East Antarctica), J. Geophys. Res., 103, 8963–8977, 1998.

Salamatin, A. N., Lipenkov, V. Y., Ikeda-Fukazawa, T., and Hondoh, T.: Kinetics of air-hydrate nucleation in polar ice sheets, Journal of Crystal Growth, 223, 285–305, 2001.

Schwander, J., Sowers, T., Barnola, J., Blunier, T., Fuchs, A., and Malaizé, B.: Age scale of the air in the Summit ice: Implication for glacial-interglacial temperature change, J. Geophys. Res., 102, 19483–19493, 1997.

Spahni, R., Schwander, J., Flückiger, J., Stauffer, B., Chappellaz, J., and Raynaud, D.: The attenuation of fast atmospheric CH_4 variations recorded in polar ice cores, Geophys. Res. Lett., 30, 1571, doi:10.1029/2003GL017093, 2003.

Uchida, T., Miyamoto, A., Shin'Yama, A., and Hondoh, T.: Crystal growth of air hydrates over 720 ka in Dome Fuji (Antarctica) ice cores: microscopic observations of morphological changes below 2000 m depth, J. Glaciol., 57, 1017–1026, 2011.

11

Mass changes in Arctic ice caps and glaciers: implications of regionalizing elevation changes

J. Nilsson[1], **L. Sandberg Sørensen**[1], **V. R. Barletta**[1,2], **and R. Forsberg**[1]

[1]Department of Geodynamics, DTU Space, Technical University of Denmark, Elektrovej 327, 2800 Lyngby, Denmark
[2]Ohio State University, 275 Mendenhall Lab, 125 S. Oval Mall, Columbus, Ohio, 43214, USA

Correspondence to: J. Nilsson (jnils@space.dtu.dk)

Abstract. The mass balance of glaciers and ice caps is sensitive to changing climate conditions. The mass changes derived in this study are determined from elevation changes derived measured by the Ice, Cloud, and land Elevation Satellite (ICESat) for the time period 2003–2009. Four methods, based on interpolation and extrapolation, are used to regionalize these elevation changes to areas without satellite coverage. A constant density assumption is then applied to estimate the mass change by integrating over the entire glaciated region.

The main purpose of this study is to investigate the sensitivity of the regional mass balance of Arctic ice caps and glaciers to different regionalization schemes. The sensitivity analysis is based on studying the spread of mass changes and their associated errors, and the suitability of the different regionalization techniques is assessed through cross-validation.

The cross-validation results shows comparable accuracies for all regionalization methods, but the inferred mass change in individual regions, such as Svalbard and Iceland, can vary up to 4 Gt a^{-1}, which exceeds the estimated errors by roughly 50 % for these regions. This study further finds that this spread in mass balance is connected to the magnitude of the elevation change variability. This indicates that care should be taken when choosing a regionalization method, especially for areas which exhibit large variability in elevation change.

1 Introduction

The most recent assessments from the Intergovernmental Panel on Climate Change (Vaughan et al., 2014) and the Snow, Water, Ice and Permafrost in the Arctic Assessment (AMAP, 2012) state that the mass loss from glaciers and ice sheets is a major contributor to sea-level rise. The use of satellite altimetry to determine the elevation change in the major ice sheets has been possible since the late 1980s and was pioneered by Zwally et al. (1987), Wingham et al. (1998) and others. In recent years this has been expanded to ice caps and glaciers using both satellite and airborne altimetry, in studies such as Gardner et al. (2011), Moholdt et al. (2010a, 2012), Abdalati et al. (2004) and Arendt et al. (2002, 2006).

The geodetic mass balance of glaciers or ice caps can be determined through the use of altimetry by measuring the temporal change in surface elevation of the glaciated area. This rate of change is then converted into volume and finally mass change by multiplying the rate of change by area and an assumed density scheme. Determining the regional rate of change in the entire glaciated area involves interpolation or extrapolation (referred to hereinafter as regionalization) of the elevation changes to unmeasured areas away from the satellite ground tracks. This regionalization might introduce a large uncertainty to the volume estimate, as the track coverage over individual ice caps and glaciers are usually limited.

The focus of this study is to determine the impact of different regionalization schemes on regional ice-mass balance estimates of Iceland, Svalbard, the Russian High Arctic and the Canadian Arctic (south and north). Studying the spread of mass change and their estimated errors allows us to judge the different regions' sensitivity to the regionalization procedure. A cross-validation allows us to identify more preferable regionalization schemes.

2 Study areas and data

In this study, we focus on five regions in the Arctic: Iceland (ICEL), Svalbard (SVLB), the Russian High Arctic (RUS), Canadian Arctic North (CAN) and Canadian Arctic South (CAS). The glacier outlines for these areas have been obtained from the "Randolph Glacier Inventory" (RGI) (Pfeffer et al., 2014).

The regional mass changes have been estimated from elevation changes obtained from the Ice, Cloud, and land Elevation Satellite (ICESat) (Schutz et al., 2005) over the time period 2003–2009. ICESat carried the Geoscience Laser Altimetry System (GLAS), which operated from 2003 to 2009 and had a repeat cycle of 96 days with a 33-day sub-cycle. The system measured the range between the satellite and a surface on the Earth, derived from the delay time between the transmitted laser pulse and the received return echo. The average ground-track sample spacing was 172 m along-track, and the ground footprint was approximately 70 m in diameter. The ICESat elevation data are obtained from the National Snow and Ice Data Center (http://nsidc.org/data/icesat/index.html) in the form of the GLA06 L1B global surface elevation data product, product release (R33).

Digital elevation models (DEMs) with a resolution of 1 km (30 arcsec) for use in the elevation-dependent regionalization are available for the five areas. The GTOPO30 DEM is used for the areas of CAS, CAN and RUS. For SVLB and ICEL, DEMs from the National Geospatial-Intelligence Agency (NGA) are used. The GTOPO30 model is estimated to have vertical accuracies of 50–200 m (http://www1.gsi.go.jp/geowww/globalmap-gsi/gtopo30/gtopo30.html). The NGA DEMs have a similar accuracy, as the data partly have common roots.

To further estimate the quality of the topography models, we have compared them to 2003–2009 surface heights obtained from ICESat by interpolating the DEMs' surface heights to the ICESat data locations using bilinear interpolation. We estimate the standard deviation of the difference between the DEM heights and the ICESat heights to judge their quality. For the GTOPO30 model we find an average standard deviation over all regions of ~ 65 m and for the NGA DEMs of ~ 45 m.

3 Data processing

In an initial step, the ICESat GLA06 product has been filtered using the quality flags and rejection parameters included in the product release. Several rejection criteria have been used in the data culling, e.g. data are only used if the flags indicated usable elevations (i_ElvuseFlg = 1) and only have one peak in the return signal (i_numPk = 1). Relevant data (i_satCorrFlg = 2) have been corrected with the provided saturation correction. Each elevation measurement has been corrected for the Gaussian centroid (GC) offsets according to Borsa et al. (2013). There exists an inter-campaign bias in the ICESat data Siegfried et al. (2011) and Hofton et al. (2013), but since this is still debated, we have not applied any bias correction in this study. The RGI glacier outlines have been used to extract only data over the glaciated areas of interest.

The ICESat ground tracks did not have perfect spatial repetition, and there could be large (up to 1°) offsets between the individual tracks from the main ground-track cluster. Tracks with large offsets have been edited out in order to produce more robust elevation change estimates. The ICESat repeat ground tracks are divided into 500 m segments to estimate surface elevation changes. The mean elevation change was estimated in each segment by least-squares regression if data from more than six campaigns were available. This method is described in detail in Sørensen et al. (2011) (referred to in that paper as the M3 method).

A cleaning procedure has been applied to the estimated elevation changes, in which elevation changes with an estimated standard deviation (estimated from the least-squares solution) outside the 95 % confidence interval of the regression errors are removed. Furthermore, a 10-point moving Hampel filter (Pearson et al., 2002) has been used to identify and remove outliers in the elevation changes. The filtering is applied in the elevation change versus elevation domain to ease outlier detection. The success of the screening was judged visually to avoid unnecessary rejection. As a last step, an along-track smoothing filter has been applied to the elevation change data. An unweighted five-point moving average filter with a corresponding physical filter distance of 2.5 km has been used. Smoothing is undertaken to remove high-frequency noise from the elevation change estimates, and to aid the fitting procedure for the extrapolation and surface fitting for the interpolation methods, which are described in Sects. 4.2.1 and 4.2.2.

4 Methods

This study uses four different methods that have been implemented to regionalize elevation change to partly unmeasured glaciated areas. The four methods can broadly be divided into two categories: interpolation and extrapolation methods. The fundamental difference between the two approaches is what main correlation dependency they use for the regionalization procedure.

The interpolation methods use the spatial correlation (horizontal) of the elevation changes to predict an elevation change value at a specific geographical location. While the extrapolation method uses the usually high altitudinal correlation of elevation change to model the elevation change at a specific elevation. The four different methods (referred to as M1–M4), based on interpolation and extrapolation, can be summarized as follows:

- M1: smooth surface fit,
- M2: spatial model,
- M3: hypsometric polynomial,
- M4: hypsometric elevation bins.

The methods introduced here will be explained in the following sections.

One of the main sources of uncertainty in the mass change estimation is the conversion to mass via a density assumption (Huss et al., 2013). A constant density of 900 kg m^{-3} is used in this study. This assumption has also been applied by others, such as Gardner et al. (2011) and Moholdt et al. (2010a, 2012), and has been used in this study to simplify comparisons to other studies and to ensure that the spread in the mass balance estimates is a result of the different regionalization schemes alone and not due to the density conversion.

4.1 Regionalization: spatial interpolation

The first regionalization method (referred to as M1) fits a smooth surface to the scattered elevation change estimates, with an along-track resolution of 500 m, using least-squares collocation (as implemented in the GRAVSOFT program GEOGRID; see Forsberg et al., 2008, and Moritz., 1978) onto a regular grid, with a grid spacing of 0.01° latitude and 0.025° longitude, corresponding to a resolution of ~ 1 km. The glaciated area of these grids have then been extracted using the RGI glacier outlines.

The least-squares collocation interpolation uses a quadrant-based nearest-neighbour search to find the N_q closest points in every quadrant around the prediction point. The data points are then interpolated by applying a second-order Markov covariance model. The covariance length is found from the data and the correlation distance is input by the user to the GEOGRID program. The correlation length has been increased until the individual satellite ground tracks are not visible on the surface. This method create a smooth continuous surface between the individual ground tracks, that usually have large cross-track spacing.

Due to data processing and data editing there is a loss of spatial coverage and thus data gaps in the along-track elevation changes. The second regionalization method (referred to as M2) tries to improve this by re-sampling the along-track data location in every track from 500 to 100 m. This to fill in data gaps and increases the along-track resolution. New along-track elevation changes are then estimated from the entire elevation change data set using the following model:

$$\dot{h} = a_0 + a_1\Phi + a_2\lambda + a_3 h + \ldots + a_N h^N, \quad (1)$$

where $\dot{h}$ is the parametrized elevation change value, a_i are the model coefficients, h is the DEM elevation, N is the model order, Φ is the latitude and λ is the longitude. The model order used for each region is the same as described in Sect. 4.1.2 for the spatial extrapolation methods. The M2 approach was chosen because it takes into account the overall spatial pattern of the elevation changes instead of just the nearest neighbours, which can in some cases be situated far away due to large across-track distances.

For the five regions of interest in this study, the number of points in each quadrant is set to $N_q = 5$, and a correlation length of 50 km gave a sufficiently smooth surface.

4.2 Regionalization: hypsometric extrapolation

The third regionalization method (M3) uses hypsometric averaging (Nuth et al., 2010; Moholdt et al., 2010a) for extrapolation of elevation change estimates to derive volume change. Hypsometric averaging is based on parametrizing elevation changes as a function of elevation using an external DEM, with the corresponding grid spacing as M1–M2. The glaciated area is divided into elevation bands or bins, and each band is assigned a representative elevation change value, estimated from the parametrized data set.

In M3, the elevation changes are parametrized by fitting a polynomial function to all the elevation change data, as in Nuth et al. (2010) and Moholdt et al. (2010a). The elevations are obtained from the glacier-masked DEMs for every region (see Sect. 2). Hypsometric averaging is then used to extrapolate the elevation changes regionally.

To determine the degree and the number of terms in the polynomial, we need a measure of how much variance the model is able to account for. The more variability that can be incorporated into the model, the better it will explain the underlying statistics of the measured data. We use the adjusted R^2 statistics as a measure of incorporated variance (see Moholdt et al., 2010a). The degree of the polynomial and the number of parameters are then increased until a convergence of this R^2 is reached. For all regions except Svalbard, a linear fit ($D = 1$) was sufficient to parametrize the relation. For Svalbard, a third-order polynomial ($D = 3$) fits the distribution best (as measured by R^2, as used by Moholdt et al., 2010a). An elevation bin range of 50 m was chosen for all regions, consistent with Gardner et al. (2011).

The fourth regionalization method (M4) also involves binning the elevation changes according to elevation (as in M3), but instead of estimating the centre bin elevation change from a continuous function we instead use the mean value of the elevation changes inside the bin. Elevation bins that do not contain any data are assigned a value from linear interpolation. DEM elevations which are not covered by the ICESat data (usually low and high elevations) are assigned a value from extrapolation of the linear function to these bins, estimated from the entire data set.

The Russian High Arctic was divided into three subregions (Novaya Zemlya, Franz Josef Land and Severnaya Zemlya) for the analysis when using regionalization methods M3 and M4, due to large geographical separation within this region.

4.3 Volume and mass change

To determine the regional volume change in the interpolated and extrapolated fields the estimated elevation changes are multiplied by their corresponding area. This procedure differs between the inter and extrapolation methods and is for that reason described below.

To estimate the volume change from the interpolation methods (M1–M2) each individual elevation change grid cell (pixel), $\dot{h}_i$, is multiplied with its corresponding area, corresponding pixel area A_p, and summed as follows to obtain the regional volume change $\dot{V}$:

$$\dot{V} = \sum(\dot{h}_i) \cdot A_\mathrm{p}. \tag{2}$$

The volume change from the extrapolated elevation changes (M3–M4) are estimated in a slightly different way. Here the estimated elevation change for each bin or band is multiplied with the total area of each band, and summed as follows to obtain the regional volume change:

$$\dot{V} = \sum\left(\dot{h}(z) \cdot A(z)\right), \tag{3}$$

where $\dot{h}$ is the specific elevation change value assigned to the elevation band/bin z, which is defined as the centre or mid-elevation of that bin (i.e. 25 m if the bin range is 0–50 m). $A(z)$ represents the total area inside the elevation band/bin at the specific binned elevation z.

The regional mass change is then estimated by multiplying the volume change by a constant density of 900 kg m^{-3}. This approach assumes that the mass changes are due to effects such as ice melt and dynamic thinning while ignoring effects like changes in accumulation rate and firn densification. This is a very simplified view and is not always valid, which makes it a large source of uncertainty.

Cross-validation

A cross-validation scheme has been employed to assess the quality of the regionalized elevation change fields from the four different methods. As the individual elevation changes are highly correlated along-track, we perform a cross-validation scheme on entire ICESat tracks of elevation changes. The individual ground tracks are assumed to be uncorrelated, and the cross-validation is performed in the following manner:

1. Remove one of the ground tracks from the original data set.
2. Use M1–M4 to regionalize the elevation changes from the reduced data set.
3. Find the estimated elevation change values from the different methods at the locations of the removed ground track.
4. Compute the difference between the estimated and original elevation changes.
5. Compute the root mean square (rms) of the residuals.
6. Repeat the procedure for all available ground tracks.

This procedure will produce one rms value for each ground track and N_rms for each method. The mean rms, $\overline{\mathrm{rms}}$, of all the ground tracks is then used to judge the quality of the different regionalization schemes.

5 Error analysis

We base the error analysis on two main concepts – the standard deviation around the mean and the standard error of the data – following the approaches of Nuth et al. (2010) and Moholdt et al. (2010a). Several studies have been dedicated to quantifying the individual point measurement errors for ICESat over ice-covered regions. Brenner et al. (2007) found that the ICESat measurement error over ice sheets varied as a function of surface slope, ranging from 0.14 m at 0.1° up to 0.5 m at 1.2°. Regions such as Svalbard have a large range in surface slope, varying between 0 and 29° at most, with a mean slope of 4.1°. Therefore, we assume a conservative error of $\varepsilon_\mathrm{icesat} = 0.17\,\mathrm{m\,a^{-1}}$, coming from a measurement error of 1 m (Nuth et al., 2010) and a measurement period of 6 years. This error does not account for the inter-campaign bias present in the ICESat data, which does not affect the spread of the regional mass balance.

To estimate the error from the elevation change estimation procedure, we use the standard deviation estimated from the least-squares solution of the elevation changes as a measure of how trustworthy the individual elevation change measurements are (Sørensen et al., 2011). ICESat elevation changes are highly correlated along-track due to distance being short between the measurements, compared to variations in the topography. In this study, we have estimated the correlation length from the semi-variogram of the elevation changes, and use the correlation length to spatially bin the elevation changes. The correlation lengths for the five regions are found to be ~ 15 km for SVLB, ~ 10 km for ICEL, ~ 20 km for CAN, ~ 20 km for CAS and ~ 10 km for RUS. For more detailed work on this topic, please see Rolstad et al (2009). The bins are then assumed to be uncorrelated, and the total number of non-empty bins is used to estimate the standard error, $\varepsilon_{\mathrm{d}h/\mathrm{d}t}$, for all four methods M1–M4:

$$\varepsilon_{\mathrm{d}h/\mathrm{d}t} = \frac{\sigma_{\mathrm{d}h/\mathrm{d}t}}{\sqrt{N}}, \tag{4}$$

where N is the number of uncorrelated bins and $\sigma_{\mathrm{d}h/\mathrm{d}t}$ is the mean standard deviation of the elevation changes. Here, $\sigma_{\mathrm{d}h/\mathrm{d}t}$ has already been reduced by a factor of $1/\sqrt{N_\mathrm{s}}$ due to the along-track smoothing, with N_s being the size of the smoothing filter.

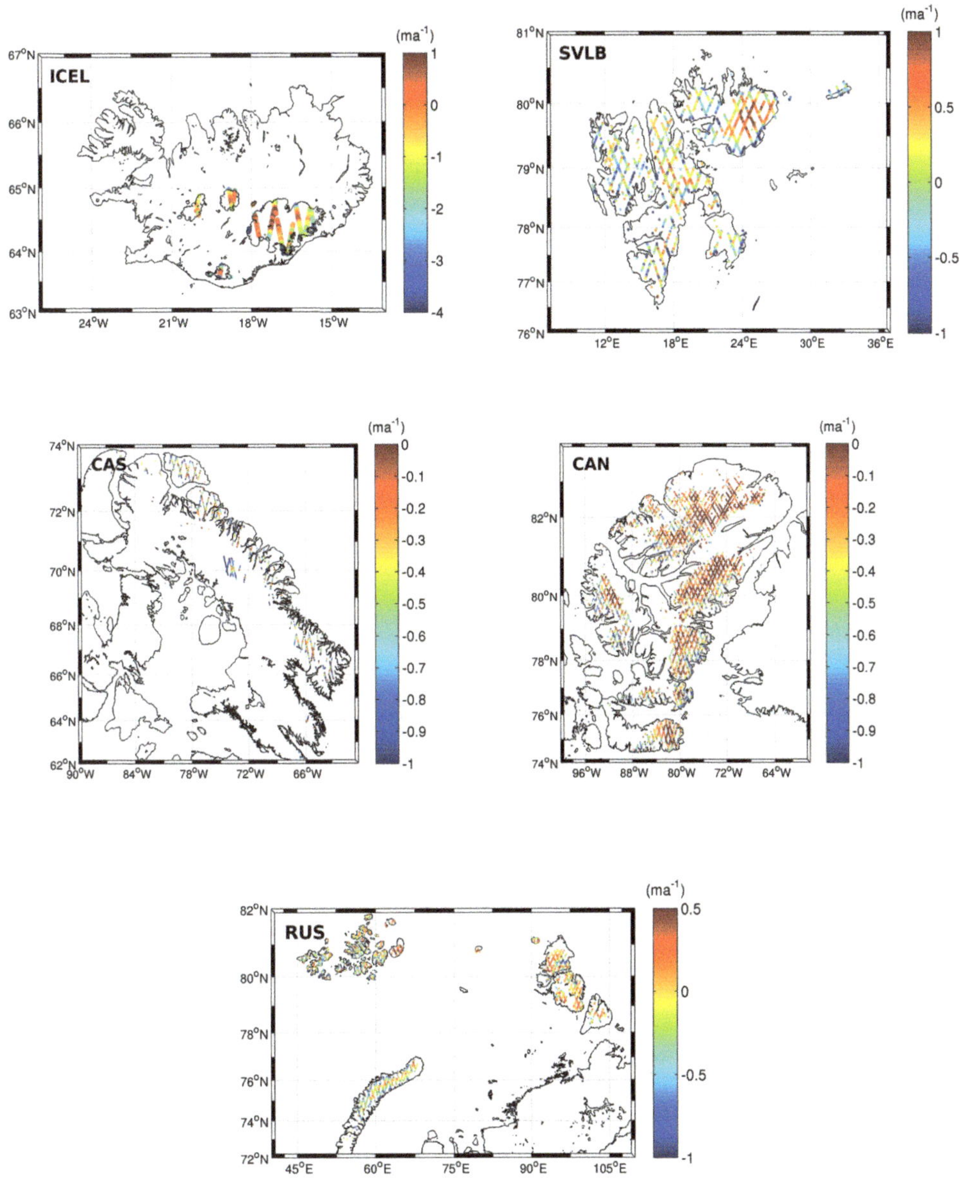

Figure 1. Spatial patterns of elevation changes of the five regions in the form of the satellite ground track coverage.

The least-squares collocation error associated with M1 and M2 is estimated by computing the standard deviation of the data around every prediction point according to Moritz. (1978). The mean value of these standard deviations is used as the interpolation error, and the standard error is computed in the same way as in the elevation change procedure:

$$\varepsilon_{\text{int}} = \frac{\sigma_{\text{int}}}{\sqrt{N}}, \tag{5}$$

where σ_{int} is the mean standard deviation from the collocation prediction of the data inside the glaciated area.

We quantify the parametrization error from the fitting of the polynomial function used in M2 and M3 by calculating the root-mean-square error (RMSE) between the original elevation change estimates and the predicted values:

$$\varepsilon_{\text{fit}} = \frac{\sigma_{\text{fit}}}{\sqrt{N-D}}, \tag{6}$$

where σ_{fit} is the RMSE between the original and predicted data, and $\sqrt{N-D}$ is the adjustment due to the degree of the polynomial. The extrapolation error, ε_{ext}, relevant for method M3, is quantified by the same approach as used in Nuth et al. (2010), with the extrapolation error being the root-sum-square (RSS) difference of the fitted error minus the elevation change error:

$$\varepsilon_{\text{ext}} = \sqrt{\varepsilon_{\text{fit}}^2 - \varepsilon_{\text{d}h/\text{d}t}^2}. \tag{7}$$

Table 1. The number of error terms present in each method. These error are then combined into a height error using RSS.

Method	Error terms
M1	$\varepsilon_{icesat}, \varepsilon_{int}, \varepsilon_{dh/dt}$
M2	$\varepsilon_{icesat}, \varepsilon_{int}, \varepsilon_{dh/dt}, \varepsilon_{fit}$
M3	$\varepsilon_{icesat}, \varepsilon_{ext}, \varepsilon_{dh/dt}$
M4	$\varepsilon_{icesat}, \varepsilon_{b}, \varepsilon_{dh/dt}$

The extrapolation error for the mean binning method is referred to as the binning error, ε_b, not to be confused with ε_{ext}. This error is associated with M4 and is defined as the standard deviation inside every elevation bin, σ_b. The standard error is then calculated by assuming that the individual bins are uncorrelated:

$$\varepsilon_b = \frac{\sigma_b}{\sqrt{N}}, \tag{8}$$

The corresponding total elevation change error, ε_{tot}, is then estimated as the RSS of the individual error sources as given in Table 1. The volumetric error, ε_{vol}, can then be estimated by multiplying the elevation change error with the total glaciated area

$$\varepsilon_{vol} = \varepsilon_{tot} \cdot A \tag{9}$$

We also include an error term, ε_ρ, to account for the simple density assumption used that ignores the fact that density is actually a function of space and time. The approach follows that of, Moholdt et al. (2010b)

$$\varepsilon_\rho = \frac{1}{2}(\rho_{ice} - \rho_{firn}), \tag{10}$$

where ρ_{ice} and ρ_{firn} are the densities of ice and firn, respectively. This error is applied to the entire glaciated region. Finally, we can estimate the mass balance error, ε_{mass}, as follows:

$$\varepsilon_{mass} = \sqrt{(\varepsilon_{vol} \cdot \rho)^2 + (\dot{V} \cdot \varepsilon_\rho)^2}, \tag{11}$$

where $\dot{V}$ is the estimated volume change.

6 Results

The along-track rates of elevation change have been derived for the five regions: ICEL, SVLB, CAS, CAN and RUS. The elevation change results are shown in Fig. 1. The regions exhibit different patterns of rates and variability in the elevation changes. To clarify these differences, a histogram of the elevation changes for the different regions is shown in Fig. 2, and the associated mean elevation change rate, standard deviation, and minimum and maximum values are presented in Table 2. ICEL shows the largest mean rate and variability in elevation change of all five regions, while RUS and CAN show the lowest variability. SVLB exhibits its own unique behaviour, with a more complex pattern of elevation change; compared to the other areas, SVLB shows the lowest mean rate of elevation change of $-0.04\,\mathrm{m\,a^{-1}}$, and the second largest variability after Iceland. The variability in the elevation changes found in the Canadian Arctic and the RUS is a factor of 2 lower than in ICEL and SVLB. The rate of elevation change is apparently a function of latitude, with lower-latitude regions like ICEL and CAS showing the largest mean rate of elevation change. This pattern is not as easily detected in the variability in the elevation change, seen in Table 2, where CAN and ICEL both show approximately the same magnitude of elevation change but a large difference in variability (63 %). Common for all five areas, though, is the negatively skewed distribution of elevation changes seen in Fig. 2.

Table 2. ICESat point statistics of elevation change for the different Arctic regions. The values are in $\mathrm{m\,a^{-1}}$ for the statistics and $\mathrm{km^2}$ for the area, and N is the number of observations.

Region	Mean	SD	Min	Max	Area	*N*
SVLB	−0.04	0.70	−6.70	2.20	33 673	4613
ICEL	−0.65	1.14	−6.20	1.79	10 989	851
CAN	−0.27	0.34	−3.00	1.44	103 990	18 022
CAS	−0.58	0.42	−3.75	1.88	40 601	3281
RUS	−0.13	0.34	−2.00	1.46	51 161	8797

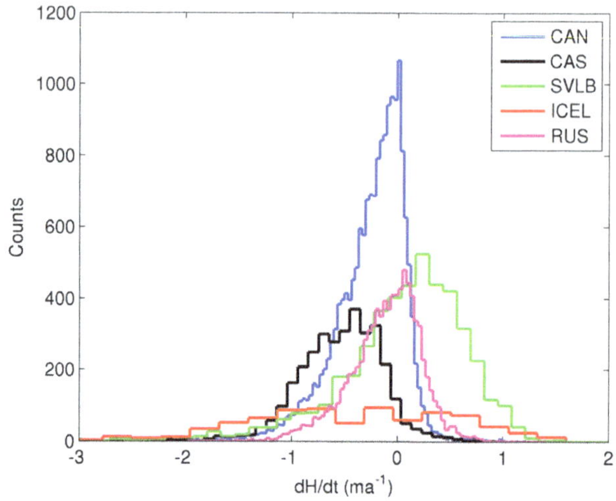

Figure 2. Histogram of elevation changes for the different Arctic regions. The Russian High Arctic (RUS) is treated as one region for visualization purposes.

The Arctic regions show a consistent pattern of large peripheral thinning and small changes in the interior regions of the ice caps (see Figs. 1 and 3). The thinning is mostly located in the low-elevation areas of the ice caps and glaciers ($h < 500$–$800\,\mathrm{m}$), and becomes less negative as the elevation increases. This pattern has also been observed in both studies of ice sheets (e.g. Pritchard et al., 2009) and glaciers (Gardner et al., 2013; Bolsh et al., 2013). The lower elevations in every region show large variability in elevation change, which are clustered around the coastal regions, in areas such as CAN and CAS.

Figure 3 shows the elevation change estimates plotted as a function of elevation, together with the estimated DEM hypsometry and the ICESat elevations averaged per 50 m elevation bin. Most regions in Fig. 3 show no evident or significant sampling bias when comparing the ICESat heights and the estimated DEM hypsometry. There are, however, some observed discrepancies in the ICESat sampling for the low elevations in both ICEL and RUS.

The mass changes estimated from all four methods and all five regions in this study are presented in Table 3, which also contains the estimated mass change error and the mean RMSE from the cross-validation procedure for each method. From Table 3 it can be seen that for regions such as CAN, CAS and RUS, only a small spread in their mass balance estimates is observed. For these three regions, the spread in mass balance estimates is well within the bounds of the estimated errors. For ICEL and SVLB, the spread of the estimated mass changes from the different methods is on the order of 50 % larger than the estimated mass change errors. The spread of the estimated mass changes for the different regions follows patterns seen in the elevation change variability (Table 2). Regions with high variability such as ICEL and SVLB show a much larger spread in the estimated mass changes than the areas with low elevation change variability.

The validity of the different regionalization schemes has been assessed though a cross-validation setup (Sect. 4.1.3). The results of the cross-validation are presented in Table 3, in the form of the mean, maximum and minimum rms for all four methods and regions. The mean RMSE follows the same pattern as detected in both the mass change estimates and elevation change variability, where areas associated with low elevation change variability and low spread in mass balance, such as CAN, CAS and RUS, show a much lower average rms ($\sim$ 65 % lower) than ICEL and SVLB. ICEL shows on average the absolute highest RMSE and also the largest spread in rms between the different methods, as much as 20 %. For the other regions, the spread in the RMSE between the different methods show much better agreement, with observed differences of up to a few percent. The maximum and minimum values obtained from the cross-validation procedure show good agreement for most areas, such as CAN, CAS and RUS. For ICEL and SVLB, a larger spread is observed in these two parameters and follows in general the difference in mass balance, at least for SVLB.

Figure 4 shows the different spatial patterns obtained from the four regionalization procedures for ICEL. Here, the M3 and M4 methods show much larger negative elevation change values at lower elevations than the results based on M1 and M2. The more negative elevation change signal can also be detected in the estimated mass balance for M3 and M4, which is approximately 26 % more negative than for M1–M2.

7 Discussion

The large degree of variability seen in the lower elevations in Fig. 3 for most regions, especially RUS and CAN, indicates that complex spatial and temporal signals have been captured in the data (ice dynamics, ablation, snow accumulation, etc.). This variability is clustered into specific coastal areas in regions such as CAN and CAS, where most of the variability is located below $h < 500$–800 m, and in areas with drainage systems. Most of the more negative elevation changes ($\dot{h} < -1.5\,\mathrm{m\,a^{-1}}$), on the tail of the elevation change distribution, are also located in these low-elevation areas.

This type of low elevation variability (excluding sampling biases) might help to explain the observed difference between the interpolation and extrapolation methods, seen in Table 3. The extrapolation methods have proven to produce more negative values in these area ($h < 500$–800 m) than the interpolation methods, because the interpolation and extrapolation regionalization schemes have two fundamental differences: (i) interpolation methods assume a spatial correlation of the elevation changes and (ii) extrapolation methods assume a vertical correlation in elevation of the elevation changes.

The interpolation approach would in theory (with satisfactory spatial coverage) capture the local spatial variability better than the extrapolation methods, as the extrapolation methods contain no spatial information. The extrapolation methods, on the other hand, make use of the usually high correlation to elevation (if the region is homogeneous enough) to produce a model that in principle is more representative of the lower elevations, given no sampling bias, because the interpolation methods might use data further away from the prediction point. These data might be located at higher elevations or may not be a good representation of the overall glacier-wide pattern, depending on how far away the data points are.

The main issue to consider for the interpolation methods is the spatial coverage of the data. If the spatial coverage is dense enough, the interpolation will be able to capture the spatial pattern of the region. The main issue to consider for the extrapolation methods is the size of the area used for the parametrization. If the area used for the parametrization is too large, important behaviour of the elevation change pattern might not be accounted for in the model. The differences between the interpolation and extrapolation methods can be reduced by dividing the extrapolation region into sub-regions before parametrization, or by including a spatial dependency in the parametrization model.

The effect of the amount of spatial coverage and homogeneity of a region can be exemplified by CAN and ICEL. CAN, for example, shows a very low spread in mass balance

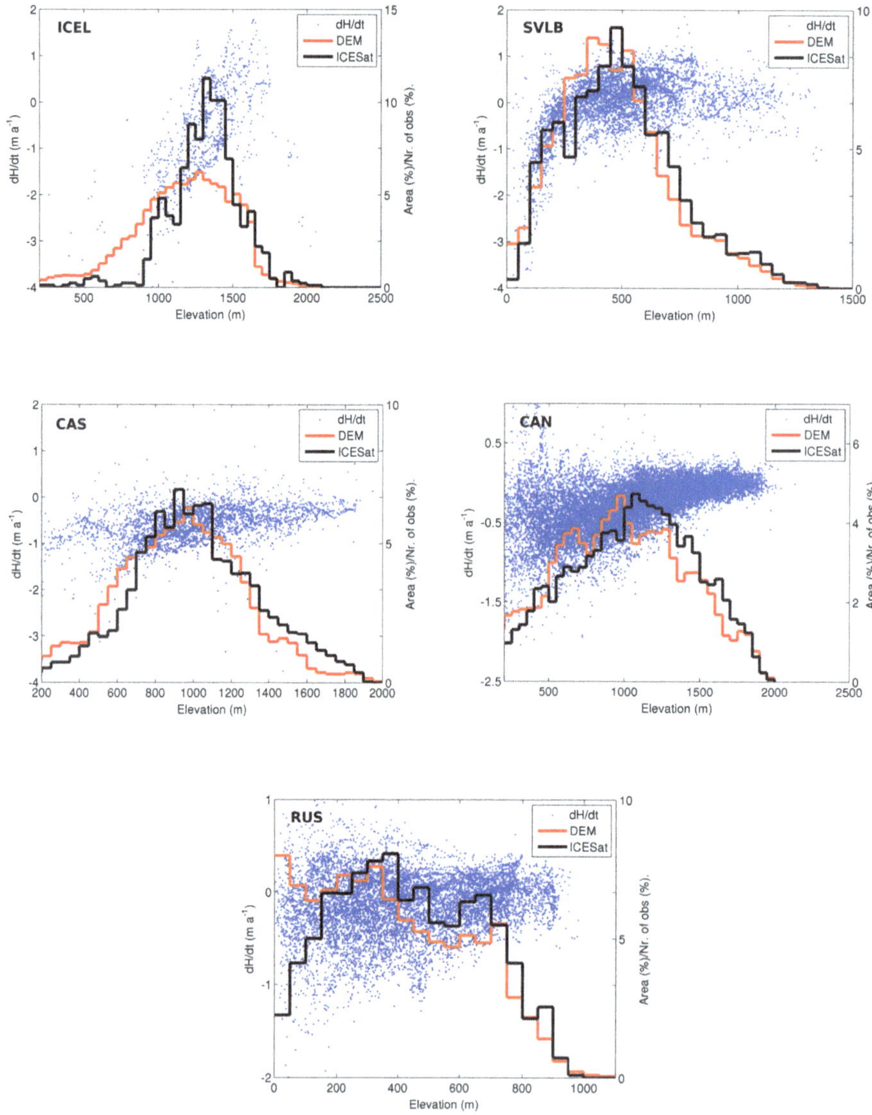

Figure 3. Elevation change (blue points) as a function of elevation for the different Arctic regions, which are used for the regional extrapolation. The black curve represents the density of ICESat's sampling and the red curve the DEM hypsometry, both per 50 m elevation bin.

compared to its large size, while ICEL shows a much larger spread in mass balance. This is due to the spatial sampling of the CAN region being dense and the elevation change variability low, compared to ICEL. This has the effect that both the interpolation and extrapolation methods can capture both the spatial and altitudinal patterns of elevation change for CAN, in contrast to, for example, ICEL with its low data density and large variability.

For most areas with a variability lower than $0.45\,\mathrm{m\,a^{-1}}$ (see Table 2), the impact of the regionalization schemes on the spread of the mass balance is small (on the order of a few percent), with a corresponding spread that falls within the mass balance error. However, for areas with much higher spatial variability and magnitude of elevation change, like ICEL and SVLB, the effect is much more prominent (Table 3). This is most certainly connected to the different types of climate regimes that the regions exhibit. Regions like CAN, CAS and RUS have a continental climate regime (dry and cold), while ICEL and SVLB are in a more maritime climate regime (wet and warm).

For ICEL, the observed difference of almost $4\,\mathrm{Gt\,a^{-1}}$ most probably arises from the low-elevation under-sampling, seen in Fig. 3, where the extrapolation methods are forced to produce more negative elevation changes. This is clearly seen in Fig. 4, where the lower elevations of Vatnajökull produced from the extrapolation methods are much more negative than their interpolation counterparts, and in the mass balance results, presented in Table 3. The estimated average

Table 3. Geodetic mass balance $\dot{m}$ from the four methods with their corresponding errors (σ), and their mean rms ($\overline{\text{rms}}$), maximum rms (rms_{max}) and minimum rms (rms_{min}) from the cross-validation procedure.

Region	Method	$\dot{m}$ [Gt a^{-1}]	σ [Gt a^{-1}]	$\overline{\text{rms}}$ [m a^{-1}]	rms_{max} [m a^{-1}]	rms_{min} [m a^{-1}]
Svalbard	M1	−3.3	2.5	0.57	3.45	0.17
	M2	−5.0	2.7	0.59	2.76	0.26
	M3	−2.5	1.6	0.59	4.10	0.18
	M4	−0.5	2.5	0.57	4.10	0.22
Iceland	M1	−7.8	1.9	1.21	4.62	0.18
	M2	−8.9	3.7	1.10	3.86	0.24
	M3	−10.9	2.7	1.03	3.25	0.40
	M4	−11.7	2.8	1.01	3.26	0.33
Canadian Arctic North	M1	−27.1	6.2	0.23	1.17	0.10
	M2	−25.4	5.8	0.27	1.18	0.12
	M3	−28.6	6.7	0.26	1.10	0.09
	M4	−28.6	7.9	0.25	1.03	0.10
Canadian Arctic South	M1	−22.0	5.7	0.39	2.46	0.15
	M2	−22.5	5.1	0.38	2.37	0.14
	M3	−22.9	5.3	0.37	2.52	0.11
	M4	−22.9	5.2	0.37	2.47	0.13
Russian High Arctic	M1	−7.1	2.0	0.25	0.79	0.04
	M2	−6.4	1.9	0.27	0.81	0.07
	M3	−7.7	3.1	0.28	0.88	0.10
	M4	−7.1	3.9	0.29	0.87	0.09

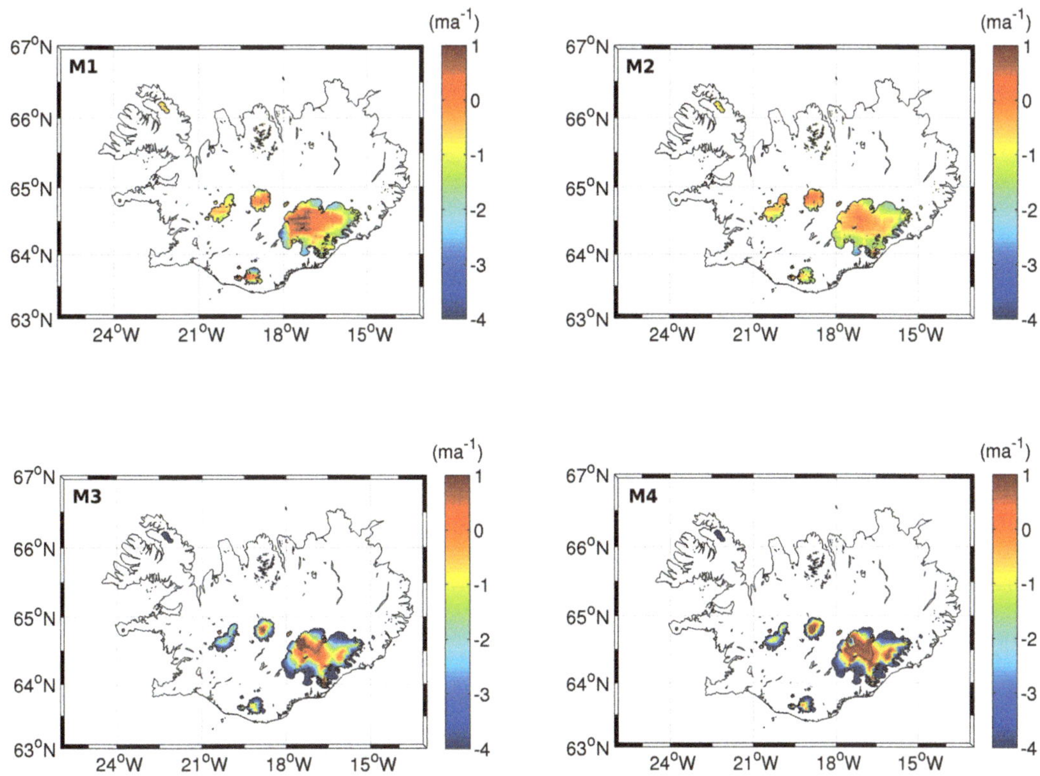

Figure 4. Regionalized elevation changes over Iceland from the M1–M4 methods showing the difference in estimated spatial patterns.

mass balance for ICEL is $-9.8 \pm 2.8\,\mathrm{Gt\,a^{-1}}$ (average of all methods) agrees well with the average contemporary mass loss of $-10 \pm 1.8\,\mathrm{Gt\,a^{-1}}$ estimated by Björnsson et al. (2013) and Gardner et al. (2013) from glaciological measurements. However, there exists an average difference of roughly 25 % between the interpolation and extrapolation methods, where the average of the interpolation methods $-8.35\,\mathrm{Gt\,a^{-1}}$ is more consistent with the results estimated by Gardner et al. (2013) of $-9 \pm 2\,\mathrm{Gt\,a^{-1}}$, while the average of the extrapolation methods $-11.3\,\mathrm{Gt\,a^{-1}}$ is more consistent with the results estimated by Björnsson et al. (2013) of $-11 \pm 1.5\,\mathrm{Gt\,a^{-1}}$.

The difference between the M1–M2 and M3–M4 methods observed in SVLB (Table 3) are most probably due to large spatial variability in the region. The regional parametrization of SVLB might not fully capture the local elevation change pattern as well as the interpolation methods. This effect can be mitigated by applying a spatial dependency, or by dividing the area into sub-regions, as previously discussed. The division into sub-regions has previously been done by Moholdt et al. (2010a), using the M3 method and a $900\,\mathrm{kg\,m^{-3}}$ density, yielding a mass balance of $-3.7\,\mathrm{Gt\,a^{-1}}$. This is in good agreement with the estimated mass balance of $-4.15\,\mathrm{Gt\,a^{-1}}$ obtained from this study by averaging the M1–M2 methods.

The estimated mass changes for CAS and RUS are on the same order as previous studies. Gardner et al. (2011) found a estimated mass loss for CAS of $-24 \pm 6\,\mathrm{Gt\,a^{-1}}$, while Moholdt et al. (2012) found a mass loss of $-9.8 \pm 1.9\,\mathrm{Gt\,a^{-1}}$ for RUS. Both results are in good agreement with the results obtained from this study of CAS of $-22.6 \pm 5.3\,\mathrm{Gt\,a^{-1}}$ and RUS of $-7.1 \pm 2.7\,\mathrm{Gt\,a^{-1}}$ by averaging methods M1–M4. The estimated mass balance for CAN, however, shows a much larger difference of roughly $10\,\mathrm{Gt\,a^{-1}}$ compared to Gardner et al. (2011), who also used ICESat. This difference can mostly be explained by the fact that there was no inter-campaign bias included in the elevation change estimation procedure. The exclusion of the inter-campaign bias gave an average mass balance for CAN of roughly $-30\,\mathrm{Gt\,a^{-1}}$. This was further reduced down to $-27.4\,\mathrm{Gt\,a^{-1}}$ when the GC offset correction (Borsa et al., 2013) was applied. As the GC offset scales with area, smaller regions are less affected by the offset, while larger regions will show a much larger difference. This is also what is observed in this study when applying the GC offset correction. The size of the mass correction introduced by the trend in the GC offset can be estimated for CAN to roughly $1.9\,\mathrm{Gt\,a^{-1}}$ by assuming a maximum trend value of $2\,\mathrm{cm\,a^{-1}}$ and the area given in Table 2. This value agrees well with the observed value of roughly $2.6\,\mathrm{Gt\,a^{-1}}$, which makes the GC offset an important correction for large-scale mass balance studies using ICESat.

To determine whether the 1 km resolution is good enough to give realistic hypsometries, we made a comparison with the ASTER GDEM (http://gdem.ersdac.jspacesystems.or.jp) over Iceland. The ASTER GDEM was re-sampled at a 150 m resolution, to make it easier to handle; binned in 50 m elevation bands; and plotted against the NGA DEM for Iceland over the glaciated areas. Iceland was chosen because the largest discrepancies between the regionalization methods were found here, and also because it exhibits the largest rate of and variability in elevation change. Even though there exists an apparent sampling bias, the calculated mass changes using the ASTER GDEM (methods M3–M4) gave only a total difference of 2 %. Thus we believe that the 1 km DEMs are of sufficient quality and resolution to give realistic hypsometries.

Table 4. The final regional and total geodetic mass balance $\dot{m}$ estimated using the results from the cross-validation procedure and in situ comparison for Iceland, with corresponding error estimates (σ).

Region	$\dot{m}$ [$\mathrm{Gt\,a^{-1}}$]	σ [$\mathrm{Gt\,a^{-1}}$]	Methods
SVLB	−4.2	2.6	M1, M2
ICEL	−9.8	2.8	M1–M4
CAN	−27.4	6.7	M1–M4
CAS	−22.6	5.3	M1–M4
RUS	−7.1	2.7	M1–M4
Total	−71.1	9.7	

The results of the cross-validation procedure, seen in Table 3, indicate that, given enough data sampling, the interpolation and extrapolation methods produce regionalized elevation change estimates of the same quality. Therefore, the interpolation methods described in this study can be used for future mass balance studies in these and other areas even with relatively sparse data sampling. This finding is in contrast to previous discussion, such as that of in Moholdt et al. (2010a), which states that the spatial sampling would usually be to sparse to allow for spatial interpolation, which is definitely true on a sub-regional basis.

Using the estimated RMSEs from the cross-validation procedure, seen in Table 3, as a guide, a combined or final geodetic mass balance was computed, which can be seen in Table 4. The final mass balance and corresponding mass error for CAN, CAS and RUS were determined using the average value of all four methods. Both the RMSE and the estimated mass balance error showed good individual agreement with each other. The final mass balance for SVLB was computed from the M1–M2 method, as these two methods showed the smallest range in the RMSE even though all four methods on average showed the same mean RMSE. Determining the final geodetic mass balance for ICEL is somewhat more arbitrary, as the low density of data points makes the cross-validation more difficult. However, here the average of all methods was chosen to determine the final mass balance of ICEL, as the average of M1–M4 shows the closest agreement with the average in situ-derived value of mass balance from Björnsson et al. (2013) and Gardner et al. (2013).

8 Conclusions

In this study, we have determined the impact of different regionalization schemes of elevation changes on the estimated mass balance of five different Arctic regions. These five regions consisted of the Canadian Arctic (north and south), the Russian High Arctic, Svalbard and Iceland. The estimated mass balance was then, in combination with a cross-validation procedure, used to determine how sensitive these regions are to different regionalization schemes of elevation change. Finally, we also estimated a mass balance budget for each region, using the results derived from the cross-validation procedure and the estimated mass errors.

The study found that the mean rates of and variability in elevation change varied extensively over the different areas in the Arctic. The rate of elevation changes showed a range of $0.6\,\mathrm{m\,a^{-1}}$ across the different regions, while the variability showed a corresponding range of $0.8\,\mathrm{m\,a^{-1}}$. Regions with large variability in elevation change showed a large spread in the estimated mass changes from the different methods, given the described setup. This spread was on average 50 % larger than the respective errors. For regions exhibiting low variability, the opposite was observed. Here, the spread of the mass changes lay well inside the estimated errors.

The statistics from the cross-validation procedure, in conjunction with the estimated mass balance results, indicate that the choice of regionalization method for regions with a variability of less than $0.5\,\mathrm{m\,a^{-1}}$ is negligible. However, if the variability exceeds $0.5\,\mathrm{m\,a^{-1}}$, caution and further analysis is required before choosing a method for mass balance studies. The results from the cross-validation further indicate that the interpolation and extrapolation methods are of the same quality for most areas. Hence the interpolation methods described in this study can also be used for mass balance studies of ice caps and glaciers with satisfactory results.

Acknowledgements. The authors would like to thank the different data contributors: the National Snow and Ice Data Center (NSIDC) for providing the ICESat data; the National Spatial-Intelligence Agency (NGA) for the Svalbard and Iceland DEM; the Ministry of Economy, Trade, and Industry (METI) of Japan and the United States National Aeronautics and Space Administration (NASA) for the ASTER GDEM; and the United States Geological Survey (USGS) for the GTOPO30 DEM. This publication is contribution no. 30 of the Nordic Centre of Excellence SVALI project "Stability and Variations of Arctic Land Ice", funded by the Nordic Top-level Research. We would like to thank the reviewers, especially Geir Moholdt, for their constructive comments and insights, which greatly improved this manuscript. This work was supported by funding from the ice2sea programme from the European Union 7th Framework Programme, grant number 226375. Ice2sea contribution number 175.

Edited by: E. Hanna

References

Abdalati, W., Krabill, W., Fredrick, E., Manizade, S., Martin, C., Sonntag, J., Swift, R., Thomas, R., Yungel, J., and Koerner, R.: Elevation changes of ice caps in the Canadian Arctic Archipelago, J. Geophys. Res., 109, F04007, doi:10.1029/2003JF000045, 2004.

AMAP: Arctic Climate Issues 2011: Changes in Arctic Snow, Water, Ice and Permafrost, SWIPA 2011 Overview Report, ISBN: 978-82-7971-073-8, 2012.

Arendt, A., Echelmeyer, K., Harrison, W., Lingle, C., and Valentine, V.: Rapid Wastage of Alaska Glaciers and Their Contribution to Rising Sea Level, Science, 297, 382–386, doi:10.1126/science.1072497, 2002.

Arendt, A. A., Echelmeyer, K., Harrison, W., Lingle, C., Zirnheld, S., Valentine, V., Ritchie, B., and Druckenmiller, M.: Updated estimates of glacier volume changes in the western Chugach Mountains, Alaska, and a comparison of regional extrapolation methods, J. Geophys. Res.-Earth Surf., 111, F03019, doi:10.1029/2005JF000436, 2006.

Bamber, J., Krabill, W., Raper, V., and Dowdeswell, J.: Anomalous recent growth of part of a large Arctic ice cap: Austfonna, Svalbard, Geophys. Res. Lett., 31, L12402, doi:10.1029/2004GL019667, 2004.

Björnsson, H., Pálsson, F., Gudmundsson, S., Magnússon, E., Adalgeirsdóóttir, G., Jóóhannesson, T., Berthier, E., Sigurdsson, O., and Thorsteinsson, T.: Contribution of Icelandic ice caps to sea level rise: Trends and variability since the Little Ice Age, Geophys. Res., 40, 1–5, doi:10.1002/grl.50278, 2013.

Bolch, T., Sandberg Sörensen, L., Simonsen, S. B., Mölg, N., Machguth, H., Rastner, P., and Paul, F.: Mass loss of Greenland's glaciers and ice caps 2003–2008 revealed from ICESat laser altimetry data, Geophys. Res. Lett., 40, 875–881, doi:10.1002/grl.50270, 2013.

Borsa, A. A., Moholdt, G., Fricker, H. A., and Brunt, K. M.: A range correction for ICESat and its potential impact on ice-sheet mass balance studies, The Cryosphere, 8, 345–357, doi:10.5194/tc-8-345-2014, 2014.

Brenner, A., DiMarzio, J. P., and Zwally, H. J.: Precision and Accuracy of Satellite Radar and Laser Altimeter Data Over the Continental Ice Sheets, IEEE Trans. Geosci. Remote Sens., 45, 321–331, doi:10.1109/TGRS.2006.887172, 2007.

Forsberg, R. and Tscherning, C. C.: An overview manual for the GRAVSOFT Geodetic Gravity Field Modelling Programs, available at: http://www.gfy.ku.dk/~cct/publ_cct/cct1936.pdf (last access: 20 January 2015), 2008.

Gardner, A., Moholdt, G., Wouters, B., Wolken, G., Burgess, D., Sharp, M., Cogley, G., Braun, C., and Labine, C.: Sharply increased mass loss from glaciers and ice caps in the Canadian Arctic Archipelago, Nature, 473, 357–360, doi:10.1038/nature10089, 2011.

Gardner, A., Moholdt, G., Cogley, J. G., Wouters, B., Arendt, A., Wahr, J., Berthier, E., Hock, R., Pfeffer, W. T., Kaser G., Ligtenberg, S. R. M., Bolch, T., Martin, J., Sharp, M. J., Hagen, J. O., van den Broeke, M. R., and Paul, F.: A Reconciled Estimate of Glacier Contributions to Sea Level Rise: 2003 to 2009, Science, 340, 852–857, doi:10.1126/science.1234532, 2013.

Hofton, M. A., Luthcke, S. B., and Blair, J. B.: Estimation of ICESat intercampaign elevation biases from comparison of lidar data in East Antarctica, Geophys. Res. Lett., 40, 852–857, doi:10.1002/2013GL057652, 2013.

Huss, M.: Density assumptions for converting geodetic glacier volume change to mass change, The Cryosphere, 7, 877–887, doi:10.5194/tc-7-877-2013, 2013.

Jacob, T., Wahr, J., Pfeffer, W. T., and Swenson, S.: Recent contributions of glaciers and ice caps to sea level rise, LETTER, Nature, 483, 514–518, doi:10.1038/nature10847, 2012.

Kaser, G., Cogley, J., Dyurgerov, M. B., Meier, M. F., and Ohmura, A.: Mass balance of glaciers and ice caps: Consensus estimates for 1961–2004, Geophys. Res. Lett., 33, L19501, doi:10.1029/2006GL027511, 2006.

Krabill, W., Abdalati, W., Fredrick E., Manizade, S., Martin, C., Sonntag, J., Swift, R., Thomas, R., Wright, W., and Yungel, J.: Greenland ice sheet: High elevation balance and peripheral thinning, Science, 289, 428–430, doi:10.1126/science.289.5478.428, 2000.

Moholdt, G., Nuth, C., Hagen, J. O., Kohler, J.: Recent elevation changes of Svalbard glaciers derived from ICESat laser altimetry, Remote Sens. Environ., 114, 2756–2767, 2010a.

Moholdt, G., Hagen, J. O., Eiken, T., and Schuler, T. V.: Geometric changes and mass balance of the Austfonna ice cap, Svalbard, The Cryosphere, 4, 21–34, doi:10.5194/tc-4-21-2010, 2010.

Moholdt, G., Wouters, B., and Gardner, A.: Recent mass change of glaciers in the Russian High Arctic, Geophys. Res., 39, L10502, doi:10.1029/2012GL051466, 2012.

Moritz, H.: Least Squares Collocation, Reviews of Geophysics and Space Physics, 16, 421–430, doi:10.1029/RG016i003p00421, 1978.

Nuth, C., Moholdt, G., Kohler, J., Hagen, J. O., and Kääb, A.: Svalbard glacier elevation changes and contribution to sea level rise, J. Geophys. Res., 115, F01008, doi:10.1029/2008JF001223, 2010.

Pearson, K. R.: Outliers in process modeling and identification, IEEE, Transa. Control Syst. Technol., 10, 55–63, doi:10.1109/87.974338, 2002.

Pfeffer, T.W., Arendt, A. A., Bliss, A., Bolch, T., Cogley, J. G., Gardner, A. S., Hagen, J. O., Hock, R., Kaser, G., Kienholz, C., Miles, E. S., Moholdt, G., Mölg, N., Paul, F., Radić, V., Rastner, P., Raup, B. H., Rich, J., Sharp, M. J.: The Randolph Consortium, J. Glaciol., 221, 537–552, doi:10.3189/2014jog13j176, 2014.

Pritchard, H. D., Arthern, R. J., Vaughan, D. G., and Edwards, L. A.: Extensive dynamics thinning on the margins of the Greenland and Antartic ice sheets, Nature, 7266, 971–975, doi:10.1038/nature08471, 2009.

Rolstad, C., Haug, T., and Denby, B.: Spatially-integrated geodetic glacier mass balance and its uncertainty based on geostatistical analysis: application to the Western Svartisen ice cap, Norway, J. Glaciol., 55, 666–680, 2009.

Schutz, B. E., Zwally, H. J., Shuman, C. A., Hancock, D., and DiMarzio, J. P.: Overview of the ICESat Mission, Geophys. Res. Lett., 32, L21S01, doi:10.1029/2005GL024009, 2005.

Siegfried, M. R., Hawley, R. L., and Burkhart, J. F.: High-resolution ground-based GPS measurements show intercampaign bias in ICESat elevation data near summit, Greenland, IEEE, Geosci. Remote Sens., 9, 3393–3400, doi:10.1109/TGRS.2011.2127483, 2011.

Sørensen, L. S., Simonsen, S. B., Nielsen, K., Lucas-Picher, P., Spada, G., Adalgeirsdottir, G., Forsberg, R., and Hvidberg, C. S.: Mass balance of the Greenland ice sheet (2003–2008) from ICESat data – the impact of interpolation, sampling and firn density, The Cryosphere, 5, 173–186, doi:10.5194/tc-5-173-2011, 2011.

Vaughan, D. G., Comiso, J. C., Allison, I., Carrasco, J., Kaser, G., Kwok, R., Mote, P., Murray, T., Paul, F., Ren, J., Rignot, E., Solomina, O., Steffen, K., and Zhang, T.: Observations: Cryosphere, in: Climate Change 2013: The Physical Science Basis. Contribution of Working Group I to the Fifth Assessment Report of the Intergovernmental Panel on Climate Change, edited by: Stocker, T. F., Qin, D., Plattner, G.-K., Tignor, M., Allen, S. K., Boschung, J., Nauels, A., Xia, Y., Bex, V., and Midgley, P. M., Cambridge University Press, Cambridge, UK and New York, NY, USA, 2013.

Wingham, D., Ridout, A., Scharroo, R., Arthern, R., and Shum, C. K.: Antartic elevation change from 1992–1996, Science, 282, 456–4458, doi:10.1126/science.282.5388.456, 1998.

Zwally, H. J., Bindschadler, R. A., Major, J. A., and Brenner, A. C.: Ice measurements by Geosat radar altimetry, Johns Hopkins APL Technical Digest (ISSN 0270-5214), Vol. 8, April–June, 251–254, 1987.

12

Influence of regional precipitation patterns on stable isotopes in ice cores from the central Himalayas

H. Pang[1], S. Hou[1], S. Kaspari[2], and P. A. Mayewski[3]

[1]Key Laboratory of Coast and Island development of Ministry of Education, School of Geographic and Oceanographic Sciences, Nanjing University, Nanjing 210093, China
[2]Department of Geological Sciences, Central Washington University, Ellensburg, WA 98926, USA
[3]Climate Change Institute and Department of Earth Sciences, University of Maine, Orono, ME 04469, USA

Correspondence to: S. Hou (shugui@nju.edu.cn)

Abstract. Several ice cores have been recovered from the Dasuopu (DSP) Glacier and the East Rongbuk (ER) Glacier in the central Himalayas since the 1990s. Although the distance between the DSP and the ER ice core drilling sites is only ~ 125 km, the stable isotopic record ($\delta^{18}O$ or δD) of the DSP core is interpreted in previous studies as a temperature proxy, while the ER core is interpreted as a precipitation proxy. Thus, the climatological significance of the stable isotopic records of these Himalayan ice cores remains a subject of debate. Based on analysis of regional precipitation patterns over the region, we find that remarkable discrepancy in precipitation seasonality between the two sites may account for their disparate isotopic interpretations. At the ER core site, the Indian summer monsoon (ISM) precipitation is dominating due to topographic blocking of the moisture from westerlies by the high ridges of Mt. Qomolangma (Everest), which results in a negative correlation between the ER $\delta^{18}O$ or δD record and precipitation amount along the southern slope of the central Himalayas in response to the "amount effect". At the DSP core site, in comparison with the ISM precipitation, the wintertime precipitation associated with the westerlies is likely more important owing to its local favorable topographic conditions for interacting with the western disturbances. Therefore, the DSP stable isotopic record may be primarily controlled by the westerlies. Our results have important implications for interpreting the stable isotopic ice core records recovered from different climatological regimes of the Himalayas.

1 Introduction

The Himalayas impact regional and even global climate due to their high elevation and large topography; however, climate research over the Himalayas is restricted owing to the lack or short duration of meteorological observations because of complex and challenging topographic conditions. Nevertheless, the Himalayas feature the largest volume of ice outside the polar regions, and can preserve high-resolution and long-term climate and environment information, ascribed to their low temperatures and relatively high snow accumulation. To reconstruct climatic and environmental information of the past in the central Himalayas, several ice cores were recently retrieved from the Dasuopu (DSP) Glacier (28°23′ N, 85°43′ E; 7200 m a.s.l.), Shishapangma (Thompson et al., 2000b; Davis et al., 2005) and the East Rongbuk (ER) Glacier (28.03° N, 86.96° E; 6518 m a.s.l.), and on Mt. Qomolangma (Everest) (Hou et al., 2000, 2003; Kaspari et al., 2007) (Fig. 1). The dating of these cores can go back to the past 2000 yr, with a time resolution from annual scales for the upper part of ice core to decadal for the lower part (Xu and Yao, 2001; Hou et al., 2004). Invaluable climatic and environmental information in the central high Himalayas has been revealed by these ice cores, such as regional warming (Thompson et al., 2000b; Hou et al., 2007), atmospheric pollution induced by anthropogenic activities since the industrial revolution (Duan et al., 2007; Kaspari et al., 2009b; Hong et al., 2009), precipitation variations over the central Himalayas (Duan et al., 2004; Kaspari et al., 2008), atmospheric dust deposition over the past several

(a)

(b)

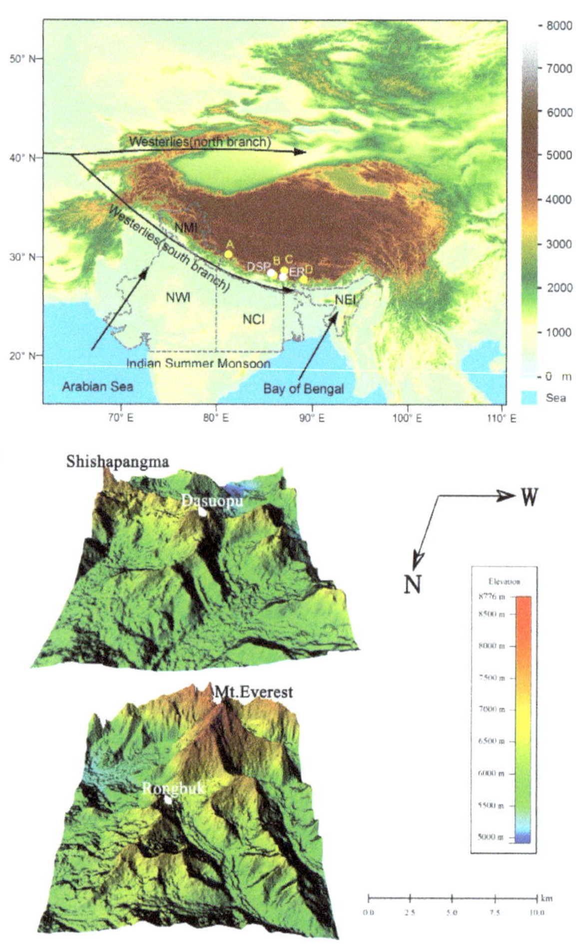

Fig. 1. Atmospheric circulation systems over the study area **(a)**, and three-dimensional topographic maps of the DSP and ER glaciers **(b)**. Filled yellow circles indicate weather stations (A: Pulan; B: Nyalam; C: Dingri; D: Pali), and the filled white circles are the DSP and ER ice core drilling sites. The grey dashed lines are boundaries of the four Indian homogenous rainfall regions north of 21° N (i.e., NEI: northeastern India; NCI: north-central India; NWI: northwestern India; NMI: northern mountainous India). Boundary lines are for reference only and may not denote actual political borders. Digital elevation model (DEM) data are from ASTER (Advanced Spaceborne Thermal Emission and Reflection Radiometer) global DEM version 2 with 30 m resolution, which is available at http://www.jspacesystems.or.jp/ersdac/GDEM/E/4.html.

centuries (Kaspari et al., 2009a; Xu et al., 2010), black carbon concentrations (Ming et al., 2008; Kaspari et al., 2011), and bacterial information (Zhang et al., 2007, 2008).

Ice core stable isotopes (δ^{18}O and δD, the water stable isotopic compositions expressed in δ units permil vs. VSMOW, the Vienna Standard Mean Ocean Water) are a fundamental and important proxy for temperature (i.e., temperature effect, a positive correlation between stable isotopic composition in precipitation and temperature) or precipitation (i.e., amount effect, a negative correlation between stable isotopic composition in precipitation and precipitation amount) (Dansgaard, 1964). Although stable isotopes in ice cores from the central Himalayas have been investigated for more than a decade, there is a discrepancy in the interpretation of the Himalayan ice core isotopic record. According to previous studies, the DSP core isotopic record is generally interpreted as a proxy for temperature (Thompson et al., 2000b; Davis et al., 2005), whereas the ER core isotopic record is often interpreted as a proxy for the Indian summer monsoon (ISM) precipitation (Zhang et al., 2005; Kaspari et al., 2007). The DSP and ER glaciers are both located in the central Himalayas (the distance between them only 125 km) and are expected to experience the same large-scale circulation systems (i.e., ISM vs. winter westerlies). Here we examine two major research questions: (1) why do the distinct interpretations of the stable isotopic records from the two cores occur? (2) Are there any undiscovered regional or local factors to justify the different interpretations?

2 Large-scale atmospheric circulation systems and local topographic features in the central Himalayas

2.1 Large-scale atmospheric circulation systems

The Himalayas, the world's highest mountains, are located on the southern edge of the Tibetan Plateau (TP), with a total length of ~2500 km from Mt. Nanga Parbat in the west to Mt. Namjagbarwa in the east. The Himalayas intercept considerable moisture due to the topographic blocking effect. During the ISM season (June to September), the ISM transports large amounts of water vapor from the Indian Ocean to the Himalayas and the southern TP via two monsoon moisture trajectories: one from the Indian Ocean across the Arabian Sea, and along the Indian River valley to the western Himalayas and TP, and the other from the Bay of Bengal moving northward to the eastern Himalayas and TP along the Brahmaputra River valley (Liu, 1989; Lin and Wu, 1990) (Fig. 1a). During the non-summer-monsoon season (October to May), the westerlies bifurcate into the northern and southern branches around the TP due to its topographic blocking (Fig. 1a), the latter affecting the Himalayas and southern TP (Wei and Gasse, 1999). Wintertime precipitation in the Himalayas is related to vapor transport by the southern branch (Yihui and Zunya, 2008).

2.2 Local topographic features

The Himalayan ranges appear as a gigantic crescent situated on the southern fringe of the TP, which can be divided into four parallel mountain belts from south to north: (1) the outer Himalayas, the sub-Himalayas, the Shivaliks hills, or the foothill zone, forming the southernmost belt of the Himalayan range bordered to the south by the Indo-Gangetic Plain, having an average elevation about 900 m a.s.l. and a width of 40–50 km; (2) the lesser Himalayas, the lower

Himalayas, or Mahabharat Range (in India it is also known as Himachal Himalayas), lying north of the sub-Himalayas, having an average elevation about 3000 m a.s.l. and a width of 60–80 km; (3) the greater Himalayas, the higher Himalayas, or the Great Himalayan Range (in India it is also known as Himadri Himalayas), the longest, highest, and most continuous belt in the Himalayan system, lying above the snow line with an average elevation of more than 6100 m a.s.l. and a width about 25 km; and (4) the trans-Himalayas or the Tibetan Himalayas, the most northerly ranges of the Himalayas in the southern part of the Tibet Autonomous Region of China, bounded by the Kailas (southwest), Nganglong Kangri (north), and Nyainqêntanglha (southeast) mountain ranges and by the Brahmaputra River (south), having an average elevation about 5300 m a.s.l. and 225 km wide in the center, narrowing to a 32 km width at the eastern and western ends (Pascoe, 1950; Menon, 1954).

From the west to east, the Himalayas are divided broadly into three mountainous regions based on general spatial differentiation of associated geographic elements: (1) the western Himalayas, extending along the mountain chain from western Nepal (west of the Kali River) through northwestern India and northern Pakistan, and then southwest along the mountains in the border region between Pakistan and Afghanistan, having a length about 880 km; (2) the central Himalayas or Nepal Himalayas, lying between the Tista River in the east and the Kali River in the west, having a length about 800 km, being characteristic of soaring heights and highest mountain peaks; and (3) the eastern Himalayas, situating between eastern Nepal (east of the Tista River) in the west and Mt. Namjagbarwa in the east, having a length about 720 km (Karan, 1966; Adams Carter, 1985).

The DSP core drilling site is located on the Mt. Shishapangma ridge, which extends in a northwest–southeast direction, facing relatively low terrains in the south and in the west (Fig. 1b). The ER core was retrieved from the East Rongbuk Col on the northeast ridge of Mt. Qomolangma. The very high west and southeast ridges of Mt. Qomolangma are in the south of the ER core (Fig. 1b).

3 Data

In 1997, three ice cores were recovered from the DSP glacier. The third one (core 3), 167.7 m long, was drilled to bedrock at the summit of the glacier (Thompson et al., 2000b). In this ice core, the cold season dust peaks nearly juxtapose with the enriched isotopic values. Based on seasonal oscillations of $\delta^{18}O$ and dust concentration as well as the beta radioactivity peak of 1963, the core was annually dated back to 1440 at 144 m, with an uncertainty of 3 yr (Thompson et al., 2000b; Davis et al., 2005). Annual $\delta^{18}O$ value is calculated by average of individual $\delta^{18}O$ values within a year. Annual accumulation rate record of this core was constructed using a two-parameter steady state flow model that takes into account the rapid thinning of annual layers near a glacier's flow divide (Bolzan, 1985; Reeh, 1988; Davis et al., 2005). The $\sim$ 560 yr records of annually averaged $\delta^{18}O$ (1450–1996) and accumulation rate (1442–1996) are available from the NOAA National Climatic Data Center for Paleoclimatology (http://www.ncdc.noaa.gov/paleo/paleo.html). The ER ice core was drilled in 2002 with a depth of 108.8 m. This ice core was annually dated to the year 1534 at the depth of 98 m by counting annual layers in terms of seasonal variations in $\delta^{18}O$ or δD, soluble ions, and trace elements (Kaspari et al., 2007). The dating was verified by the beta radioactivity peak of 1963 and major volcanic horizons. Dating uncertainties are estimated to be ± 0 yr at 1963 (20 samples per year) and ± 5 yr at 1534 (4 samples per year). Annual $\delta^{18}O$ value is also calculated by average of individual $\delta^{18}O$ values within a year. For correcting annual ice layer thinning with increasing the depth, Kaspari et al. (2008) developed a flow model to construct the ER core annual accumulation rates based on the annual-layer thickness data. Details about this core – including sampling processing, laboratory analysis, dating, and ice flow model for constructing annual accumulation rates – can be found elsewhere (Kaspari et al., 2007, 2008).

To reveal precipitation seasonality along the Himalayas, monthly precipitation amount data from four Himalayan meteorological stations (Pali, Dingri, Nyalam, and Pulan stations; Fig. 1a) are used, available from the China Meteorological Data Sharing Service System. In order to compare the Himalayan ice records with the ISM rainfall data, the instrumental summer monsoon rainfall series of four Indian homogenous rainfall regions north of 21° N (i.e., NMI: northern mountainous India; NWI: northwestern India; NCI: north-central India; NEI: northeastern India) (Fig. 1a) are used since these monsoon-impacted regions are directly adjacent to the Himalayas. The ISM rainfall data during 1813–2006 are available from the Indian Institute of Tropical Meteorology at http://www.tropmet.res.in (Sontakke and Singh, 1996; Sontakke et al., 2008). For comparing with the Indian summer rainfall series over a same period, the DSP and ER ice-core $\delta^{18}O$ and accumulation rate records beginning from 1813 are used in this paper. Lastly, in order to analyze the influence of atmospheric circulation systems on the Himalayan ice core records, the monthly reanalysis data are used, including the NCEP/NCAR reanalysis data with $2.5° \times 2.5°$ resolution beginning from 1948 (Kistler et al., 2001), the Global Precipitation Climatology Centre (GPCC) monthly precipitation data set beginning from 1951 with $1.0° \times 1.0°$ resolution, and the 20th century reanalysis data (V2) with $2.0° \times 2.0°$ resolution beginning from 1871 (Compo et al., 2011). The NCEP/NCAR and the 20th century reanalysis data are obtained from the NOAA/ESRL PSD (National Oceanic and Atmospheric Administration/Earth System Research Laboratory, Physical Sciences Division) at http://www.esrl.noaa.gov/psd/data/gridded/. The GPCC precipitation data are available at http://gpcc.dwd.de.

4 Results and discussion

4.1 Differences between the DSP and ER ice-core records

Previous studies have hypothesized that the DSP core site receives most precipitation during the ISM season, and ascribed the DSP $\delta^{18}O$ variations at interannual or longer timescales to air temperature changes at mean condensation level (Thompson et al., 2000b; Davis et al., 2005). For the ER core, Zhang et al. (2005) found there are no correlations between the annual ER $\delta^{18}O$ record and air temperature data from more than 30 meteorological stations distributed in the north and south of the Himalayas, which suggests that the ER $\delta^{18}O$ record is not influenced by temperature. Later, Kaspari et al. (2007) found the annual ER δD is inversely correlated with the ISM season precipitation over the nearby region of Mt. Qomolangma, which verifies that the ER ice core isotopic composition is controlled by the amount effect. To check whether a relationship exists between the Himalayan ice core records ($\delta^{18}O$ and accumulation rate) and the ISM rainfall, we calculated correlation coefficients between the Himalayan ice core records and the Indian monsoon rainfalls over the four Indian homogenous regions in north India (NMI, NWI, NCI, and NEI), as shown in Table 1. From Table 1, we find a weak but significant positive correlation between the ER accumulation rate and the summer monsoon rainfall over the NWI region, whereas there are no correlations between the DSP accumulation rate and the Indian summer rainfalls.

In order to further discern differences between the DSP and ER ice core records, time series of the two ice cores $\delta^{18}O$ and accumulation rate records are compared, as shown in Fig. 2. The trend in the DSP accumulation rate is opposite to that of the ER and the average accumulation rate at the DSP site (1.28 m i.e. yr^{-1}) is almost 3 times the rate at the ER site (0.44 m i.e. yr^{-1}). In addition, the DSP and ER accumulation rates show step changes: higher values (the average 1.66 m i.e. yr^{-1}) before 1880, and lower values (the average 1.07 m i.e. yr^{-1}) with a significantly decreasing trend beginning from 1880 at the DSP site (Fig. 2f), but lower values (the average 0.33 m i.e. yr^{-1}) with small variation amplitude before 1938, and higher values (the average 0.65 m i.e. yr^{-1}) with a significantly increasing trend beginning from 1938 at the ER site (Fig. 2c). We choose the year 1880 (1938) as a boundary of the DSP (ER) accumulation step change because the DSP (ER) accumulation rate exhibited a remarkable decrease (increase) beginning from 1880 (1938). Otherwise, the increasing linear trend of the DSP $\delta^{18}O$ is significant at 99 % confidence level (Fig. 2g), which was interpreted as an indicator of rising temperature (Thompson et al., 2000b; Yao et al., 2006). However, such an increasing trend is not significant for the ER $\delta^{18}O$ record (Fig. 2d). Furthermore, the ER $\delta^{18}O$ record even shows a decreasing trend to more negative $\delta^{18}O$ beginning from 1938 (Fig. 2d). The remarkable difference in the snow accumulation rate reconstructed from the DSP and ER cores suggests that the major precipitation origins of the two sites may differ, which could account for the different interpretations of the DSP and ER isotopic records.

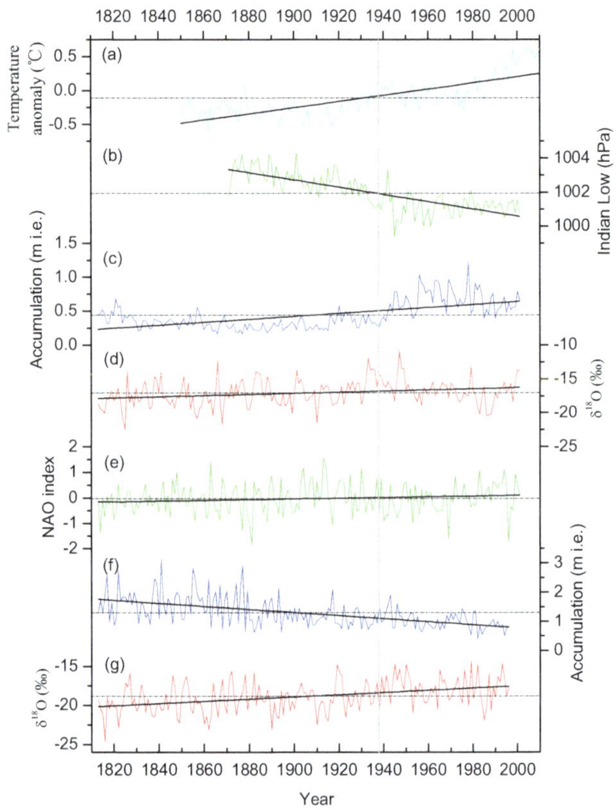

Fig. 2. Variations of **(a)** Northern Hemisphere annual temperature anomaly (Jones et al., 2013), **(b)** the Indian Low intensity, **(c)** the ER annual accumulation rate, **(d)** the annual mean ER $\delta^{18}O$, **(e)** the non-monsoon-season (October to May) NAO index (Luterbacher et al., 2002), **(f)** the DSP annual accumulation rate, and **(g)** the annual mean DSP $\delta^{18}O$ since 1813. The short dash-dotted lines are the averages of each series, and the bold lines are their linear trends. The vertical grey line indicates the boundary year 1938.

4.2 Regional precipitation patterns and their impacts on the Himalayan ice core stable isotopic record

4.2.1 Different precipitation seasonality along the Himalayas

To discern the precipitation difference between the DSP and ER sites, seasonal distribution of precipitation at four weather stations – Pali (27°44′ N, 89°05′ E, 4300 m a.s.l.), Dingri (28°38′ N, 87°05′ E, 4300 m a.s.l.), Nyalam (28°11′ N, 85°58′ E, 3810 m a.s.l.), and Pulan (30°17′ N, 81°15′ E, 4900 m a.s.l.) – along the Himalayas (Fig. 1a) is presented in Fig. 3. The precipitation distribution over the eastern (western) Himalayas is unimodal (bimodal). Most precipitation over the eastern Himalayas occurs during

Table 1. Pearson correlation coefficients between the Himalayan ice core records ($\delta^{18}O$ and accumulation rate) and the summer monsoon rainfall of the four Indian homogenous rainfall regions north of 21° N (i.e., NEI: northeastern India; NCI: north-central India; NWI: north-western India; NMI: northern mountainous India) beginning from 1813.

	NEI	NCI	NWI	NMI	ER-accum	DSP-accum	ER-$\delta^{18}O$	DSP-$\delta^{18}O$
NEI	1							
NCI	0.19[b]	1						
NWI	−0.17[b]	0.46[a]	1					
NMI	0.01	0.54[a]	0.62[a]	1				
ER-accum	−0.10	0.02	0.20[b]	0.00	1			
DSP-accum	0.01	0.01	−0.09	0.02	−0.31[a]	1		
ER-$\delta^{18}O$	−0.12	0.03	0.02	0.01	0.01	0.10	1	
DSP-$\delta^{18}O$	−0.06	−0.25[a]	−0.02	−0.15[c]	0.28[a]	−0.39[a]	0.16[b]	1

Note: two-tailed test of significance is used.
[a] 99 % confidence level, [b] 95 % confidence level, and [c] 90 % confidence level. ER: East Rongbuk; DSP: Dasuopu.

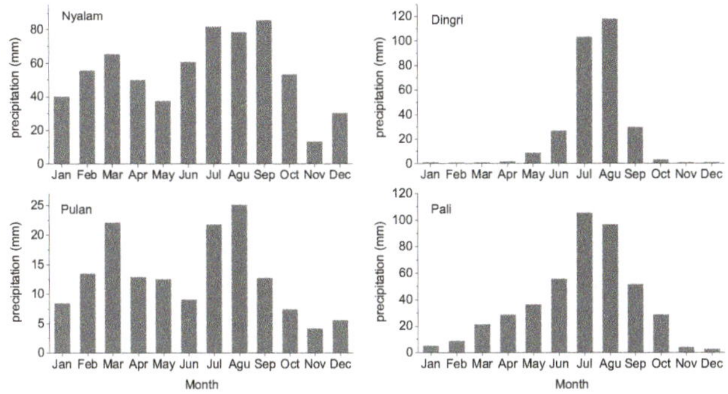

Fig. 3. Monthly long-term mean of precipitation at four weather stations along the Himalayas (Nyalam, Pulan, Dingri, and Pali). Seasonal distribution of precipitation is calculated based on the meteorological data observed from January 1973 to January 2011.

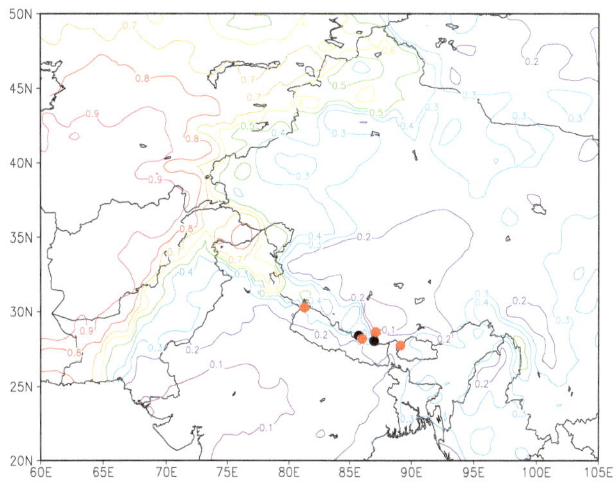

Fig. 4. Map showing the non-monsoon season precipitation ratio to annual precipitation (October–May/annual) along the Himalayas. Reanalysis data are from the monthly long-term mean (1981–2010) precipitation data with a 0.5° × 0.5° resolution from the Global Precipitation Climatology Centre (GPCC). Filled red circles are the four weather stations along the Himalayas (from the left to right: Pulan, Nyalam, Dingri, and Pali). The non-monsoon precipitation ratios at the Himalayan stations are generally consistent with those calculated from the GPCC data. Filled black circles are DSP and ER ice cores sites (left: DSP; right: ER).

the ISM season, whereas precipitation over the western Himalayas takes place during the winter–spring and summer seasons.

To further verify precipitation seasonality along the Himalayas, spatial distribution of the non-monsoon season (October to May) precipitation ratio to annual precipitation over the study area is calculated based on the monthly long-term mean (1981–2010) precipitation data with a 0.5° × 0.5° resolution from the GPCC (available at http://www.esrl.noaa.gov/psd/data/gridded/data.gpcc.html), as shown in Fig. 4. It is clear that the non-monsoon season precipitation ratio over the western high Himalayas (40–80 %) is higher than that over the southern and northern slopes of the Himalayas (≤ 20 %), suggesting local capture of moisture from the westerlies by the western high Himalayas. Furthermore, the non-monsoon season precipitation ratio seems to gradually decrease from the western Himalayas to Mt. Qomolangma due to moisture wastage by sequential condensation of moisture from the westerlies during its being transported eastward. Interestingly, the highest peak Mt. Qomolangma seems to be a boundary of the non-monsoon season precipitation ratio because west of it the non-monsoon season precipitation is high, whereas east of it the non-monsoon season precipitation ratio is rather low (Fig. 4). From this point, we can divide the Himalayas into western and eastern Himalayas by Mt. Qomolangma. According to this division, the DSP core is in the western Himalayas, while the ER core is in the eastern Himalayas, although the distance between them is only about 125 km.

The difference in precipitation seasonality between the western and eastern Himalayas must have an important impact on the Himalayan ice core stable isotopic records, because the stable isotopic composition in summer monsoonal precipitation is controlled by the "amount effect", whereas the stable isotopic ratio in non-summer-monsoon precipitation is dominated by the "temperature effect" (Tian et al., 2003, 2007). The non-monsoon season precipitation associated with the westerlies at Nyalam station accounts for 53 % of its total annual precipitation. This site is nearby the DSP glacier; thus we speculate that the proportion of non-monsoon season precipitation at the DSP core is also considerable. As a result, we conjecture that the stable isotopic record in the DSP core is influenced substantially by winter westerlies. On the other hand, the proportion of summertime precipitation at Dingri station accounts for more than 90 % of its annual precipitation. Dingri is nearby the ER glacier; therefore we suggest that summer monsoonal precipitation dominates at the ER glacier and the ER stable isotopic record is mainly controlled by the ISM. Thus, the different seasonal distribution of precipitation between the western and eastern Himalayas is likely the main reason for the different climatological interpretations of the DSP and ER ice-core stable isotopic records. In the following two sections (Sects. 4.2.2 and 4.2.3), we further investigate how the winter westerlies and the ISM impact the DSP and ER ice-core isotopic records, respectively.

4.2.2 Influence of winter westerlies on the DSP isotopic record

To attest the control of winter westerlies on the DSP $\delta^{18}O$, we first performed a composite analysis of moisture flux during the winter–spring season (February–April) using the monthly NCEP/NCAR reanalysis data (wind vector and specific humidity). Here, we choose February–April as the winter–spring season because the non-monsoon precipitation peak occurs during these months (Fig. 3). To define the higher and lower DSP accumulation rate years, the DSP accumulation rate record was standardized by subtracting the mean and dividing by the standard deviation (SD). A year with a value > 1.0 SD is defined as a year with higher accumulation rate, while a value < -1.0 SD is defined as a year with lower accumulation rate. Based on this division, there are 9 years with higher DSP accumulation rate (1952, 1953, 1968, 1970, 1972, 1975, 1976, 1980, and 1981) and 8 years with lower DSP accumulation rate (1954, 1956, 1957, 1965, 1979, 1982, 1984, and 1994). Figure 5 shows the composite analysis result of the winter–spring season mean moisture flux at 400 hPa level between years with higher and lower DSP accumulation rate (higher minus lower). It is clear that moisture transport from the westerlies – originating in the northern Atlantic Ocean passing through northern Africa, Arabia, Iran, Afghanistan, Pakistan, and the Tibetan Plateau – is stronger when the DSP accumulation rate is higher, suggesting that

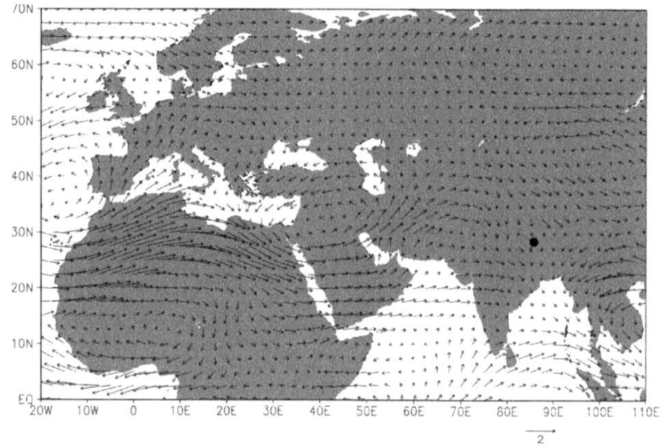

Fig. 5. Composite analysis of moisture flux (multiplying by wind vector and specific humidity) at 400 hPa level during the winter–spring season (February to April) between years with higher and lower Dasuopu accumulation rate (higher minus lower). The filled black circle indicates the DSP core drilling site.

the winter–spring season precipitation at the DSP site is significantly influenced by moisture transport from the westerlies. Moreover, we calculated the correlation between the DSP accumulation rate and the winter–spring season precipitation amount derived from the GPCC monthly precipitation data beginning from 1951. However, no significantly positive correlation was found over the western Himalayas (figure not shown), which may be due to the recently increasing contribution of the ISM precipitation to the DSP precipitation or sparse precipitation observations over the high Himalayas.

In order to verify the increasing impact of the ISM circulation on the DSP accumulation rate and $\delta^{18}O$ record beginning from 1938, which we suggest in Sect. 4.4, the correlations of the DSP accumulation rate vs. summer monsoon season precipitation and the DSP $\delta^{18}O$ record vs. the monsoonal precipitation were also calculated based on the GPCC monthly precipitation data beginning from 1951, as shown in Fig. 6. As we expected, we find a weak but significant positive correlation between the DSP accumulation rate and the monsoonal precipitation over the north-central India over the period of 1951–1996 (Fig. 6a). Additionally, we also find a significantly negative correlation between the DSP $\delta^{18}O$ record and the summer monsoon rainfall along the southern slope of the Himalayas (Fig. 6b). Concerning the increasing contribution of the ISM precipitation to the DSP precipitation during recent decades, this will be discussed in detail in Sect. 4.4.

It is well known that the midlatitude westerlies are closely related to North Atlantic Oscillation (NAO). NAO is defined by the difference in pressure between the permanent high-pressure system located over the Azores (Azores High) and the permanent low-pressure system located over Iceland (Icelandic Low). When the NAO index is high, a large pressure difference between the two places induces stronger

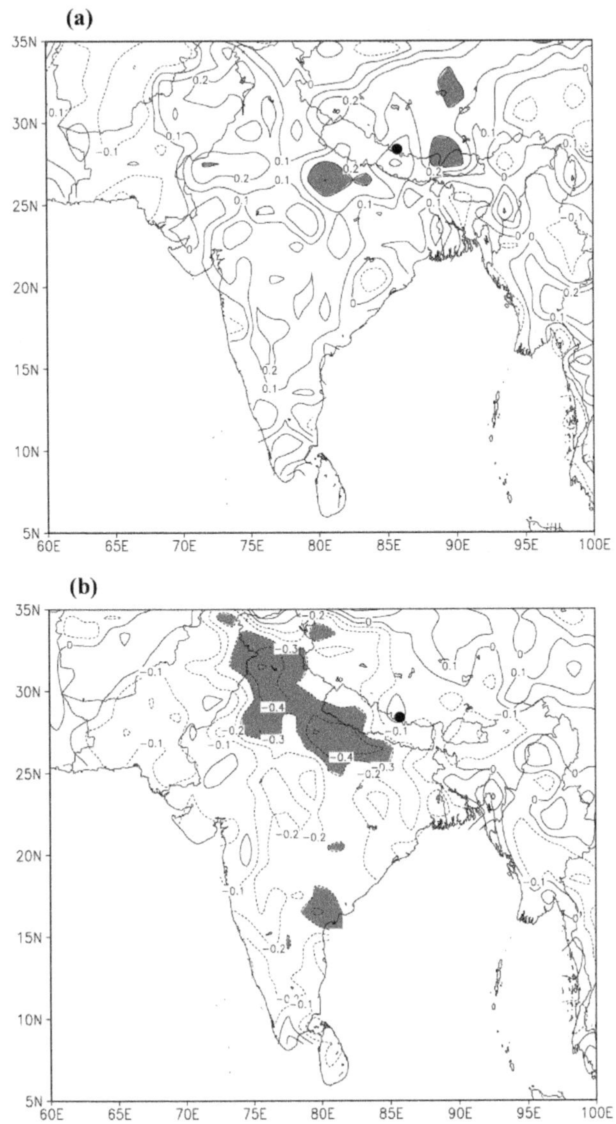

Fig. 6. Correlation coefficients of the DSP accumulation rate vs. summer monsoon season precipitation **(a)** and the DSP $\delta^{18}O$ record vs. the monsoonal precipitation **(b)** over the period of 1951–1996. Grey shading indicates correlation significance at 95 % confidence level. The filled black circle indicates the DSP core drilling site. Summer monsoon precipitation data are from the GPCC monthly precipitation data set from 1951–present with 1.0° × 1.0° resolution (available at http://gpcc.dwd.de).

westerly flow over the mid-high latitudes, leading to storm tracks shifting northward and the mid-high-latitude Europe undergoing milder winters but more frequent rainfall. When the NAO index is low, storm tracks shift southward, and the westerlies enhance over the mid-low latitudes, which results in colder winters but more rainfall in southern Europe, the Mediterranean, and northern Africa. Hence, stronger westerlies over the mid-low latitudes transport more and colder air masses eastward into the western Himalayas and bring higher snowfall there, and vice versa. We calculated the correlation between the DSP accumulation rate and an NAO index (Luterbacher et al., 2002) during the non-monsoon season beginning from 1813 and find a good negative correlation ($r = -0.216$, $N = 184$, $p = 0.003$) (Fig. 2e and f), proving the control of the westerlies on the DSP precipitation.

As a result, stronger (weaker) westerlies over the mid-low latitudes not only bring higher (lower) precipitation but also lead to a colder (warmer) temperature environment in the western Himalayas because of more (less) and colder air mass invasion. These temperature variations could lead to $\delta^{18}O$ variations if the temperature effect was at play as suggested by previous studies (Thompson et al., 2000b; Davis et al., 2005). Although most researchers have suggested that the temperature effect at the DSP site is significant (Thompson et al., 2000b; Davis et al., 2005), they have not delimited a specific region where the temperature is positively correlated with the DSP $\delta^{18}O$ record and only presented a good correlation between the DSP $\delta^{18}O$ record and the Northern Hemisphere temperature (Thompson et al., 2000b; Davis et al., 2005). In order to determine where the temperature is correlated with the DSP $\delta^{18}O$ record, we calculated the correlation between the DSP $\delta^{18}O$ record and mean air temperatures during the non-monsoon season at different pressure levels deriving from the 20th century reanalysis (V2) data. Surprisingly, there is no correlation between the DSP $\delta^{18}O$ record and air temperature over the regions adjacent to the DSP site. Nevertheless, we find a good positive correlation region over the Azores High area and its adjacent regions between the DSP $\delta^{18}O$ record and air temperatures in the mid-low troposphere (from 850 to 300 hPa level). Moreover, the positive correlation region is stable at different pressure levels. Because the DSP core elevation is very high (7200 m a.s.l.), we only present the correlation coefficients at 400 hPa level as a demonstration, shown in Fig. 7. The higher (lower) air temperature over the Azores High region may imply the position of Azores High shifting northward (southward) in response to the northward (southward) shifting of the westerlies. Thus, such a positive correlation region may indicate a close relationship between the NAO and the DSP $\delta^{18}O$ record. According to previous studies, there is a good positive correlation between the NAO index and the Northern Hemisphere temperature (Hurrell, 1996; Gimeno et al., 2003), which may explain why there is a good correlation between the DSP $\delta^{18}O$ record and the Northern Hemisphere temperature found by previous studies (Thompson et al., 2000b; Davis et al., 2005). However, the link between westerlies and temperature does not prove that DSP $\delta^{18}O$ should respond to temperature.

As a result, changes of moisture transport trajectory of winter westerlies associated with the NAO may be more important than the local air temperature in controlling the DSP $\delta^{18}O$. When the mid-low-latitude winter westerlies are stronger (corresponding to a low NAO index), more vapor originating from ocean (the Atlantic Ocean) will be transported into the western Himalayas, leading to higher

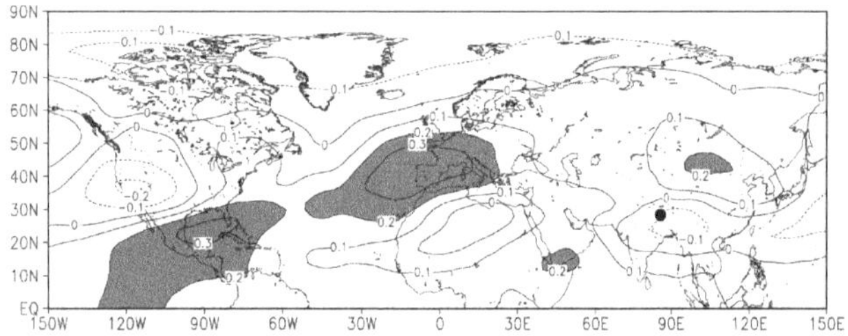

Fig. 7. Correlation coefficients between the annual mean DSP δ^{18}O and mean air temperature during the non-monsoon season (October to May) at 400 hPa level beginning from 1871. The monthly air temperature data with 2.0° × 2.0° resolution are from the 20th century reanalysis (V2). Grey shading indicates correlation significance at 95 % confidence level. The filled black circle indicates the DSP core drilling site.

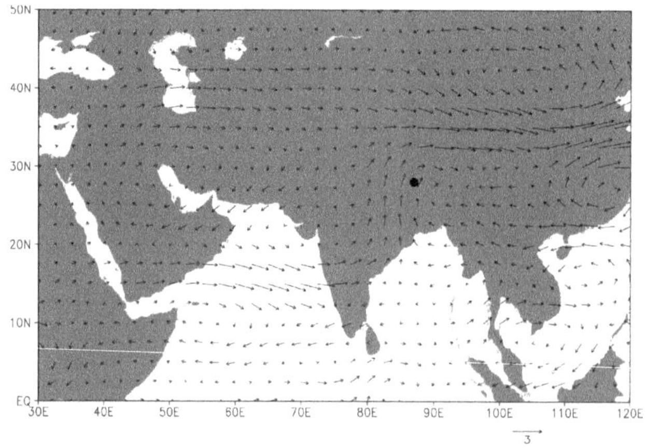

Fig. 8. Composite analysis of summer mean (June to September) moisture flux (multiplying by wind vector and specific humidity) at 400 hPa level between years with higher and lower ER accumulation (higher minus lower). The filled black circle indicates the ER core drilling site.

accumulation rate and lower DSP δ^{18}O. Conversely, when the mid-low-latitude winter westerlies are weaker (corresponding to a high NAO index), less oceanic moisture and more continental recycling vapor from northern Africa, the Mediterranean, and western Asia will be transported into the western Himalayas, resulting in lower accumulation rate and higher DSP δ^{18}O.

The influence of moisture from winter westerlies on isotopic composition in precipitation at the DSP site is also supported by the second-order isotopic parameter, deuterium excess, defined as d-excess= δD-8δ^{18}O (Dansgaard, 1964). The moisture from westerlies originating in the Atlantic Ocean is enhanced by the Mediterranean evaporation during its being transported eastward. The d-excess value in the mixing vapor (Atlantic vapor plus the vapor evaporated from the Mediterranean) is generally high due to the isotopic disequilibrium between the moisture in the atmosphere and that of the Mediterranean Sea water body resulting from strong temperature contrast between them (Gat and Carmi, 1970; Rindsberger et al., 1983; Gat et al., 2003). For instance, the average value of d-excess in the 50 vapor samples collected over the Mediterranean Sea during winter of 1995 is as high as 21 ‰ (Gat et al., 2003), which is close to the average value of d-excess (20 ‰) in winter precipitation at Nyalam station nearby the DSP core site (Tian et al., 2005, 2007). Previous investigations suggest that the d-excess values of snow pit samples from the DSP glacier as high as 17.3 ‰ (Tian et al., 2001). The high d-excess values in precipitation at or nearby the DSP core site suggest that the influence of the westerlies is significant at the DSP site.

4.2.3 Influence of Indian summer monsoon on the ER isotopic record

To further verify the dominant ISM control on the ER δ^{18}O, we also first performed a composite analysis of moisture flux during the monsoon season using the monthly NCEP/NCAR reanalysis data. Similarly, we defined the higher and lower ER accumulate rate years based on the same method as used for the DSP accumulation rate. There are 10 years with higher ER accumulation rate (1956, 1957, 1963, 1964, 1965, 1969, 1970, 1978, 1981, and 1983) and 8 years with lower ER accumulation rate (1953, 1954, 1955, 1968, 1979, 1985, 1987, and 1990). Figure 8 shows the composite analysis result of the summer monsoon season mean moisture flux at 400 hPa level between years with higher and lower ER accumulation rate (higher minus lower). It is clear that the ISM moisture flux from the Arabian Sea, via northern central India, to the central Himalayas is stronger when the ER accumulation rate is higher, and vice versa, indicating that the precipitation at the ER core site is significantly influenced by the ISM moisture transport. In addition, the correlation between the ER accumulation rate and summer monsoon season precipitation is also analyzed based on the GPCC monthly precipitation data beginning from 1951, as shown in Fig. 9a. It is clear that strongest positive correlation occurs over the northwestern region of India (i.e., the core region of the India Low), which is in agreement with the correlation analysis between the ER accumulation rate and the summer monsoon rainfalls over the four northern sub-India regions (Table 1). This suggests that precipitation at ER core site is controlled by the large-scale ISM circulation.

In order to check the influence of the ISM circulation system on precipitation at the ER core site on a longer timescale, we calculated the Indian Low intensity as the summer (June to September) mean sea level pressure over the core region of the Indian Low (20–30° N, 60–90° E) based on the monthly 20th century reanalysis sea level pressure data beginning from 1871, as shown in Fig. 2b. It is clear that the Indian Low intensity is completely opposite to the ER accumulation rate: lower Indian Low pressure (in response to stronger ISM circulation) corresponds well with higher ER accumulation

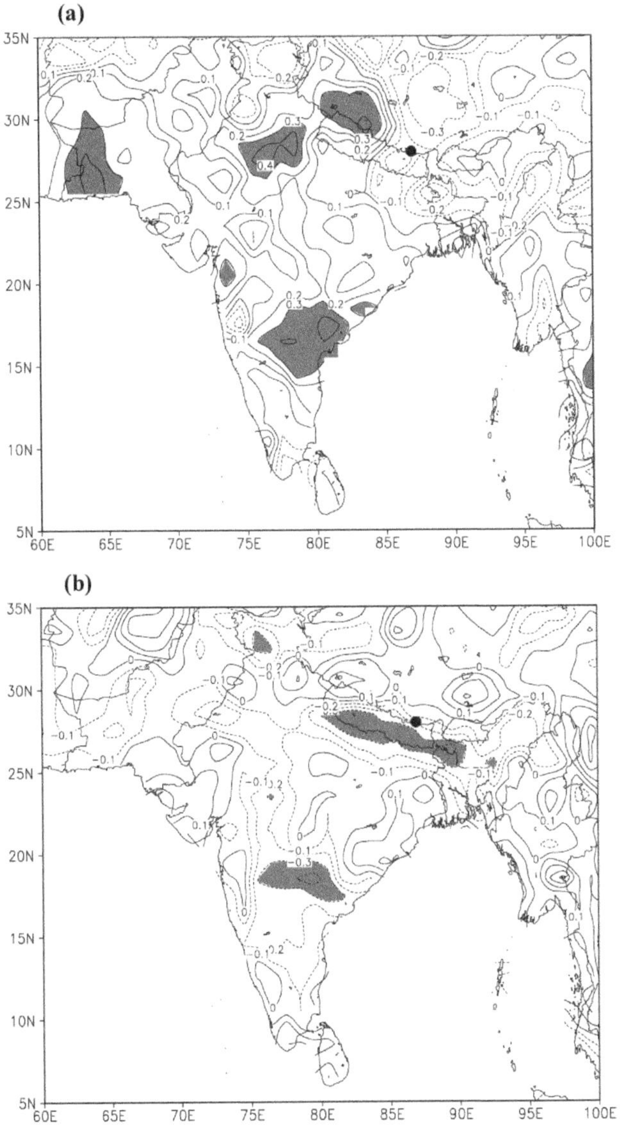

Fig. 9. Correlation coefficients of the ER accumulation rate vs. summer monsoon season precipitation **(a)** and the ER $\delta^{18}O$ record vs. the monsoonal precipitation **(b)** over the period of 1951–2001. Grey shading indicates correlation significance at 95 % confidence level. The filled black circle indicates the ER core drilling site. Summer monsoon precipitation data are from the GPCC monthly precipitation data set from 1951-present with 1.0° × 1.0° resolution (available at http://gpcc.dwd.de).

rate, and vice versa (Fig. 2b and c). The negative correlation between them is very significant ($r = -0.663$, $N = 131$, $p < 0.001$), which suggests that the influence of the ISM circulation on precipitation at the ER site is dominating.

Finally, we calculated the correlation between the ER $\delta^{18}O$ record and summer monsoon season rainfall based on the GPCC monthly precipitation data beginning from 1951, as shown in Fig. 9b. We find a significantly negative correlation between the ER $\delta^{18}O$ record and summer monsoon rainfall along the southern slope of the Himalayas (a long, narrow region) (Fig. 9b). Interestingly, we also find a similar such long, narrow correlation belt along the southern slope of the Himalayas between the DSP $\delta^{18}O$ record and the monsoonal rainfall over the period of 1951–1996 (Fig. 6b). Note that the narrow-band negative correlation region along the southern slope of the Himalayas between the ER (DSP) $\delta^{18}O$ record and the summer precipitation does not overlap with the positive correlation region in the northwest (north-central) of India between the ER (DSP) accumulation rate and the summer precipitation. This suggests that summer monsoon precipitation $\delta^{18}O$ over the high Himalayas is probably controlled by precipitation processes associated with deep convective activities over the southern slope of Himalayas due to its very steep topographic gradient, where the heavy isotopes in vapor are washed out strongly by intense precipitation processes.

Relative to high d-excess values in precipitation at the DSP site, the averaged d-excess value of the ER core over the period of 1813–2001, 9.3 ‰, is much lower. This lower d-excess in the ER core is probably attributed to the high humidity over the ISM moisture source region. Recently, Pang et al. (2012) also indicated that the interannual variation of d-excess in the ER core is primarily controlled by the ISM moisture transport.

4.3 Potential mechanisms for regional precipitation patterns

4.3.1 Unique geographical features associated with different precipitation seasonality along the Himalayas

The weather systems responsible for wintertime precipitation in the western Himalayas are the western disturbances (Gupta et al., 1999). They are westerly upper-tropospheric synoptic-scale waves, originating as cut-off lows (i.e., a closed low that has become completely displaced (cut off) from and moves eastward independent of the basic westerly current) over the region adjoining the Caspian Sea (Rao, 2003). These systems intrude into the western Himalayas via the notch (30–32.5° N, 70–75° E) formed by the Himalayas and the Hindu Kush mountains (Lang and Barros, 2004). Many disturbances are split into two or more secondary systems by interaction with the mountains in the western Himalayas when they migrate eastward. Snowstorms are generally related to the terrain-locked low-pressure disturbances. As a result, most moisture may precipitate over the western Himalayas because of the orographic capture of snowstorms, leading to more wintertime precipitation over the western Himalayas than over the eastern Himalayas, as shown in Figs. 3 and 4.

A numerical simulation of a representative snow event suggests that significant wintertime precipitation over the central Himalayas only occurs when the western disturbances evolve to a favorable geometry with respect to the mountains (Lang and Barros, 2004). The DSP core is located on the Mt. Shishapangma ridge, which extends in a northwest–southeast direction. The DSP core faces relatively low terrains in the south and west (Fig. 1b), which provides a broad space for the western disturbances invading and developing. Furthermore, the northwest–southeast ridge of Mt. Shishapangma is diagonal to the flow of the westerlies, which is favorable for interacting with the western disturbances. As a consequence, the local favorable orographic conditions may result in significant wintertime precipitation at the DSP core site. On the other hand, the ER core was retrieved from the East Rongbuk Col on the northeast ridge of Mt. Qomolangma. The very high west and southeast ridges of Mt. Qomolangma (Fig. 1b) may constrain the western disturbances in the southern slope of Mt. Qomolangma ridges, which thus leads to little wintertime precipitation at the ER core site. In other words, the ER core site is equivalently situated in the leeward slope (rain shadow region) of the western disturbances. Therefore, the DSP core site is heavily influenced by the western disturbances whereas the ER core is not. The much greater accumulation rate at the DSP core site than that at the ER core site supports our conclusion.

4.3.2 Altitudinal differences in westerly vs. monsoon moisture sources

Previous studies have shown that most annual precipitation at low altitudes of the central Himalayas falls during the ISM season (Shrestha, 2000; Lang and Barros, 2004), while Lang and Barros (2004) found that high elevations (> 3000 m a.s.l.) in the central Himalayas can receive up to 40 % of their annual precipitation during winter. The result seems to imply that wintertime precipitation associated with the western disturbances in the central Himalayas increases with rising elevation. The local advantageous topographic conditions may cause considerable wintertime precipitation to fall at the DSP core site (7200 m a.s.l.). However, the local unfavorable orographic conditions may result in little wintertime precipitation at the ER core site (6518 m a.s.l.), as mentioned above.

According to investigation on altitudinal distribution of the ISM rainfall in the central Himalayas, Dhar and Rakhecha (1981) indicated that the zones of maximum rainfall occur near the foothills at an elevation of 2000 to 2400 m a.s.l.. Beyond the elevation, summer monsoon rainfall decreases continuously as elevation increases until the Great Himalayan Range is reached. The result suggests that the ISM precipitation above the elevation of 2400 m a.s.l. decreases significantly with rising elevation in the central Himalayas. Nevertheless, the 700 m elevation difference between the DSP and ER ice core sites may cause little difference in summer monsoon precipitation between the two sites because the convective activity associated with the ISM precipitation can reach a high elevation.

As a result, with elevation rising in the central Himalayas, the potential increase in wintertime precipitation associated with the westerlies and the decrease in summertime precipitation related to the ISM probably lead to larger proportion of wintertime precipitation than the summertime precipitation at the DSP core site.

4.4 Variations of winter westerlies and Indian summer monsoon during the past two centuries recorded by the Himalayan ice core accumulation

Good correlation between the ER accumulation rate and the Indian Low intensity over the period of 1813–2001 indicates that the high ridges of Mt. Qomolangma are effective topographical barriers between the westerlies' precipitation and the ER core site. Low ER accumulation rate before 1938 corresponds to the weak Indian Low intensity and high ER accumulation rate beginning from 1938 is in response to the Indian Low intensifying (Fig. 2b and c). Furthermore, the weak (strong) ISM/low (high) ER accumulation rate seems to correspond to the relatively lower (higher) Northern Hemisphere temperature based on their trends (Fig. 2a, b, and c). This suggests that the ISM intensity over the central Himalayas may weaken in a cold climate and intensify in a warm climate. The overall increase trend of the ER accumulation rate during the past 200 yr may reflect the ISM circulation intensifying from the Little Ice Age (LIA) cold climate to the today's global warming climate. Global circulation models also predict an increase of the summer monsoon precipitation in High Asia as a consequence of global warming (Meehl and Washington, 1993; Hu et al., 2000).

In comparison with the contribution of ISM precipitation to precipitation at the DSP site, the contribution of winter westerlies may be more important. Although there exists a 700 m elevation difference between the DSP and ER sites, the difference in summer monsoon precipitation between the two sites may be small, as the convective system associated with the ISM precipitation can reach a high elevation. We therefore suppose the summer precipitation amount at the DSP site equals that at the ER site. In addition, we assume the non-monsoon precipitation ratio to annual precipitation at the ER site is 20 % based on the result in Fig. 4. Based on the above assumptions, we can calculate the contribution of the westerlies to precipitation at the DSP site using the ER accumulation rate data. For instance, the average DSP accumulation rate is 1.66 m i.e. yr^{-1} and the average ER accumulation rate is 0.34 m i.e. yr^{-1} over the period of 1813–1879; correspondingly, the proportion of precipitation associated with the westerlies at the DSP site is about 84 %. Similarly, the proportion is 78 % over the period of 1880–1937, and 48 % over the period of 1938–1996. It is noteworthy that the proportion of the westerlies' precipitation at the DSP site during the period 1938–1996 is comparable with the non-monsoon precipitation ratio to annual precipitation (53 %) at Nyalam weather station, suggesting that our estimation of the contribution of the westerlies to DSP precipitation is likely reliable. The very high DSP accumulation rate/proportion of the westerlies' precipitation over the period of 1813–1879 indicates that the westerlies' intensity over the central Himalayas is strong under the condition of the latter part of the LIA cold climate (Fig. 2a and f). The relatively low DSP accumulation rate and evident decrease of the proportion of the westerlies' precipitation over the period of 1938–1996 suggest that the westerlies' intensity over the central Himalayas is weak under the condition of the today's global warming climate (Fig. 2a and f). The general decrease trend of the proportion of the westerlies' precipitation at the DSP site over the past two centuries may reflect the gradual decrease of the westerlies' intensity over the central Himalayas, which may correspond to gradually northward shifting of the midlatitude westerlies due to global warming since the end of the LIA. Observed significant increases in high-latitude cyclone frequency and intensity (McCabe et al., 2001) as well as the upward trend in the NAO index (Thompson et al., 2000a) during the past several decades both indicate the northward shifting of the midlatitude westerlies due to recently significant global warming. The ECHAM4–OPYC3 coupled general circulation model also produces a significantly strengthened annual-mean meridional pressure gradient over the North Atlantic associated with projected global warming, also indicating the northward shift of the midlatitude westerlies (Hu and Wu, 2004).

The negative correlation between the DSP and ER accumulation rates (Table 1) and their opposite linear trends (Fig. 2c and f) during the past two centuries are likely indicative of interplay between the winter westerlies and the ISM, as indicated by An et al. (2012). More recently, Joswiak et al. (2013) found evidence of interplay between the westerlies and the ISM based on an ice core d-excess record from the Tanggula Mountains in the central Tibetan Plateau. They indicated that a relatively greater contribution of moisture from the westerlies prior to the 1940s and an increase in the contribution of the ISM moisture after the 1940s, which are consistent with our results. In addition, Yao et al. (2012) found that the precipitation in the Himalayas associated with the ISM has been decreasing during the past three decades (1979–2010), while the precipitation in the eastern Pamir related to the westerlies has been strengthening. In Fig. 2c, the ER accumulation rate also shows a significantly decreasing trend beginning from the late 1970s. However, the DSP accumulation rate does not exhibit an increasing trend beginning from the late 1970s, which is likely due to the evident decrease of the ISM precipitation over this region during this period.

5 Conclusions

Local favorable terrain conditions cause the DSP core site receive more precipitation from winter westerlies. Stronger winter westerlies over the mid-low latitudes transport more and colder vapor into the western Himalayas and lead to higher DSP accumulation rate and lower $\delta^{18}O$ value, and vice versa. However, the ER core site receives more precipitation from the ISM due to moisture from the westerlies being shielded by the high ridges of Mt. Qomolangma, which causes the ER $\delta^{18}O$ record to show a significant amount effect. As a result, the past different interpretations of the stable isotopic records from the DSP and ER core sites may not be incompatible, which can be explained by the contrasting precipitation seasonality at the two sites. Nevertheless, the conclusions of the paper are only representative of the two specific Himalayan ice core locations, which are not applicable to other low-latitude ice cores.

Acknowledgements. This work was jointly supported, in China, by the National Natural Science Foundation of China (41330526, 41171052, and 41321062), National Basic Research Program of China (2010CB951401), the Chinese Academy of Sciences (XDB03030101-4), and the Fundamental Research Funds for the Central Universities (1082020904), and, in United States, by the US National Science Foundation (ATM 0139491 and ATM0754644). We are grateful to many scientists, technicians, students, and porters for their hard work in the field. Special thanks are given to anonymous reviewers for their constructive comments. We also thank Y. Wang for drawing the three-dimensional topographic maps of the Dasuopu and East Rongbuk glaciers.

Edited by: A. N. LeGrande

References

Adams Carter, H.: Classification of the Himalaya, Am. Alp. J., 27, 109–141, 1985.

An, Z., Colman, S. M., Zhou, W., Li, X., Brown, E. T., Timothy Jull, A. J., Cai, Y., Huang, Y., Lu, X., Chang, H., Song, Y., Sun, Y., Xu, H., Liu, W., Jin, Z., Liu, X., Cheng, P., Liu, Y., Ai, L., Li, X., Yan, L., Shi, Z., Wang, X., Wu, F., Qiang, X., Dong, J., Lu, F., and Xu, X.: Interplay between the Westerlies and Asian monsoon recorded in Lake Qinghai sediments since 32 ka, Scientific Reports, 2, 619, doi:10.1038/srep00619, 2012.

Bolzan, J. F.: Ice flow at the Dome C ice divide based on a deep temperature profile, J. Geophys. Res., 90, 8111–8124, 1985.

Compo, G. P., Whitaker, J. S., Sardeshmukh, P. D., Matsui, N., Allan, R. J., Yin, X., Gleason, B. E., Vose, R. S., Rutledge, G., Bessemoulin, P., Brönnimann, S., Brunet, M., Crouthamel, R. I., Grant, A. N., Groisman, P. Y., Jones, P. D., Kruk, M., Kruger, A. C., Marshall, G. J., Maugeri, M., Mok, H. Y., Nordli, Ø., Ross, T. F., Trigo, R. M., Wang, X. L., Woodruff, S. D., and Worley, S. J.: The Twentieth Century Reanalysis Project, Q. J. Roy. Meteor. Soc., 137, 1–28, doi:10.1002/qj.776, 2011.

Dansgaard, W.: Stable isotopes in precipitation, Tellus, 16, 436–468, 1964.

Davis, M. E., Thompson, L. G., Yao, T., and Wang, N.: Forcing of the Asian monsoon on the Tibetan Plateau: Evidence from high-resolution ice core and tropical coral records, J. Geophys. Res., 110, D04101, doi:10.1029/2004JD004933, 2005.

Dhar, O. N. and Rakhecha, P. R.: The effect of elevation on monsoon rainfall distribution in the central Himalayas, in Monsoon dynamics, edited by: Lighthill, J. and Pearce, R. P., New York, Cambridge University Press, 253–260, 1981.

Duan, K., Yao, T., and Thompson, L. G.: Low-frequency of southern Asian monsoon variability using a 295-year record from the Dasuopu ice core in the central Himalayas, Geophys. Res. Lett., 31, L16209, doi:10.1029/2004GL020015, 2004.

Duan, K., Thompson, L. G., Yao, T., Davis, M. E., and Mosley-Thompson, E.: A 1000 year history of atmospheric sulfate concentrations in southern Asia as recorded by a Himalayan ice core, Geophys. Res. Lett., 34, L01810, doi:10.1029/2006GL027456, 2007.

Gat, J. R. and Carmi, I.: Evolution of the isotopic composition of atmospheric waters in the Mediterranean Sea area, J. Geophys. Res., 75, 3039–3048, 1970.

Gat, J. R., Klein, B., Kushnir, Y., Roether, W., Wernli, H., Yam, R., and Shemesh, A.: Isotope composition of air moisture over the Mediterranean Sea: An index of the air-sea interaction pattern, Tellus B, 55, 953–965, 2003.

Gimeno, L., Torre, L., Nieto, R., García, R., Hernández, E., and Ribera, P.: Changes in the relationship NAO-Northern hemisphere temperature due to solar activity, Earth Planet. Sc. Lett., 206, 15–20, 2003.

Gupta, A., Rathore, L. S., Singh, S. V., and Mendiratta, N.: Performance of a global circulation model in predicting the winter systems and associated precipitation over North West India during 1994–97, in: The Himalayan Environment, edited by: Dash, S. K. and Bahadur, J., New Age International, 123–138, 1999.

Hong, S., Lee, K., Hou, S., Hur, S., Ren, J., Burn, L., Rosman, K., Barbante, C., and Boutron, C.: An 800-year record of atmospheric As, Mo, Sn, and Sb deposition in central Asia in high-altitude ice cores from Mt. Qomolangma (Everest), Himalayas, Environ. Sci. Technol., 43, 8060–8065, 2009.

Hou, S., Qin, D., Wake, C. P., Mayewski, P. A., Ren, J., and Yang, Q.: Climatological significance of ice core net-accumulation record at Mt. Qomolangma (Everest), Chinese Sci. Bull., 45, 259–264, 2000.

Hou, S., Qin, D., Zhang, D., Kang, S., Mayewski, P. A., and Wake, C. P.: A 154a high resolution ammonium record from the Rongbuk Glacier, north slope of Mt. Qomolangma (Everest), Tibet-Himal region, Atmos. Environ., 37, 721–729, 2003.

Hou, S., Qin, D., Jouzel, J., Masson-Delmotte, V., von Grafenstein, U., Landais, A., Caillon, N., and Chappellaz, J.: Age of Himalayan bottom ice cores, J. Glaciol., 50, 467–468, 2004.

Hou, S., Chappellaz, J., Jouzel, J., Chu, P. C., Masson-Delmotte, V., Qin, D., Raynaud, D., Mayewski, P. A., Lipenkov, V. Y., and Kang, S.: Summer temperature trend over the past two millennia using air content in Himalayan ice, Clim. Past, 3, 89–95, doi:10.5194/cp-3-89-2007, 2007.

Hu, Z. Z. and Wu, Z.: The intensification and shift of the annual North Atlantic Oscillation in a global warming scenario simulation, Tellus A, 56, 112–124, 2004.

Hu, Z. Z., Latif, M., Roeckner, E., and Bengtsson, L.: Intensified Asian summer monsoon and its variability in a coupled model forced by increasing greenhouse gas concentrations, Geophys. Res. Lett., 27, 2681–2684, 2000.

Hurrell, J. W.: Influence of variations in extratropical wintertime teleconnections on Norther Hemisphere temperature, Geophys. Res. Lett., 23, 665–668, 1996.

Jones, P. D., Parker, D. E., Osborn, T. J., and Briffa, K. R.: Global and hemispheric temperature anomalies-land and marine instrumental records, in: Trends: A Compendium of Data on Global Change. Carbon Dioxide Information Analysis Center, Oak Ridge National Laboratory, US Department of Energy, Oak Ridge, Tenn., USA. doi:10.3334/CDIAC/cli.002, 2013.

Joswiak, D. R., Yao, T., Wu, G., Tian, L., and Xu, B.: Ice-core evidence of westerly and monsoon moisture contributions in the central Tibetan Plateau, J. Glaciol., 59, 56–66, doi:10.3189/2013JoG12J035, 2013.

Karan, P. P.: Geographic regions of the Himalayas, Bulletin of Tibetology, 3, 5–25, 1966.

Kaspari, S., Mayewski, P. A., Kang, S., Sneed, S., Hou, S., Hooke, R., Kreutz, K., Introne, D., Handley, M., Maasch, K., Qin, D., and Ren, J.: Reduction in northward incursions of the South Asian monsoon since 1400 AD inferred from a Mt. Everest ice core, Geophy. Res. Lett., 34, L16701, doi:10.1029/2007GL030440, 2007.

Kaspari, S., Hooke, R., Mayewski, P. A., Kang, S., Hou, S., and Qin, D.: Snow accumulation rate on Mt. Everest: synchroneity with sites across the Tibetan Plateau on 50-100 year timescales, J. Glaciol., 54, 343–352, 2008.

Kaspari, S., Mayewski, P. A., Handley, M., Kang, S., Hou, S., Sneed, S., Maasch, K., and Qin, D.: A High-Resolution Record of Atmospheric Dust Composition and Variability Since 1650 AD from a Mt. Everest Ice Core, J. Climate, 22, 3910–3925, doi:10.1175/2009JCLI2518.1, 2009a.

Kaspari, S., Mayewski, P. A., Handley, M., Kang, S., Sneed, S., Hou, S., and Qin, D.: Recent Increases in Atmospheric Concentrations of Bi, U, Cs, S and Ca from a 350-Year Mt. Everest Ice Core Record, J. Geophys. Res., 114, D04302, doi:10.1029/2008JD011088, 2009b.

Kaspari, S., Schwikowski, M., Gysel, M., Flanner, M. G., Kang, S., Hou, S., and Mayewski, P. A.: Recent increase in black carbon concentrations from a Mt. Everest ice core spanning 1860–2000 AD, Geophy. Res. Lett., 38, L04703, doi:10.1029/2010GL046096, 2011.

Kistler, R., Kalnay, E., Collins, W., Saha, S., White, G., Woollen, J., Chelliah, M., Ebisuzaki, W., Kanamitsu, M., Kousky, V., van den Dool, H., Jenne, R., and Fiorino, M.: The NCEP-NCAR 50-year reanalysis: monthly means CD-ROM and documentation, B. Am. Meteorol. Soc., 82, 247–268, 2001.

Lang, T. J. and Barros, A. P.: Winter storms in the central Himalayas, J. Meteorol. Soc. Jpn., 82, 829–844, 2004.

Lin, Z. and Wu, X.: A preliminary analysis about the tracks of moisture transport on the Qinghai-Xizang Plateau, Geogr. Res., 9, 33–40, 1990 (in Chinese).

Liu, G.: Hydrometeorological characteristics of the Tibet Plateau, IAHS Publ. No. 179, Atmospheric Deposition (Proceedings of the Baltimore Symposium, May 1989), 267–280, 1989.

Luterbacher, J., Xoplaki, E., Dietrich, D., Jones, P. D., Davies, T. D., Portis, D., Gonzalez-Rouco, J. F., von Storch, H., Gyalistras, D., Casty, C., and Wanner, H.: Extending North Atlantic Oscillation Reconstructions Back to 1500, Atmos. Sci. Lett., 2, 114–124, doi:10.1006/asle.2001.0044, 2002.

McCabe, G. J., Clark, M. P., and Serreze, M. C.: Trends in Northern Hemisphere surface cyclone frequency and intensity, J. Climate, 14, 2763–2768, 2001.

Meehl, G. A. and Washington, W. M.: South Asian summer monsoon variability in a model with doubled atmospheric carbon dioxide concentration, Science, 260, 1101–1104, 1993.

Menon, A. G. K.: Fish geography of the Himalayas, Proc. Nat. Inst. Sci. India, 22, 467–493, 1954.

Ming, J., Cachier, H., Xiao, C., Qin, D., Kang, S., Hou, S., and Xu, J.: Black carbon record based on a shallow Himalayan ice core and its climatic implications, Atmos. Chem. Phys., 8, 1343–1352, doi:10.5194/acp-8-1343-2008, 2008.

Pang, H., Hou, S., Kaspari, S., Mayewski, P. A., Introne, D., Masson-Delmotte, V., Jouzel, J., Li, Z., He, Y., and Qin, D.: Atmospheric circulation change in the central Himalayas indicated by a high-resolution ice core deuterium excess record, Clim. Res., 53, 1–12, 2012.

Pascoe, E. H.: A manual of the geology of India and Burma, Vol. 1 (3rd Edn.), Government of India Press, Calcutta, 485 pp., 1950.

Rao, P. L. S.: Circulation characteristics over the Himalayas during winter season, Meteorol. Atmos. Phys., 83, 19–33, 2003.

Reeh, N.: A flow-line model for calculating the surface profile and the velocity, strain-rate, and stress fields in an ice sheet, J. Glaciol., 34, 46–54, 1988.

Rindsberger, M., Magaritz, M., Carmi, I., and Gilad, D.: The relation between air mass trajectories and the water isotope composition of rain in the Mediterranean Sea area, J. Geophys. Res., 88, 43–46, 1983.

Shrestha, M. L.: Interannual variation of summer monsoon rainfall over Nepal and its relation to Southern Oscillation index, Meteorol. Atmos. Phys., 75, 21–28, 2000.

Sontakke, N. A. and Singh, N.: Longest instrumental regional and all-India summer monsoon rainfall series using optimum observations: Reconstruction and update, Holocene, 6, 315–331, 1996.

Sontakke, N. A., Singh, N., and Singh, H. N.: Instrumental period rainfall series of the Indian region (1813–2005): revised reconstruction, update and analysis, Holocene, 18, 1055–1066, 2008.

Thompson, D. W. J., Wallace, J. M., and Hegerl, G. C.: Annular modes in the extratropical circulation. Part II: trends, J. Climate, 13, 1018–1036, 2000a.

Thompson, L. G., Yao, T., Mosley-thompson, E., Davis, M. E., Henderson, K. A., and Lin, P. N.: A high-resolution millennial record of the south Asian monsoon from Himalayan ice cores, Science, 289, 1916–1919, 2000b.

Tian, L., Yao, T., Sun, W., Stievenard, M., and Jouzel, J.: Relationship between δD and $\delta^{18}O$ in precipitation from north to south of the Tibetan Plateau and moisture cycling, Sci. China Ser. D, 44, 789–796, 2001.

Tian, L., Yao, T., Schuster, P. F., White, J. W. C., Ichiyanagi, K., Pendall, E., Pu, J., and Yu, W.: Oxygen-18 concentrations in recent precipitation and ice cores on the Tibetan Plateau, J. Geophys. Res., 108, 4293, doi:10.1029/2002JD002173, 2003.

Tian, L., Yao, T., White, J. W. C., Yu, W., and Wang, N.: Westerly moisture transport to the middle of Himalayas revealed from the high deuterium excess, Chinese Sci. Bull., 50, 1026–1030, 2005.

Tian, L., Yao, T., MacClune, K., White, J., Schilla, A., Vaughn, B., Vachon, R., and Ichiyanagi, K.: Stable isotopic variations in West China: A consideration of moisture sources, J. Geophys. Res., 112, D10112, doi:10.1029/2006JD007718, 2007.

Vuille, M., Werner, M., Bradley, R. S., and Keimig, F.: Stable isotopes in precipitation in the Asian monsoon region, J. Geophys. Res., 110, D23108, doi:10.1029/2005JD006022, 2005.

Wei, K. and Gasse, F.: Oxygen isotopes in lacustrine carbonates of West China revisited: implications for post glacial changes in summer monsoon circulation, Quat. Sci. Rev., 18, 1315–1334, 1999.

Xu, B. and Yao, T.: Dasuopu ice core record of atmospheric methane over the past 2000 years, Sci. China Ser. D, 44, 689–695, 2001.

Xu, J., Hou, S., Qin, D., Kaspari, S., Mayewski, P. A., Petit, J. R., Delmonte, B., Kang, S., Ren, J., Chappellaz, J., and Hong, S.: A 108.83-m ice-core record of atmospheric dust deposition at Mt. Qomolangma (Everest), Central Himalaya, Quat. Res., 73, 33–38, 2010.

Yao, T., Guo, X., Thompson L. G., Duan, K., Wang, N., Pu, J., Xu, B., Yang, X., and Sun, W.: $\delta^{18}O$ record and temperature change over the past 100 years in ice cores on the Tibetan Plateau, Sci. China Ser. D, 49, 1–9, doi:10.1007/s11430-004-5096-2, 2006.

Yao, T., Thompson, L., Yang, W., Yu, W., Gao, Y., Guo, X., Yang, X., Duan, K., Zhao, H., Xu, B., Pu, J., Lu, A., Xiang, Y., Kattel, D. B., and Joswiak, D.: Different glacier status with atmospheric circulations in Tibetan Plateau and surroundings, Nat. Clim. Change, 2, 663–667, doi:10.1038/nclimate1580, 2012.

Yihui, D. and Zunya, W.: A study of rainy seasons in China, Meteorol. Atmos. Phys., 100, 121–138, doi:10.1007/s00703-008-0299-2, 2008.

Zhang, D., Qin, D., Hou, S., Kang, S., Ren, J., and Mayewski, P. A.: Climatic significance of $\delta^{18}O$ records from an 80.36 m ice core in the East Rongbuk Glacier, Mount Qomolangma (Everest), Sci. China Ser. D, 48, 266–272, 2005.

Zhang, S., Hou, S., Ma, X., Qin, D., and Chen, T.: Culturable bacteria in Himalayan glacial ice in response to atmospheric circulation, Biogeosciences, 4, 1–9, doi:10.5194/bg-4-1-2007, 2007.

Zhang, S., Hou, S., Wu, Y., and Qin, D.: Bacteria in Himalayan glacial ice and its relationship to dust, Biogeosciences, 5, 1741–1750, doi:10.5194/bg-5-1741-2008, 2008.

13

Seasonal changes of ice surface characteristics and productivity in the ablation zone of the Greenland Ice Sheet

D. M. Chandler[1], J. D. Alcock[1], J. L. Wadham[1], S. L. Mackie[2], and J. Telling[1]

[1]Bristol Glaciology Centre, School of Geographical Sciences, Bristol University, Bristol, UK
[2]School of Earth Sciences, Bristol University, Bristol, UK

Correspondence to: D. M. Chandler (davemchandler@gmail.com)

Abstract. Field and remote sensing observations in the ablation zone of the Greenland Ice Sheet have revealed a diverse range of ice surface characteristics, primarily reflecting the variable distribution of fine debris (cryoconite). This debris reduces the surface albedo and is therefore an important control on melt rates and ice sheet mass balance. Meanwhile, studies of ice sheet surface biological processes have found active microbial communities associated with the cryoconite debris, which may themselves modify the cryoconite distribution. Due to the considerable difficulties involved with collecting ground-based observations of the ice surface, our knowledge of the physical and biological surface processes, and their links, remains very limited. Here we present data collected at a field camp established in the ice sheet ablation zone at 67° N, occupied for almost the entire melt season (26 May–10 August 2012), with the aim of gaining a much more detailed understanding of the physical and biological processes occurring on the ice surface. These data sets include quadrat surveys of surface type, measurements of ice surface ablation, and in situ biological oxygen demand incubations to quantify microbial activity. In addition, albedo at the site was retrieved from AVHRR (Advanced Very High Resolution Radiometer) remote sensing data. Observations of the areal coverage of different surface types revealed a rapid change from complete snow cover to the "summer" (summer study period) ice surface of patchy debris ("dirty ice") and cryoconite holes. There was significant correlation between surface albedo, cryoconite hole coverage and surface productivity during the melt season, but microbial activity in "dirty ice" was not correlated with albedo and varied widely throughout the season. While this link suggests the potential for a remote-sensing approach to monitoring cryoconite hole biological processes, very wide seasonal and spatial variability in net surface productivity demonstrates the need for caution when extrapolating point measurements of biological processes to larger temporal or spatial scales.

1 Introduction

The ablation zone of the Greenland Ice Sheet displays a remarkable variety of surface characteristics that vary widely in time and space. The varying distribution of ice, snow, melt water and sediment across the surface strongly influences the surface energy balance, and therefore the melt rate and mass balance (e.g. Konzelmann and Braithwaite, 1995; Knap and Oerlemanns, 1996; Bøggild et al., 2010). Recent studies have also shown links between sediment distribution and biological activity, indicative of feedback between microbial processes and albedo (e.g. Anesio et al., 2009; Uetake et al., 2010).

Due to the logistical demands of observing the ice sheet surface on foot, which is particularly problematic in the ablation zone where rivers, slush and crevasses make surface travel difficult, much of our present knowledge of its surface properties and processes has been gained through short 1–2 day field visits, automated data acquisition (primarily at automatic weather stations, some of which have now been operation for several years – for example those on the K-transect: http://www.projects.science.uu.nl/iceclimate/aws/greenland.php) and satellite remote sensing. Consequently there are very few data sets comprising manual observations of seasonal changes in surface characteristics, and none on seasonal changes in biological activity on the ice sheet. Many more observations are needed to accurately estimate the contribution of observed biological activity to global chemical cycles, ideally leading to a method by which such activity can be monitored remotely.

In 2012 a field camp was established centrally in the ice sheet ablation zone at 1030 m elevation, 35 km from the western margin, with the aim of acquiring a continuous 10-week time series of data relating to surface characteristics and processes. The camp was occupied continuously from 26 May to 10 August, covering most of the melt season. Resulting data on surface ablation, surface type and surface productivity (as quantified by biological oxygen demand (BOD) experiments) are compared with a satellite remote-sensing time series of surface albedo for the same location. Using these data sets, we identify the characteristics of physical and biological surface processes at the field site and assess the extent to which these can be quantified remotely in surface albedo retrievals. We finally examine in more detail the seasonal changes in surface productivity, which are presently unknown in Greenland. To our knowledge, this is the longest record of ground-based surface observations yet to be collected on the ice sheet.

1.1 Ablation zone ice surface characteristics

During the melt season, the ice sheet surface in the ablation zone displays a wide range of features commonly found on terrestrial glaciers, including crevasses, rivers, lakes, cryoconite holes and snow remnants. A variable amount of fine surface debris is also present, for example as dispersed particles, as a thin layer in cryoconite holes, or as thick deposits (Fig. 1). This fine debris, usually referred to as cryoconite, is mostly composed of fine mineral particles with a small (< 10 %) organic matter content (Bayley, 1891; Hodson, 2010; Wientjes et al., 2011). Typical particle sizes are of order 10^{-5} to 10^{-4} m (data from several studies reported in Table 1 of Hodson et al., 2010). During the summer, debris causes some parts of the ablation zone to appear notably darker than the surrounding ice in satellite images (van de Wal and Oerlemans, 1994; Wientjes et al., 2011). These dark zones can be clearly observed each year during the melt season in true-colour MODIS AERONET imagery of west Greenland (http://lance-modis.eosdis.nasa.gov/imagery/subsets/?project=aeronet&subset=Kangerlussuaq). The lower albedo of the dark region in the ablation zone has previously been interpreted as arising from the accumulation of surface melt water (Oerlemans and Vugts, 1993; Knap and Oerlemans, 1996). Whether debris- or water-related, this zone of lower albedo has a direct impact on surface melt rates and therefore on ice sheet mass balance.

As a consequence of the variable surface characteristics and debris cover, the ice surface albedo can range widely, from > 80 % (fresh snow) to typically 40–60 % (bare ice) and as low as 10 % for cryoconite debris (Knap and Oerlemans, 1996; Bøggild et al., 2010). However, bare ice containing

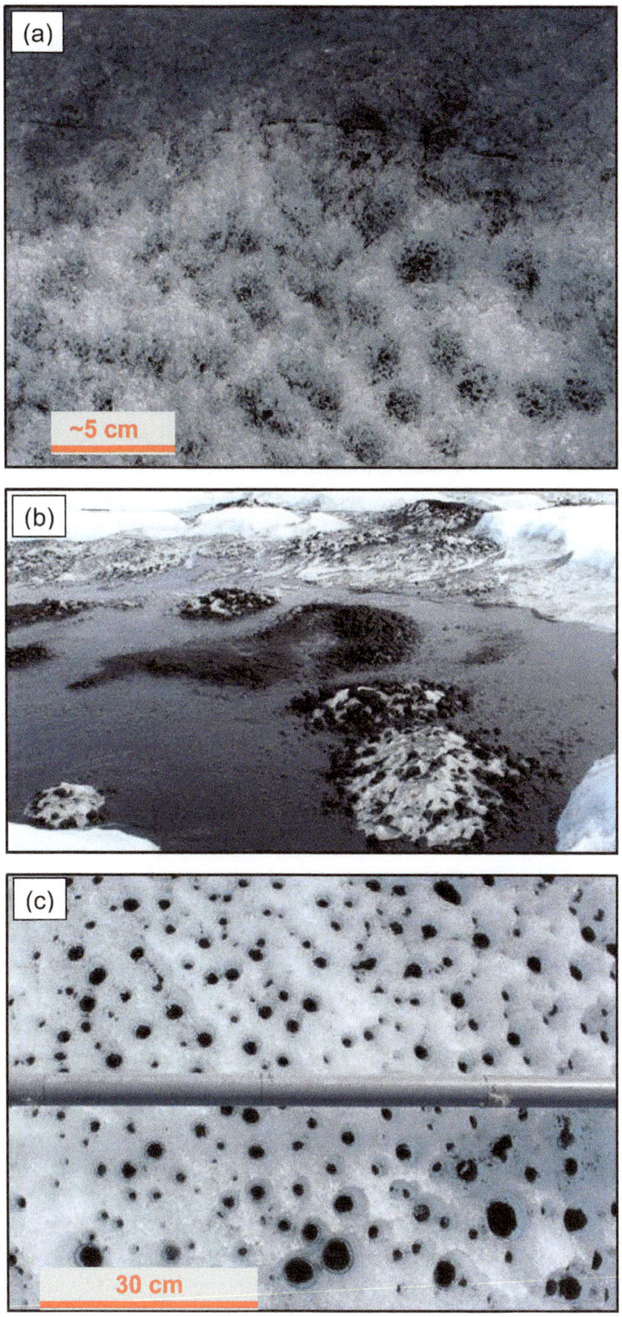

Figure 1. Debris on the ice sheet surface at approximately 67.0° N, 49.3° W, 1030 m, in the 2012 melt season. **(a)** Dispersed debris. The foreground shows debris released by the melting-out of cryoconite holes while the more uniform cover in the background is typical "dirty ice". The linear feature oriented right to left is a healed fracture. The image width at the surface is ~ 50 cm. **(b)** Thick deposits in a slowly draining lake bed, here reaching ~ 6 cm thick. **(c)** Cryoconite holes. The plastic pole is marked at 30 cm intervals. These are relatively small cryoconite holes; some reach over 1 m in diameter.

Figure 2. The ice sheet surface at 67.0° N, 50.0° W, 580 m (7 km from the margin), early in the melt season (10 June 2011). The foreground is a recently drained pond with cryoconite debris and "spikes" of clean ice that have melted much more slowly due to their higher albedo. The "hills" in the background have heights of 3–5 m and very low debris cover.

cryoconite holes has been found to have a higher albedo than bare ice with a lower mass loading of dispersed debris, because the depth of the cryoconite holes keeps the debris partially hidden at most incidence angles (Bøggild et al., 2010). Seasonal changes in albedo have been identified in the ablation zone, where a sharp then gradual decrease in albedo in the early melt season is associated with the melting of highly reflective seasonal snow and the subsequent increase in surface water and/or exposure of surface debris (Knap and Oerlemans, 1996; Stroeve et al., 2001; Wientjes et al., 2011). Although none of these studies has presented detailed time series lasting longer than a year, consistency between the different years and locations investigated suggests the reported seasonal patterns are widespread and persistent.

In addition to the different ice types and patchy debris cover, the ice surface is also characterised by its small-scale topography. For example, the variable action of surface melt leads to the formation of the "lumpy" topography in the lower ablation zone, with peak-to-peak height differences between valleys and ridges increasing down-glacier and reaching 5 m close to the margin in west Greenland (Fig. 2). This topography includes the characteristic "domes", "hummocks" and "gullies" described by Smeets and van den Broeke (2008) towards the end of the melt season at 67° N. Development of these topographical features and the change from snow to bare ice lead to large spatial and seasonal changes in the aerodynamic roughness length of several orders of magnitude. This affects the sensible and latent heat fluxes into (or out of) the ice but has been rarely measured in the ablation zone (Duynkerke and van den Broeke, 1994; Smeets and van den Broeke, 2008). Remote sensing of surface roughness has been shown to be possible using a Multi-angle Imaging SpectroRadiometer (MISR) in the Sermeq Kujalleq (Jacobshavn) region of west Greenland (Nolin et al., 2002; Nolin and Payne, 2007) but seasonal changes in roughness measured using this technique have not yet been reported. Roughness is not considered further in the present paper but should be addressed in future studies.

1.2 Ablation zone ice surface biological processes

The lack of any obvious plant or animal life on the ice sheet gives the impression that the surface is devoid of biological activity. However, studies of surface debris on the Greenland Ice Sheet have revealed the existence of active microbial communities including bacteria, cyanobacteria and green algae (Steinböck, 1936; Hodson et al., 2010; Stibal et al., 2012). Much of the activity was originally believed to be associated with the fine debris concentrated in a thin layer on the bottom of cryoconite holes (Anesio et al., 2009), but activity in algae present over much of the remaining surface in the ablation zone has recently been reported (Cook et al., 2012).

Cryoconite holes are widespread in the ablation zone of the Greenland Ice Sheet as well as in many other glaciated areas worldwide, and may form an important component of regional chemical cycles (e.g. Hodson et al., 2007; Anesio et al, 2009; Cook et al., 2012; Telling et al, 2012b). Since the Greenland Ice Sheet's melt zone has (potentially) a much greater surface area of microbial communities than that of any other terrestrial ice mass outside Antarctica, the role of the ice sheet's surface debris in the global carbon cycle has attracted increasing attention. Measurements of ecosystem productivity by in situ incubation experiments (Table 1) have revealed active communities in the few field sites visited on the ice sheet to date. Incubation experiments have now seen widespread use on many other glaciers, where similarly high levels of biological activity have been observed (Telling et al., 2010). The most detailed study in Greenland collected productivity measurements along a 70-km transect across the ablation zone in west Greenland at 67° N (Stibal et al., 2012; Cook et al., 2012), close to the field site studied in the present paper. Stibal et al. (2012) found a peak in cryoconite hole carbon storage rate ($22 \pm 4.8\,\mu$gC day^{-1} per unit mass of debris) at 11 km from the margin (761 m a.s.l.) and subsequent decrease to near zero by 65 km (1446 m a.s.l.). The most active part of the transect in terms of carbon storage rate per unit area was ~ 7 mgC m^{-2} day^{-1} at the ice margin (399 m a.s.l.) where debris cover is very high (Stibal et al., 2012). Interestingly, measurements of surface algae productivity on the debris-poor ice between holes suggested the carbon storage by surface algae might be considerably higher than the above figures for cryoconite holes (Yallop et al., 2012), but the different method (^{14}C uptake; 1 h incubation rather than 24 h) introduced considerable uncertainty in the estimates. Attempts have been made to calculate ice sheet-scale carbon fluxes from Greenland's cryoconite holes and surface algae by extrapolating existing data (Hodson, 2010; Cook, 2012); however, the great variability in carbon fluxes measured at different field sites (Table 1) and a lack of knowledge regard-

Table 1. Cryoconite hole carbon flux measurements in Greenland.

Source	Method	Carbon storage rate[1] μgC [g or L]$^{-1}$ d^{-1}	Notes
Anesio et al. (2009)	^{14}C	Water 54 ± 60 Sediments 115 ± 56	Frøya Glacier, east Greenland, 74° N, date not specified
Hodson et al. (2010)	TDIC, O_2	Sediments Mean -3.9[2] P: 12–30 R: 4.8–29	South-west margin, 67° N, 2-km transect at margin (~ 2 % of melt zone), 23–30 August 2008
Stibal et al. (2012); Cook et al. (2012)	TDIC	Sediments Mean: 5.4[3] Range from below detection limit to 22 ± 4.8	South-west margin, 67° N, 65-km transect with nine stations from margin to slush zone, August 2010.

[1] Units μgC g^{-1} d^{-1} or μgC L^{-1} d^{-1} represent the carbon flux per gram of dry sediment per day or the carbon flux per litre of water per day. [2] Mean NEP calculated from O_2 and TDIC incubation experiments BI1 to BI5 reported in Table 2 of Hodson et al. (2010). [3] Mean NEP of nine stations along the transect using results reported in their Table 3.

ing seasonal variability requires many more measurements to be made before realistic extrapolation is possible.

In addition to its biogeochemical interest, biological activity in cryoconite debris can modify the characteristics of the debris itself, for example by generating dark organic material or by causing small grains to clump together into larger, less mobile particles. These processes alter the albedo of the cryoconite holes (Langford et al., 2010; Takeuchi, 2002), causing a corresponding reduction in the ice surface albedo, one of the key factors controlling ice sheet mass balance. We speculate that microbial processes will also affect the mobility of debris within the surface drainage system, and although this question remains open at present, it could be addressed using the time-lapse technique employed recently to observe cryoconite particle motion in Svalbard (Irvine-Fynn et al., 2011).

Given the importance of ice surface characteristics in controlling surface melt (and thus mass balance), and the increasing evidence for links between biological activity and surface characteristics, there is clearly a need to gain a better understanding of ice sheet surface processes in the ablation zone. In this paper, changes in the areal coverage and associated biological activity over most of a melt season are described for different surface types identified in the ablation zone of the western Greenland Ice Sheet at 67° N. The resulting time series reveal the seasonal patterns in much greater detail than has been achieved in previous field studies.

2 Methods

Field data were collected during the 2012 summer melt season at 66.97° N, 49.27° W, in the ablation zone of the ice sheet at 1030 m altitude and approximately 35 km from the western margin (Fig. 3). The field site is on the western edge of the "dark" band of the ablation zone identified by Wientjes et al. (2011). Site locations for the surveys and incubation experiments (see below) were all located down-stream from, and on the opposite side of, a surface melt water stream passing adjacent to the field camp, to minimise any interference or contamination from human activity.

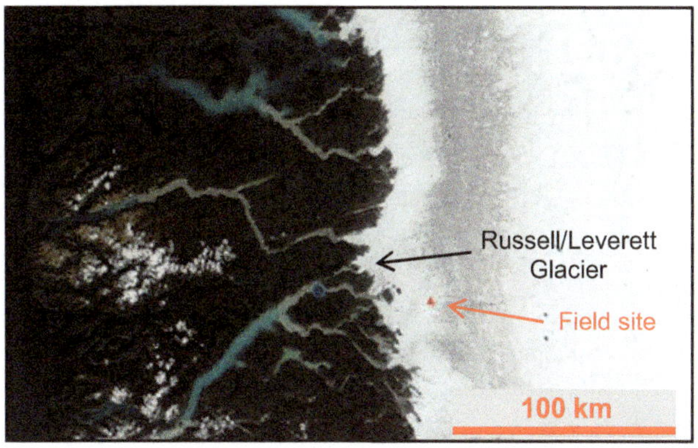

Figure 3. Field site location (red triangle) at ~ 67° N in west Greenland, with the band of dark ice oriented north-south clearly visible in the ablation zone. The blue circle west of the field site marks Kangerlussuaq. The background image is a MODIS image acquired on 17 July 2012.

2.1 Surface characteristics and ablation

Daily ice ablation was recorded by measuring the distance between the ice surface and the tops of five 40 cm plastic poles installed in holes drilled 50–100 cm into the ice. Installing the stakes into the ice below the surface avoided the lateral melting that normally occurs around poles left protruding from the ice. Stakes were arranged in a "+" shape with four outer stakes and one in the centre, at ~ 2 m spacing. The ice surface was often very rough, so for consistency

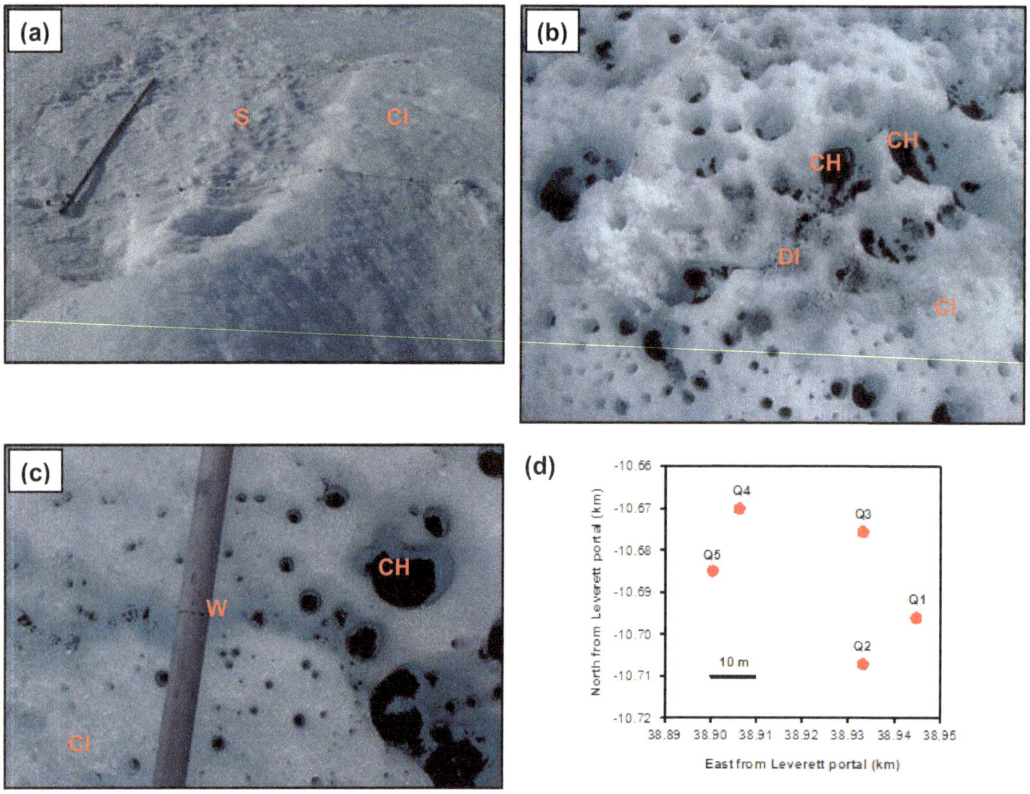

Figure 4. (**a**) Q2, 1 June 2012, showing the quadrat ropes and pole. Surface cover is 55 % snow (S) and 45 % clean ice (CI). (**b**) Lower left corner of Q4, 13 July 2012, with clean ice (CI), dirty ice (DI) and cryoconite holes (CH; Note that CS incubations sampled both the sediment and overlying water from these holes; CW incubations sampled only the water). (**c**) Q4 on 29 July, with a very small melt water channel linking cryoconite holes under the mark on the pole. (**d**) Locations of quadrats on 1 June 2012, co-ordinates are relative to the portal from which melt water draining from the field site emerges at the ice margin.

each measurement was taken to be the vertical distance between the stake top and a 25 cm hacksaw blade laid centrally across the top of the hole in an east–west orientation. The initial surface slope at the site was flat (less than 10°), but by August the site had become a lump with the central and northern stakes on its flat top and the eastern, southern and western stakes on sides sloping outwards at 10–20°. Stakes were re-drilled before the stake had started to protrude from the hole. This method yields ablation expressed as an elevation change, rather than the more commonly encountered unit of mm water equivalent (mm w.e.). Conversion to mm w.e. from measured elevation change requires the densities of water and ice. In practice the density of ice close to the surface is very difficult to measure owing to the porous nature of the melt crust which has a density much lower than that of solid glacier ice. If this crust were to maintain a constant thickness, change in surface height would still yield ablation in mm w.e.; however, the melt crust characteristics (e.g. thickness, density) were very variable, and therefore we have not attempted to make this conversion.

Surface characteristics were monitored frequently from 1 June to 11 August (total 18 surveys) in five 2.7×2.7 m quadrats situated within a region of roughly 60×40 m (Fig. 4a and d). Each quadrat comprised a 10×10 grid with 30 cm spacing. Quadrat locations were marked by drilling ~ 50 cm deep holes in the ice at each corner of a 3×3 m square, to ensure the same area of ice was surveyed each time. To carry out a survey, two nylon ropes (both marked with the positions of the hole edges and then at 30 cm intervals) were laid out in parallel between each pair of holes. The ropes were kept tight by attaching steel weights that could hang freely in the drilled holes. A rigid 3 m pole, also marked at 30 cm intervals, was laid between the first pair of markers on the ropes. The surface type was recorded immediately in front of each mark along the pole, before moving the pole to the next pair of rope markers and recording the second row of surface types, and so on. Six surface types were present at the field site, of which five types were recorded in the quadrats (Table 2; Fig. 4). All surveys were carried out by the same pair of observers, to ensure consistency in the classification.

From a biological aspect, it would have been preferable to distinguish between superimposed ice and englacial ice (here both classified as CI) since they may well contain different abundances and/or species of micro-organisms. In practice it

Table 2. List of surface types identified at the field site. All except Type F were present in the quadrat surveys.

Surface Type	Code	Description
Snow	S	Seasonal snow cover, including a coarse-grained granular material that is probably rotting superimposed ice (Fig. 4a)
Water	W	Surface water that is not covering cryoconite debris (Fig. 4c)
Clean ice	CI	Englacial ice with no visible debris (Fig. 4a, b, c)
Dirty ice	DI	Englacial ice with debris partially or completely covering the surface (Figs. 1a, b, 4b)
Cryoconite hole	CH	Water-filled holes with a layer of debris in the bottom (Figs. 1c, 4b, c)
Fractures, moulins	F	Open fractures, crevasses and moulins were all present close to the field site but were not recorded in the quadrat surveys

would have been difficult to confidently classify the ice surface in the field and in some cases they may even be mixed (for example, in a layer ~ 10 cm thick where spring snow melt has penetrated the previous summer's permeable melt crust and then refrozen). Sometimes there were clues in the shape of the crystal remnants but classification would nevertheless be uncertain and subjective. Therefore, we do not make the distinction between these ice types.

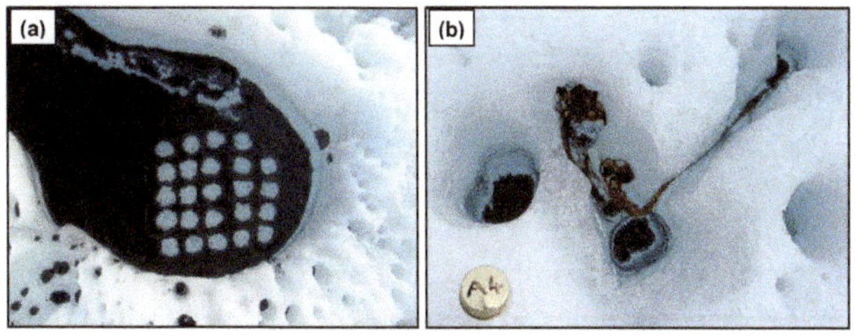

Figure 5. **(a)** Sample collection for the cryoconite water plus sediment incubation (13 July) left 24 round spaces, each about the diameter of the BOD bottle (~ 65 mm). **(b)** Orange-green material sampled for the 4th and 9th July incubations. The lid diameter is 27 mm.

2.2 Productivity (biological oxygen demand)

Net ecosystem production (NEP) and respiration were quantified within each of the different surface environments recorded in the quadrat surveys, at approximately weekly intervals, by measuring the rate of O_2 production in samples collected from the ice surface. The method is described in detail by Telling et al. (2010). One sample was collected for each surface type present, and divided into six glass 250 mL biological oxygen demand (BOD) bottles. Ice samples were melted before pouring into the BOD bottles, by collecting the ice in a Whirl-pak bag and immersing the bag in lukewarm water. These melting samples were left in a black plastic box, typically for 2–3 h, until completely melted (partially melted samples were not used). For samples representing cryoconite holes, debris from the base of the hole was collected using a 100 mL syringe fitted with a ~ 20 cm plastic tube, and the area of debris removed from the hole was selected to match the area of the BOD bottle base (Fig. 5a). This ensured that the debris thickness in the bottle was representative of that in the hole. These bottles were then filled with water collected from the same hole as the debris. The debris typically settled out onto the base of the bottle within 30 s.

The temperature and dissolved oxygen content of the water in each BOD bottle were measured using a PreSens Fibox3 fibre-optic oxygen meter (manufacturer's stated accuracy: ±1 %), and the bottle was then immediately sealed with a glass stopper. Bottles were checked carefully to ensure no air bubbles were present. Three out of the six bottles for each sample type were wrapped in foil to eliminate light (therefore measuring respiration only); the other three remained unwrapped to allow photosynthesis as well as respiration (to measure NEP). All bottles were placed in a cryoconite hole sufficiently deep to ensure complete immersion, in order to replicate typical light and temperature conditions as well as to prevent the bottle freezing and cracking. After 24±1 h, the bottles were recovered and the dissolved oxygen content and temperature were re-measured.

In addition to these standard weekly incubation experiments, we carried out four additional experiments to better characterise temporal and spatial variability and errors in NEP rates, as follows. (1) To better quantify the variance in the four common environments (water, clean ice, dirty ice, cryoconite hole) we carried out additional incubations with 12 light and 12 dark replicates of just one sample type. (2) Temporal variability in cryoconite hole productivity within the 24 h incubation time was measured from 4–5 August by filling 32 bottles (16 light, 16 dark) with cryoconite hole debris and water, then recovering sets of bottles at 3 h intervals for 24 h. (3) Spatial variability and effects of cryoconite hole diameter were investigated on 6–7 August by choosing five sites and incubating light/dark pairs of samples from three holes of varying size (nominally "small", "medium" and "large") at each site. (4) Two standard incubations (4 and 9 July) also included samples of orange-green material (Fig. 5b) sampled from the former bed of a shrinking lake (66.97° N, 49.21° W; 3 km east of the main field site). This material was only observed at one location, in a strip a few tens of metres wide extending for less than 500 m along the lake edge.

The rate of oxygen production in the BOD bottles, expressed in $mg\,L^{-1}\,h^{-1}$, is equal to the difference between the start and end point dissolved oxygen concentrations, divided by the incubation time. In this experiment, two rates of oxygen production were measured for each sample type: $\Delta O_2^{(light)}/\Delta t$ and $\Delta O_2^{(dark)}/\Delta t$ for the light and dark bottles, respectively. We assume that respiration rates were independent of light conditions and therefore the same in the light and dark bottles. The rate of oxygen production by photosynthesis in the light bottles, representing that in the original environment, is therefore $P = (\Delta O_2^{(light)} - \Delta O_2^{(dark)})/\Delta t$.

To convert oxygen production into carbon storage, we assume photosynthesis can be represented as

$$nCO_2 + nH_2O \rightarrow (CH_2O)_n + nO_2. \tag{1}$$

Respiration is taken to be the reverse reaction. In photosynthesis, this equation predicts that n moles of CO_2 are stored for each n moles of O_2 released. This assumption of a 1 : 1 stoichiometric ratio has been adopted in previous studies of cryoconite holes (Hodson et al., 2007, 2010; Anesio et al., 2009) but may not be exact: for example in Svalbard, Telling et al. (2010) found 1.24 ± 0.20 moles of O_2 were released for each mole of CO_2 consumed during photosynthesis, while 0.80 ± 0.17 moles of CO_2 were released for each mole of O_2 consumed during respiration. Neglecting this uncertainty given the error bounds on the above ratios, a 1 : 1 molar ratio will yield 1.375 g of CO_2 (or 0.375 g of C) consumed per 1 g of O_2 released.

Carbon storage rates in the BOD bottles are more useful after conversion to a surface flux (e.g. $mgC\,m^{-2}\,d^{-1}$) for each sample type. To achieve this we required the bottle volume V_B, bottle radius R_B, sample thickness (for ice samples) H_S, incubation duration T, water density ρ_W and sample density ρ_S. For water and cryoconite hole samples (CW, CS), the area of ice sheet A_S sampled to fill the bottle is $A_S = \pi R_B^2$. For ice samples, $A_S = V_B \rho_W / H_S \rho_S$. The C flux $F_{C,X}$ from each ice type X is then $F_{C,X} = \Delta O_2^{(light)} \times 0.375 V_B/(A_S T)$ and the total carbon flux into the surface at the field site is the sum of $A_X F_{C,X}$ over all the ice types X, weighted for their respective fractional areas A_X as determined in the quadrat surveys. Parameter values were as follows: $V_B = 250$ mL, $R_B = 32$ mm, $H_S \approx 30$ mm, $\rho_S = 900\,kg\,m^{-3}$ (ice), $\rho_S = 300\,kg\,m^{-3}$ (snow), $\rho_W = 1000\,kg\,m^{-3}$. Quadrat surveys and BOD experiments did not always coincide, so both data sets were linearly interpolated to daily resolution to allow calculation of daily values for the product $A_X F_{C,X}$ and for the net C flux.

2.3 Surface albedo

A surface broadband albedo product (AVHRR-SAL, CM-59), derived from data collected by the AVHRR (Advanced Very High Resolution Radiometer) instrument onboard the NOAA polar orbiting MetOp platforms, is available from the Climate Monitoring Satellite Application Facility (CM SAF) for the Arctic region. Monthly and weekly averaged products are available on a 15 × 15 km grid, and an instantaneous product on a 1 × 1 km grid. We use the latter product here because of its higher resolution. A summary of the product follows; for a full description see Karlsson et al. (2012). Albedo is calculated for each AVHHR-imaged pixel ($\sim 1.1\,km^2$), using satellite and sun viewing geometry and recorded top-of-atmosphere reflectances. In addition, land cover data, a cloud mask, surface pressure and water vapour from Deutsche Wetter Dienst are used. Cloud shadow effects are reduced by the removal of not only all cloud pixels, but also of all pixels neighbouring a cloud pixel. A topographical correction is applied to the observed reflectances and used to assign a reflectance to hidden slopes. Water vapour and surface pressure data from Deutsche Wetter Dienst are used under the assumption of constant ozone content and aerosol optical depth of the atmosphere to convert top-of-atmosphere reflectances to surface reflectances. These surface reflectances are used to calculate black sky surface albedo, following Schaepman-Strub et al. (2006) for radiation of 0.25 to 2.5 μm; this is almost equivalent to full broadband surface albedo since shortwave infrared contributions to the full broadband albedo are generally very small, and we treat it here as equivalent. Distance-weighted interpolation was used to calculate the surface broadband albedo for the field site location from the 1 × 1 km grid in which the CM SAF product is provided.

Only those images where all four closest pixels were assigned an albedo value were retained for further analysis. We also restricted the image acquisition time to within 6 h of local solar noon to avoid images acquired at high zenith angles. Finally, remaining images were grouped into daily bins before calculating the mean and standard deviation albedo value for each day.

2.4 Meteorological data

Meteorological data were obtained from an automatic weather station installed on the nearby K-transect (Greuell et al., 2001). These data were kindly provided by the Institute for Marine and Atmospheric Sciences, Utrecht University (IMAU). Further details of the field sites and instrumentation can be found at http://www.projects.science.uu.nl/iceclimate/aws/home.html. There are several AWSs (Automatic Weather Stations) on the transect, the closest of which in terms of both location and altitude is S6 (67°04 N, 49°23 W, 1030 m; 11 km NNW from our field site). The measurements of most importance to this paper are the air temperature and relative humidity (both at 2 m above the

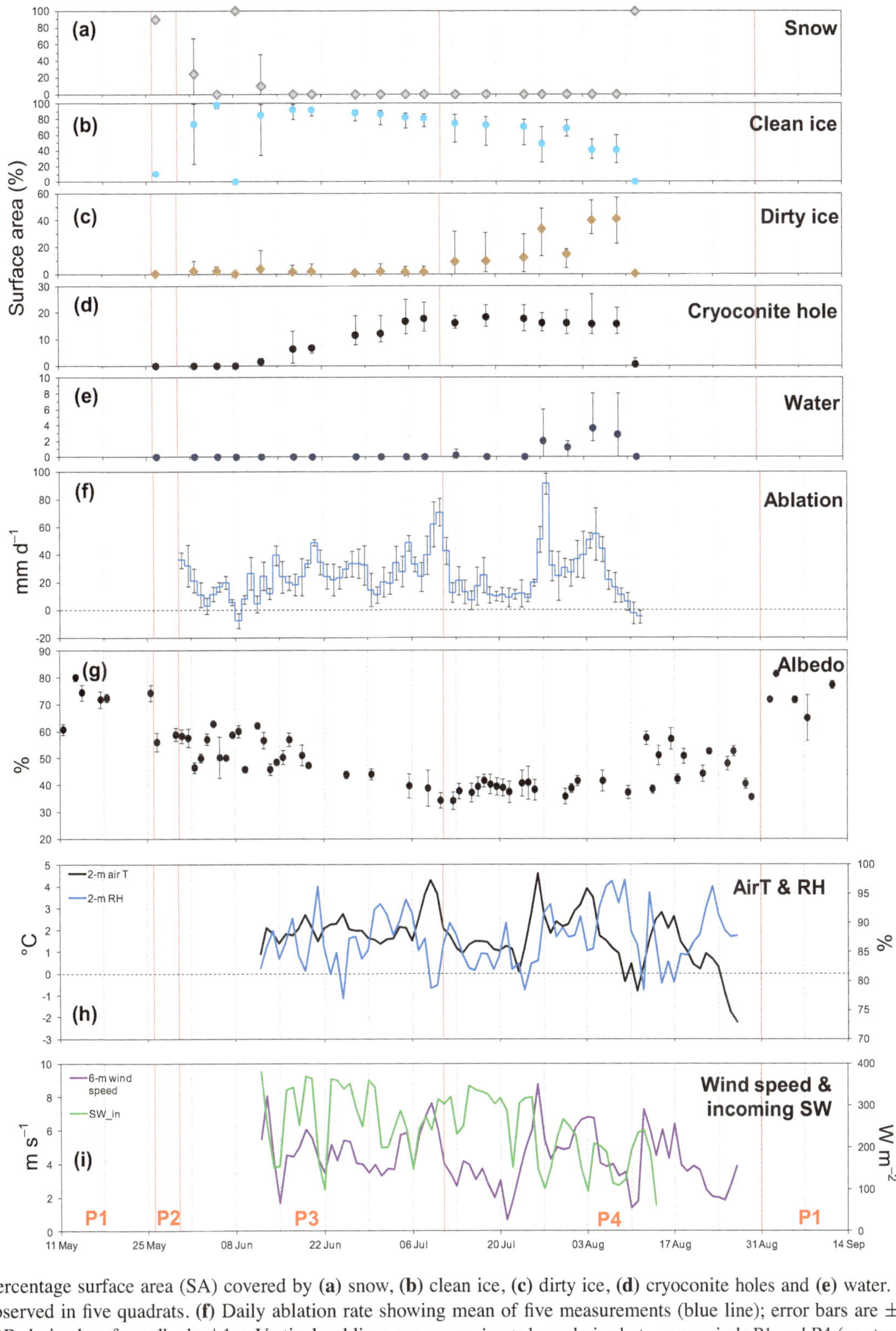

Figure 6. Percentage surface area (SA) covered by **(a)** snow, **(b)** clean ice, **(c)** dirty ice, **(d)** cryoconite holes and **(e)** water. Error bars are the range observed in five quadrats. **(f)** Daily ablation rate showing mean of five measurements (blue line); error bars are $\pm 1\sigma$. **(g)** Daily mean AVHRR-derived surface albedo $\pm 1\sigma$. Vertical red lines are *approximate* boundaries between periods P1 and P4 (see text). **(h–i)** Daily averages of 2 m air temperature, relative humidity, 6 m wind speed and incoming short-wave radiation intensity at AWS S6 located 11 km from our field site.

surface), wind speed (6 m above the surface) and incoming short-wave radiation. Each of these are logged at 30 min intervals.

To assess whether photosynthetic activity is dependent on light intensity, we used the AWS S6 incoming shortwave radiation data recorded by a Kipp and Zonen CNR1 radiometer (305 to 2800 nm sensitivity; accuracy 10 %). The SW data for S6 are missing before 12 June and were therefore estimated using data from AWS S5. Estimates were based on linear regression between SW data from the two AWSs S5 and S6 for the period 12 June–15 August, which yielded $SW_{S6} = 0.984SW_{S5} + 15.17$ [$W\,m^{-2}$]; $r^2 = 0.89$; $n = 3088$. This provided a better fit than that using data from AWS S9 ($r^2 = 0.71$).

3 Results

Time series of ice surface types, ablation rate, albedo and BOD are plotted in Figs. 6 and 7. Mean $\pm 1\sigma$ ablation during the observation period (averaged over the five stakes) was $25 \pm 2\,mm\,d^{-1}$; a peak of $91 \pm 7\,mm\,d^{-1}$ occurred on 27 July. Total ablation from 29 May to 11 August was 1.8 m (average), ranging from 1.6 to 2.0 m. Accumulation (negative ablation) was only recorded on 8 June, and 10–11 August, following fresh snow falls.

3.1 Surface characteristics

A total of 18 quadrat surveys were completed between 1 June and 11 August. The extensive snow cover present on arrival at the site (estimated at 90 % cover on 26 May) rapidly melted to slush during heavy rain on 27 May and had reduced to 24 % when the first quadrat survey was completed on 1 June (Fig. 6a). No snow was left in the quadrats by 5 June, although numerous, deep channels filled with slush and snow were still common close to the field site after this time. The retreating snow uncovered a smooth, mostly debris-free ice surface that lacked any evidence of the cryoconite holes and debris which were observed to be widespread at this site in late August 2011. Hence, we assume that this upper ice surface layer was composed of superimposed ice formed by the refreezing of percolating snow melt in spring/early summer. After a brief period of fresh snowfall on 7–8 June, the 2011 ice surface became increasingly exposed and cryoconite holes were first observed in the quadrats on 13 June (Fig. 6d). The area covered by cryoconite holes gradually increased, reaching ~ 18 % on 8 July, before remaining steady at this percentage coverage until the final survey on 11 August, when they were mostly hidden by fresh snow.

Very little dirty ice (DI) was present early in the season: areal coverage remained below 4 % until 13 July (Fig. 6c), when a period of warm, cloudy and very windy weather resulted in rapid surface melting (reaching $71 \pm 10\,mm\,d^{-1}$ on 10 July). Many of the small, shallow cryoconite holes melted

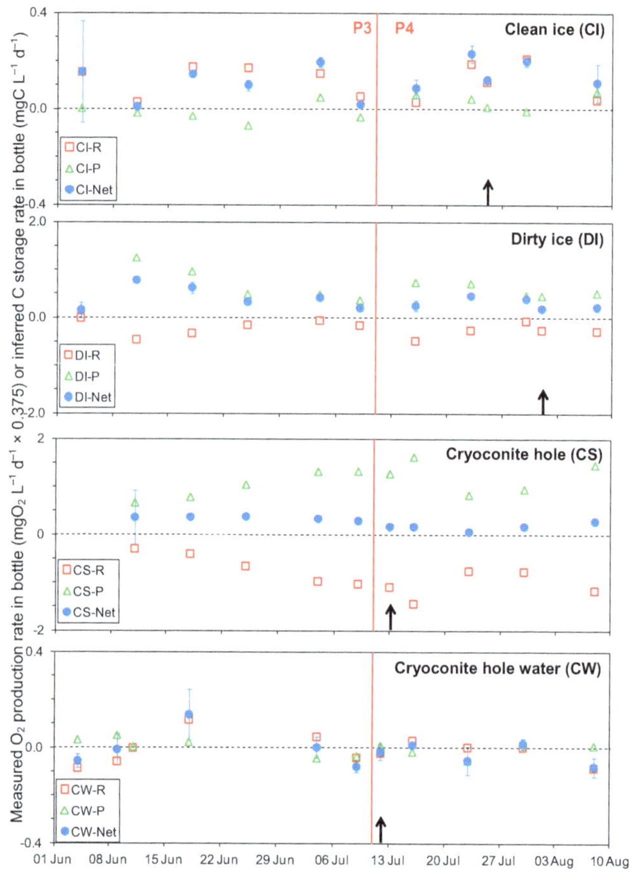

Figure 7. Seasonal variation in productivity as measured by O_2 production or inferred C storage in the BOD incubation experiments for **(a)** clean ice; **(b)** dirty ice; **(c)** cryoconite holes (cryoconite debris plus water); and **(d)** cryoconite hole water. Symbols are respiration (R; red squares) and net carbon storage (N; blue circles), both measured directly, and inferred photosynthesis (P; green triangles) calculated by P = N–R. Note the different scales on the vertical axes. Arrows mark the times of 24-bottle single-sample-type variance experiments (see Fig. 10) and red lines mark the approximate change from period P3 to P4 (see text).

out completely in this period, and the resulting dispersal of their debris lead to the first notable increase in DI cover between 8 and 13 July. This is illustrated by the patchy debris in Fig. 1a. DI coverage became more spatially variable after this time, as indicated by the wide error bars in Fig. 6c. A further increase between 24 and 27 July coincided with a shorter but more intense melt event. Excavating dirty ice surfaces with an ice axe showed that the debris was confined to the upper melt crust, consistent with findings in a field study of the ice surface in northeast Greenland (Bøggild et al., 2010).

Surface water outside of cryoconite holes covered very little of the quadrat surfaces (maximum 3.6 %; 4 August; Fig. 6e). Although many streams and lakes/ponds exist in this part of the ice sheet, the wide spacing of narrow channels results in a low percentage surface area. Much of the local

Table 3. Summary of productivity data for the regular (~ weekly) incubation experiments, presented as mean $\pm 1\sigma$ of carbon storage rates in n incubation experiments. P, R and net are results for inferred photosynthesis only, inferred respiration only, and both processes combined. C storage is the carbon storage rate in the bottle. C flux is the seasonal mean of C storage when converted to carbon flux, based on the surface area sampled to fill the bottle; C flux (total) is the C flux weighted by the fractional surface area of that type, with [5, 95 %] percentiles of 66 interpolated daily values. Observation periods for productivity were 4 June–8 August 2012 except snow (9 June only) and orange-green material (4 and 9 July only).

Surface type	n		C storage rate $mgC\,L^{-1}\,d^{-1}$	C flux $mgC\,m^{-2}\,d^{-1}$	C flux (total) $mgC\,m^{-2}\,d^{-1}$
Clean ice (CI)	10	Net	0.13 ± 0.08	5.7 ± 3.4	2.4 [0.30, 4.1]
		P	0.002 ± 0.048		
		R	0.12 ± 0.08		
Dirty ice (DI)	10	Net	0.40 ± 0.20	18 ± 9.0	0.89 [0.07, 3.0]
		P	0.64 ± 0.31		
		R	-0.24 ± 0.17		
Cryoconite hole water (CW)	10	Net	0.004 ± 0.010	0.5 ± 8.1	−0.015 [−0.16, 0.001]
		P	-0.010 ± 0.035		
		R	0.014 ± 0.16		
Cryoconite hole water & debris (CS)	8	Net	0.24 ± 0.11	30 ± 14	2.5 [0, 4.7]
		P	1.0 ± 0.33		
		R	-0.76 ± 0.47		
Snow (SN)	1	Net	-0.025 ± 0.027	-0.23 ± 0.24	−0.010 [−0.060, 0]
		P	-0.003 ± 0.035		
		R	-0.022 ± 0.022		
Orange-green material (OG)	2	Net	0.002 ± 0.064	0.3 ± 8.1	–
		P	1.7 ± 0.22		
		R	-1.7 ± 0.3		

(< 10 m-scale) melt water drainage appeared to flow through a permeable melt crust, which reached several cm thickness after long periods of sunny weather, before eventually joining the larger, visible drainage channels. The melt crust is formed by the penetration of solar radiation into the ice, which permits melting within a relatively thick surface layer and creates a wide variety of ice formations (Fig. 8). Exceptions to this pattern were periods of warm, windy weather (Fig. 6), when the surface melt crust was quickly stripped away and replaced by a hard, slippery and impermeable surface; these times correspond to the higher (but still small) incidence of surface water in the quadrats in late July and early August.

The ice surface at the field site was initially quite smooth, mostly snow with a few ice patches and lumps of less than 50 cm in height (similar to the photo in Fig. 9a, taken on 1 June). As melt progressed, the surface became increasingly uneven with a mix of bumps up to approximately 150 cm high and melt channels up to 3 m deep (similar to the photo in Fig. 9b, taken on 31 July).

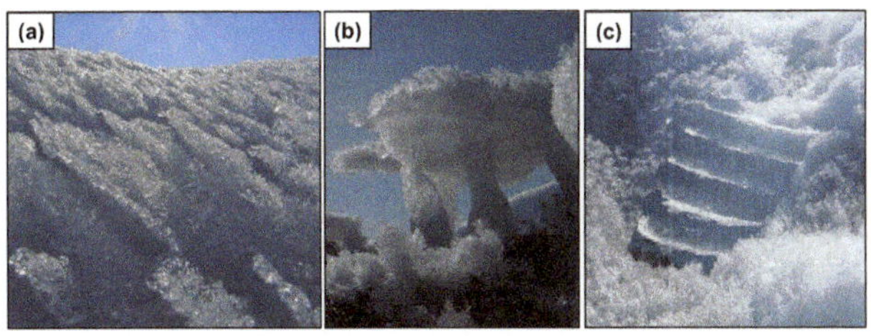

Figure 8. Ice formations arising from radiative melt. **(a)** "Feathers" ca. 5–10 cm long, overlying a permeable melt crust a further several centimetres thick. These needed several days of cold, sunny weather to form. **(b)** "Mushrooms", approximately 20 cm diameter and 15 cm tall, which we believe formed from remnants of the previous year's cryoconite holes. **(c)** "Stacks" of cryoconite hole lids, formed when the water in the cryoconite hole freezes overnight. Each lid represents the water level that night, then the water level drops during the day as the surrounding ice surface ablates.

3.2 Productivity

In total, 17 incubation experiments were carried out between 4 June and 8 August (Fig. 7; Table 3). Results averaged over the regular weekly experiments in the light bottles showed changes in O_2 significantly different from zero ($p < 0.05$) in CI, DI and CS, and no significant change in CW, SN and OG. Significant inferred photosynthesis and respiration, as estimated from the combined light and dark bottle results (see Sect. 2.2), was found in DI and CS as well as in the orange-green material (OG) collected from the former lake bed 3 km from the site (Table 3). In each of these three sample types, light bottles gained O_2 (stored C) while dark bottles lost O_2

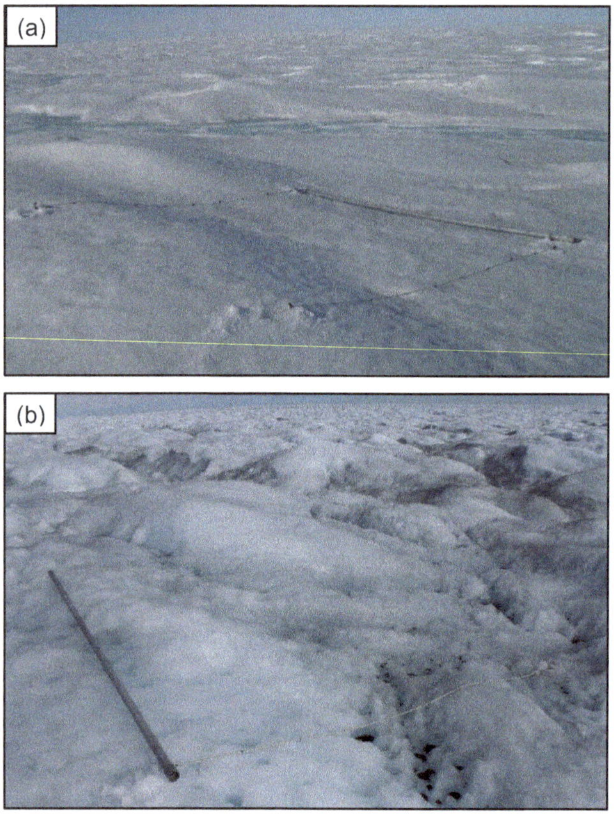

Figure 9. Quadrat Q3 on **(a)** 1 June 2012, with the pole on Row 12 and **(b)** 31 July 2012, with the pole on Row 1. The photos were taken looking approximately northeast. Note the darker and rougher surface on 31 July compared with 1 June.

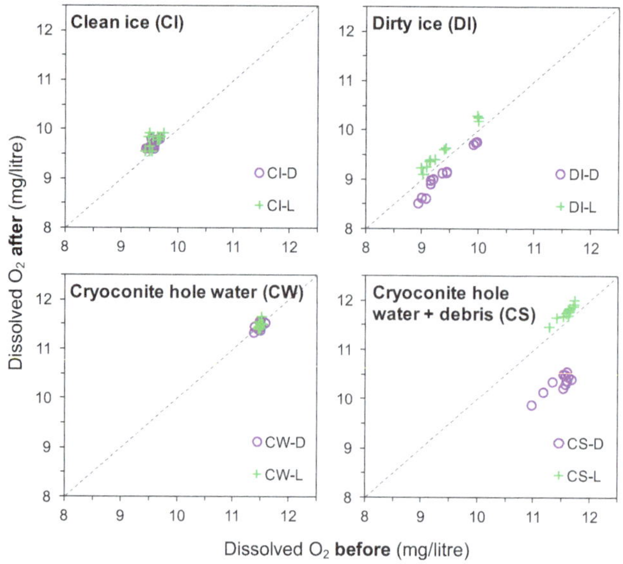

Figure 10. Dissolved O_2 concentrations before and after incubation in experiments with 12 light (purple circles, "-L") and 12 dark bottles (green crosses, "-D") of each sample type. Samples plotting ABOVE the dashed line GAINED oxygen. Light samples indicate the balance between rates of respiration and photosynthesis, while dark samples reflect respiration rates only.

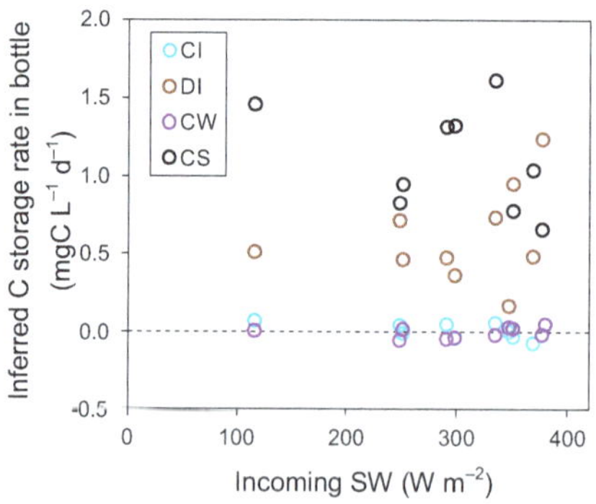

Figure 11. Relationship between incident SW radiation intensity and inferred C storage rate by photosynthesis, for clean ice (CI, blue), dirty ice (DI, brown), cryoconite holes (CH, black) and cryoconite hole water (CW, purple).

(respired C). In contrast, CI samples significantly gained a very small but significant amount of O_2 in both the light and dark bottles, by amounts that were not significantly different from each other (Fig. 7a; $p > 0.05$). The greatest activity in the bottles was in OG, where mean photosynthesis and respiration rates were almost equal at $1.7\,\mathrm{mgC\,L^{-1}\,d^{-1}}$. Since the OG material was present in such small amounts, and observed at just one location, its contribution to carbon storage on the ice sheet is taken as negligible.

A better estimate of the variance within the main four sample types was provided by the additional incubation experiments with 24 bottles of the same type (12 light, 12 dark: Fig. 10). Although there was a wide spread of initial O_2 concentrations, especially in the debris-rich CS and DI samples, the variability of changes in dissolved O_2 after 24 h (which are the important results indicating the level of biological oxygen demand and inferred C storage) was much smaller. Specifically, standard deviations of ΔO_2 expressed as a fraction of the mean were 0.32, 0.27, 3.1 and 0.30 in light bottles containing CI, DI, CW and CS, respectively, and 0.49, 0.32, 1.5 and 0.13 in the dark bottles, respectively. Each of these means is significantly different from zero ($p < 0.05$) except for those for CW.

Seasonal variations in C storage rate within the bottles, as inferred from measured increases in dissolved O_2, are shown in Fig. 7 for the four main surface types. There was no significant correlation in activity between activity between any two pairs of surface types (Spearman's rank correlation; $p > 0.05$) and no clear seasonal pattern (Fig. 7), except for an apparently steady increase in the rates of photosynthesis and respiration in cryoconite hole (CS) samples from 11 June to 16 July. There was no significant correlation between incident SW radiation and C storage rate over the study period (Fig. 11); however, we found a strong link between incident SW intensity and productivity in the sub-24 h activity exper-

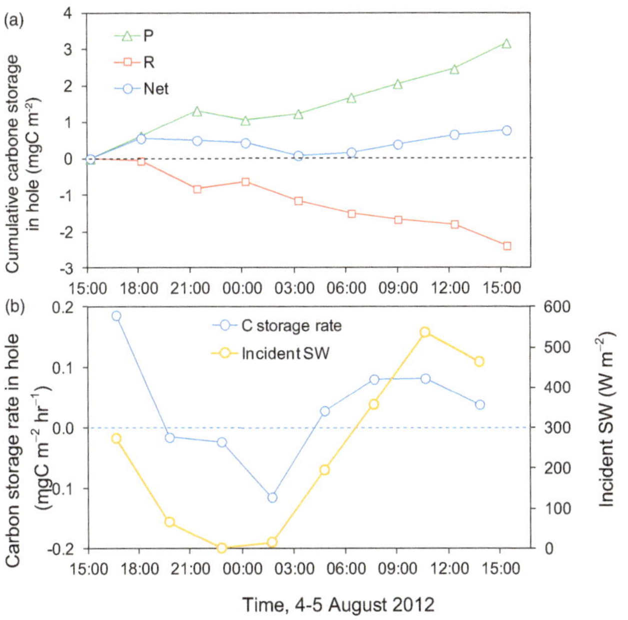

Figure 12. Changes in cryoconite hole C storage at sub-diurnal time scales. **(a)** Cumulative C storage as measured in bottles incubated for 3 to 24 h. **(b)** Equivalent C flux in the same bottles as part **(a)**, plotted alongside incident shortwave radiation intensity (3-h averages of 30-min data).

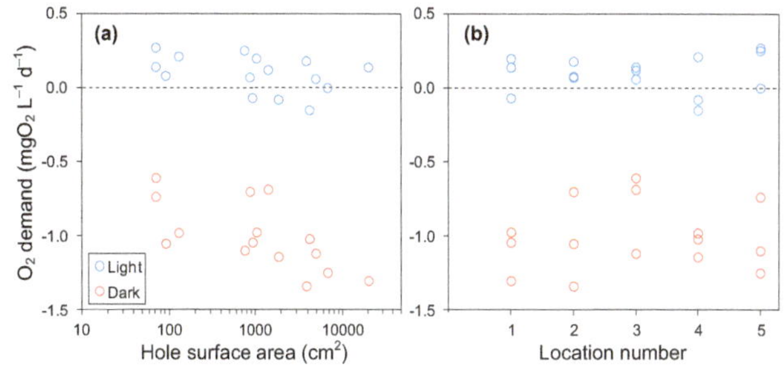

Figure 13. Variation in biological oxygen demand with **(a)** cryoconite hole surface area and **(b)** sampling location, 6–7 August 2012. To convert the measured O_2 demand to inferred C storage in $mgC\,m^{-2}\,d^{-1}$, multiply the vertical axis values by 29.1.

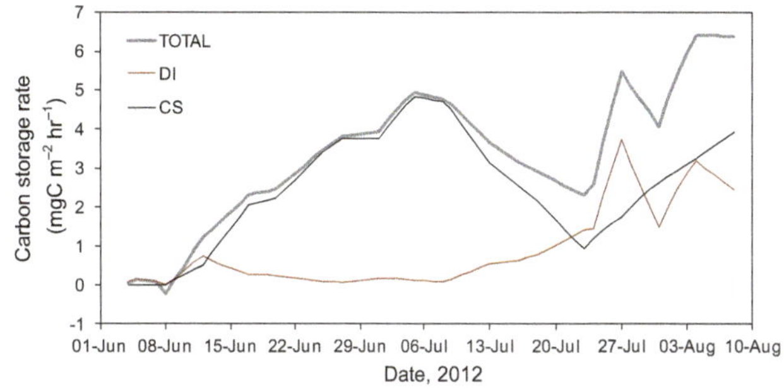

Figure 14. Seasonal variation in C fluxes from the field site, showing individual contributions from DI and CS surfaces. Daily fluxes are plotted, which were calculated by linear interpolation of ~ weekly incubation measurements and quadrat surveys.

iment on 4–5 August (Fig. 12), which is in contrast to the seasonal results. These short incubations showed a change from O_2 production (C storage) to O_2 storage (C release) during the darkest hours of the experiment (20:00–03:00 LT), but with an overall gain in dissolved O_2 (net C storage) after 24 h consistent with the standard experiments.

Variations in productivity with cryoconite hole diameter and location were investigated in an experiment on 6–7 August, when samples from 15 holes in 5 areas were incubated (Fig. 13). Note that hole depth was not investigated in detail, because holes are typically encountered in rough, sloping ices surfaces such that the hole top is difficult to identify; the hole bases are also often sloping, and the water level relative to the ice surface can vary on a day-to-day basis depending on the melt crust thickness. Spearman's correlation between hole diameter and oxygen demand showed weak negative correlation between hole diameter and net productivity in the light bottles ($r = -0.52$; $p = 0.027$) and stronger negative correlation between hole diameter and respiration rate in the dark bottles ($r = -0.65$; $p = 0.005$) (Fig. 13a). There was no apparent difference in productivity between the five locations sampled (Fig. 13b).

Biological oxygen demand in the bottles was converted to net inferred C flux for each surface type, using estimates of the sampled ice area in each bottle and the contribution of each surface type to the total surface area as measured in the quadrat surveys. This yielded a seasonal pattern of C flux from each surface type (Fig. 14; Table 3) and an estimate of the total C flux at the field site during the summer study period of $3.4\,mgC\,m^{-2}\,d^{-1}$. This mean was calculated from 66 daily values, each based on linearly interpolated quadrat and productivity data. The main contributions were $2.5\,mgC\,m^{-2}\,d^{-1}$ from CS and $0.9\,mgC\,m^{-2}\,d^{-1}$ from DI, while contributions from SN and CW were negligible. Note that CI was excluded from the total because the small gain in O_2 appeared to be an artefact of the method (see Sect. 4.3).

3.3 Albedo

Daily albedo data satisfying the quality control criteria (Sect. 2.3) were available for 69 out of the 153 days from 1 May to 30 September (Fig. 6g). Albedo (reported as mean $\pm 1\sigma$) from 1 to 25 May was 74 ± 6 % with no significant trend. After a sharp drop to 57 ± 3 % on 26 May, albedo fell unsteadily until early July, before stabilising at 39 ± 2 % from 8 July to 9 August. Following fresh snow on 10 August, albedo was variable in the ca. 40–60 % range before rising sharply on 1 September and remaining steady throughout that month at 73 ± 3 %.

4 Discussion

4.1 Surface characteristics

Results from the quadrat surveys, albedo retrieval and visual observations summarised above clearly demonstrate the great temporal and spatial variability in surface conditions at the field site. First, using observed patterns in surface classification and albedo we can divide the "summer" study period into four distinct periods, as follows.

- *Period 1:* pre-melt. Very little exposed ice before the winter snow has melted; cold, dry snow; no surface water; high albedo (> 70 %). In 2012, the field site was established on 26 May only a few hours before the onset of heavy rain marking the end of this period.
- *Period 2:* snow melt. 26–30 May. A rapid transition from dry snow to wet snow, slush, melt ponds and bare ice over just a few days, with a corresponding sharp drop in albedo. A similarly rapid transition was observed by D. Chandler at a lower site 7 km from the margin (~ 600 m above sea level) in June 2011. Period 2 is likely to be relatively short in this part of the ice sheet, due to the very low precipitation in comparison with other parts of the west coast (Burgess et al., 2010).
- *Period 3:* early summer ice surface. 30 May–11 July. Superimposed ice and remaining snow patches gradually melt to reveal the dirty ice surface and cryoconite holes from the previous year. The area covered by cryoconite holes slowly increases to reach ~ 17 %; albedo slowly decreases to a minimum (34 %).
- *Period 4:* late summer ice surface. 11 July–31 August. Cryoconite hole cover remains steady while dirty ice coverage varies widely, depending on prevailing weather conditions (warm/cloudy or cold/sunny); albedo remains low but becomes more variable. This period ends with an unsteady transition back to Period 1 conditions as evidenced by the snowfall on 10–11 August and similar events observed in the albedo record as peaks after the field campaign had ended.

Once cryoconite hole coverage had stabilised (Period 4), the mean areal coverage of 16.7 % (range 12–27 % in individual quadrat surveys) was much greater than that measured by Cook et al. (2012) on their nearby ablation-zone transect of sites at ~ 67° N, where maximum coverage was reported as ~ 7 % on 2–3 August 2010, but it is similar to coverage reported by Bøggild et al. (2010) along a transect of the very narrow ablation zone in northeast Greenland. Cook et al. (2012) analysed photographs of nine 1 m × 1 m quadrats to determine cryoconite hole extent at each site, therefore sampling smaller areas (noting the high spatial variability indicated by error bars in Fig. 6). We argue that our strategy of five 2.7 m ×2.7 m quadrats (total 500 measurements) will have yielded a more spatially representative sample at our field site. Photographic analysis should be used with caution since many of the deep but narrow cryoconite holes quickly become obscured as the viewing angle deviates from vertical: for example, a typical cryoconite hole 2 cm in diameter and 10 cm deep will be completely hidden in an image with a viewing angle of only $\tan^{-1}(2/10) = 11°$ from vertical.

Despite the warm, cloudy conditions causing cryoconite holes to melt out and release their debris, resulting in the sharp increases in DI coverage following ablation peaks in July (Fig. 6), there was little corresponding reduction in CH coverage. This is probably because only the small, shallow holes melted out, and the debris contained in just a few holes is sufficient to create a much larger area of DI once it has been dispersed. The same conditions that lead to hole melt-out also caused the surface to become hard and impermeable, making the rapid dispersion of debris much easier than would have been the case over the permeable, rough surface typical of sunny conditions. Another possible process that could increase DI coverage is input of windblown dust – however, it is unclear why this dust would immediately settle in patches rather than being generally distributed over the ice surface, so do we do not believe this process made a significant contribution. Overall, the combination of an early-summer increase in CH and late-summer increase in DI caused a steady decrease in the area of CI from 98 % on 5 June to a minimum of 41 % on 8 August.

The increasing areal coverage of cryoconite holes through the melt season was mainly due to the re-emergence of the previous year's holes, which we believe became exposed by the melting of superimposed ice, rather than the formation of new holes. We infer this because (1) the increasing CH area did not occur at the expense of DI, (2) excavation of DI surfaces revealed no visible debris below the ice surface, (3) there was no visible evidence of significant input of wind blown dust (most clean ice remained clean), and (4) because of our observations of thick (> 10 cm) ice lids over newly emerging holes in June and July. These lids might have formed by refreezing of water in the hole over winter, by superimposed ice forming in snow above the hole, or both. In either case it seems likely that many holes can persist for more than one season, becoming reactivated in the spring as has been observed elsewhere: for example, the Dry Valleys, Antarctica (Fountain et al., 2004). Some emerging holes had lids with a radial crystal structure characteristic of re-freezing processes, and were easy to spot in the rotting ice surface; others had little or no surface expression and were found only after the lid collapsed under foot pressure (we speculate that these lids are more likely to consist of superimposed ice; their remains are shown in Fig. 8b). The large number of cryoconite holes apparently present beneath the surface early in the season suggests that most holes persisted through the winter. Interannual observations are still needed to confirm this. At present it is uncertain how long the holes are active for in the spring before losing their winter ice lids, and this early season activity has yet to be measured.

Field data for this study were collected in 2012, which was notable for its unusually warm conditions across the whole ice sheet: in particular regarding the record-breaking melt extent and surface mass balance (Tedesco et al., 2013). Nevertheless, similarities between the seasonal changes in albedo observed in the present and previous studies (for example Knap and Oerlemans, 1996; Stroeve et al., 2001) suggest the pattern discussed above is persistent between years, including this extreme melt year. Extended visits to the same site in 2011 (25 June–8 July; 31 July; 12–20 August) have also allowed qualitative visual comparison of the ice surface between the two summers. Many features were recognised in both seasons: the gradual exposure of cryoconite holes from under ice lids; patchiness of dirty ice coverage; increasing amplitude of the surface lumpiness as the melt season progresses. The only noted difference of significance was the frequent occurrence of warm, cloudy and windy conditions in July and August 2012 (Fig. 6), which created a hard and slippery ice surface and caused melt-out of some of the cryoconite holes – these events were not observed in 2011, when conditions were predominantly cooler and sunnier. This suggests a greater contribution of sensible or latent heat fluxes to surface melt in 2012.

The presence of the melt crust has interesting implications for ice ablation observations based solely on surface height, when these are used to determine ice mass loss. If the melt crust were to maintain constant characteristics (density, thickness) then surface height change is directly correlated with ice mass loss, usually expressed in mm water equivalent. However, changes in the melt crust will introduce erroneous ablation estimates when using this approach: most significantly a strong overestimate of ablation at the onset of warm conditions, when the melt crust is stripped to leave a solid ice surface. Similarly, following the return to cold, sunny conditions, there will be an artificial dip in observed ice mass loss as the melt crust reforms, since ice is being lost from a layer several cm thick with little change in surface height.

4.2 Ice surface albedo

The mean albedo of 45 % reported by Knap and Oerlemans (1996) for Zone II of the ablation zone on 25 July 1991 is similar to that retrieved from the 2012 AVHRR data for our field site once the albedo had stabilised early in Period 4 (39 ± 2 % from 8 July to 9 August). The seasonal changes in albedo (Fig. 6g) followed a similar pattern to observations from this part of the ice sheet in previous remote sensing and AWS studies, in particular the gradual decrease in Period 3: for example at Site S6 on the K-transect, located at the same altitude (1030 m) as the present field site (Knap and Oerlemans, 1996). However, while Knap and Oerlemans (1996) attributed albedo changes solely to variations in surface melt, the gradual decrease in albedo observed in our Period 3 occurred when there was an overall decrease in stored surface water and slush as the surface drainage system developed. Therefore, while the initial sharp drop at the end of May was due to a transition from loose, dry snow encountered on arrival at the site to water ponds and wet snow, the subsequent gradual decrease in albedo between 29 May and 10 July was more likely due to increasing exposure of cryoconite holes than to increasing melt water coverage (Fig. 6). This is consistent with the recent work of Wientjes et al. (2011). These two distinctly different responses demonstrate the ability of surface albedo retrievals to provide information on both surface melt and surface cryoconite hole coverage.

4.3 Surface productivity

Throughout the season, dirty ice (DI) and cryoconite water-plus-sediment (CS) incubations consistently showed O_2 loss in dark bottles and slight but significant O_2 gain in light bottles (Fig. 7b and c) corresponding to net C storage (see Sect. 3.2) indicative of active, weakly autotrophic biological communities as have been observed previously in cryoconite holes in Greenland and elsewhere (Anesio et al., 2009; Cook et al., 2012). In contrast, there was slight O_2 gain in both light and dark clean ice (CI) incubations (Fig. 7a), with O_2 gains found to be insignificantly different between the light and dark bottles. This O_2 gain was therefore a light-independent process and was most likely an artefact of the method, for example a small amount of O_2 may have been absorbed by the water when the oxygen probe was inserted to take the post-incubation reading. This problem would have had a greater effect in melted ice samples (CI, DI) than water samples (CW, CS) as melted ice samples were under-saturated with O_2 following melting: compare CI, DI starting O_2 levels with those of CW, CS in Fig. 10. Correcting the DI results for this same artefact would reduce the total O_2 gain by about 30 %, from 1.1 down to 0.8 $mgO_2\,L^{-1}\,d^{-1}$. We also note that the measured DI productivity may not be a true representation of in situ rates since we were using melted samples. Firstly, the solar radiation may have been weaker in the incubation environment than at the ice surface owing to the shading effect of the hole edges; secondly, the nutrient concentrations may have been altered during melting, for example with some chemical species becoming more diluted when water within the grain boundaries in the rotting surface ice was mixed with bulk melt from the surrounding crystals. In situ measurements in ice would be extremely challenging, due to the open nature of the system (ice permeable to water and gases) and ice-water exchange of solutes during melting-refreezing processes. Nevertheless, our measurements represent the potential for activity in this surface environment.

In contrast to the clear seasonal changes in the ice surface characteristics, there was very little evidence for any seasonal signal in the results of productivity experiments, except for a general increase in O_2 consumption in cryoconite holes during Period 3 (Fig. 7c). This trend did not significantly alter the net productivity so was presumably balanced by increasing photosynthesis, suggesting a strong link between the rates of respiration and photosynthesis similarly to results reported by Hodson et al. (2010) and Telling et al. (2012a) but contrary to others (for example, net respiration: Stibal and Tranter, 2007). In our case, increasing respiration may reflect the increasing exposure time of holes to direct solar radiation following melting of the ice lids (i.e. increasing cell count), or a thicker accumulation of sediment. Unfortunately the sediment was always too thin (up to a few mm) to allow accurate thickness measurements. Interestingly, both the surface area and activity of cryoconite holes increased during Period 3, then remained steady in early Period 4, mirroring the surface albedo (decreasing in Period 3, steady in early Period 4). Total C flux into cryoconite holes increased during this time due to an increase in the area of exposed holes rather than changes in net productivity (Fig. 14).

Combining results from the quadrat and BOD experiments yielded total C fixation over the course of the experiment (4 June–9 August) of $0.22\,\mathrm{gC\,m^{-2}}$, of which DI and CS contributed 26 and 74 %, respectively. The melt season continued beyond the end of the experiment, probably by ~ 3 weeks according to the albedo data, so the total flux for the complete season was most likely greater (simple extrapolation would yield $0.29\,\mathrm{gC\,m^{-2}}$). However, the annual net C flux remains uncertain since the summer is relatively short and for much of the year there is either no daylight or the ice surface is covered by highly reflective snow which greatly reduces light intensity reaching the microbial communities beneath. Since high rates of respiration were recorded in dark bottles it is evident that any C stored during the summer might be subsequently released over the winter. The effect of this on annual fluxes will be dependent on the continuation of microbial activity in the extended periods without sunlight. Whilst much of this period is very cold, solute-rich water remains in the grain boundaries surrounding the ice crystals, which could allow microbial activity albeit at much reduced rates. This could have a significant effect on carbon fluxes and should clearly be addressed in future studies in Greenland.

The short term experiment with cryoconite hole sediment demonstrates the necessity of using incubations lasting for multiples of 24 h to minimise the effects of diurnal variations in activity driven by solar radiation intensity (Fig. 12). However, despite the clear dependency of radiation intensity on productivity at diurnal time scales, there was no clear relationship between activity and radiation intensity at seasonal time scales (Fig. 11), so we assume the incoming solar radiation intensity did not contribute significantly to the seasonal variability in productivity. Instead, other factors such as hole age, hole depth, debris composition, debris thickness or water chemistry may have been important. Interestingly, Stibal et al. (2012) found slope angle was an important factor along their ablation zone transect at 67° N. However, their slopes were computed from GPS positions separated by many kilometres and are unlikely to reflect the local slope angle owing to the very rough ice sheet surface in the ablation zone at the horizontal scale of 1 to 100 m (compare local slopes in Figs. 2 and 9 with the calculated topographic slope of less than 1°). The correlation reported by Stibal et al. (2011) is therefore unlikely to be a direct causal relationship.

In the same region of the ice sheet as our field site, Hodson et al. (2010) reported mean C storage rates in cryoconite holes of $-3.9\,\mu\mathrm{gC\,g^{-1}\,d^{-1}}$, which is equivalent to $-8.0\,\mathrm{mgC\,m^{-2}\,d^{-1}}$ using their value of $0.206\,\mathrm{g\,cm^{-2}}$ of sediment in cryoconite holes. At Transect Site 7, representative of our field site, Stibal et al. (2011) measured NEP, photosynthesis, respiration C storage rates of 1.1 ± 0.5, -4.9 ± 1.8, $3.8 \pm 2.0\,\mathrm{mgC\,m^{-2}\,d^{-1}}$ (based on results reported in $\mu\mathrm{gC\,m^{-2}\,d^{-1}}$ converted to $\mathrm{mgC\,g^{-1}\,d^{-1}}$ using their measured surface debris cover of $180\,\mathrm{g\,m^{-2}}$). By comparison, C fluxes from cryoconite holes in our study ranged from 5 to $33\,\mathrm{mgC\,m^{-2}\,d^{-1}}$ (mean $24\,\mathrm{mgC\,m^{-2}\,d^{-1}}$). Our data and those of previous studies highlight the great variability in measured C fluxes associated with microbial activity; even the direction of the flux (C storage or C release) varies between studies.

4.4 Linking albedo and productivity

Since there is a clear response of the AVHRR-derived albedo to the snow melt and increasing cryoconite hole exposure (Fig. 6; Periods P3–P4), it is likely that a relationship could be established between retrieved albedo and cryoconite hole coverage. This is illustrated in Fig. 15, which shows a clear link between albedo and cryoconite hole coverage as well as between albedo and net surface productivity. A robust relationship based on remotely sensed albedo would enable a much more wide-ranging assessment of cryoconite hole coverage than has been possible in this or previous field studies, and could also be used to predict surface C fluxes either directly (Fig. 15, bottom panel) or indirectly via the hole coverage and hole productivity. The scatter evident in Fig. 15 shows there would be a large uncertainty associated with any prediction of C flux based on albedo from this data set alone, but this uncertainty is nevertheless smaller than that associated with extrapolation of point measurements collected in isolation and could be reduced either with further observations or by analysis of data from previous studies given that AVHRR data are available for periods covering the studies listed in Table 1.

The albedo product used here appeared less sensitive to DI coverage, but it is worth noting that the sudden increases in DI during high melt events occurred exclusively in periods of cloudy weather, when albedo data are not available from satellite remote sensing, so it does not seem likely that AVHRR data will be useful for assessing DI coverage. Further, there was great spatial variability in DI cover (Fig. 6c), making it difficult to correlate albedo and DI cover with a relatively small data set. The contribution of DI to the mean

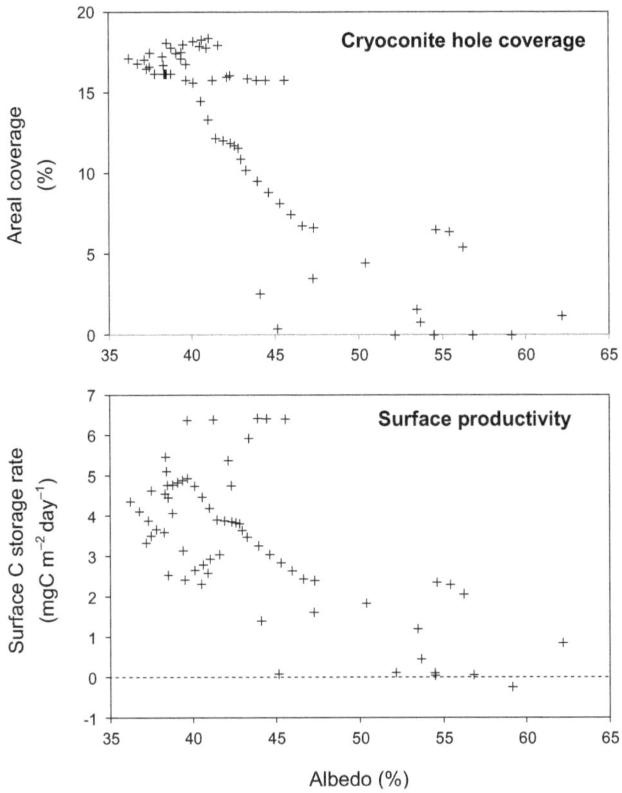

Figure 15. Correlations between albedo and cryoconite hole coverage (top) and between albedo and surface productivity (bottom). Each point is a daily value, linearly interpolated from the respective time series since observations of each quantity were generally not made simultaneously.

C flux was approximately one-third that of cryoconite holes; therefore, an alternative approach to measuring DI coverage remotely may be needed if the DI contribution is included in this approach to estimating ice sheet C fluxes.

5 Conclusions

A time-series of ice surface characteristics and productivity from 26 May to 10 August 2012 has provided a detailed insight into seasonal changes in ice surface processes during the melt season. The field site was located about half way between the ice margin and equilibrium line in west Greenland, near the edge of the so-called "dark zone" of the Greenland Ice Sheet's ablation zone (Wientjes et al., 2011). Specific findings of interest are as follows:

1. Based on changes in ice surface characteristics observed in ground observations and satellite AVHRR imagery, the melt season was classified into four distinct periods. These are: winter snow cover (P1); snow melt (P2); melting of superimposed ice and exposure of cryoconite holes (P3); and the mid- to late-summer ice surface (P4).

2. There was a marked surface response to major melt events, most notably expressed as loss of the several centimetre-thick, permeable melt crust to expose the impermeable underlying ice, and the dispersion of debris from shallow cryoconite holes.

3. In situ incubation experiments revealed active microbial communities, as have been reported in limited previous studies in Greenland. Rates of activity in the different ice surface environments were very variable both between environments and between incubations in the same environment. There was no clear seasonal signal, aside from apparent correlations between albedo, cryoconite hole coverage and cryoconite hole productivity in melt season Periods 3 and 4. These correlations were mostly attributed to changes occurring in Period 3. On average over the full study period, we found significant net O_2 production (inferred as net C storage) in dirty ice and cryoconite holes. There was no significant O_2 gain or loss in snow or water samples.

While there is good evidence from this and previous studies that seasonal changes in ice surface characteristics are broadly persistent between years, the very variable productivity results obtained at this field site and elsewhere suggest that there remains great uncertainty in the magnitude of biologically mediated carbon fluxes into the ablation zone of the ice sheet. However, the link between cryoconite hole productivity, cryoconite hole areal coverage and ice surface albedo shows there is potential for the large-scale estimation of biological processes and associated carbon fluxes, which could aid in extrapolating point observations up to the ablation zone scale. However, the areas of dirty ice also contribute significantly to carbon fluxes but a similar correlation between dirty ice coverage and albedo was not apparent.

Acknowledgements. This work was funded by NERC grant no. NE/H023879/1. M. van den Broeke (IMAU) kindly provided the meteorological data, and AVHRR imagery was supplied by the EUMETSAT Climate Monitoring Satellite Application Facility (CM SAF). We thank the three reviewers for their detailed comments on the original version of the paper.

Edited by: E. Larour

References

Anesio, A. M., Hodson, A. J., Fritz, A., Psenner, R., and Sattler, B.: High microbial activity on glaciers: importance to the global carbon cycle, Glob. Change Bio., 15, 955–960, doi:10.1111/j.1365-2486.2008.01758.x, 2009.

Bayley, W. S.: Minerology and petrography, Am. Natur., 25, 138–146, 1891.

Bøggild, C. E., Brandt, R. E., Brown, K. J., and Warren, S. G.: The ablation zone in northeast Greenland: ice types, albedos and impurities, J. Glaciol., 56, 101–113, 2010.

Burgess, E. W., Forster, R. R., Box, J. E., Mosley-Thompson, E., Bromwich, D. H., Bales, R. C., and Smith, L. C.: A spatially calibrated model of annual accumulation rate on the Greenland Ice Sheet (1958–2007), J. Geophys. Res., 115, F02004, doi:10.1029/2009JF001293, 2010.

Cook, J. M., Hodson, A. J., Anesio, A. M., Hanna, E., Yallop, M., Stibal, M., Telling, J., and Huybrechts, P.: An improved estimate of microbially mediated carbon fluxes from the Greenland Ice Sheet, J. Glaciol., 58, 1098–1108, doi:10.3189/2012JoG12J001, 2012.

Duynkerke, P. G. and van den Broeke, M. R.: Surface energy balance and katabatic flow over glacier and tundra during GIMEX-91, Global Planet. Change, 9, 17–28, 1994.

Fountain, A. G., Tranter, M., Nylen, T. H., Lewis, K .J., and Mueller, D. R.: Evolution of cryoconite holes and their contribution to runoff from glaciers in the McMurdo Dry Valleys, Antarctica, J. Glaciol., 50, 35–45, 2004.

Greuell, W., Denby, B., Van der Wal, R. S. W., and Oerlemans, J.: 10 years of mass-balance measurements along a transect near Kangerlussuaq, central West Greenland, J. Glaciol., 47, 157–158, 2010.

Hodson, A., Anesio, A. M., Ng, F., Watson, R., Quirk, J., Irvine-Fynn, T., Dye, A., Clark, C., McCloy, P., Kohler, J., and Sattler, B,: A glacier respires: Quantifying the distribution and respiration CO_2 flux of cryoconite across an entire Arctic supraglacial ecosystem, J. Geophys. Res.-Biogeo., 112, G04S36, doi:10.1029/2007JG000452, 2007.

Hodson, A., Bøggild, C., Hanna, E., Huybrechts, P., Langford, H., Cameron, K., and Houldsworth, A.: The cryoconite ecosystem on the Greenland Ice Sheet, An. Glaciol., 51, 123–129, 2010.

Irvine-Fynn, T. D. L., Bridge J. W., and Hodson, A.: In situ quantification of supraglacial cryoconite morphodynamics using time-lapse imaging: an example from Svalbard, J. Glaciol., 57, 651–657, 2011.

Karlsson, K-G., Riihelä, A., Müller, R., Meirink, J. F., Sedlar, J., Stengel, M., Lockhoff, M., Trentmann, J., Kaspar, F., Hollmann, R., and Wolters, E.: CLARA-A1: CM SAF Clouds, Albedo and Radiation dataset from AVHRR data – Edition 1 – Monthly Means/Daily Means/Pentad Means/Monthly Histograms, Satellite Application Facility on climate Monitoring, doi:10.5676/EUM_SAF_CM/CLARA_AVHRR/V001, 2012.

Knap, W. H. and Oerlemans, J.: The surface albedo of the Greenland ice sheet: satellite-derived and in situ measurements in the Søndre Strømfjord area during the 1991 melt season, J. Glaciol., 42, 364–374, 1996.

Konzelmann, T. and Braithwaite, R. J.: Variations of ablation, albedo and energy balance at the margin of the Greenland ice sheet, Kronprins Christian Land, eastern north Greenland, J. Glaciol., 41, 174–182, 1995.

Langford, H., Hodson, A., Banwart, S., and Bøggild, C.: The microstructure and biogeochemistry of Arctic cryoconite granules, Ann. Glaciol., 51, 87–94, 2010.

Nolin, A. W. and Payne, M. C.: Classification of glacier zones in western Greenland using albedo and surface roughness from the Multi-angle Imaging SpectroRadiometer (MISR), Remote Sens. Environ., 107, 264–275, 2007.

Nolin, A. W., Fetterer, F. M., and Scambos, T. A.: Surface roughness characterizations of sea ice and ice sheets: Case studies with MISR data, IEEE T. Geosci. Remote, 40, 1605–1615, 2002.

Oerlemans, J. and Vugts, H. F.: A meteorological experiment in the melting zone of the Greenland ice sheet. Bull. Am. Meteorol. Soc., 74, 355–365, 1993.

Schaepman-Strub, G., Schaepman, M. E., Painter, T. H., Dangel, S., and Martonchik, J. V.: Reflectance quantities in optical remote sensing – definitions and case studies, Remote Sens. Environ., 103, 27–42, 2006.

Smeets, C. J. P. P. and van den Broeke, M. R.: Temporal and spatial variations in the aerodynamic roughness length in the ablation zone of the Greenland Ice Sheet, Bound-Lay. Meteorol., 128, 315–338, doi:10.1007/s10546-008-9291-0, 2008.

Steinböck, O.: Über Kryokonitlöcher und ihre biologische Bedeutung, Z. Gletscherkd., 24, 1–21, 1936.

Stibal, M. and Tranter, M.: Laboratory investigation of inorganic carbon uptake by cryoconite debris from Werenskioldbreen, Svalbard, J. Geophys. Res., 112, G04S33, doi:10.1029/2007JG000429, 2007.

Stibal, M., Telling, J., Cook, J., Mak, K. M., Hodson, A., and Anesio, A. M.: Environmental controls on microbial abundance and activity on the Greenland Ice Sheet: a multivariate analysis approach, Microb. Ecol., 63, 74–84, doi:10.1007/s00248-011-9935-3, 2012.

Stroeve, J.: Assessment of Greenland albedo variability from the advanced very high resolution radiometer Polar Pathfinder data set, J. Geophys. Res., 106, 33989–34006, 2001.

Takeuchi, N.: Optical characteristics of cryoconite (surface dust) on glaciers: the relationship between light absorbency and the property of organic matter contained in the cryoconite, Ann. Glaciol., 34, 404–419, 2002.

Tedesco, M., Fettweis, X., Mote, T., Wahr, J., Alexander, P., Box, J. E., and Wouters, B.: Evidence and analysis of 2012 Greenland records from spaceborne observations, a regional climate model and reanalysis data, The Cryosphere, 7, 615–630, doi:10.5194/tc-7-615-2013, 2013.

Telling, J., Anesio, A. M., Hawkins, J., Tranter, M., Wadham, J. L., Hodson, A. J., Irvine-Fynn, T., and Yallop, M. L.: Measuring rates of gross photosynthesis and net community production in cryoconite holes: a comparison of field methods, Ann. Glaciol., 51, 153–162, 2010.

Telling, J., Anesio, A. M., Tranter, M., Stibal, M., Hawkings, J., Irvine-Fynn, T., Hodson, A., Butler, C., Yallop, M., and Wadham, J. L.: Controls on the autochthonous production and respiration of organic matter in cryoconite holes on High Arctic glaciers, J. Geophys. Res.-Biogeo., 117, G01017, doi:10.1029/2011JG001828, 2012a.

Telling, J., Stibal, M., Anesio, A. M., Tranter, M., Nias, I., Cook, J., Bellas, C., Lis, G., Wadham, J. L., Sole, A., Nienow, P., and Hodson, A.: Microbial nitrogen cycling on the Greenland Ice

Sheet, Biogeosciences, 9, 2431–2442, doi:10.5194/bg-9-2431-2012, 2012b.

Uetake, J., Naganuma, T., Hebsgaard, M. B., Kanda, H., and Koshima, S.: Communities of algae and cyanobacteria on glaciers in west Greenland, Polar Science, 4, 71–80, 2010.

van de Wal, R. S. W. and Oerlemans, J.: An energy balance model for the Greenland Ice Sheet, Global Planet. Change, 9, 115–131, 1994.

Wientjes, I. G. M., Van de Wal, R. S. W., Reichart, G. J., Sluijs, A., and Oerlemans, J.: Dust from the dark region in the western ablation zone of the Greenland ice sheet, The Cryosphere, 5, 589–601, doi:10.5194/tc-5-589-2011, 2011.

Yallop, M. L., Anesio, A. M., Perkins, R. G., Cook, J., Telling, J., Fagan, D., MacFarlane, J., Stibal, M., Barker, G., Bellas. C., Hodson, A., Tranter, M., Wadham, J., and Roberts, N. W.: Photophysiology and albedo-changing potential of the ice algae community on the surface of the Greenland ice sheet, ISME Journal, 6, 2302–2313, 2012.

14

Stratigraphy of Lake Vida, Antarctica: hydrologic implications of 27 m of ice

H. A. Dugan[1,2], P. T. Doran[1], B. Wagner[3], F. Kenig[1], C. H. Fritsen[4], S. A. Arcone[5], E. Kuhn[4], N. E. Ostrom[6], J. P. Warnock[7], and A. E. Murray[4]

[1]Department of Earth and Environmental Sciences, University of Illinois at Chicago, Chicago, IL, USA
[2]Center for Limnology, University of Wisconsin-Madison, Madison, WI, USA
[3]Institute of Geology and Mineralogy, University of Cologne, Cologne, Germany
[4]Division of Earth and Ecosystem Sciences, Desert Research Institute, Reno, NV, USA
[5]US Army Cold Regions Research and Engineering Laboratory, Hanover, NH, USA
[6]Department of Zoology, Michigan State University, East Lansing, MI, USA
[7]Department of Geology and Environmental Geosciences, Northern Illinois University, DeKalb, IL, USA

Correspondence to: H. A. Dugan (hilarydugan@gmail.com)

Abstract. Lake Vida, located in Victoria Valley, is one of the largest lakes in the McMurdo dry valleys and is known to contain hypersaline liquid brine sealed below 16 m of freshwater ice. For the first time, Lake Vida was drilled to a depth of 27 m. Below 21 m the ice is marked by well-sorted sand layers up to 20 cm thick within a matrix of salty ice. From ice chemistry, isotopic composition of $\delta^{18}O$ and $\delta^{2}H$, and ground penetrating radar profiles, we conclude that the entire 27 m of ice formed from surface runoff and the sediment layers represent the accumulation of surface deposits. Radiocarbon and optically stimulated luminescence dating limit the maximum age of the lower ice to 6300 ^{14}C yr BP. As the ice cover ablated downwards during periods of low surface inflow, progressive accumulation of sediment layers insulated and preserved the ice and brine beneath, analogous to the processes that preserve shallow ground ice. The repetition of these sediment layers reveals hydrologic variability in Victoria Valley during the mid- to late Holocene. Lake Vida is an exemplar site for understanding the preservation of subsurface brine, ice, and sediment in a cold desert environment.

1 Introduction

Little is known about the habitability of cold liquid environments sealed off from the atmosphere, be it the subglacial lakes of Antarctica (Siegert et al., 2013) or beneath the icy shell of Europa (McKay, 2011). Located in the McMurdo dry valleys of Antarctica, Lake Vida has the thickest ice of any subaerial lake on Earth and is one of the few "ice-sealed" ecosystems known to support a diverse and active microbial population in a cold, anoxic, aphotic brine (Murray et al., 2012; Dugan et al., 2015). The brine was first discovered 16 m below the surface of Lake Vida and is hypothesized to have been sealed from the atmosphere for several millennia (Doran et al., 2003). At present, the lake level (i.e., the ice surface) of Lake Vida is rising, which implies that the brine is progressively getting farther from the surface. In this paper, we present evidence for how and when Lake Vida formed to further understand the structure and evolution of the existing brine system beneath Lake Vida.

On most lakes in the dry valleys, the thickness of ice ranges from 3–6 m (Wharton Jr. et al., 1992). This thickness is maintained by energy loss at the surface (conduction and ablation) and energy gained at the bottom of the ice cover (freezing) (McKay et al., 1985). Constant ablation at the ice surface and freezing at the bottom of these floating ice covers limit the maximum age of the surface ice to ~5 years (Dugan et al., 2013). However, the Lake Vida ice cover is at least partially grounded (Doran et al., 2003), so the ice does not turn over in the same way. Water that is flowing to the lake is trapped on the surface of the ice where it freezes and is later ablated or buried by subsequent ice buildup. In this way, the thick ice on Lake Vida may record past hydrological changes similar to a glacier; however, unlike a glacier, intermittent accumulation and ablation may lead to large discontinuities in the ice cover during prolonged cold/dry periods.

In Victoria Valley, it is theorized that the Lake Vida basin was occupied by a 200 m deep glacial lake 8600 ^{14}C yr BP (Hall et al., 2002), after which lake levels began to decline. This lake is inferred to have had a water column (likely with an ice cover) that permitted sedimentation and led to the lacustrine deposits seen on the landscape today. A deep water column would preclude the presence of bottom ice, and therefore it is improbable that any of the observed ice existed during this time. Therefore, the 27 m of ice currently on Lake Vida was formed subsequent to 8600 yr BP.

During mid- to late Holocene, it is possible that Lake Vida was influenced by events similar to the repeated lake level drawdowns and complete desiccation events recorded in lacustrine sediment cores and geochemical diffusion profiles of the large lakes of Taylor and Wright valleys. For instance, it is speculated that Taylor Valley underwent a valley-wide desiccation event at 1000–1200 yr BP (Lyons et al., 1998), and Lake Fryxell had lowstands at 6400, 4700, 3800, and around 1600 yr BP (Wagner et al., 2006; Whittaker et al., 2008). Lake Vanda, in Wright Valley, underwent a lowstand at 1200 yr BP (Wilson, 1964), or prior to 2000 yr BP (Gumbley et al., 1974). In these closed basin lakes, desiccation is thought to be the result of climatic changes and not a result of large drainage events. If Lake Vida was influenced by similar climatic patterns, some or all of these events may be recorded in the ice cover.

The aim of this study is to reconstruct the history of the ice cover on Lake Vida. We examine the isotopic and ion geochemistry, sediment characteristics, and diatom composition of a 27 m ice core, as well as ground penetrating radar (GPR) profiles, to extrapolate the strata in single cores. Both radiocarbon dating and optically stimulated luminescence (OSL) dating are employed to establish the time of deposition of sediment layers, and ice core stratigraphy is used as a means of establishing periods of lake level drawdown.

2 Study site

Lake Vida (77°23′ S, 161°56′ E), situated in Victoria Valley, Antarctica, is one of the largest (6.8 km^2) and highest (350 m a.s.l. – above sea level) lakes in the McMurdo dry valleys (Fig. 1). The lake is endorheic (closed basin) and receives inflow via streams originating from the Victoria Upper, Victoria Lower, and Clark glaciers. Annual precipitation is < 35 mm (Fountain et al., 2010). Lake Vida occupies a unique climatological niche where summer temperatures can rise slightly above 0 °C to generate stream flow, yet unusually cold winters (compared to the other major valleys in the region) maintain a thick ice cover on the lake. From 1995 to 2000, the mean annual air temperature at Lake Vida (−27.4 °C) was 7 to 10 °C lower than valley bottom temperatures in Taylor and Wright Valleys, but mean summer temperatures were similar (Doran et al., 2002).

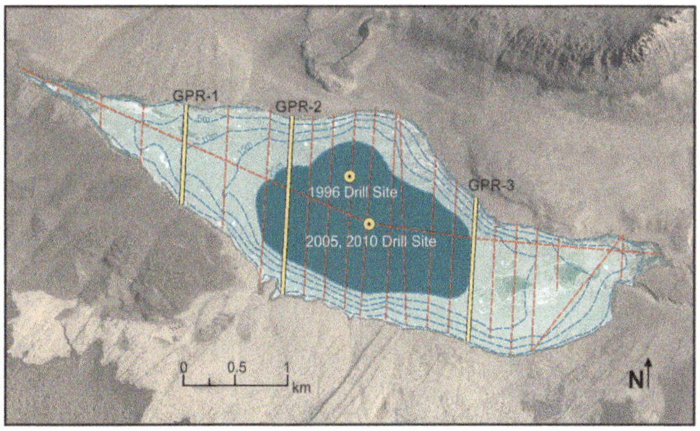

Figure 1. Location of drill sites and GPR transects (red dashed line) on Lake Vida in central Victoria Valley ©DigitalGlobe (2011). Highlighted GPR transects 1–3 are presented in the paper. Bathymetric lines 0–20 m were digitized and interpolated from GPR profiles. The dark blue area below 20 m is of unknown depth.

From drilling in 2010, it is known that the ice on Lake Vida extends to at least 27 m (Murray et al., 2012). A unique feature of Lake Vida is the presence of liquid brine within the ice cover, which infiltrates the drill hole at approximately 16 m and rises to 10.5 m below the surface. The brine is anoxic, with salinity of 195 g L^{-1} and temperature of 13.4 °C (Murray et al., 2012). It is hypothesized that the brine is contained within small fractures or channels in the ice and rises to 10.5 m when the confining layer of freshwater ice in the upper 16 m is breached.

3 Methods

An electric 15 cm diameter SideWinder drill (Kyne and McConnell, 2007) was used to retrieve a 27 m and a 20 m long ice core located 6 m apart in the center of Lake Vida in November 2010. The 27 m ice core was split in half for archival purposes and subsampled into 5 cm lengths. Where recovery was incomplete for the 27 m core (between 16 and 20 m), the 20 m ice core was subsampled. Longitudinal thick-sections (~0.5 cm thick) were cut from the ice core face and viewed under cross-polarized light for ice crystal fabric analyses. Subsamples were washed with deionized MilliQ water to remove possible brine contamination and allowed to completely melt for processing on a Dionex 1500 ion chromatograph for major ion analysis and a Los Gatos Research

Table 1. Summary of analyses and analytical errors.

Analysis	Material	Analytical error
TC and TIC	Ice	±0.16 %
Anions	Ice	±2.8 µm (SO_4^{2-}), ±0.7 µm (Cl^-)
Cations	Ice	±1.7 µm (Na^+), ±0.4 µm (K^+)
		±1.0 µm (Mg^{2+}), ±1.0 mum (Ca^{2+})
δ^2H	Ice/sediment pore water	±0.8 ‰
δ^2H	Brine	±2 ‰
$\delta^{18}O$	Ice/sediment pore water	±0.1 ‰
$\delta^{18}O$	Brine	±0.2 ‰
OSL	Bulk sediment	$\pm \leq 100$ yr BP
^{14}C	Sediment (insoluble organic fraction)	$\pm \leq 58$ yr BP
^{14}C	Sediment (carbonate)	$\pm \leq 49$ yr BP

liquid water isotope analyzer for isotopic composition of ^2H and ^{18}O (analytical errors provided in Table 1). When necessary, samples were diluted to near seawater salinity prior to analysis. Isotopic values are reported with respect to the VS-MOW international standard. Salinity is reported as the sum of concentration of total ions.

Sediment layers in the 27 m ice core were subsampled in duplicate 1 cm segments, which were freeze-dried or allowed to melt in order to extract pore water by centrifugation. For grain size analyses, 2 g samples were sieved through a 1000 µm sieve and pretreated with 30 % H_2O_2 for 18 h in a 50 °C water bath. Following pretreatment, samples were shaken following the addition of 1 mL of 30 mg L^{-1} Graham's salt ($Na_4P_2O_7$) as a dispersant and analyzed on a Micromeritics Saturn DigiSizer 5200 particle size analyzer (detection limit 0.1–1000 µm). Sand/silt classifications are based on the Udden–Wentworth scale (Wentworth, 1922). Sediment layers were also subsampled to evaluate diatom assemblages and absolute abundance (valves/g dry weight) via light microscopy. Preparation methods followed standard techniques (Warnock and Scherer, 2015). A known mass of freeze-dried sediment was reacted with 10 % HCl and 10 % H_2O_2 to remove carbonates and organics. Abundance per gram was extrapolated from diatom counts on coverslips in a beaker of known area.

Freeze-dried samples were analyzed for total carbon (TC) and total inorganic carbon (TIC) with an elemental analyzer (Dimatec Co.). Total organic carbon was calculated from the difference in TC and TIC. Six samples with the highest organic carbon content (0.4–2.1 %) were chosen for radiocarbon dating of the organic fraction. Samples were prepared by removing carbonates and humic acids by acid–alkali–acid extraction prior to graphite conversion in an automated graphitization system. This required an overnight treatment with 1 % HCl to remove carbonates, followed by 4-hour humic acid extraction with 1 % NaOH and a secondary overnight treatment of 1 % HCl to eliminate any CO_2 that may have been absorbed during the NaOH treatment (Grootes et al., 2004). Two samples at 23.90 and 26.43 m with 1.3–2.3 % TIC were selected for radiocarbon dating of carbonates and were treated with dilute sulphuric acid. Two 5 cm long sediment sections at 21.51 and 25.54 m were cut by band saw under red light for OSL dating (UIC Luminescence Dating Research Laboratory) and remained frozen until analysis. The advantage of frozen sediment for OSL dating is that the water content is expected to have remained constant since freezing took place (Demuro et al., 2008; Arnold and Roberts, 2011). OSL dating of quartz grains from sediment layers was performed using single aliquot regeneration protocols (Murray and Wintle, 2003).

To confirm the continuity of horizons and sediment layers noted in the ice cores, 55 km of GPR transects was recorded over the surface of Lake Vida in 2010 (Fig. 1) using a GSSI SIR-3000 acquisition unit equipped with a 400 MHz antenna. Transects were recorded at 400 ns time range and 2048 16 bit samples per trace, with five manual gain points at −20, 0, 25, 30, and 50. A dielectric constant of 3.15 was initially chosen for depth calibration, but this was altered based on known characteristics of the ice cores. In post-processing, radar profiles were triple stacked and passed through a 200 and 500 MHz triangle FIR filter to remove high- and low-frequency noise. Lake levels were annually surveyed from benchmarks tied into historical optical survey transects conducted by New Zealand Antarctic Program and were recorded in m a.s.l.

4 Results

A water column of brine was not encountered 20 m below the surface of Lake Vida as previously hypothesized (Doran et al., 2003). Rather, wet ice and sediment continued below this depth. After four thick (> 10 cm) sediment layers were encountered below 21 m, the drill became lodged in what was almost certainly a sediment-rich layer (based on the slow progress of drilling) at 27.01 m. The last sample obtained was an ice layer from 26.62 to 26.81 m.

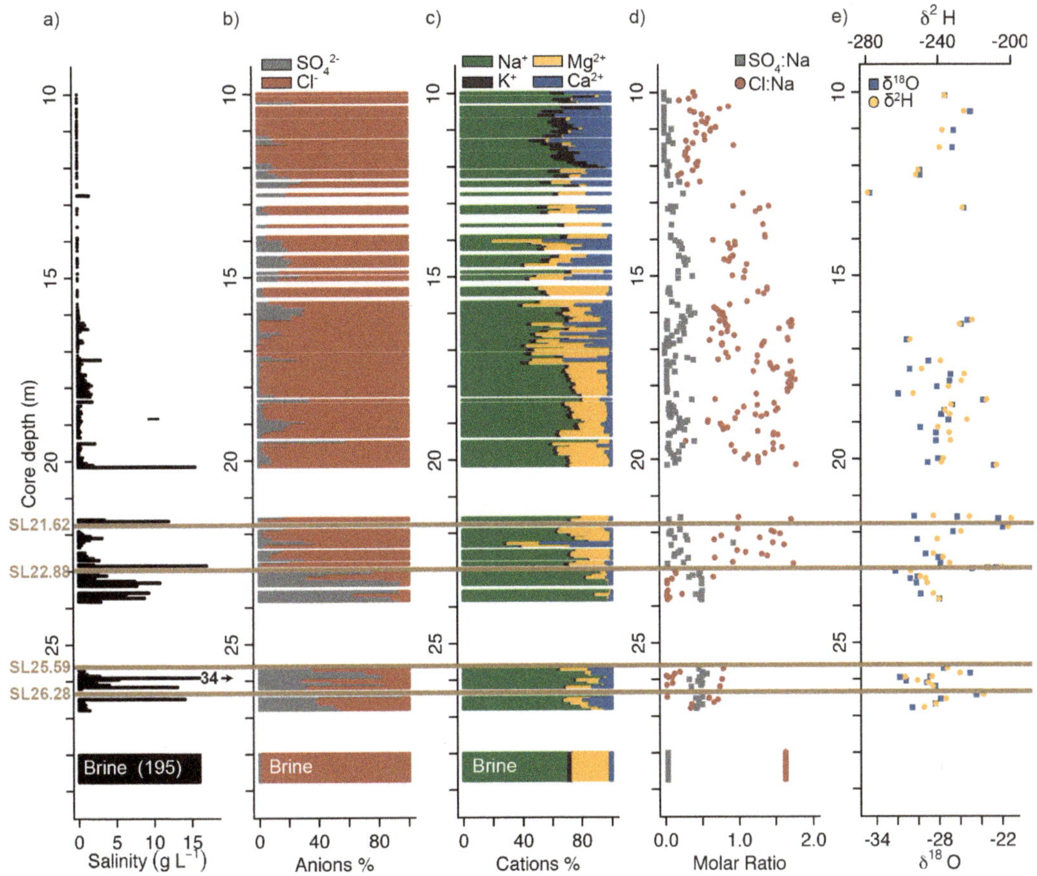

Figure 2. **(a)** Salinity of ice samples below 10 m subsampled from the Lake Vida ice cores. Salinities ranged from < 1 to 34 g L^{-1}. Brine salinity is 195 g L^{-1}. SL21.62, SL22.88, SL25.59, and SL26.28 represent sediment layers > 10 cm thick present in the 27 m ice core. **(b, c)** Total percentage of major anions and cations. Ion percentages in the brine are noted at the base of the figure. **(c)** Ratio of $Na^+ : Cl^-$ and $Na^+ : SO_4^{2-}$ (mol : mol). **(d)** Stable isotope composition (δ^2H and δ^{18}O) of ice samples and sediment pore water.

There was an overall trend of increasing salinity in the ice core with depth as well as a shift at 23 m from Cl^- to SO_4^{2-} as the dominant anion and an accompanying decrease in Mg^{2+} (Fig. 2). Below 21 m, the ice salinity was variable from < 1 to 34 g L^{-1}. During extraction, all cores below 16 m depth were in contact with the brine. If samples were contaminated by brine in the drill hole, we would expect the percentage of major anions/cations and the ratio of $Na^+ : Cl^-$ and $Na^+ : SO_4^{2-}$ in the ice to be similar to the brine (Fig. 2). As the ice chemistry is distinct from the brine, we conclude the ice was not substantially contaminated by brine in the drill hole. The fabric of the ice also changed with depth, from large individual ice crystals with c-axes oriented upwards, which is typical of freshwater ice, to ice composed of randomly oriented small crystals that appear to have recrystallized over time (Fig. 3).

Throughout the cores, there are many small pockets of sediments and thin sediment layers (Fig. 4). All sediments sampled from the ice core were predominately sand (grain size: 62.5 to 2000 μm), with only 4 out of 79 samples having a mean grain size < 100 μm or a percentage of silt and clay (< 62.5 μm) > 6 % of the total volume. In the 27 m core, the four thick sediment layers below 21 m will be referred to by their depth in the core: SL21.62, SL22.88, SL25.59, and SL26.28 (Fig. 3). These layers had water contents < 10 % and total thicknesses of 19, 15, 11, and 19 cm, respectively. At the base of SL26.28, mean grain size began to decrease with a concomitant increase in TC (Fig. 5). Microscopy of the lower sediment layers and ice revealed abundant diatom frustules of *Luticola gaussi* and the genera *Pinnularia*, specifically *P. deltaica* and *P. quaternaria*, only in SL26.28. These species are commonplace in the sediments and water column of Taylor Valley lakes, streams, and cryoconite holes (Stanish et al., 2012, 2013). *L. gaussi* is considered a freshwater species and cosmopolitan to the Antarctic diatom flora (Kopalová et al., 2013), as are *P. deltaica* and *P. quaternaria*. However, no studies have been done to specifically test the salinity tolerance of these species.

The upper sediment layers noted in two ice cores retrieved in 2010 correspond to those cored in 1996 (Doran et al., 2003) and 2005 (Taylor, 2009) (Fig. 4). The radiocarbon

Figure 3. **(a)** Images of thick sections of Lake Vida ice between two sheets of polarized film. At 12.40–12.47 m, the entire section is a single ice crystal. With depth the average grain size of ice crystals decreases to < 1 cm. **(b)** Photographs of sediment sections SL21.62, SL22.88, SL25.59, and SL26.28.

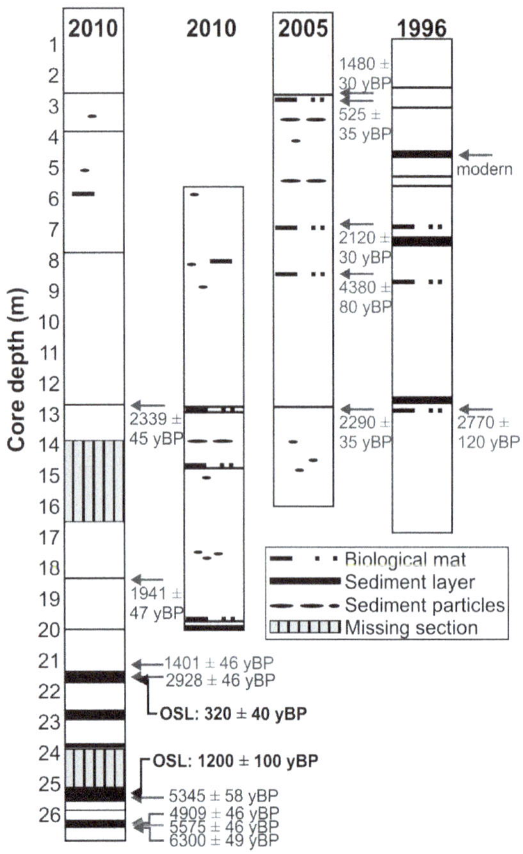

Figure 4. Lake Vida ice core logs from 1996, 2005 and 2010. Two ice cores were drilled in 2010. Where samples were missing in the 27 m core (between 16 and 20 m), the 20 m ice core was subsampled. For core locations in 1996 and 2005, see Doran et al. (2003) and Murray et al. (2012). Surface heights are adjusted to the 2010 lake elevation.

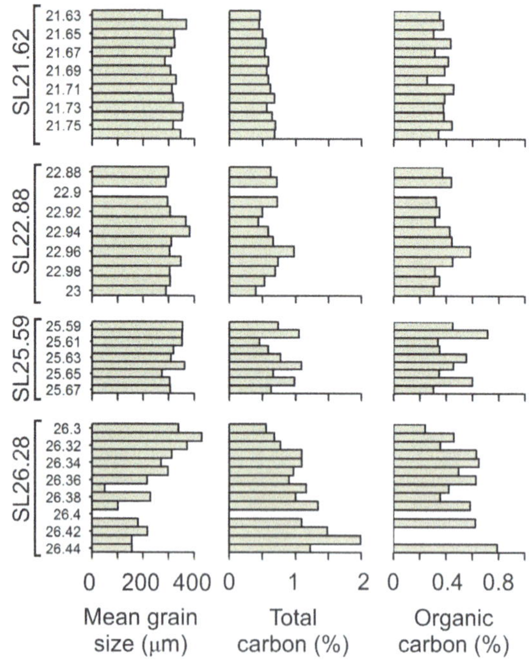

Figure 5. Mean grain size (μm), percentage of total carbon (%), and percentage of organic carbon (%) in sediment sections removed from the 27 m ice core from Lake Vida.

dates vary significantly, especially in the upper ice where there is no correlation between age and depth. The dates from SL26.28 do show increasing age with depth (4909 ± 46 to 6300 ± 49 ^{14}C yr BP). It is noted that δ^{13}C values are unreliable and not reported due to high fractionation during accelerator mass spectrometry measurements.

OSL samples showed no evidence of exposure to sunlight during collection and returned dates of 320 ± 40 and 1200 ± 100 yr BP for SL21.62 and SL25.59, respectively, dates which are younger than the radiocarbon ages from the respective horizons.

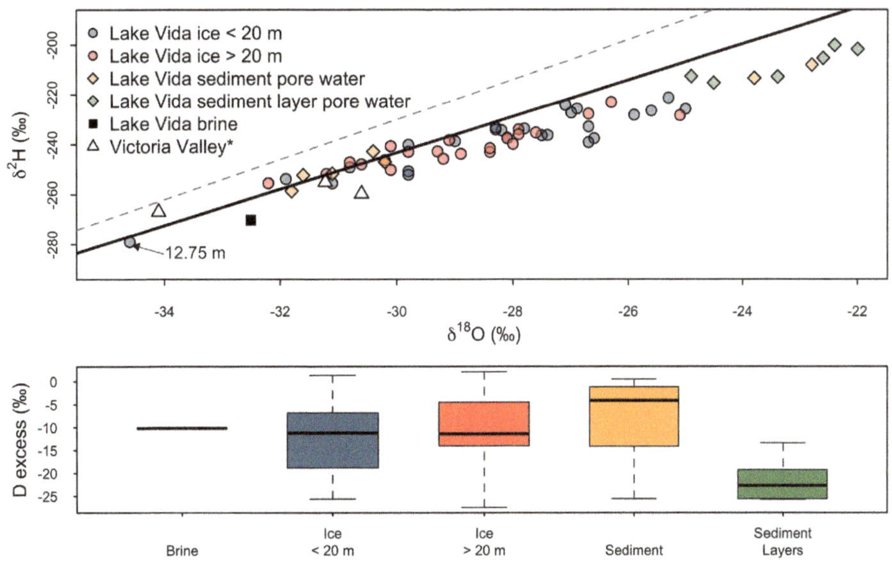

Figure 6. **(a)** Stable isotope composition of hydrogen (δ^2H) and oxygen (δ^{18}O) in Lake Vida ice, pore water in thin sediment pockets and thick sediment layers, and brine. Published values for Victoria Valley surface water and snow are denoted by black triangles (Hagedorn et al., 2010). The local meteoric water line of glacial water (solid line, Gooseff et al., 2006) and global meteoric water line (dashed line) are plotted for reference. **(b)** Box and whisker plot of deuterium excess values for brine ($n = 1$), ice samples above 20 m ($n = 25$), ice samples below 20 m ($n = 26$), thin sediment pockets ($n = 7$), and sediment layers ($n = 6$). The thick black line represents the median values, and box edges represent the 25th and 75th quantiles. Whisker lines extend to the extreme values.

Stable isotope (δ^{18}O and δ^2H) values of both the ice and sediment pore water fell on or below the local meteoric water line but on a slope consistent with the sublimation of ice (Sokratov and Golubev, 2009; Hagedorn et al., 2010; Lacelle et al., 2011, 2013) and isotopic values reported around the dry valleys (Gooseff et al., 2006; Harris et al., 2007) (Fig. 6). Most of the ice and sediment pore water samples have δ^{18}O between −25 and −32 ‰. There are, however, two notable exceptions. The first is the sample at 12.75 m that is significantly depleted in ^{18}O versus all other ice samples. The second is the pore water of the four thick sediment layers below 21 m, as well as some thinner sediment layers above, that are relatively enriched in ^{18}O in comparison to the ice.

GPR profiles distinctly map the edge of the lake basin until an impenetrable basal reflector around 21 m (Fig. 7). Along the basin edges there are features resembling ancient terraces, especially at 8 m (Fig. 7). From 8 to 12 m, synchronous, wavy reflectors are spaced approximately 1 m apart. In all profiles, ice and sediment layers appear to be continuous across the lake.

5 Discussion

The low ion concentration, absence of large sediment layers, and clear GPR returns in the top 8 m of ice suggest that the upper ice has formed recently under a positive water balance. The level of Lake Vida has risen 3.5 m in the last 40 years (Doran, 2015) and has a hydrologic history similar to Lake Bonney (Fig. 8), which has been documented to have risen ~16 m from 1903 to 2010 (Chinn, 1993; Doran, 2015). If a linear extrapolation is applied to the Vida record based on the correlation of volumetric change to Lake Bonney, the surface of Lake Vida would have risen ~7.7 m from 1903 to 2010 when our cores were collected. Therefore, the 8 m contour in Fig. 1 may be an approximate representation of the lake shore in 1903. This is an indication of the rapidity with which the level of Lake Vida can change over time.

5.1 Upper ice

The ice between 9 and 13 m contains almost no sediment. In addition, the ice sample at 12.75 m has a salinity of 3.5 g L^{-1} and is heavily depleted in ^{18}O and ^{2}H versus all other ice samples (Fig. 6). Both an increase in salinity and the relative depletion of heavy isotopes are a signature of freshwater freezing. During freezing at ice–water interface, equilibrium isotope fractionation preferentially retains heavy isotopes in the ice, which leads to a depletion in heavy isotopes in the unfrozen water (Gallagher et al., 1989; Horita, 2009). Under equilibrium conditions, the isotopic composition for the fraction of water that remains unfrozen during the freezing of ice downward (δ_f) can be approximated by Rayleigh fractionation (Eq. 1), where α is the fractionation factor between ice and water (measured at 0 °C), f is the fraction of water that remains unfrozen, and δ_o is the original isotopic value of the water (Miller and Aiken, 1996; Fritz et al., 2011).

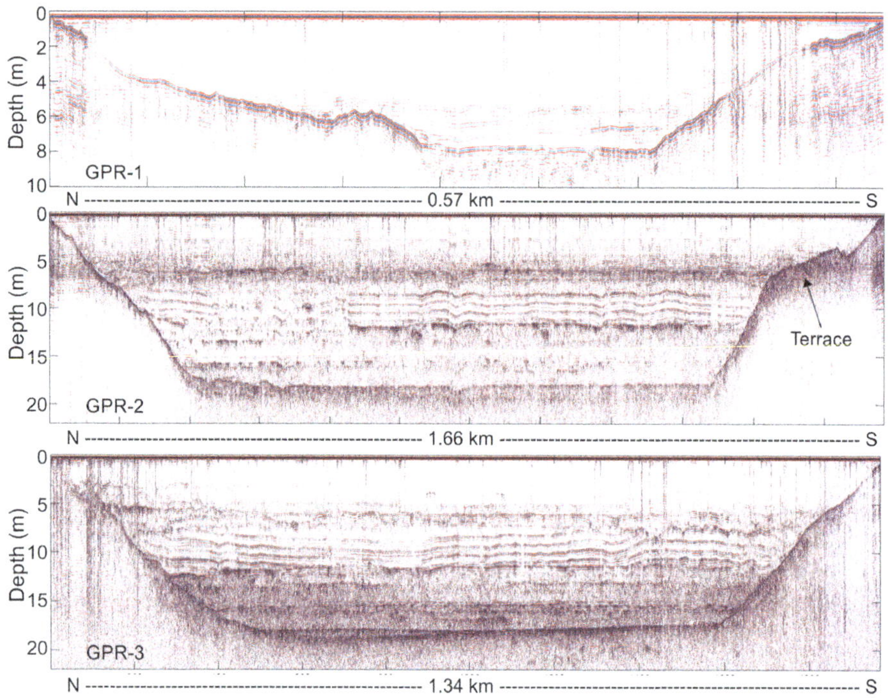

Figure 7. GPR profiles recorded north to south across Lake Vida (Fig. 1).

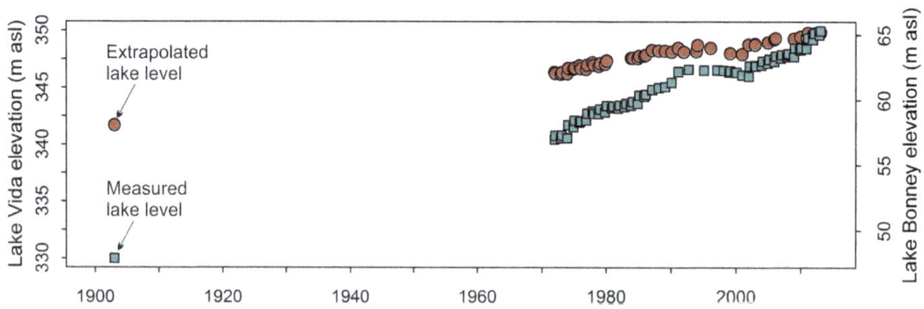

Figure 8. Lake Vida and Lake Bonney surface elevations. The 1903 Lake Bonney elevation is inferred from a measurement at Lake Bonney narrows by Robert Falcon Scott (Chinn, 1993). The 1903 Lake Vida elevation is extrapolated from the correlation between the two lakes for 1971–2010.

$$(\delta_f - \delta_o) \cong 10^3(\alpha - 1)(\ln f) \qquad (1)$$

As a simple model for the depletion of stable isotopes during freezing, we employ Eq. (1) with $\delta_o = -25$ and -30 ‰ (within the range of $\delta^{18}O$ of Lake Vida ice) and $\alpha = 1.0029$ (average value from Horita, 2009) to test if the high-salinity and low-isotopic values at 12.75 m could have resulted from the downward freezing of 3 to 4 m of ice above. To obtain the observed $\delta^{18}O$ at 12.75 m of -34.6 ‰, parameter f must equal 0.04 and 0.20 for an original source δ_o of -25.0 and -30, respectively. This illustrates that the relatively depleted ^{18}O at 12.75 m could easily be generated through freezing processes. This, along with the high salinity and lack of sediment particles, indicates a high likelihood of downward freezing in this section of the ice core. A 3 to 4 m layer of water on the surface of the lake could easily result from the combination of a large surface flood and the melt generated at the water/ice contact. Particles settling out from this event may have generated the sediment seen between 12 and 13 m (Fig. 4). This record may be affirmation that anomalous warming events, such as the flood year of 2001/2002 (Barrett et al., 2008; Doran et al., 2008), are not unprecedented in the dry valleys.

In the GPR profiles, the continuity of the horizontal reflectors across the lake validates the extrapolation of ice core records across the entire lake body. The noticeable undulations in the reflectors between 8 to 12 m are interpreted as density contrasts in ice layers (Arcone and Kreutz, 2009), which may have been formed from 3 to 4 m of liquid water freezing downwards, as discussed previously in regards to isotopic composition. Initially horizontal, these bands were later forced into their present configuration by pressure due to freezing below. Also evident in the radar profiles is a preserved paleo-terrace at 8 m depth along the south end of the lake (Fig. 7). This may be evidence that Lake Vida maintained an elevation of 8 m below present for a prolonged period. The lack of downcutting along the lake margins below 8 m (Fig. 7) further suggests the lake level has mostly risen since this time, which, based on our previous calculation, encompasses the past 100 years.

5.2 Lower ice

In all GPR profiles the radar signal is attenuated below 21 m. Doran et al. (2003) interpreted this horizontal reflector as the top of a large brine body. From the drilling detailed in this study, we now interpret this impenetrable basal reflector to be the SL21.62 sediment layer, as the thickness and salt content likely inhibited radar penetration (Frolov, 2003).

Below 16 m in the ice core, salinity increases and the ice appears to have recrystallized. Recrystallization is induced by temperature changes, stress or strain on the ice, and/or the presence of debris (Samyn et al., 2008). There are three processes by which the lower ice and sediment layers may have formed: (1) the repeated freezing of surface water and deposition of sediment layers (increased salinity is generated from the concentration of salts through evaporation/sublimation), (2) the formation of segregated ice in lake sediments from the freezing of brine from beneath, or (3) the lower ice is remnant glacial ice.

The following were a priori hypotheses:

1. Victoria Valley has not been occupied by a valley glacier since the Miocene (Fountain et al., 1998). Combined with the presence of a glacial lake, there should be no remnant glacial ice near the surface of Victoria Valley.
2. If the ice were segregated ice, we would expect:
   a. a gradient in isotopic composition throughout the sediment and ice layers, as discussed in French and Harry (1990) and observed in closed system freezing of massive ground ice (Fritz et al., 2011);
   b. diatoms present in the sediment layers, which were originally lake sediments;
   c. the lower ice to have similar ionic ratios to the underlying brine.

However, we must note that segregated ice has not been researched in a lacustrine context.

3. If the sediment layers were formed from surface deposition, we would expect the layers to be roughly horizontal.

The following results support that the lower lake ice was formed from surface deposition and is not segregated ice formed from brine seeping upwards through lake sediments.

1. Only the deepest sediment layer, SL26.28, contains abundant diatom frustules. This layer also begins to fine downward in grain size and increases in total carbon content (Fig. 5). This is the only layer that resembles lake sediments and yet is underlain by ice with similar chemical composition to all ice below 21 m.
2. The GPR profiles reveal a lake-wide horizontal layer at 21 m (Fig. 7).
3. The lower ice is chemically distinct from the brine (Fig. 2).
4. The sediment layers are relatively enriched in the stable isotopes of oxygen and hydrogen versus the surrounding ice (Fig. 6).

Therefore, we hypothesize the entire 27 m of ice and sediment was formed from surface processes, and the brine that enters the drill holes is sourced from below 27 m. Here, we explore processes that might have led to the formation of interspersed sediment layers in 27 m of lake ice.

If the lower ice formed from surface inflow rather than brine, the sediment layers in the lower ice are unusual. Field observations at Lake Vida during the 1990s, when lake levels were static, and during the 2000s, when lake levels were rising, reveal that the surface of the lake is largely flat and free of surface sediment. Aeolian transport of sediment is common in Victoria Valley (Speirs et al., 2008); however, aeolian deposition onto a dry, flat ice cover has a high probability of further redistribution by wind and may not be readily entrained into the ice column. This conclusion is supported by observations that windblown sediment largely does not get trapped on the frozen ice surface of Lake Vida but saltates across the lake. Therefore, we propose that sediment is mainly delivered onto the surface of the lake through fluvial transport as melt streams flood the lake surface. A saturated lake surface provides a mechanism for sediment to both infiltrate cracks in the ice and freeze beneath a new layer of water.

In January 2002, anomalously high stream discharge flooded the surface of Lake Vida with turbid water, yet only a thin band of sediment < 1 cm is evident in the upper 3 m of the ice core. Therefore, it is improbable that sediment layers 20 cm in thickness formed from a single surface flooding event. Additionally, it has been noted in Taylor Valley (where lake ice is formed at the bottom of the ice cover, not the surface) that during the austral summer the low albedo of surface sediment can cause it to warm and move downward in the ice cover (Hendy, 2010). This movement tends to aggregate sediment into layers and pockets approximately 2 m below the surface (Priscu et al., 1998). From this, it is hypothesized that the thick sediment layers in Lake Vida may have formed from repeated deposition rather than individual events and were amassed during periods of ice cover ablation (negative water balance).

The hypothesis that sediment layers were formed from long-term evaporation and sublimation is further supported by the isotopic enrichment and low deuterium excess of water contained within the sediment layers (Fig. 6). When water evaporates from a water body to the atmosphere, the remaining water becomes enriched in the heavy isotopes of oxygen and hydrogen (Horita, 2009). The same has been shown for the sublimation of snow and ice (Neumann et al., 2008; Sokratov and Golubev, 2009; Lacelle et al., 2011). Likewise, negative deuterium excess values suggest that evaporative fractionation has more strongly modified oxygen isotopes over hydrogen isotopes.

During periods of significant lake drawdown, the sediment layers may have been visible near the surface of the lake and would have been analogous to the ice-cemented permafrost found at higher elevations in the Victoria Valley. At 450 m a.s.l. in Victoria Valley, the ice-rich permafrost in a 1.6 m soil profile had similar sediment characteristics to the lower sediment layers in Lake Vida, with a median grain size range of 357–510 μm and water content $< 13\,\%$ (Hagedorn et al., 2007). The dry permafrost/ice-cemented contact was found at 22 cm below the surface, which was also the approximate maximum depth of the 0 °C isotherm during the 3 years of study (Hagedorn et al., 2007). At the edge of Lake Vida, soil temperatures at 10 cm depth rise only slightly above 0 °C during the short summer (Fountain, 2015). On Lake Vida, we propose that similar thermal conditions preserved the ice found below and between the thick sediment layers. As sediment layers on the ice thickened to almost 20 cm, the amassed sediment provided insulation for the ice beneath and allowed the ice to remain below freezing temperatures during the summer. This processes allowed for the repeated deposition and preservation of sediment and ice layers.

Only in SL26.28 does the mean grain size significantly decrease toward the base of the core (Fig. 5). The occurrence of silts is not common in the dry valleys but is found in the sediment beneath ice-covered lakes (Wagner et al., 2006, 2011). The increase in TC concurrent with the decrease in grain size, as well as the presence of freshwater diatom frustules, points to the possible occurrence of more open water conditions during this time.

5.3 Constraining the age of the lower ice

Isolating individual dates of deposition or burial of the sediment layers is challenging. Radiocarbon dates in the dry valleys can often be erroneously old due to a reservoir effect (Doran et al., 1999; Berger and Doran, 2001), where an inherited age can result from the direct input of old carbon into lakes (Doran et al., 2014). A residence age can similarly result from limited atmospheric exchange of lake water due to permanent ice covers, strong salinity gradients, or the inclusion of old organic material reworked into modern stream water (Doran et al., 1999; Hendy and Hall, 2006). In Victoria Valley, water travels more than 1 km from glacial sources to Lake Vida, which should allow waters to equilibrate with the atmosphere before reaching the lake. Moat waters too should contain mostly modern carbon, although Doran et al. (2014) found moat waters in Taylor Valley lakes with apparent dissolved inorganic carbon (DIC) ages as high as 3790 ^{14}C yr BP. Only lakes with large open water moats, like Lake Fryxell, seem to have modern DIC. Hall and Henderson (2001) found that at Lake Vida, lacustrine carbonates along the shorelines with uranium/thorium ages of 9600 yr BP had a ^{14}C reservoir age of +3600 yr BP. They concluded that old CO_2 was likely input into the system with meltwater from the Ross Ice Shelf.

Of the eight radiocarbon samples taken from the upper 13 m of ice in Lake Vida, only one sample returned a concentration indicative of modern carbon (Fig. 4). All indications point to erroneously old carbon dates as we assume that at least the upper 7.7 m of ice formed during the last century. However, the radiocarbon dates impart a maximum age constraint on the ice, as it is almost impossible that samples were contaminated by young carbon, and indicate the top 27 m of ice formed after 6300 ^{14}C yr BP. This aligns with the geomorphic reconstructions that indicate that Victoria Valley was filled with a deep (> 200 m) glacial lake prior to 8600 yr ^{14}C BP (Hall et al., 2002, 2010), and the Ross Ice Shelf retreated from the mouth of the dry valleys between 6500 and 8340 ^{14}C yr BP (Hall and Denton, 2000). Furthermore, it may be that the ice cover originated much later, as radiocarbon dates of the dissolved organic carbon fractions in the Lake Vida brine date between 2955 and 4150 ^{14}C yr BP (Murray et al., 2012).

OSL dates at SL21.62 and SL25.59 represent the date at which the minerals in the sediment layers were last exposed to solar radiation or, more specifically, an interval when the ice cover was thin enough to allow sunlight to penetrate to the dated sediment layer, followed by a period of ice growth or further sediment burial that extinguished the light source to the sediment layer. To constrain the amount of time between sediment deposition and burial, we draw on SL26.28, where the relative difference in radiocarbon ages along the length of the layer suggests formation spanned 1400 years. Using this inference and the assumption that light could only penetrate a few centimeters into a layer, we assume burial lag of < 300 years. Therefore, the OSL dates indicate a lake level drawdown and rebound at 1200 (+300) and 320 (+300) yr BP. A lowering at 1200 years matches previous paleolimnological studies of lake levels in Taylor and Wright Valleys (Wilson, 1964; Lyons et al., 1998). Also, if our previous interpolation of a 7.7 m lake level rise in 103 years is further hindcast, it suggests that a 21.62 m lake level rise is not improbable over 320 years.

All errors associated with the quality of dating methods are subsumed when radiocarbon and OSL techniques are considered together. The discrepancy between the two dating techniques has been documented before in the dry valleys (Berger et al., 2010, 2013). Contaminated OSL samples yield artificially young dates and therefore results are considered a minimum age, whereas contaminated radiocarbon samples tend to yield artificially old dates and therefore are considered maximum ages. When viewed together, the two dating techniques constrain evaporation events between 6300 and 320 yr BP and suggest that the current Lake Vida system is a few millennia in age.

From the ages available in this study, there is no discernible correlation between lake highstands/lowstands and the temperature proxy record from neighboring Taylor Dome (Mayewski et al., 1996; Stager and Mayewski, 1997; Steig et al., 2000). This lack of synchronicity between runoff/lake level and temperature has been noted in other lake level reconstructions (Whittaker et al., 2008) and throughout the instrumental climate record (Levy et al., 2013).

6 Conclusions

Our analyses point toward the formation of sediment layers in the Lake Vida ice cover from the accumulation of sediment entrained in the ice cover accumulating during periods of ice ablation and lake level drawdown. The capacity of Lake Vida to integrate watershed processes presents a fundamental framework for understanding hydrological and climatological shifts over time.

The inconsistency in radiocarbon dates makes a full reconstruction of the history of the Lake Vida ice cover challenging. However, several conclusions are gained from this ice/sediment record:

1. A hydrologically variable climate is not unique to recent times. Lake Vida has experienced major drawdowns that led to the accumulation of four thick sediment layers in the lower ice cover. These drawdowns may have occurred as early as 6300 ^{14}C yr BP, but OSL ages and a presumed reservoir effect in radiocarbon ages suggest these events were likely constrained to the last 1–3 millennia.

2. 27 m of ice was produced from glacial streams flooding the ice surface. In the ice collected to date, there is no freezing from the bottom of the ice cover downwards as suggested by Doran et al. (2003).

3. The brine that entered the drill holes at 16 m and rose to 10.5 m in both 2005 and 2010 appears to be hydrologically connected and sourced from below 27 m.

Lake Vida represents a unique lacustrine system that has recorded a hydrologic history in the growing ice cover. As climate is projected to change (Thompson et al., 2011), Lake Vida may provide an ideal environment for tracking the influence of climate on hydrology in the dry valleys.

Acknowledgements. This research was supported by National Science Foundation grants 0739698 and 0739681 and NASA grant NNX10AN23H. We would like to thank field team members Peter Glenday, Jay Kyne, Seth Young, and Brian Glazer and the RPSC Field Safety and Training Program, the United States Antarctic Program, and PHI helicopters. Graduate student funding to H. Dugan was assisted by the Natural Sciences and Engineering Research Council of Canada and the University of Illinois at Chicago. Graduate student funding to E. Kuhn was supported by the Desert Research Institute Division of Earth and Ecosystem Sciences and the Fulbright CAPES-Brazil grant 2163-08-8. Two anonymous reviewers significantly improved this paper.

Edited by: J. Boike

References

Arcone, S. A. and Kreutz, K.: GPR reflection profiles of Clark and Commonwealth Glaciers, Dry Valleys, Antarctica, Ann. Glaciol., 50, 121–129, doi:10.3189/172756409789097531, 2009.

Arnold, L. J. and Roberts, R. G.: Paper I – Optically stimulated luminescence (OSL) dating of perennially frozen deposits in north-central Siberia: OSL characteristics of quartz grains and methodological considerations regarding their suitability for dating, Boreas, 40, 389–416, doi:10.1111/j.1502-3885.2011.00209.x, 2011.

Barrett, J. E., Virginia, R. A., Wall, D. H., Doran, P. T., Fountain, A. G., Welch, K. A., and Lyons, W. B.: Persistent effects of a discrete warming event on a polar desert ecosystem, Global Change Biol., 14, 2249–2261, doi:10.1111/j.1365-2486.2008.01641.x, 2008.

Berger, G. W. and Doran, P. T.: Luminescence-dating zeroing tests in Lake Hoare, Taylor Valley, Antarctica, J. Paleolimnol., 25, 519–529, doi:10.1023/A:1011144502713, 2001.

Berger, G. W., Doran, P. T., and Thomsen, K. J.: Single-grain and multigrain luminescence dating of on-ice and lake-bottom deposits at Lake Hoare, Taylor Valley, Antarctica, Quatern. Geochronol., 5, 679–690, 2010.

Berger, G. W., Doran, P. T., and Thomsen, K.: Micro-hole and multigrain quartz luminescence dating of Paleodeltas at Lake Fryxell, McMurdo Dry Valleys (Antarctica), and relevance for lake history, Quatern. Geochronol., 18, 119–134, doi:10.1016/j.quageo.2013.04.002, 2013.

Chinn, T. J.: Physical hydrology of the Dry Valley lakes, in: Phys. Biogeochem. Process Antarct. Lakes. Antarct. Res. Ser., edited by: Green, W. and Friedmann, E. I., vol. 59, American Geophysical Union, Washington, D.C., 1–52, 1993.

Demuro, M., Roberts, R., Froese, D., Arnold, L., Brock, F., and Ramsey, C.: Optically stimulated luminescence dating of single and multiple grains of quartz from perennially frozen loess in western Yukon Territory, Canada: Comparison with radiocarbon chronologies for the late Pleistocene Dawson tephra, Quatern. Geochronol., 3, 346–364, doi:10.1016/j.quageo.2007.12.003, 2008.

DigitalGlobe: Worldview-2 scene 10300100097A2E00, Level Standard 1B, DigitalGlobe, Longmont, Colorado, 28 January 2011.

Doran, P. T.: McMurdo Dry Valley lake levels, Long Term Ecological Research Network, doi:10.6073/pasta/112812a4c1579a00b6bb6f7f26ab39fd, 2015.

Doran, P. T., Berger, G. W., Lyons, W. B., Wharton Jr., R. A., Davisson, M. L., Southon, J., and Dibb, J. E.: Dating Quaternary lacustrine sediments in the McMurdo Dry Valleys, Antarctica, Palaeogeogr. Palaeoclim., 147, 223–239, 1999.

Doran, P. T., McKay, C. P., Clow, G. D., Dana, G. L., Fountain, A. G., Nylen, T., and Lyons, W. B.: Valley floor climate observations from the McMurdo dry valleys, Antarctica, 1986–2000, J. Geophys. Res., 107, 4772, doi:10.1029/2001jd002045, 2002.

Doran, P. T., Fritsen, C. H., McKay, C. P., Priscu, J. C., and Adams, E. E.: Formation and character of an ancient 19-m ice cover and underlying trapped brine in an "ice-sealed" east Antarctic lake, P. Natl. Acad. Sci., 100, 26–31, doi:10.1073/pnas.222680999, 2003.

Doran, P. T., McKay, C. P., Fountain, A. G., Nylen, T., McKnight, D. M., Jaros, C., and Barrett, J.: Hydrologic response to extreme warm and cold summers in the McMurdo Dry Valleys, East Antarctica, Antarct. Sci., 20, 299–509, 2008.

Doran, P. T., Kenig, F., Knoepfle, J., Mikucki, J., and Lyons, W.: Radiocarbon distribution and the effect of legacy in lakes of the McMurdo Dry Valleys, Antarctica, Limnol. Oceanogr., 59, 811–826, 2014.

Dugan, H. A., Obryk, M. O., and Doran, P. T.: Lake ice ablation rates from permanently ice covered Antarctic lakes, J. Glaciol., 59, 491–498, 2013.

Dugan, H. A., Doran, P. T., Tulaczyk, S., Mikucki, J. A., Arcone, S. A., Auken, E., Schamper, C., and Virginia, R. A.: Subsurface imaging reveals a confined aquifer beneath an ice-sealed Antarctic lake, Geophys. Res. Lett., 42, 96–103, doi:10.1002/2014GL062431, 2015.

Fountain, A. G.: Lake Vida meteorological station measurements, Long Term Ecological Research Network, doi:10.6073/pasta/897a5537defe0a7e0f0d8343641a067e, 2015.

Fountain, A. G., Dana, G. L., Lewis, K. J., Vaughn, B., and Mcknight, D. M.: Glaciers of the McMurdo Dry Valleys, Southern Victoria Land, Antarctica, in: Ecosystem Dynamics in a Polar Desert: the McMurdo Dry Valleys, Antarctica, edited by: Priscu, J. C., Antarctic Research Series, American Geophysical Union, Washington, D.C., 65–75, 1998.

Fountain, A. G., Nylen, T. H., Monaghan, A., Basagic, H. J., and Bromwich, D. H.: Snow in the McMurdo Dry Valleys, Antarctica, Int. J. Climatol., 30, 633–642, doi:10.1002/joc.1933, 2010.

French, H. M. and Harry, D. G.: Observations on buried glacier ice and massive segregated ice, western arctic coast, Canada, Permafrost Periglac. Process., 1, 31–43, doi:10.1002/ppp.3430010105, 1990.

Fritz, M., Wetterich, S., Meyer, H., Schirrmeister, L., Lantuit, H., and Pollard, W. H.: Origin and characteristics of massive ground ice on Herschel Island (western Canadian Arctic) as revealed by stable water isotope and Hydrochemical signatures, Permafrost Periglac. Process., 22, 26–38, doi:10.1002/ppp.714, 2011.

Frolov, A. D.: A review of the nature and geophysical studies of the thick permafrost in Siberia: Relevance to exploration on Mars, J. Geophys. Res., 108, 8039, doi:10.1029/2002JE001881, 2003.

Gallagher, J. B., Burton, H. R., and Calf, G. E.: Meromixis in an antarctic fjord: a precursor to meromictic lakes on an isostatically rising coastline, Hydrobiologia, 172, 235–254, 1989.

Gooseff, M. N., Lyons, W. B., McKnight, D. M., Vaughn, B., Fountain, A. G., and Dowling, C.: A stable isotopic investigation of a polar desert hydrologic system, McMurdo Dry Valleys, Antarctica, Arct. Antarct. Alp. Res., 38, 60–71, doi:10.1657/1523-0430(2006)038[0060:asiioa]2.0.co;2, 2006.

Grootes, P. M., Nadeau, M., and Rieck, A.: ^{14}C-AMS at the Leibniz-Labor: radiometric dating and isotope research, Nucl. Instrum. Methods Phys. Res. Sect. B, 223–224, 55–61, doi:10.1016/j.nimb.2004.04.015, 2004.

Gumbley, J., Wilson, A. T., Hendy, C. H., and Nelson, C. S.: Sedimentology of shallow cores from Lake Vanda, Antarct. J. United States, 9, 135–137, 1974.

Hagedorn, B., Sletten, R. S., and Hallet, B.: Sublimation and ice condensation in hyperarid soils: Modeling results using field data from Victoria Valley, Antarctica, J. Geophys. Res., 112, F03017, doi:10.1029/2006JF000580, 2007.

Hagedorn, B., Sletten, R. S., Hallet, B., McTigue, D. F., and Steig, E. J.: Ground ice recharge via brine transport in frozen soils of Victoria Valley, Antarctica: Insights from modeling δ^{18}O and δD profiles, Geochim. Cosmochim. Acta, 74, 435–448, 2010.

Hall, B. L. and Denton, G. H.: Radiocarbon Chronology of Ross Sea Drift, Eastern Taylor Valley, Antarctica: Evidence for a Grounded Ice Sheet in the Ross Sea at the Last Glacial Maximum, Geogr. Ann. Ser. A, 82, 305–336, doi:10.1111/j.0435-3676.2000.00127.x, 2000.

Hall, B. L. and Henderson, G. M.: Use of uranium-thorium dating to determine past ^{14}C reservoir effects in lakes: examples from Antarctica, Earth Planet. Sc. Lett., 193, 565–577, 2001.

Hall, B. L., Denton, G. H., Overturf, B., and Hendy, C. H.: Glacial Lake Victoria, a high-level Antarctic Lake inferred from lacustrine deposits in Victoria Valley, J. Quaternary Sci., 17, 697–706, doi:10.1002/jqs.691, 2002.

Hall, B. L., Denton, G. H., Fountain, A. G., Hendy, C. H., and Henderson, G. M.: Antarctic lakes suggest millennial reorganizations of Southern Hemisphere atmospheric and oceanic circulation, P. Natl. Acad. Sci., 107, 21355–21359, doi:10.1073/pnas.1007250107, 2010.

Harris, K. J., Carey, A. E., Lyons, W. B., Welch, K. A., and Fountain, A. G.: Solute and isotope geochemistry of subsurface ice melt seeps in Taylor Valley, Antarctica, Geol. Soc. Am. Bull., 119, 548–555, doi:10.1130/B25913.1, 2007.

Hendy, C. H.: The role of polar lake ice as a filter for glacial lacustrine sediments, Geogr. Ann., 82, 271–274, 2010.

Hendy, C. H. and Hall, B.: The radiocarbon reservoir effect in proglacial lakes: examples from Antarctica, Earth Planet. Sc. Lett., 241, 413–421, 2006.

Horita, J.: Isotopic evolution of saline lakes in the low-latitude and polar regions, Aquat. Geochem., 15, 43–69, doi:10.1007/s10498-008-9050-3, 2009.

Kopalová, K., Nedbalová, L., Nývlt, D., Elster, J., and Van de Vijver, B.: Diversity, ecology and biogeography of the freshwater diatom communities from Ulu Peninsula (James Ross Island, NE Antarctic Peninsula), Polar Biol., 36, 933–948, doi:10.1007/s00300-013-1317-5, 2013.

Kyne, J. and McConnell, J.: The SideWinder for powering a handcoring auger in drilling and lifting, Ann. Glaciol., 47, 101–104, doi:10.3189/172756407786857721, 2007.

Lacelle, D., Davila, A. F., Pollard, W. H., Andersen, D., Heldmann, J., Marinova, M., and McKay, C. P.: Stability of massive ground ice bodies in University Valley, McMurdo Dry Valleys of Antarctica: Using stable O-H isotope as tracers of sublimation in hyperarid regions, Earth Planet. Sc. Lett., 301, 403–411, 2011.

Lacelle, D., Davila, A. F., Fisher, D., Pollard, W. H., DeWitt, R., Heldmann, J., Marinova, M. M., and McKay, C. P.: Excess ground ice of condensation-diffusion origin in University Valley, Dry Valleys of Antarctica: evidence from isotope geochemistry and numerical modeling, Geochim. Cosmochim. Acta, 120, 280–297, doi:10.1016/j.gca.2013.06.032, 2013.

Levy, J. S., Fountain, A. G., Dickson, J. L., Head, J. W., Okal, M., Marchant, D. R., and Watters, J.: Accelerated thermokarst formation in the McMurdo Dry Valleys, Antarctica, Sci. Rep., 3, 2269, doi:10.1038/srep02269, 2013.

Lyons, W. B., Tyler, S. W., Wharton Jr., R. A., McKnight, D. M., and Vaughn, B. H.: A Late Holocene desiccation of Lake Hoare and Lake Fryxell, McMurdo Dry Valleys, Antarctica, Antarct. Sci., 10, 247–256, 1998.

Mayewski, P. A., Twickler, M. S., Whitlow, S. I., Meeker, L. D., Yang, Q., Thomas, J., Kreutz, K., Grootes, P. M., Morse, D. L., Steig, E. J., Waddington, E. D., Saltzman, E. S., Whung, P.-Y., and Taylor, K. C.: Climate change during the last deglaciation in Antarctica, Science, 272, 1636–1638, doi:10.1126/science.272.5268.1636, 1996.

McKay, C. P.: The search for life in our Solar System and the implications for science and society, Philos. T. Roy Soc. A, 369, 594–606, doi:10.1098/rsta.2010.0247, 2011.

McKay, C. P., Clow, G. D., Wharton Jr., R. A., and Squyres, S. W.: Thickness of ice on perennially frozen lakes, Nature, 313, 561–562, 1985.

Miller, L. G. and Aiken, G. R.: Effects of glacial meltwater inflows and moat freezing on mixing in an ice-covered Antarctic lake as interpreted from stable isotope and tritium distributions, Limnol. Oceanogr., 41, 966–976, 1996.

Murray, A. E. and Wintle, A.: The single aliquot regenerative dose protocol: potential for improvements in reliability, Radiat. Meas., 23, 497–500, 2003.

Murray, A. E., Kenig, F., Fritsen, C. H., McKay, C. P., Cawley, K. M., Edwards, R., Kuhn, E., Mcknight, D. M., Ostrom, N. E., Peng, V., Ponce, A., Priscu, J. C., Samarkin, V., Townsend, A. T., Wagh, P., Young, S. A., To, P. T., and Doran, P. T.: Microbial life at $-13\,^{\circ}C$ in the brine of an ice-sealed Antarctic lake, P. Natl. Acad. Sci., 109, 20626–20631, doi:10.1073/pnas.1208607109, 2012.

Neumann, T., Albert, M., Lomonaco, R., Engel, C., Courville, Z., and Perron, F.: Experimental determination of snow sublimation rate and stable-isotopic exchange, Ann. Glaciol., 49, 1–6, doi:10.3189/172756408787814825, 2008.

Priscu, J. C., Fritsen, C. H., Adams, E. E., Giovannoni, S. J., Paerl, H. W., McKay, C. P., Doran, P. T., Gordon, D. A., Lanoil, B. D., and Pinckney, J. L.: Perennial Antarctic lake ice: An oasis for life in a polar desert, Science, 280, 2095–2098, doi:10.1126/science.280.5372.2095, 1998.

Samyn, D., Svensson, a., and Fitzsimons, S. J.: Dynamic implications of discontinuous recrystallization in cold basal ice: Taylor Glacier, Antarctica, J. Geophys. Res., 113, F03S90, doi:10.1029/2006JF000600, 2008.

Siegert, M. J., Kennicutt, M. C., and Bindschadler, R. A.: Antarctic Subglacial Aquatic Environments, John Wiley & Sons, Washington, D.C., 2013.

Sokratov, S. A. and Golubev, V. N.: Snow isotopic content change by sublimation, J. Glaciol., 55, 823–828, doi:10.3189/002214309790152456, 2009.

Speirs, J. C., McGowan, H. A., and Neil, D. T.: Meteorological controls on sand transport and dune morphology in a polar-desert: Victoria Valley, Antarctica, Earth Surf. Proc. Land., 33, 1875–1891, doi:10.1002/esp.1739, 2008.

Stager, J. C. and Mayewski, P. A.: Abrupt early to mid-Holocene climatic transition registered at the equator and the poles, Science, 276, 1834–1836, doi:10.1126/science.276.5320.1834, 1997.

Stanish, L. F., Kohler, T. J., Esposito, R. M., Simmons, B. L., Nielsen, U. N., Wall, D. H., Nemergut, D. R., and McKnight, D. M.: Extreme streams: flow intermittency as a control on diatom communities in meltwater streams in the McMurdo Dry Valleys, Antarctica, Can. J. Fish. Aquat. Sci., 69, 1405–1419, doi:10.1139/f2012-022, 2012.

Stanish, L. F., Bagshaw, E. A., McKnight, D. M., Fountain, A. G., and Tranter, M.: Environmental factors influencing diatom communities in Antarctic cryoconite holes, Environ. Res. Lett., 8, 045006, doi:10.1088/1748-9326/8/4/045006, 2013.

Steig, E. J., Morse, D. L., Waddington, E. D., Stuiver, M., Grootes, P. M., Mayewski, P. A., Twickler, M. S., and Whitlow, S. I.: Wisconsinan and Holocene climate history from an ice core at Taylor Dome, Western Ross Embayment, Antarctica, Geogr. Ann. Ser. A, 82, 213–235, doi:10.1111/j.0435-3676.2000.00122.x, 2000.

Taylor, A. B.: History of a perennially ice-sealed lake using ice geochemistry, MSc thesis, University of Illinois at Chicago, Chicago, 2009.

Thompson, D. W. J., Solomon, S., Kushner, P. J., England, M. H., Grise, K. M., and Karoly, D. J.: Signatures of the Antarctic ozone hole in Southern Hemisphere surface climate change, Nat. Geosci., 4, 741–749, doi:10.1038/ngeo1296, 2011.

Wagner, B., Melles, M., Doran, P. T., Kenig, F., Forman, S. L., Pierau, R., and Allen, P.: Glacial and postglacial sedimentation in the Fryxell basin, Taylor Valley, southern Victoria Land, Antarctica, Palaeogeogr. Palaeoclim., 241, 320–337, 2006.

Wagner, B., Ortlepp, S., Doran, P. T., Kenig, F., Melles, M., and Burkemper, A.: The Holocene environmental history of Lake Hoare, Taylor Valley, Antarctica, reconstructed from sediment cores, Antarct. Sci., 23, 307–319, doi:10.1017/S0954102011000125, 2011.

Warnock, J. P. and Scherer, R. P.: A revised method for determining the absolute abundance of diatoms, J. Paleolimnol., 53, 157–163, doi:10.1007/s10933-014-9808-0, 2015.

Wentworth, C.: A scale of grade and class terms for clastic sediments, J. Geol., 30, 377–392, 1922.

Wharton Jr., R. A., McKay, C. P., Clow, G. D., Andersen, D. T., Simmons, G. M., and Love, F. G.: Changes in ice cover thickness and lake level of Lake Hoare, Antarctica – Implications for local climatic-change, J. Geophys. Res., 97, 3503–3513, 1992.

Whittaker, T. E., Hall, B. L., Hendy, C. H., and Spaulding, S. A.: Holocene depositional environments and surface-level changes at Lake Fryxell, Antarctica, Holocene, 18, 775–786, doi:10.1177/0959683608091797, 2008.

Wilson, A. T.: Evidence from chemical diffusion of a climatic change in the McMurdo Dry Valleys 1,200 years ago, Nature, 201, 176–177, doi:10.1038/201176b0, 1964.

Snow-cover reconstruction methodology for mountainous regions based on historic in situ observations and recent remote sensing data

A. Gafurov[1], S. Vorogushyn[1], D. Farinotti[1,*], D. Duethmann[1], A. Merkushkin[2], and B. Merz[1]

[1]Department of Hydrology, GFZ German Research Centre for Geosciences, Potsdam, Germany
[2]Uzbek Hydrometeorological Service (Uzhydromet), Tashkent, Uzbekistan
[*]now at: Swiss Federal Institute for Forest, Snow and Landscape Research WSL, Birmensdorf, Switzerland

Correspondence to: A. Gafurov (gafurov@gfz-potsdam.de)

Abstract. Spatially distributed snow-cover extent can be derived from remote sensing data with good accuracy. However, such data are available for recent decades only, after satellite missions with proper snow detection capabilities were launched. Yet, longer time series of snow-cover area are usually required, e.g., for hydrological model calibration or water availability assessment in the past. We present a methodology to reconstruct historical snow coverage using recently available remote sensing data and long-term point observations of snow depth from existing meteorological stations. The methodology is mainly based on correlations between station records and spatial snow-cover patterns. Additionally, topography and temporal persistence of snow patterns are taken into account. The methodology was applied to the Zerafshan River basin in Central Asia – a very data-sparse region. Reconstructed snow cover was cross validated against independent remote sensing data and shows an accuracy of about 85 %. The methodology can be used in mountainous regions to overcome the data gap for earlier decades when the availability of remote sensing snow-cover data was strongly limited.

1 Introduction

Water resources from remote mountain catchments play a crucial role in the development of regions in or in the vicinity of mountain ranges (Pellicciotti et al., 2012). Seasonal snow is an important water resource in many of Earth's semi-arid regions (Durand et al., 2008). Particularly in Central Asia, seasonal snowmelt decisively contributes to the total runoff volume (Ososkova et al., 2000; Unger-Shayesteh et al., 2013).

Information on snow cover and snow depth and ideally on snow water equivalent in Central Asian catchments is crucial for seasonal forecasts of water availability and for calibration and validation of hydrological models. However, the available sparse station-based data are insufficient to represent the snow-cover variability over the large and remote mountain areas (Erickson et al., 2005). The development of remote sensing techniques during recent decades allows the derivation of snow cover spatially (Liu et al., 2012). Widely used remotely sensed snow-cover products are from Advanced Very High Resolution Radiometer (AVHRR), Landsat and Moderate Resolution Imaging Spectroradiometer (MODIS) missions. Whereas AVHRR (launched 1978) and Landsat (launched 1972) offer remote sensing data for a longer period, MODIS is available only after 2000. However, snow cover from Landsat and AVHRR needs to be derived by the end user themselves, whereas MODIS offers already-compiled snow-cover product. The above-mentioned snow products are extremely useful to study snow cover worldwide; however, they are strongly limited by the presence of clouds. Recently, the reconstruction of snow-cover time series from AVHRR data for Central Asia has been reported by Zhou et al. (2013), but they are also limited in time starting in 1986 at earliest.

Several studies used remotely sensed snow cover either as input to hydrological models (Tekeli et al., 2005; Immerzeel et al., 2008; Li et al., 2008; Wang et al., 2010) or for calibration and validation purposes (Parajka and Blöschl, 2008a; Corbari et al., 2009; Liu et al., 2012; Duethmann et al., 2014).

Table 1. Uzhydromet snow observation stations with indication of elevation (Elev), and snow predictability index (SPI) and land predictability index (LPI) values for the study area (see Sect. 4). The entries "records on snow reconstructed days" indicate whether a station was snow covered (0) or snow free (1) during a day for which snow-cover reconstruction was conducted and Landsat scenes were available for validation.

				Records on snow reconstructed days			
Name	Elev (m.a.s.l)	SPI (%)	LPI (%)	10 Apr 1998	20 Nov 1998	29 Apr 1999	15 Nov 1999
Dukant	1984	12.4	3.7	1	0	0	1
Kamchik	2145	10.0	9.1	1	0	0	1
Kul	2161	23.2	6.5	0	0	0	1
Minchukur	2136	22.6	4.2	0	0	0	0
Oigaing	2187	2.2	12.3	1	1	1	1
Pskem	1256	2.9	0.8	1	0	0	1
Chimgan	1677	10.2	9.0	1	0	0	1

Particularly for hydrological model calibration, spatially distributed snow-cover data offer high information content required to constrain model parameters (Finger et al., 2011; Duethmann et al., 2014).

In Central Asia, continuous hydrometeorological records are widely available from the 1960s and earlier until the collapse of the Soviet Union in 1991. In contrast, continuous remote sensing snow-cover data from MODIS are readily available after 2000, when station data are very scarce. We present a methodology which enables reconstructing historical snow-cover pattern using long-term, point-based observations from existing meteorological stations and recent remotely sensed snow-cover data. By merging high-resolution spatial satellite data with long-term station data, snow-cover patterns can be reconstructed for several decades into the past.

Only a limited number of studies on snow-cover reconstruction have been conducted in the past that use long-term station observations and recent remote sensing data (Robinson, 1991; Brown, 2000; Frei et al., 1999; Brown and Robinson, 2011). These studies are, however, conducted at the continental scale under conditions of dense station network availability and neglecting the effect of topography. Robinson (1991) and Frei et al. (1999) conducted reconstruction of snow cover based on regression analysis between snow characteristics and snow-cover area (SCA) derived from AVHRR satellite observations. As snow characteristics both studies used snow-cover duration derived from interpolated station records. Another study by Brown (2000) conducted reconstruction of snow cover for "pre-satellite era" interpolating snow-depth data from station network. For grid cells of nearly 200 km, the interpolation of snow cover was done using different thresholds for snow depth and compared against NOAA snow-cover extent during "satellite era". The calibration showed 2 cm to be most appropriate snow depth for accurate snow-cover reconstruction based on station data. Brown and Robinson (2011) updated and extended the snow-cover reconstruction of Brown (2000) to the period 1922–2010 and used these data for trend analysis of snow-cover extent in the Northern Hemisphere. These studies can be helpful in assessing climate-related variations of snow cover but are hardly transferable to smaller catchment scale with moderate resolution and limited station data availability.

Different to those studies mentioned above, we present a methodology for snow-cover reconstruction (1) with moderate spatial resolution (500 m), (2) suitable for catchment scale hydrological studies, (3) accounting topography and (4) delivering spatially distributed snow-cover maps. The methodology takes advantage of the strong control of topography on the spatial snow-cover distribution. Hence, measurements from snow observation stations at different elevations can be interpreted as representative sites to predict snow-cover patterns. The methodology consists of five successive steps which make use of topographic information and correlations between station records and spatial snow-cover patterns. In order to test the presented methodology, snow-cover reconstruction was conducted for 4 days (Table 1) for which independent Landsat data were available.

2 Study area

The methodology for snow-cover reconstruction was developed and tested for the area containing the upper Zerafshan River basin, Central Asia (Fig. 1).

The upper Zerafshan River basin is located in the Gissaro-Alai mountain range. Elevation ranges from 658 to 5402 m a.s.l. and basin area is about 12 000 km^2. The Zerafshan River basin is currently an endorheic basin in the inner Central Asia that no longer contributes to the Amu Darya River. It originates in Tajikistan and flows towards Uzbekistan, where its water is used for agricultural production in the oases of Samarkand and Bukhara. The flow regime is strongly dominated by snow and glacier melt, as can be inferred from the temperature, precipitation, SCA and flow

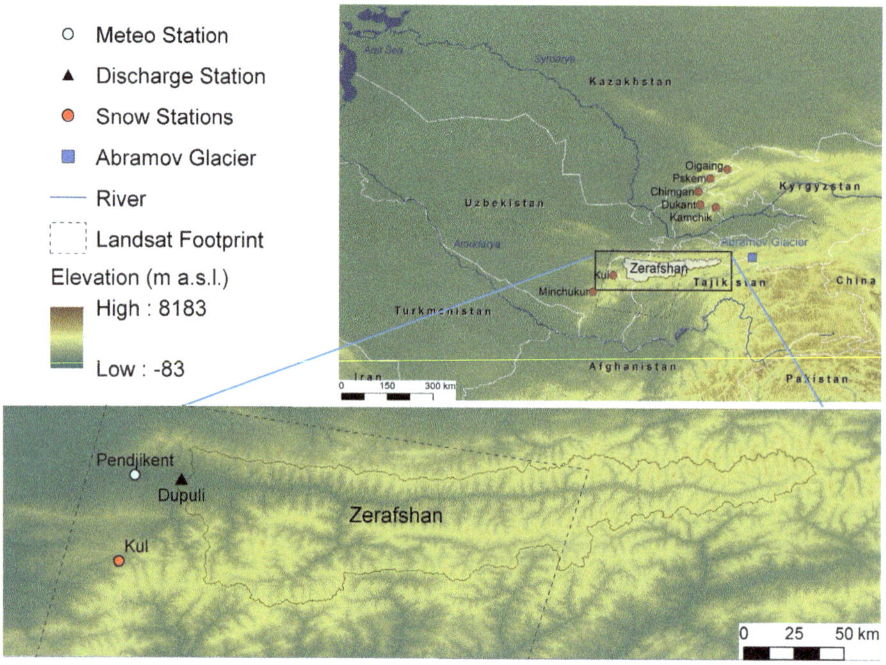

Figure 1. Location of the upper Zerafshan River basin in the Gissaro-Alai mountain range, Central Asia. Snow-cover reconstruction was conducted for the entire area of Fig. 1b and validated for the area with Landsat footprint.

regimes in Fig. 2. The highest precipitation is brought by westerly flows during winter and spring, with a clear minimum during summer and early autumn (Aizen et al., 1995). The highest runoff, however, occurs during summer months and is driven by snow and glacier melt. According to MODIS Landcover product (MCD12Q1) from 2009, the main land cover types in the study area are grasslands (60 %), croplands (9 %), open shrublands (7 %), woody savannahs (6 %) and permanent snow- and ice-covered areas (5 %).

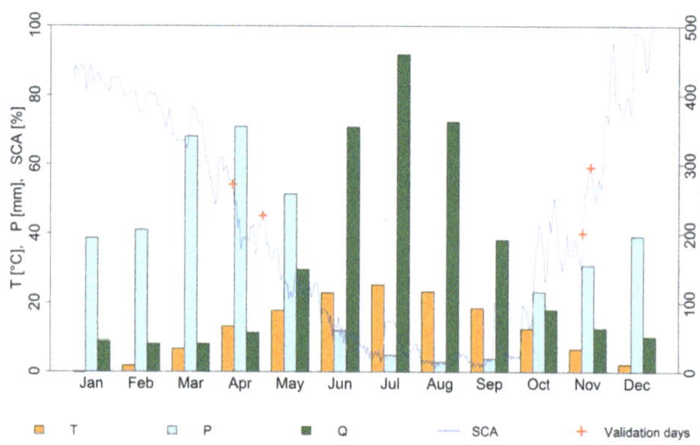

Figure 2. Monthly average air temperature (T), cumulative precipitation (P), discharge (Q) and SCA dynamics in the Zerafshan basin. T, P and Q means are based on data for the period from 1930 until 2008. Daily SCA is for 2004 obtained from MODIS and cloud eliminated using Gafurov and Bárdossy (2009). Temperature (0 °C in January) and precipitation data are from the Pendjkent station (1016 m a.s.l.), and discharge is measured at the Dupuli gauge (see Fig. 1).

3 Data

We used (1) daily in situ snow-depth data, (2) daily MODIS snow-cover data, (3) a digital elevation model (DEM) and (4) Landsat data. The first three data sets were used for snow-cover reconstruction whereas Landsat data were used as an independent data set to validate the results.

3.1 In situ snow-depth data

Daily snow-depth data in the period from 1964 to 2012 were available for seven climate stations located at different elevations (Fig. 1, Table 1). These data contain continuous snow-depth measurements including records on no-snow conditions. Snow depth in these stations are recorded at 1 cm threshold depth. The data were provided by Uzbek Hydrometeorological Service (Uzhydromet).

3.2 MODIS snow-cover data

MODIS daily snow-cover data from the Terra satellite with 500 m spatial resolution (MOD10A, version V005) were employed for the time period of 2000 to 2012. We used MODIS Terra snow-cover data due to the longer time series compared to the Aqua satellite, which delivered snow-cover data only after 2002. The MODIS snow-cover product is based on the

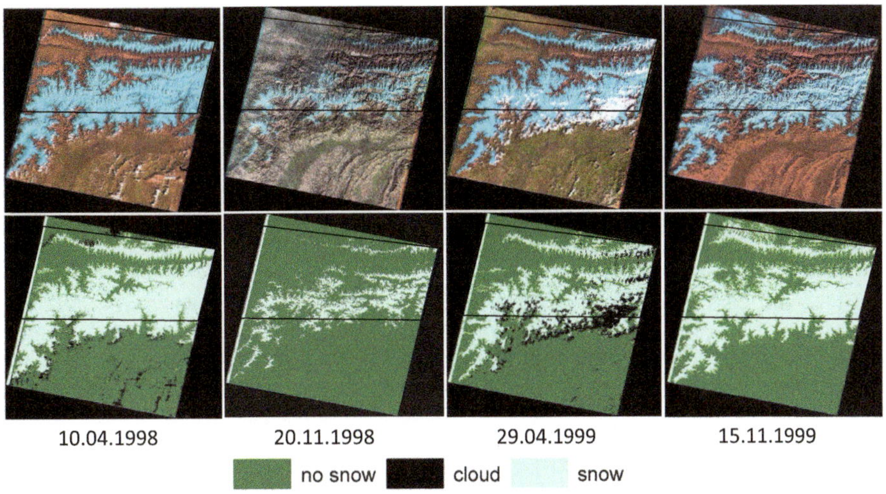

Figure 3. Original Landsat scenes (top row) and derived snow-cover maps used for validation (bottom row). Black outlines show the validation domain for Zerafshan basin.

normalized difference snow index (NDSI) algorithm (Hall et al., 2002). Its accuracy was tested in different parts of the world showing good agreement with in situ data (Klein and Barnett, 2003; Tekeli et al., 2005; Parajka et al., 2006; Ault et al., 2006; Wang et al., 2008; Liang et al., 2008; Huang et al., 2011; Gafurov et al., 2013; Parajka et al., 2012). The main drawback of MODIS snow-cover data is the limitation due to cloud cover. There have been several studies on filtering methods for reducing cloud cover or even removing it completely (e.g., Parajka and Blöschl, 2008b; Gafurov and Bárdossy, 2009; Tong et al., 2009; Hall et al., 2010; Lòpez-Burgos et al., 2013). We used original MODIS snow-cover data to exclude any uncertainty that may be introduced by cloud filtering. The data were obtained from the National Aeronatics and Space Administration (NASA) Earth Observving System Data and Information System (EOSDIS) Reverb platform. MODIS data are distributed as tiles with the size of $10° \times 10°$, which makes up a total of 36 horizontal (h) and 18 vertical (v) tiles covering the entire globe. In this study, the tile h23v05, which covers the study area completely, was used.

3.3 Digital elevation model

The void-filled DEM with 90 m spatial resolution from NASA Shuttle Radar Topography Mission (SRTM) was used. SRTM DEM data were obtained from the CGIAR CSI (Consultative Group on International Agricultural Research, Consortium for Spatial Information) database (Jarvis et al., 2008). To have the same resolution as the MODIS data (500 m), the 90 m SRTM DEM was aggregated to 500 m.

3.4 Landsat data

Optical remote sensing data from the Landsat Thematic Mapper sensor were used to validate the reconstructed snow-cover maps. The Landsat data have a spatial resolution of 30 m and a temporal resolution of 16 days. Landsat data from 4 nearly clear-sky days in the snow season (10 April 1998, 20 November 1998, 20 April 1999 and 15 November 1999) were used for validation purposes. Snow-cover maps for the Landsat footprint (see Fig. 1) were prepared using the NDSI methodology. For a detailed description of the algorithm used for deriving snow-cover maps from Landsat refer to Gafurov et al. (2013). Figure 3 shows raw and processed Landsat snow-cover maps for the study area.

Since Landsat has a spatial resolution of 30 m and snow reconstruction was performed for 500 m pixels based on MODIS resolution, the processed Landsat snow-cover maps were spatially aggregated to 500 m resolution. This was done by classifying each of the 500 m pixels as snow covered or snow free, based on the majority of the 30 m Landsat pixels within the 500 m pixel.

4 Methodology

The methodology presented hereafter is based on similarity of different locations in terms of presence or absence of snow at a given time. The idea is to use the information about the presence of snow from one location in order to estimate the presence of snow at another location. The similarity between different locations was assessed using both observed snow cover at meteorological stations (i.e., records of snow depth > 0) and MODIS snow-cover data. This similarity was quantified using the concept of conditional probability (CP). The conditional probability gives the probability of one event (e.g., snow cover at a pixel) to occur given that another event (e.g., snow depth > 0 at station) has already occurred. Additionally to the similarity between two locations, temporally persistent monthly snow-cover patterns and elevation-based classifications were used. The methodology consists of five successive steps in which each step estimates a certain fraction of snow cover. This leads to a complete snow-cover reconstruction after step 5. Once the similarities and temporally persistent snow fields are established using existing stations and remote sensing data, the methodology can be applied to other time periods and is solely based on snow records at meteorological stations. The following five steps are detailed in the next sections:

1. pixel to station CP fields
2. temporally persistent monthly probability fields
3. pixel to pixel CP fields
4. usage of elevation information
5. pixel to station CP for CP < 1.

4.1 Pixel-to-station conditional probability

In the first step, we consider the CP of each pixel, given the observed data from a set of snow stations. We compute the CP of each pixel as follows:

$$P^{\mathrm{s}}(S_{x,y}|S_n) = \frac{\sum(1-\mathrm{ABS}(S_{x,y,t}-S_{n,t}))}{N_{x,y}} \quad \forall \quad S_{n,t}=1, \tag{1}$$

$$P^{\mathrm{l}}(S_{x,y}|S_n) = \frac{\sum(1-\mathrm{ABS}(S_{x,y,t}-S_{n,t}))}{N_{x,y}} \quad \forall \quad S_{n,t}=0, \tag{2}$$

where $P^{\mathrm{s}}(S_{x,y}|S_n)$ ($P^{\mathrm{l}}(S_{x,y}|S_n)$) is the CP of a pixel with coordinates xy to be covered by snow (land) given that the station n also records snow depth > 0 ($= 0$) at the same time. $S_{x,y,t}$ and $S_{n,t}$ are binary variables indicating the presence ($S = 1$) or absence ($S = 0$) of snow at pixel xy and station n for day t, respectively. $N_{x,y}$ is the number of observations simultaneously available at pixel xy (excluding cloud-covered days) and station n over the 12 years (2000–2012) for which station n showed snow ($S = 1$) or snow-free ($S = 0$) conditions. The value of $P^{\mathrm{s}}(S_{x,y}|S_n)(P^{\mathrm{l}}(S_{x,y}|S_n))$ varies from 0 to 1, with $P^{\mathrm{s}}(S_{x,y}|S_n) = 1$ ($P^{\mathrm{l}}(S_{x,y}|S_n) = 1$) indicating that a pixel at xy was always observed as snow covered (snow free) in the MODIS data when station n measured snow depth > 0 ($= 0$), whilst $P^{\mathrm{s}}(S_{x,y}|S_n) = 0$ ($P^{\mathrm{l}}(S_{x,y}|S_n) = 0$) indicates an opposite relationship, i.e., that the MODIS product at xy always showed snow-free (snow-covered) conditions when station n had snow depth > 0 ($= 0$).

CPs were computed for each MODIS pixel in the study area (total of 169 776 pixels) using over 12 years of available MODIS data and observed snow-depth measurements. Hence, the daily snow-cover maps from MODIS are treated as snow observations for each 500 m grid cell, giving rise to a very dense "observation network". An example for a CP map for snow and land conditions for Chimgan station is given in Fig. 4. In total, 14 maps were derived (two maps for every of the seven stations: one for $P^{\mathrm{s}}(S_{x,y}|S_n)$ and one for $P^{\mathrm{l}}(S_{x,y}|S_n)$).

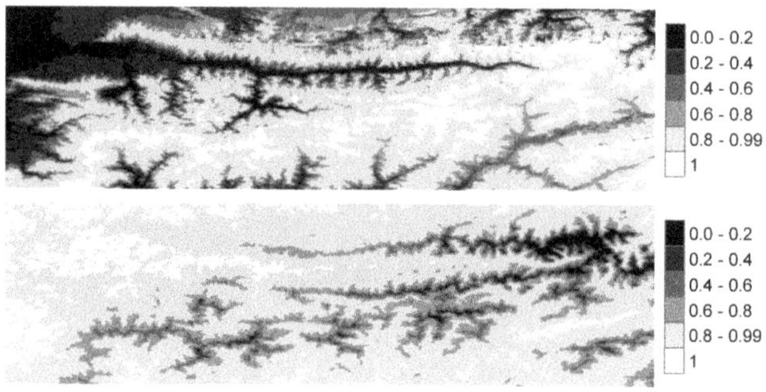

Figure 4. CP maps for snow (top) and land (bottom) conditions of Chimgan station (see Fig. 1) for the study area. The figure shows the same domain as Fig. 1b.

The number of pixels with $P^{\mathrm{s}}(S_{x,y}|S_n) = 1$ ($P^{\mathrm{l}}(S_{x,y}|S_n) = 1$) varies from station to station. The higher the number of pixels with $P^{\mathrm{s}}(S_{x,y}|S_n) = 1$, the higher the predictive power of the station for snow classification is. Similarly, the higher the number of pixels with $P^{\mathrm{l}}(S_{x,y}|S_n) = 1$, the higher the predictive power of the station to predict snow-free conditions is. In order to quantify the predictive power of each station, we introduce two terms: snow predictability index (SPI) and land predictability index (LPI). These terms give the fractions of the reconstruction domain with $P^{\mathrm{s}}(S_{x,y}|S_n) = 1$ for a given station for snow and land conditions, respectively:

$$\mathrm{SPI}_n = \frac{\sum(P^{\mathrm{s}}(S_{x,y}|S_n) - 1)}{N} \cdot 100 \quad [\%], \tag{3}$$

$$\mathrm{LPI}_n = \frac{\sum(P^{\mathrm{l}}(S_{x,y}|S_n) = 1)}{N} \cdot 100 \quad [\%], \tag{4}$$

where SPI_n and LPI_n are the snow predictability index and the land predictability index of station n, respectively. N is the total number of pixels (169 776) in the entire study area. Pixels with $P^{\mathrm{s}}(S_{x,y}|S_n) = 1$ in Fig. 4 (top) add up to 10.2 %, which is the SPI value of the Chimgan station for the entire domain. This means that when the Chimgan station shows a snow depth above zero, 10.2 % of the study area can be classified as snow covered as well. Pixels with $P^{\mathrm{l}}(S_{x,y}|S_n) = 1$ in Fig. 4 (bottom) add up to 9.0 %, which means that when the Chimgan station shows snow-free conditions, 9.0 % of the study area can be assigned as snow free.

Figure 4 shows that $P^{\mathrm{s}}(S_{x,y}|S_n) = 1$ occurs mainly at high elevations (see Fig. 1), whilst $P^{\mathrm{l}}(S_{x,y}|S_n) = 1$ occurs mainly at low elevations. This is not surprising since pixels with an

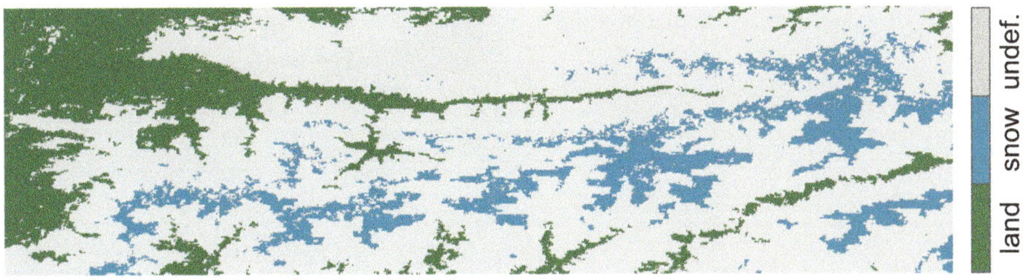

Figure 5. Temporally persistent spatial snow ($MP^s = 1$) and land ($MP^l = 1$) patterns for April in the study area shown in Fig. 1b. Pixels with "undef" indicate $MP^s < 1$ or $MP^l < 1$, for which classification is not possible in this step.

elevation far higher than the elevation of Chimgan station tend to be snow covered if Chimgan station records positive snow depth, whilst pixels with an elevation far below the elevation of Chimgan station tend to be snow free if Chimgan station records snow depth of zero. Table 1 shows the SPI and LPI values for each station. The stations Kul and Minchukur near the Zerafshan basin (see Fig. 1) have the highest SPI values (23.2 and 22.6 %, respectively). Other stations, located further away from the catchment, have smaller SPI values. Noticeably, Oigaing station, located farthest away from the Zerafshan basin, has the highest LPI value (12.3 %). This can be explained by the high elevation of the station. When the station indicates snow-free conditions, pixels with significantly lower elevation are likely to be snow free as well. Assuming that the dependencies remain stable in time, the computed CPs of each pixel can be used to classify individual pixels for any arbitrary day prior to the availability of MODIS data (before 2000) for which station records are available:

$$S_{x,y,t} = 1 \quad \text{if} \quad \left(P^s(S_{x,y}|S_n) = 1 \text{ and } S_{n,t} = 1\right), \tag{5}$$

$$S_{x,y,t} = 0 \quad \text{if} \quad \left(P^l(S_{x,y}|S_n) = 1 \text{ and } S_{n,t} = 0\right). \tag{6}$$

This step leads to a partially reconstructed snow-cover map which is further enhanced in the next steps.

4.2 Monthly probability fields

Snow-cover extent is a seasonally variable parameter. Accordingly, the probability of a certain pixel to be covered by snow or land varies with time. The second step for reconstructing snow cover is based on the observation that during different months, certain pixels are snow covered or snow free with high confidence. The spatial distributions of such temporally persistent patterns can be identified using the available MODIS snow-cover data in the period 2000–2012. A monthly probability (MP) of each pixel to be covered by snow or land in a certain month is computed according to

$$MP^s_{x,y,m} = \frac{\sum(S_{(x,y,t),\ (t\in m)} = 1)}{N_{x,y,m}}, \tag{7}$$

$$MP^l_{x,y,m} = \frac{\sum(S_{(x,y,t),\ (t\in m)} = 0)}{N_{x,y,m}}, \tag{8}$$

where $MP^s_{x,y,m}$ and $MP^l_{x,y,m}$ are the probabilities of pixel xy to be covered by snow or land in month m, respectively. $S_{(x,y,t)}$ indicates the coverage (snow or land) of pixel xy on day t. $N_{x,y,m}$ is the total number of MODIS observations of pixel xy and month m in the period 2000–2012. The maximum value of $MP^s_{x,y,m}$ ($MP^l_{x,y,m}$) is 1, meaning that the pixel xy was covered always by snow (land) in month m during the cloud-free days in the period 2000–2012. Computation of MPs for every pixel in the study area leads to MP maps for all 12 months as illustrated exemplarily in Fig. 5 for April.

Pixels with $MP^s_{x,y,m} = 1$ in Fig. 5, i.e., pixels that were always snow covered in April, add up to 12.7 % of the whole area. This means that 12.7 % of the study area can be classified as snow covered in April. Pixels with $MP^l_{x,y,m} = 1$ add up to 14.9 %, i.e., 14.9 % of the domain can be classified as snow free in April. To remain consistent with the terminology used in the first step, we call the sum of pixels with $MP^s_{x,y,m} = 1$ ($MP^l_{x,y,m} = 1$) the monthly SPI (LPI) value for snow (land). Monthly SPI and LPI are defined in a similar way as in step 1 (Eqs. 3 and 4).

The main idea in this step is to transfer these temporally persistent monthly spatial snow/land patterns (SPI_m/LPI_m) to the past to reconstruct historical snow cover. However, the validity of these temporally persistent spatial snow/land patterns over a longer time in the past is not assured due to, e.g., potential warmer/cooler or wetter/dryer climate conditions. In order to account for possible climatic variability, we introduce a buffer as vertical elevation shift from month-specific minimum snow and maximum land lines. We define a month-specific minimum snow line ($H^s_{\min,m}$) and a month-specific maximum land line ($H^l_{\max,m}$) as

$$H^s_{\min,m} = \min\left(H_{x,y}\right) \quad \forall \quad MP^s_{x,y,m} = 1, \tag{9}$$

$$H^l_{\max,m} = \max\left(H_{x,y}\right) \quad \forall \quad MP^l_{x,y,m} = 1, \tag{10}$$

where $H_{x,y}$ is the elevation of the pixel xy. $H^l_{\max}$ is thus the maximum elevation of all pixels with $MP^l_{x,y,m} = 1$ in month

Table 2. Monthly SPI, LPI, $H^{s}_{min,m}$ and $H^{l}_{max,m}$ values for the study region.

	SPI		LPI	
Month	Fraction (%)	H^{s}_{min}	Fraction (%)	H^{l}_{max}
1	10.1	1544	0.0	658
2	11.4	1164	0.0	2233
3	24.0	1658	0.4	2923
4	12.7	2354	14.9	3323
5	1.1	2900	32.7	3707
6	0.0	3568	49.5	4387
7	0.0	5402	66.3	4434
8	0.0	5402	78.3	4520
9	0.0	4285	51.2	4330
10	0.2	3405	25.3	4089
11	0.9	2750	0.4	2581
12	8.5	2373	0.0	1707

m. Note that below the altitude H^{l}_{max} not all pixels necessarily have $MP^{l}_{x,y,m} = 1$. Similarly, H^{s}_{min} is the minimum elevation of all pixels with $MP^{s}_{x,y,m} = 1$ in month m. Again, H^{s}_{min} does not necessarily represent the elevation above which all pixels have $MP^{s}_{x,y,m} = 1$. Table 2 lists monthly SPI and LPI values as well as $H^{s}_{min,m}$ and $H^{l}_{max,m}$ values for the selected area.

These $H^{s}_{min,m}$, $H^{l}_{max,m}$ and monthly SPI/LPI maps were used to further reconstruct the snow cover resulting from step 1:

$$S_{x,y,t} = 1 \quad (11)$$
$$\text{if } \left(H_{x,y} > H^{s}_{min,m} + \text{buffer and } MP^{s}_{x,y,m} = 1\right) \quad t \in m,$$

$$S_{x,y,t} = 0 \quad (12)$$
$$\text{if } \left(H_{x,y} < H^{l}_{max,m} - \text{buffer and } MP^{l}_{x,y,m} = 1\right) \quad t \in m,$$

where "buffer" is a parameter accounting for the possible vertical shift in snow line. In order to account for climate variability not represented by the period for which satellite observations are available, "buffer" was set to 500 m. Due to the absence of historical data on snow-line variations in the region, the buffer was estimated corresponding to the maximum observed variation in the equilibrium line altitude (ELA) of Abramov glacier (Table 3, see Fig. 1 for location) for the period 1972–1998 (WGMS, 2001; Pertziger, 1996) and is thus a conservative estimate for the variations in snow line for the study area.

4.3 Pixel-to-pixel conditional probability

In step 1, CPs of each pixel in accordance to station records were computed, and any pixel that had $P^{s}(S_{x,y}|S_n) = 1$ was classified according to the station record. The idea behind the third step is similar, but CPs of each pixel in accordance to other pixels are computed this time. In such a way, the state of different pixels is used as a predictor for snow cover elsewhere. We define the CP of any pixel with coordinates xy to be covered by snow (land) given that another pixel ij is covered by snow (land) as follows:

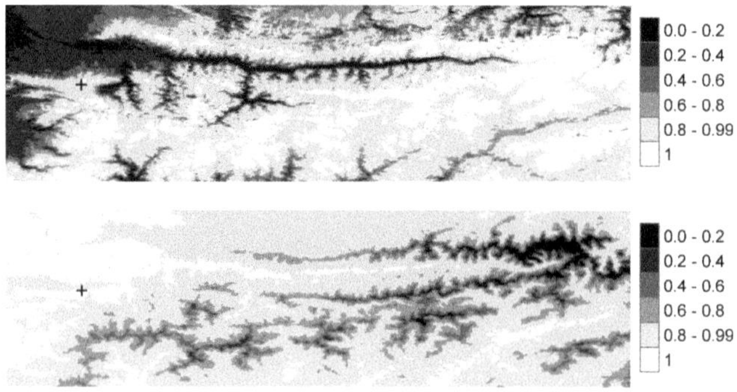

Figure 6. CP maps for snow (top) and land (bottom) conditions for the pixel $x = 100$, $y = 100$ (black cross) with elevation 2206 m a.s.l.

$$P^{s}(S_{i,j}|S_{x,y}) = \frac{\sum(1 - \text{ABS}(S_{i,j,t} - S_{x,y,t}))}{N_{i,j}} \quad \forall \quad S_{x,y,t} = 1, \quad (13)$$

$$P^{l}(S_{i,j}|S_{x,y}) = \frac{\sum(1 - \text{ABS}(S_{i,j,t} - S_{x,y,t}))}{N_{i,j}} \quad \forall \quad S_{x,y,t} = 0, \quad (14)$$

where $S_{i,j,t}$ indicates whether the pixel with coordinates ij is snow covered ($S = 1$) or snow free ($S = 0$) for a given day t and $N_{i,j}$ is the total number of valid observations (clear sky, no cloud) at pixel ij simultaneously available for a given condition ($S = 1$ or $S = 0$) in the period 2000–2012.

The computation of $P^{s}(S_{i,j}|S_{x,y})$ and $P^{l}(S_{i,j}|S_{x,y})$ according to Eqs. (13) and (14) is repeated in an "all-versus-all" procedure, which means that all possible combinations of (xy) and (ij) are considered. For the region of interest, this yields at maximum 339 552 (double the total number of pixels) CP maps (i.e., two maps for every pixel: for snow and land condition). However, not all of these maps were used for snow reconstruction since some pixels may have no perfect dependence (no pixels with CP = 1) to any other pixel in the study area. An example of the CP maps for snow and snow-free conditions for the pixel located at $x = 100$ and $y = 100$ is given in Fig. 6.

The pixel with coordinates $x = 100$, $y = 100$ has $P^{s}(S_{i,j}|S_{100,100}) = 1$ for 30 684 (18 %) other pixels in the study area, which means that when the particular pixel was snow covered during 2000–2012, 30 684 other pixels were always snow covered as well. Thus, the SPI value of that

Table 3. ELA records of Abramov glacier (WGMS, 2001; Pertziger, 1996).

Year	1972	1977	1987	1988	1989	1990	1991	1992	1993	1994	1995	1996	1997	1998
ELA (m a.s.l)	4020	4393	4130	4170	4200	4220	4242	4110	4120	4250	4240	4163	4440	4130

pixel is 18 %, and this can be interpreted as the predictive power for snow of the pixel for the entire study area. Analogously, 23 770 pixels (14 %) have $P^{\mathrm{l}}(S_{i,j}|S_{100,100}) = 1$ with this pixel and the LPI value of this pixel is 14 %. The SPI and LPI values of each pixel are derived through Eqs. (3) and (4) and illustrated in Fig. 7 for all pixels in the study area.

The maximum SPI value (Fig. 7, top) is 46 %, meaning that, according to the observations of the period 2000–2012, 46 % (78 189 pixels) of the study area was always snow covered when that particular pixel was snow covered. The maximum LPI value (Fig. 7 bottom) is 88 %, meaning that this particular pixel is able to predict snow-free conditions for 88 % (149 685 pixels) of the basin. These two pixels with maximum SPI and LPI values are located within an area which has high predictive power for snow and land, respectively. When interpreting Fig. 7, three further features are worth noting: (1) pixels with SPI = 0 or LPI = 0 exist as well, but these pixels have no predictive power and are therefore not used in the snow-cover reconstruction; (2) SPI and LPI maps generally reflect the topography of the catchment: lower elevation pixels have higher SPI values and pixels at higher elevations have higher LPI values; (3) snow-free pixels are easier to predict than snow-covered ones.

The SPI and LPI maps were used for classifying pixels that are still undefined after the previous steps:

$$S_{i,j,t} = 1 \quad \text{if} \quad \left(P^{\mathrm{s}}(S_{i,j}|S_{x,y}) = 1 \text{ and } S_{x,y,t} = 1\right), \tag{15}$$

$$S_{i,j,t} = 0 \quad \text{if} \quad \left(P^{\mathrm{l}}(S_{i,j}|S_{x,y}) = 1 \text{ and } S_{x,y,t} = 0\right). \tag{16}$$

Since in this step SPI and LPI maps were generated for every pixel in the basin, this step tends to classify a significantly larger area than the first step where only seven stations were used for constructing CPs.

4.4 Snow-cover estimation using elevation information from neighboring pixels

This step is adapted from Gafurov et al. (2009) and is based on the information of neighboring pixels. Let us consider a pixel that has not been classified as snow covered or snow free in any of the previous steps. If any of the adjacent eight pixels is covered by snow and the elevation of that snow-covered pixel is lower than the pixel that is still undefined, then the undefined pixel is classified as snow covered. The same idea is applied for snow-free pixels. Hence, this step

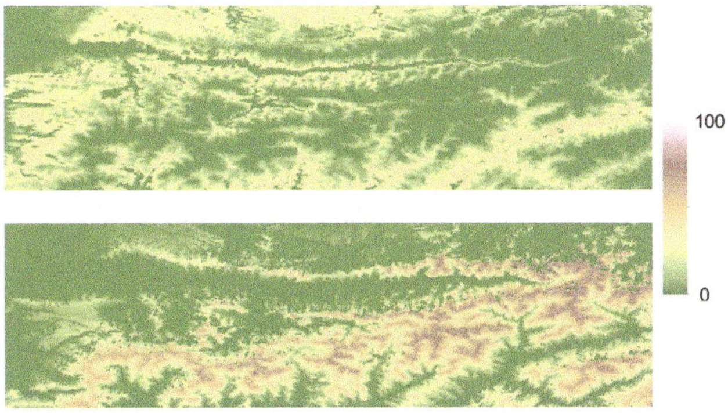

Figure 7. SPI (top) and LPI (bottom) values of each pixel (in %) in the study area defined in Fig. 1b.

can be formalized as follows:

$$S_{x,y,t} = 1 \quad \text{if} \quad S_{x+k,y+k,t} = 1 \quad \text{and } H_{x+k,y+k} < H_{x,y} \quad k \in (-1, 1), \tag{17}$$

$$S_{x,y,t} = 0 \quad \text{if} \quad S_{x+k,y+k,t} = 0 \quad \text{and } H_{x+k,y+k} > H_{x,y} \quad k \in (-1, 1). \tag{18}$$

This step takes only elevation of neighboring pixel into account. However, in areas where factors others than elevation have an influence on neighboring pixel condition (e.g., pixels located near to water surfaces or forests), additional information such as a land cover map could be introduced into this step.

4.5 Snow-cover estimation with CP < 1

In the last step, the $P^{\mathrm{s}}(S_{x,y}|S_n)$ and $P^{\mathrm{l}}(S_{x,y}|S_n)$ values calculated in step 1 are used again. Whereas in step 1 only CP= 1 conditions were considered, this step considers CP< 1 relations (see Fig. 4) to classify still-undefined pixels. Still-undefined pixels are classified according to the highest CP value found amongst all 14 CP values available (seven CPs for snow and seven CPs for land). However, since the stations are located at different elevations with varying snow-covered (snow-free) days, their computed CPs can have varying levels of confidence depending on the number of observations used to construct probabilities. To account for this in the reconstruction, we compute lower bound confidence intervals

Table 4. Contingency table (in %) for the reconstructed snow-cover maps validated against four aggregated Landsat snow-cover images. Four cases are distinguished: SS, LL, SL and LS. The first (second) letter indicates the classification according to the presented algorithm (Landsat). "S" stands for snow and "L" for land. "Total" indicates the percentage of pixels classified after each step. Results refer to the Landsat domain (dashed line) shown in Fig. 1b.

Day	Step	SS	LL	SL	LS	Total
10 Apr 1998	1	8.5	7.2	0.0	0.0	15.7
	2	13.0	14.9	0.1	0.1	28.1
	3	17.5	27.1	0.8	0.1	45.5
	4	20.6	29.3	1.3	0.2	51.4
	5	43.4	42.3	12.5	1.8	100.0
20 Nov 1998	1	0.1	11.2	0.0	0.0	11.3
	2	0.2	11.3	0.0	0.0	11.5
	3	0.5	30.5	0.0	0.0	31.0
	4	1.4	33.4	0.0	0.1	34.9
	5	18.1	67.0	2.2	12.7	100.0
29 Apr 1999	1	0.1	13.9	0.0	0.0	14.0
	2	7.6	21.5	0.0	0.5	29.6
	3	9.0	35.2	0.1	0.7	45.0
	4	11.0	38.1	0.3	0.9	50.3
	5	24.5	58.8	14.3	2.3	100.0
15 Nov 1999	1	18.4	4.1	0.0	0.4	22.9
	2	18.4	4.2	0.0	0.4	23.0
	3	24.8	15.7	0.4	0.8	41.7
	4	28.5	18.3	0.7	1.3	48.9
	5	41.7	42.0	11.2	5.1	100.0

(CI) of each CP. As the CP estimates follow a binomial distribution, we compute lower bound CI of CP according to

$$\mathrm{CI}^{\mathrm{low}}_{(P^{\mathrm{s}}(S_{x,y}|S_n))} = P^{\mathrm{s}}\left(S_{x,y}|S_n\right) - z\sqrt{\frac{1}{N_{x,y}}P^{\mathrm{s}}(S_{x,y}|S_n)(1-P^{\mathrm{s}}(S_{x,y}|S_n))}, \tag{19}$$

$$\mathrm{CI}^{\mathrm{low}}_{(P^{\mathrm{l}}(S_{x,y}|S_n))} = P^{\mathrm{l}}\left(S_{x,y}|S_n\right) - z\sqrt{\frac{1}{N_{x,y}}P^{\mathrm{l}}(S_{x,y}|S_n)(1-P^{\mathrm{l}}(S_{x,y}|S_n))}, \tag{20}$$

where $\mathrm{CI}^{\mathrm{low}}_{(P^{\mathrm{s}}(S_{x,y}|S_n))}$ and $\mathrm{CI}^{\mathrm{low}}_{(P^{\mathrm{l}}(S_{x,y}|S_n))}$ are lower bound of 95 % CI of CP for snow and land conditions, respectively. z is the constant for 95 % confidence level (1.96). The computed lower bound CI for each CP will help to classify still-undefined pixel coverage based on highest confidence level for $P\left(S_{x,y}|S_n\right) < 1$ case among 14 CPs. Thus, we use the highest lower bound CI among all CPs of this particular pixel to be decisive for reconstruction.

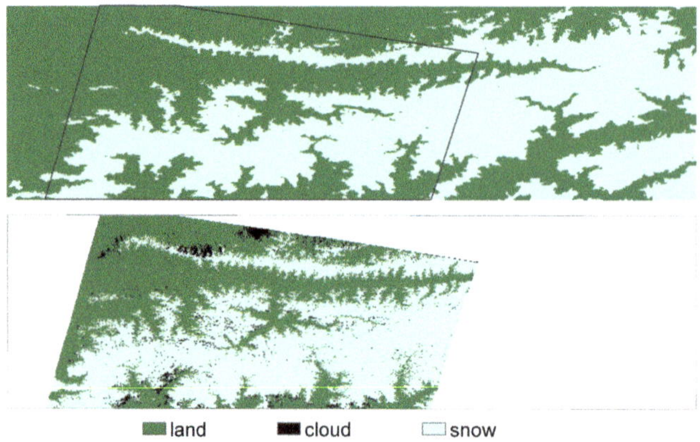

Figure 8. Reconstructed (top) and Landsat (bottom) snow-cover maps for 10 April 1998.

$$S_{x,y,t} = 1 \quad \text{if} \quad \max\left(\mathrm{CI}^{\mathrm{low}}_{(P^{\mathrm{s}}(S_{x,y}|S_n))}\right) > \max\left(\mathrm{CI}^{\mathrm{low}}_{(P^{\mathrm{l}}(S_{x,y}|S_n))}\right) \quad n \in 1:7 \tag{21}$$

$$S_{x,y,t} = 0 \quad \text{if} \quad \max\left(\mathrm{CI}^{\mathrm{low}}_{(P^{\mathrm{l}}(S_{x,y}|S_n))}\right) > \max\left(\mathrm{CI}^{\mathrm{low}}_{(P^{\mathrm{s}}(S_{x,y}|S_n))}\right) \quad n \in 1:7 \tag{22}$$

Taking maximum lower bound CI values for still-undefined pixels in the last step allows us to complete the classification for all pixels. However, since in this step $P(S_{x,y}|S_n) < 1$ was considered, the reconstruction is subject to uncertainty that stems from non-perfect agreement between station records and a pixel in the period 2000–2012.

5 Results and discussion

Applying the five steps described above, snow-cover maps for the area containing Zerafshan basin were reconstructed for 4 days in 1998 and 1999. The maps contain binary information showing whether a given pixel was covered by snow or not. The accuracy of the reconstructed snow-cover maps was assessed by comparing against independent snow maps derived from four Landsat images from the same days. The validation could be done only for recent years (1998 and 1999) due to the availability of cloud-free Landsat images during the snow period only for those days. The comparison was performed on a pixel-to-pixel basis, and the accuracy was assessed in a contingency table (Table 4). In Table 4, the sum of percentages of "SS" and "LL" columns represent the degree of accuracy after each reconstruction step, related to the total share of reconstructed pixels. Accordingly, the sum of "SL" and "LS" indicate the error in relation to the total percentage.

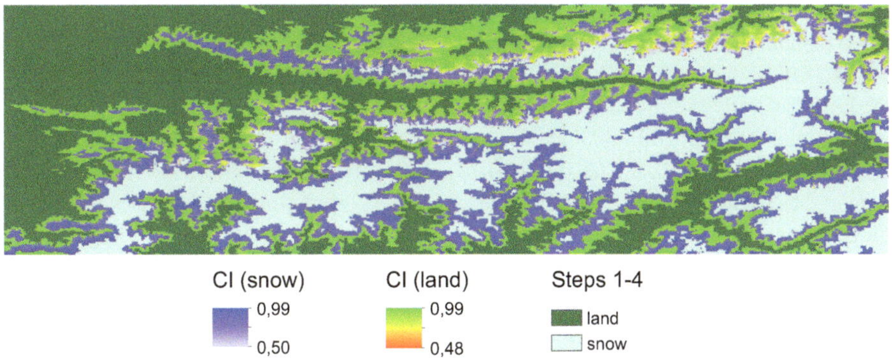

Figure 9. Fraction of reconstruction in steps 1–4 and maximum CI values for snow and land in step 5 for the study area illustrated in Fig. 1b.

As an example, Fig. 8 shows the reconstructed and Landsat-derived snow-cover maps for 10 April 1998. The comparison of these maps results in 85.7 % of correct reconstruction (cases SS + LL in Table 4) and 14.3 % of erroneous reconstruction (SL + LS). Steps 1–4 show high accuracy with only little erroneous reconstruction (ER) whereas step 5 has the lowest accuracy in all validation days. However, the reconstruction fraction (RF) is very high in step 5 compared to previous steps. Note that ER may also be enhanced by erroneous snow-cover estimation from raw Landsat data and due to the spatial aggregation of Landsat 30 m original resolution to 500 m. Another potential bias may come from similar approaches (NDSI) used to map snow cover both for MODIS snow-cover maps, which are used to assess CPs between station and pixels, and Landsat snow-cover maps, which are used to validate reconstructed snow-cover maps. However, different threshold values than MODIS were used to map snow cover from Landsat, assuring best visual validation of snow and snow-free surface cover.

In order to better illustrate snow reconstruction in step 5, Fig. 9 shows the areal fraction for which the reconstruction was performed in steps 1–4 and maximum lower bound CI obtained in step 5 under CP < 1 condition for the validation day of 10 April. Most of the still-unclassified pixels after steps 1–4 have CI values close to 1 and only few pixels have a lower CI value (reddish and light blue colors in Fig. 9). Figure 10 illustrates the trade-off between RF and ER as a function of lower bound CI in step 5. For example, for the validation day of 10 April, ER from steps 1–4 adds up to 1.5 % (Table 4) and RF to 51.4 % (snow and land classes in Fig. 9). With decreasing CI, RF increases but at the cost of an increased ER. However, Fig. 10 also shows that RF is relatively high until about CI = 0.9 with increasing ER. On all 4 days used for validation, an almost complete reconstruction is achieved with CI > 0.9.

Figures 8, 9 and 10 also demonstrate that the methodology provides two types of results for snow-cover reconstruction: deterministic and probabilistic snow-cover maps. Deterministic maps result from the complete classification of pixels (Fig. 8) with binary information (snow/snow free), taking

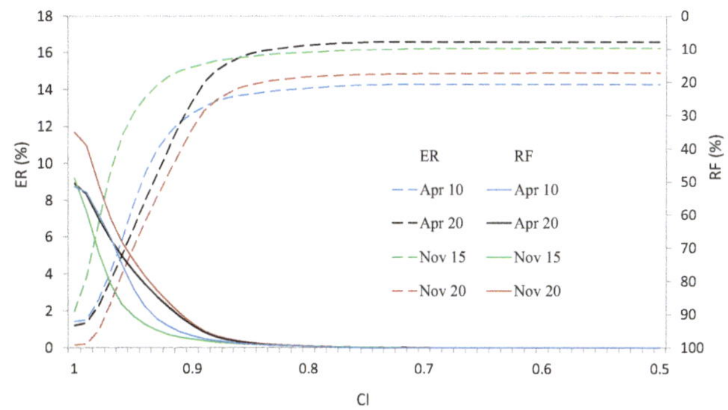

Figure 10. Trade-off between erroneous reconstruction (ER, dashed lines) and reconstruction fraction (RF, solid lines) after step 4 as a function of CI.

CI < 1 in step 5 into account at the expense of the overall accuracy. However, the accuracy is still quite high with a range of 83.3–85.7 % for the 4 validation days and is only slightly less than the accuracy of the MODIS snow-cover product in Central Asia (ca. 92 %) when compared to Landsat snow information (Gafurov et al., 2013). Alternatively, probabilistic snow-cover maps (Fig. 9) deliver a partial snow-cover reconstruction with high accuracy resulting from steps 1–4 and, as result of step 5, a probability statement for snow cover for the remaining pixels.

The validation days for this study were chosen deliberately from the snowmelt and snow accumulation season (transition period) when snow-cover estimation is particularly challenging. For the time outside the snowmelt or snow accumulation period, higher accuracies can be expected since a higher fraction can be reconstructed in the first four steps already. During the snow transition period, snow-cover conditions such as ephemeral snow cover can occur which exacerbates snow-cover estimation. However, in the reconstruction process using steps 1, 2 and 3 the conditions with ephemeral snow cover (in the period 2000–2012) are accounted for as well. Under such conditions, station or MODIS data may see different snow cover than the reality. In such cases (e.g., MODIS sees "land" although there is ephemeral snow, whilst the station sees "snow" since it is a manual point recording with a certain threshold), and CP and MP of the pixel produce the value of < 1 as they do not show the same event and are not used in the first three steps for reconstruction. Only distinct snow-cover records from both station and MODIS are used to identify snow-covered areas in these steps with CP = 1 (MP = 1). Reduced CP values that may partly be due to ephemeral snow cover are, however, used in step 5 in order to classify areas still undefined in the steps 1–4 and may thus contribute to the accuracy loss in step 5.

The validation of reconstructed snow-cover maps were done using independent Landsat data in this study. Alternatively, the AVHRR snow-cover data, which are also available beyond the MODIS data availability in the past, can be used for validation purposes. However, AVHRR snow-cover data have a coarser spatial resolution ($\sim$ 1.1 km) than the resolution (500 m) used in this study. Unfortunately, processed AVHRR snow-cover data were not available at the time of manuscript writing and remain alternative data to be used for validation.

6 Limitations of the methodology

The predictive power of the observations at meteorological stations for snow-cover reconstruction is limited by the elevation range of the stations. If all meteorological stations are located at high elevations, they will be good predictors during summer for snow-free conditions but will perform poorly when predicting snow-covered areas during winter due to their elevation and correspondingly lower SPI values. Conversely, low-elevation stations are better indicators for snow-covered pixels at higher altitudes than they are for snow-free ones. Hence, a wide spread in station elevation is optimal for accurate snow-cover reconstruction. In our case study, the application of the presented methodology suffered from the small number of station data (only seven stations). A higher number of stations would lead to a higher number of SPI and LPI maps and would allow us to reconstruct a larger areal fraction of snow cover in the first four steps with high accuracy. Noticeably, the stations do not need to be located inside the area of interest.

Reconstruction of the snow cover for the past is based on the assumptions that (1) the calibration period, i.e., the MODIS data period, is representative for the past period, and (2) the relationship between station records and spatial snow patterns derived from MODIS data is stationary, i.e., does not significantly change in time. A calibration period which lacks extreme conditions, e.g., snow-rich or snow-scarce years, might lead to larger errors in the reconstruction. A longer calibration period is expected to lead to more robust relationships for reconstructing snow cover.

The problem of representativity of the MODIS period in the reconstruction step 2 is tackled by the introduction of the elevation buffer to capture the effect of interannual temporal variability of snow-line elevation. For this the temporal variability of the recorded ELA from the neighboring Abramov glacier was used as a proxy. Through changes in climatic conditions of the calibration period going beyond temporal variability of the snow-line elevation in the reconstruction period, the relationships between station records and some pixels (step 1) and between pixels (step 3) may become non-representative. This occurs if the snow line in the future/calibration period more often separates the station of the pixels compared to the reconstruction period. Hence, an analysis of temperature and precipitation trends and comparison of climatology between calibration and reconstruction periods may provide some confidence on representativeness of the relationships used.

The statistical relationship (CP) between point measurements and aerial patterns computed in this study highly depend on topography. Since the Zerafshan basin has a very heterogeneous topography with high elevation range, good predictive power (SPI and LPI) of individual stations could be obtained. This is important to estimate initial snow cover in the first step, which is a base input for next steps (except step 2). Thus, we can conclude that the methodology is well applicable for mountainous areas where high SPI and LPI values can be obtained. However, it might be difficult to exploit statistical relationships between point measurements and aerial pattern in lowland areas; this is a subject to be tested.

7 Conclusion

In this study, a methodology for reconstructing past snow cover using historical in situ snow-depth data, recent remote sensing snow-cover data and topographic data was presented. The methodology is based on (1) constructing relationships between station observations and remote sensing data, (2) estimating the monthly variation of snow cover from remote sensing data, (3) deriving pixel-to-pixel relationships using remote sensing data and (4) using neighborhood relations. Once the dependence between individual pixels and station records is derived, this dependence is used to reconstruct past snow cover based solely on station records.

The methodology was applied to a study area containing the Zerafshan River basin – a basin with high topographic gradients – in Central Asia and showed correct classification in the range of 83.3–85.7 % when compared to four Landsat snow-cover scenes. This high agreement is noteworthy, given that only seven stations and 12 years of remote sensing data were available. Moreover, snow-cover reconstruction was done for snowmelt and onset season when snow classification is challenging compared to outside snowmelt and onset seasons when large areas are easy to classify as snow or land. The agreement is only slightly less than that of original MODIS snow-cover product with accuracy of about

92 % for Central Asia when compared to Landsat-derived snow-cover maps (Gafurov et al., 2013). Just 12 years of MODIS data was sufficient to extract stable patterns of snow cover and relate them to station records in the Zerafshan basin with heterogeneous topography. Hence, we conclude that the developed methodology is suitable to derive past snow cover in remote mountainous regions such as the Zerafshan basin with very limited data availability. Reconstructed snow-cover patterns can be used for hydrological model calibration/validation and for understanding snow-cover dynamics over large areas prior to the age of satellite observations. The performance of methodology presented here for non-mountainous areas remains an open question.

Acknowledgements. This work was carried out within the framework of the CAWa (Water in Central Asia) project (http://www.cawa-project.net, contract no. AA7090002), funded by the German Federal Foreign Office as part of the "Berlin Process". Doris Duethmann was supported by the SuMaRiO project (Sustainable Management of River Oases along the Tarim River/China), funded by the BMBF (German Ministry for Education and Research) — funding measure "Sustainable Land Management", reference no. LLA2-02.

Edited by: R. Brown

References

Aizen, V., Aizen, E., and Melack, J.: Climate snow cover, glaciers and runoff in the Tien Shan, Central Asia, Water Resour. Bull., 31, 1113–1129, 1995.

Ault, T., Czajkowski, K., Benko, T., Coss, J., Struble, J., Spongberg, A., Templin, M., and Gross, C.: Validation of the MODIS snow product and cloud mask using student and NWS cooperative station observations in the Lower Great Lakes Region, Remote Sens. Environ., 105, 341–353, 2006.

Brown, R.: Northern Hemisphere Snow Cover Variability and Change, 1915–1997, J. Climate, 13, 2339–2355, 2000.

Brown, R. D. and Robinson, D. A.: Northern Hemisphere spring snow cover variability and change over 1922–2010 including an assessment of uncertainty, The Cryosphere, 5, 219–229, doi:10.5194/tc-5-219-2011, 2011.

Corbari, C., Ravazzani, G., Martinelli, J., and Mancini, M.: Elevation based correction of snow coverage retrieved from satellite images to improve model calibration, Hydrol. Earth Syst. Sci., 13, 639-649, doi:10.5194/hess-13-639-2009, 2009.

Duethmann, D., Peters, J., Blume, T., Vorogushyn, S., and Güntner, A.: The value of satellite-derived snow cover images for calibrating a hydrological model in snow-dominated catchments in Central Asia, Water Resour. Res., 50, 2002–2021, doi:10.1002/2013WR014382, 2014.

Durand, M., Molotch, N., and Margulis, A.: A Bayesian approach to snow water equivalent reconstruction, J. Geophys. Res., 113, D20117, doi:10.1029/2008JD009894, 2008.

Erickson, T., Williams, M., and Winstral, A.: Persistence of topographic controls on the spatial distribution of snow in rugged mountain terrain, Colorado, United States, Water Resour. Res., 41, W04014, doi:10.1029/2003WR002973, 2005.

Finger, D., Pellicciotti, F., Konz, M., Rimkus, S., and Burlando, P.: The value of glacier mass balance, satellite snow cover images and hourly discharge for improving the performance of a physically based distributed hydrological model, Water Resour. Res., 47, W07519, doi:10.1029/2010WR009824, 2011.

Frei, A., Robinson, D., and Hughes, M.: North American Snow Extent: 1900–1994, Int. J. Climatol., 19, 1517–1534, 1999.

Gafurov, A. and Bárdossy, A.: Cloud removal methodology from MODIS snow cover product, Hydrol. Earth Syst. Sci., 13, 1361–1373, doi:10.5194/hess-13-1361-2009, 2009.

Gafurov, A., Kriegel, D., Vorogushyn, S., and Merz, B.: Evaluation of remotely sensed snow cover product in Central Asia, Hydrol. Res., 44, 506–522, doi:10.2166/nh.2012.094, 2013.

Hall, D., Riggs, G., and Salomonson, V.: Development of methods for mapping global snow cover using Moderate Resolution Imaging Spectroradiometer (MODIS) data, Remote Sens. Environ., 83, 181–194, 2002.

Hall, D., Riggs, G., Foster, J., and Kumar, S.: Development and evaluation of a cloud-gap-filled MODIS daily snow-cover product, Remote Sens. Environ., 114, 496–503, 2010.

Huang, X., Liang, T., Zhang, X., and Guo, Z.: Validation of MODIS snow cover products using Landsat and ground measurements during the 2001–2005 snow seasons over northern Xinjiang, China, Int. J. Remote Sens., 32, 133–152, 2011.

Immerzeel, W., Droogers, P., Jong, S., and Bierkens, M.: Large-scale monitoring of snow cover and runoff simulation in Himalayan river basins using remote sensing, Remote Sens. Environ., 113, 40–49, 2008.

Jarvis, A., Reuter, H., Nelson, A., and Guevara, E.: Hole-filled SRTM for the globe Version 4, available at the CGIAR-CSI SRTM 90m Database: http://srtm.csi.cgiar.org (last access: August 2013), 2008.

Klein, A. and Barnett, A.: Validation of daily MODIS snow cover maps of the Upper Rio Grande River Basin for the 2000–2001 snow year, Remote Sens. Environ., 86, 162–176, 2003.

Liang, T., Zhang, X., Xie, H., Wu, C., Feng, Q., Huang, X., and Chen, Q.: Toward improved daily snow cover mapping with advanced combination of MODIS and AMSR-E measurements, Remote Sens. Environ., 112, 3750–3761, 2008.

Li, X. and Williams, M.: Snowmelt runoff modeling in an arid mountain watershed, Tarim Basin, China, Hydrol. Process., 22, 3931–3940, 2008.

Liu, T., Willems, P., Feng, X., Li, Q., Huang, Y., Bao, A., Chen, X., Veroustraete, F., and Dong, Q.: On the usefulness of remote sensing input data for spatially distributed hydrological modeling: case study of the Tarim River basin in China, Hydrol. Process., 26, 335–344, 2012.

López-Burgos, V., Gupta, H. V., and Clark, M.: Reducing cloud obscuration of MODIS snow cover area products by combining spatio-temporal techniques with a probability of snow approach, Hydrol. Earth Syst. Sci., 17, 1809–1823, doi:10.5194/hess-17-1809-2013, 2013.

Ososkova, T., Gorelkin, N., and Chub, V.: Water resources of Central Asia and adaptation measures for climate change, Environ. Monit. Assess., 61, 161–166, 2000.

Parajka, J. and Blöschl, G.: Validation of MODIS snow cover images over Austria, Hydrol. Earth Syst. Sci., 10, 679–689, doi:10.5194/hess-10-679-2006, 2006.

Parajka, J. and Blöschl, G.: The value of MODIS snow cover data in validating and calibrating conceptual hydrological models, J. Hydrol., 358, 240–258, 2008a.

Parajka, J. and Blöschl, G.: Spatio-temporal combination of MODIS images – potential for snow cover mapping, Water Resour. Res., 44, 1–13, 2008b.

Parajka, J., Holko, L., Kostka, Z., and Blöschl, G.: MODIS snow cover mapping accuracy in a small mountain catchment – comparison between open and forest sites, Hydrol. Earth Syst. Sci., 16, 2365–2377, doi:10.5194/hess-16-2365-2012, 2012.

Pellicciotti, F., Buegrl, C., Immerzeel, W., Konz, M., and Shrestha, A.: Challenges and uncertainties in hydrological modeling of remote Hindu-Kush-Karakoram-Himalayan (HKH) Basins: Suggestions for calibration strategies, Mt. Res. Dev., 39, 39–50, 2012.

Pertziger, F.: Abramov Glacier Data Reference Book: Climate, Runoff, Mass Balance, SANIIGMI, Tashkent, Uzbekistan, 279 pp., 1996.

Robinson, D.: Merging operational satellite and historical station snow cover data to monitor climate change, Palaeogeogr. Palaeocl., 90, 235–240, 1991.

Tekeli, A., Akyurek, Z., Sorman, A., Sensoy, A., and Sorman, A.: Using MODIS snow cover maps in modeling snowmelt runoff process in the eastern part of Turkey, Remote Sens. Environ., 97, 216–230, 2005.

Tong, J., Déry, S. J., and Jackson, P. L.: Interrelationships between MODIS/Terra remotely sensed snow cover and the hydrometeorology of the Quesnel River Basin, British Columbia, Canada, Hydrol. Earth Syst. Sci., 13, 1439–1452, doi:10.5194/hess-13-1439-2009, 2009.

Unger-Shayesteh, K., Vorogushyn, S., Farinotti, D., Gafurov, A., Duethmann, D., Mandychev, A., and Merz, B.: What do we know about past changes in the water cycle of Central Asian headwaters? A review, Global Planet. Change, 110, 4–25, doi:10.1016/j.gloplacha.2013.02.004, 2013.

Wang, J., Li, H., and Hao, X.: Responses of snowmelt runoff to climatic change in an inland river basin, Northwestern China, over the past 50 years, Hydrol. Earth Syst. Sci., 14, 1979–1987, doi:10.5194/hess-14-1979-2010, 2010.

Wang, X., Xie, H., and Liang, T.: Evaluation of MODIS snow cover and cloud mask and its application in Northern Xinjiang, China, Remote Sens. Environ., 112, 1497–1513, 2008.

World Glacier Monitoring Service (WGMS): Glacier Mass Balance Bulletin No. 6 (1998–1999), edited by: Haeberli, W., Frauenfelder, R., and Hoelzle, M., IAHS (ICSI)/ UNEP/UNESCO, World Glacier Monitoring Service, Zurich, Switzerland, available at: http://www.wgms.ch/mbb/mbb6/wgms_2001_gmbb6.pdf (last access: 4 January 2013), 2001.

Zhou, H., Aizen, E., and Aizen, V.: Deriving long term snow cover extent dataset from AVHRR and MODIS data: Central Asia case study, Remote Sens. Environ., 136, 146–162, doi:10.1016/j.rse.2013.04.015, 2013.

16

Lidar snow cover studies on glaciers in the Ötztal Alps (Austria): comparison with snow depths calculated from GPR measurements

K. Helfricht[1,2], M. Kuhn[2], M. Keuschnig[1,3], and A. Heilig[4,5]

[1]alpS – Centre for Climate Change Adaptation, Grabenweg 68, 6020 Innsbruck, Austria
[2]Institute of Meteorology and Geophysics, University of Innsbruck, Innrain 52, 6020 Innsbruck, Austria
[3]Department of Geography and Geology, University of Salzburg, Hellbrunnerstrasse 34, 5020 Salzburg, Austria
[4]Commission for Geodesy and Glaciology, Bavarian Academy of Sciences and Humanities, Alfons-Goppel-Str. 11, 80539 Munich, Germany
[5]Institute of Environmental Physics, University of Heidelberg, Im Neuenheimer Feld 229, 69120 Heidelberg, Germany

Correspondence to: K. Helfricht (helfricht@alps-gmbh.com)

Abstract. The storage of water within the seasonal snow cover is a substantial source of runoff in high mountain catchments. Information about the spatial distribution of snow accumulation is necessary for calibration and validation of hydro-meteorological models. Generally, only a small number of precipitation measurements deliver precipitation input for modelling in mountain areas. The spatial interpolation and extrapolation of measurements of precipitation is still difficult. Multi-temporal application of lidar techniques from aircraft, so-called airborne laser scanning (ALS), provides surface elevations changes even in inaccessible terrain. These ALS surface elevation changes can be used to derive changes in snow depths of the mountain snow cover for seasonal or subseasonal time periods. However, since glacier surfaces are not static over time, ablation, densification of snow, densification of firn and ice flow contribute to surface elevation changes. ALS-derived surface elevation changes were compared to snow depths derived from 35.4 km of ground penetrating radar (GPR) profiles on four glaciers. With this combination of two different data acquisitions, it is possible to evaluate the effect of the summation of these processes on ALS-derived snow depth maps in the high alpine region of the Ötztal Alps (Austria). A Landsat 5 Thematic Mapper image was used to distinguish between snow covered area and bare ice areas of the glaciers at the end of the ablation season. In typical accumulation areas, ALS surface elevation changes differ from snow depths calculated from GPR measurements by −0.4 m on average with a mean standard deviation of 0.34 m. Differences between ALS surface elevation changes and GPR derived snow depths are small along the profiles conducted in areas of bare ice. In these areas, the mean absolute difference of ALS surface elevation changes and GPR snow depths is 0.004 m with a standard deviation of 0.27 m. This study presents a systematic approach to analyze deviations from ALS generated snow depth maps to ground truth measurements on four different glaciers. We could show that ALS can be an important and reliable data source for the spatial distribution of snow depths for most parts of the here investigated glaciers. However, within accumulation areas, just utilizing ALS data may lead to systematic underestimation of total snow depth distribution.

1 Introduction

In alpine catchments, the so-called glacio–nival runoff regime is caused by the storage of water in the seasonal snow cover and in glaciers (Aschwanden et al., 1986; Kuhn and Batlogg, 1998; Kuhn, 2000). High flow rates caused by snow- and ice melt in spring and summer are of interest in terms of water resources management for reservoirs and flood forecasting (e.g. Weingartner et al., 2003; Kirnbauer et al., 2009; Schöber et al., 2010).

For calculating reliable runoff amounts in these mountain catchments, precipitation data in high elevations are needed and the heterogeneous distribution of the snow cover has to be considered in hydro-meteorological models (e.g. Kuhn, 2003; Huss et al., 2008). However, measurements of solid precipitation are known to be affected by considerable errors (e.g. Sevruk, 1985; Lundberg et al., 2010) and only insufficient data are available for model calibration and validation of the snow depth distribution on the catchment scale. Thereby the spatial representativeness of individual measurements of snow depths and snow densities derived from direct measurements in terms of snow probings and snow pits (Fierz et al., 2009) and from automatic measurement systems (Lundberg et al., 2010) has to be questioned (Grünewald and Lehning, 2011).

Whereas various hydro-meteorological models have been tested versus snow cover extent from satellite remote sensing data of MODIS (Moderate Resolution Imaging Spectroradiometer) or Landsat scenes (e.g. Rott and Markl, 1989; Hall et al., 2002; Schöber et al., 2010), to date no snow depth or SWE can be obtained from satellite data in a high spatial resolution of at least 100 m (Lundberg et al., 2010; Nolin, 2011), which is required to capture the spatial variability of the mountain snow cover (Blöschl, 1999). In contrast, laser altimetry was recognized to be useful to derive small scale snow depth distributions in mountain areas (Deems et al., 2013). Several snow cover studies based on the application of lidar from aircraft (airborne laser scanning – ALS) have been performed in mountain areas, mostly in unglacierized terrain (e.g. Deems et al., 2008; Melvold and Skaugen, 2013), but also in glacierized catchments (e.g. Dadic et al., 2010; Helfricht et al., 2012; Sold et al., 2013). Since 2001, ALS surveys of annual and seasonal surface elevation changes (Δz_{ALS}) were performed in glacierized catchment of the Ötztal Alps (Austria, 46°50′ N, 10°50′ E) (Geist and Stötter, 2007; Helfricht et al., 2012). In 2010/11 an ALS survey of the whole mountain range of the Ötztal Alps was conducted to estimate seasonal snow depths.

In this study Δz_{ALS} are investigated for the accumulation season ranging roughly from the beginning of October (t_1) to the end of April respective the beginning of May (t_2). Figure 1 illustrates the processes causing surface elevation changes on glaciers in this period. The elevation z_1' at the time t_2 corresponds to the primary surface z_1 at the time t_1. The elevation z_2 corresponds to the actual surface at the time t_2. Δz_{ALS} is the result of changes in snow depth (Δz_{HS}), densification of snow and firn (Δz_{DSF}), ablation (Δz_{ABL}) and vertical ice flow (Δz_{ICE})

$$\Delta z_{ALS} = \Delta z_{HS} + \Delta z_{DSF} + \Delta z_{ABL} + \Delta z_{ICE} = z_2 - z_1, \quad (1)$$

whereas Δz_{HS} at the time t_2 is defined as the elevation difference between z_2 and z_1'

$$\Delta z_{HS} = z_2 - z_1', \quad (2)$$

and the summation of densification, ablation and vertical ice flow is the elevation difference between $z_{1'}$ and z_1

$$\Delta z_{DSF} + \Delta z_{ABL} + \Delta z_{ICE} = z_1' - z_1. \quad (3)$$

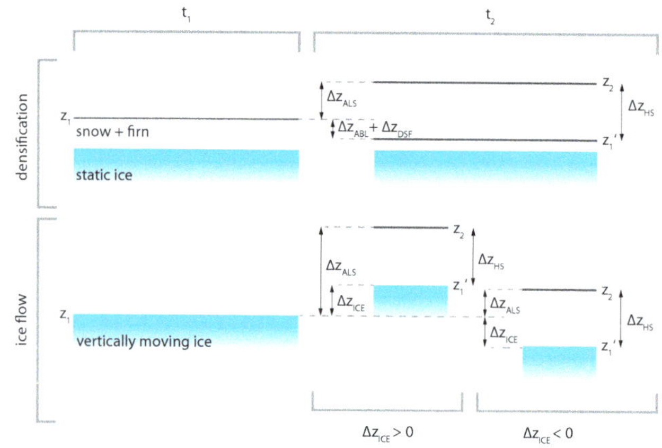

Fig. 1. Changes in surface elevation due to snow depth (Δz_{HS}), ablation (Δz_{ABL}) and densification of snow and LSC (Δz_{DSF}) resulting in ALS surface elevation changes Δz_{ALS} between the two ALS flights (t_1) and (t_2). Upper part: firn and snow on top of static ice. Lower part: firn and snow on top of a vertically moving ice surface (Δz_{ICE}).

The densification of snow and firn layers existing at t_1 leads to an underestimation of Δz_{HS} in the case of snow and firn layers on a static ice body at t_2. Changes of the reference elevation due to vertical ice flow can have both, negative and positive values. Submergence ice flow ($\Delta z_{ICE} < 0$) leads to an underestimation of Δz_{HS} by Δz_{ALS}. Emergent ice flow ($\Delta z_{ICE} > 0$) causes overestimation of Δz_{HS} by Δz_{ALS}. Hence, "ground truth" data are necessary to evaluate Δz_{ALS} for snow cover and mass balance studies on glacier surfaces (e.g. Huss et al., 2009; Koblet et al., 2010; Fischer, 2011).

The aim of the present study is to show the consequence of differences between Δz_{ALS} and actual snow depths on the application of multi-temporal ALS to derive seasonal snow depths on the investigated glaciers. The questions is whether and where differences between Δz_{ALS} and actual snow depths on glacier surfaces are of magnitude, such that they have to be considered in snow-hydrological studies based on ALS surveys. Therefore, we compared Δz_{ALS} to snow depths calculated from ground penetrating radar (GPR) measurements and analyzed differences larger than the combined uncertainties of ALS and GPR measurements. The presented results will support the application of ALS surveys for snow cover studies on glaciers in this mountain area and in Alpine catchments with comparable glacier behaviour in terms of ice flow velocities and accumulation area extent. Systematic underestimation or overestimation of snow depths by Δz_{ALS} can be attributed to specific areas for all here investigated glaciers.

After the introduction an overview of methods and data used to process, the GPR and lidar data is given in Sect. 2. Conditions on the glaciers during ALS surveys are treated in more detail in Sect. 3. In Sect. 4 the results are presented followed by the discussion in Sect. 5. Finally, conclusions are made about the applicability of surface elevation changes derived from ALS for snow cover studies in glacierized catchments.

2 Data and methods

2.1 Study sites

The study sites are located next to the Alpine main ridge in the Ötztal Alps (46°48′ N, 10°46′ E; Fig. 2). Four of the largest glaciers in this mountain range, namely Hintereisferner, Kesselwandferner, Gepatschferner and Vernagtferner were chosen to evaluate Δz_{ALS} on glacier surface for snow cover studies. All of these glaciers have been subject to scientific research for decades: Hintereisferner (HEF, 46°48′ N, 10°46′ E) has one of the world's longest mass balance series starting in 1952 (Hoinkes, 1970; Fischer, 2010; Fischer et al., 2012). Based on these data sets, a variety of models with different complexity were developed to compute glacier surface mass balance (Kuhn et al., 1999; Escher-Vetter et al., 2009). Ice thickness was measured by Span et al. (2005). Since 2001 a series of ALS surveys enabled the comparison of geodetic and direct glaciological measurements of mass balance on Hintereisferner (Geist and Stötter, 2007; Fischer, 2011). However, since the last acceleration of ice flow on Hintereisferner in the 1970s, horizontal velocities have slowed down dramatically to less than $10\,\mathrm{m\,a^{-1}}$ (Span et al., 1997; Span and Kuhn, 2003). They found that emergence velocities measured on the glacier tongue of Hintereisferner decreased to about $1\,\mathrm{m\,a^{-1}}$at the beginning of the 1990s. Haberkorn (2011) observed a depression of the ice surface for a certain area along the tongue of Hintereisferner. Measured horizontal velocities at ablation stakes were $4.6\,\mathrm{m\,a^{-1}}$ at most. Vertical velocities could not be derived due to small absolute values with changing signs. In 2010/2011 ablation stakes along the flow line on the glacier tongue of Hintereisferner were georeferenced with DGPS (H. Schneider, personal communication, 2013). In combination with the ablation measurements at these positions, vertical velocities were calculated. The mean annual vertical velocities increased from zero in the area of the GPR profile H3 (Fig. 8) to $0.55\,\mathrm{m\,a^{-1}}$ of annual emergence velocity at about 2 km from the terminus. Annual vertical ice flow measured at an ablation stake in the central flow line near the GPR profile H5 (Fig. 8) was $0.85\,\mathrm{m\,a^{-1}}$ in 2010/2011.

Kesselwandferner (KWF, 46°50′ N, 10°48′ E) was subject to scientific studies investigating density profiles and deformation of firn layers (Ambach and Eisner, 1966). Mass balance measurements from Kesselwandferner using the direct glaciological method started in the hydrological year 1952/53 (Fischer et al., 2011) and geodetic mass balances are available since 1964. In Abermann et al. (2007) annual vertical and horizontal velocities at several stake locations along the central flow line are presented. After the glacier's advance in the 1980s, vertical velocities decreased to $1.5\,\mathrm{m\,a^{-1}}$ at most, both for submergence and emergent flow. Since the measurements of Abermann et al. (2007), annual surveys of the ablation stakes and the accumulation stakes have been continued. For the last 5 yr, mean annual vertical velocities along the flow line remained similar at up to $1\,\mathrm{m\,a^{-1}}$(H. Schneider, personal communication, 2013).

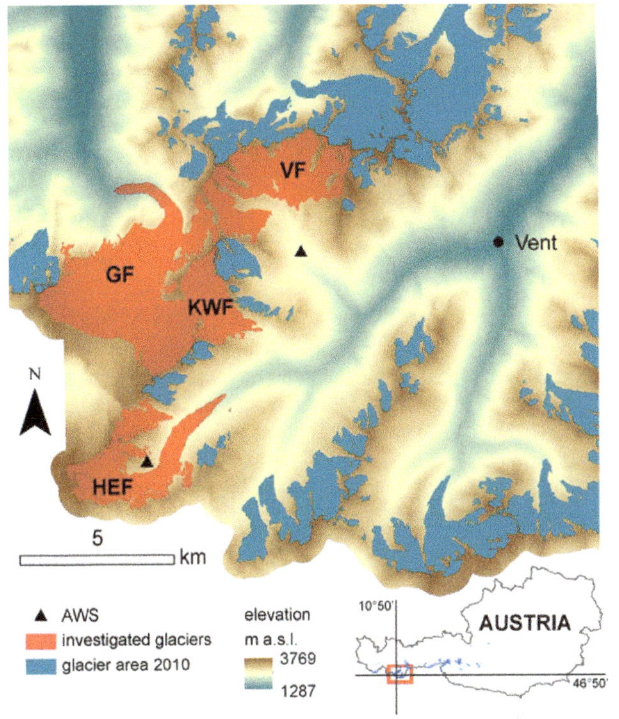

Fig. 2. Map of the investigation area next to Vent in the Ötztal Alps showing the locations of the glaciers Hintereisferner (HEF), Gepatschferner (GF), Kesselwandferner (KWF) and Vernagtferner (VF) and the automatic weather-stations (triangles).

Gepatschferner (GF, 46°51′ N, 10°45′ E) is the largest glacier in this region. Gepatschferner is the only glacier with an area of more than $10\,\mathrm{km^2}$ in the Ötztal Alps. Its changes in volume and area in the last decades are summarized by Abermann et al. (2009) and ice thickness was measured by Fischer et al. (2007). Glacier velocities have not been investigated on Gepatschferner. Recently a network of ablation stakes was set up to investigate the dynamical behaviour of the glacier tongue. Highest velocities are expected to occur at the crevassed area before the inflow to the glacier tongue. First results show maximum horizontal velocities of up to $50\,\mathrm{m\,a^{-1}}$ with a vertical ice flow of approx. $5\,\mathrm{m\,a^{-1}}$ measured below the ice fall. Velocities decrease towards the glacier tongue, which itself decays very rapidly. No velocity data exist from the large plateau of the accumulation area.

Since 1964 the mass balance of Vernagtferner (VF, 46°52′ N, 10°49′ E) is measured with the direct glaciological method by the Commission for Glaciology, Bavarian

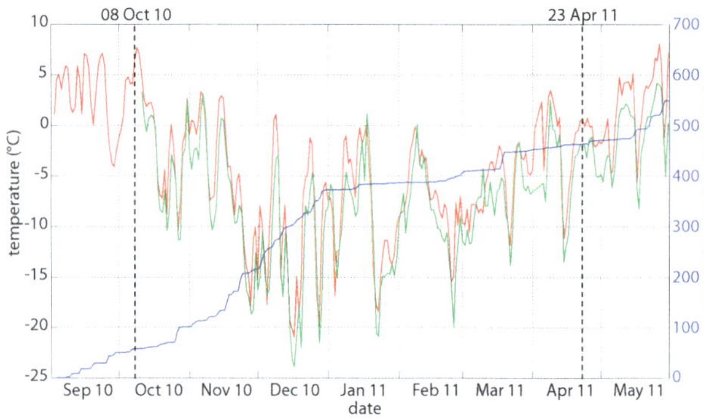

Fig. 3. Mean daily air temperature measured at the automatic weather station (AWS, Fig. 2) Hintereisferner (3027 m a.s.l.; green curve) and the AWS Vernagtferner (2640 m a.s.l.; red curve) for the accumulation season 2010/2011. Cumulative precipitation is shown for the AWS Vernagtferner (blue curve). Dates of ALS surveys are marked as dotted lines.

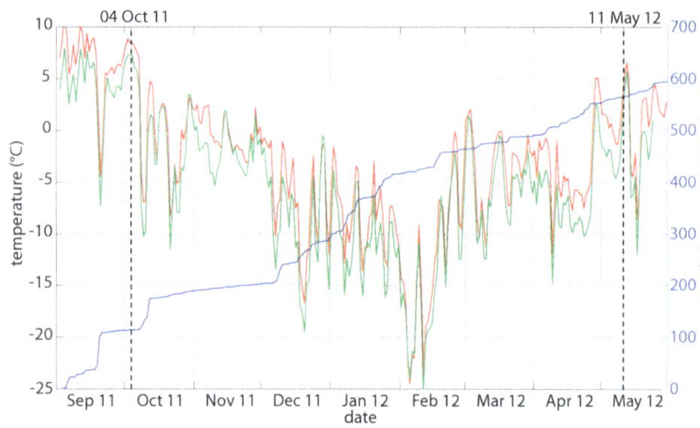

Fig. 4. Mean daily air temperature measured at the automatic weather station (AWS, Fig. 2) Hintereisferner (3027 m a.s.l.; green curve) and the AWS Vernagtferner (2640 m a.s.l.; red curve) for the accumulation season 2011/2012. Cumulative precipitation is shown for the AWS Vernagtferner (blue curve). Dates of ALS surveys are marked as dotted lines.

Academy of Sciences and Humanities, Munich. In combination with a runoff gauge in front of the glacier tongue, a series of hydro-meteorological models were calibrated for energy balance and hydrological studies (Escher-Vetter et al., 2009). Mayer et al. (2013) showed that the glacier slowed down dramatically.

Apart from the long time series of mass balance and ice flow velocity measurements on these glaciers, no measurements of ice flow in the accumulation season exist. With respect to recently observed slow ice velocities in this region, the influence of ice flow on seasonal surface elevation changes can be assumed to be small. However, even 1 m of submergence would have an enormous effect for the interpretation of Δz_{ALS} in terms of snow depth without applying corrections for ice dynamics.

For this study, glacier outlines of the four investigated glaciers derived from the Austrian glacier inventory of 2006

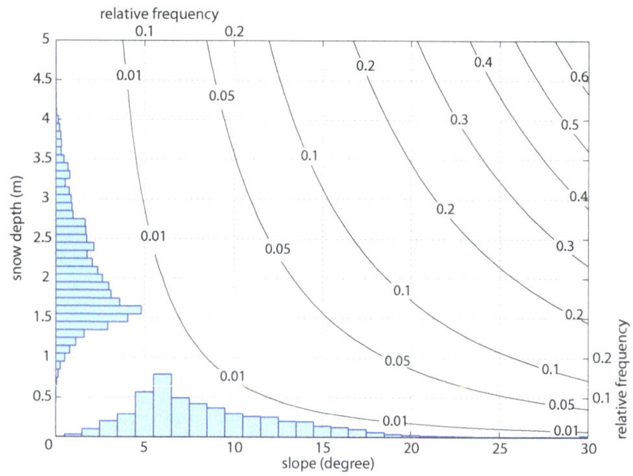

Fig. 5. Contour lines of the correction of GPR snow depth as a function of slope and snow depth. The relative frequency distribution of GPR snow depths h_{v} and the frequency distribution of the occurring slopes are presented in horizontal and vertical bars, respectively.

(Abermann et al., 2009) were updated based on ALS data of 2010 according to Abermann et al. (2010).

2.2 Ground penetrating radar

Ground penetrating radar (GPR) is an active sensing technique to locate targets or internal layers within ground, ice and snow (Daniels, 2004). A transmitted electromagnetic wave is partly reflected at boundaries between two layers with different relative dielectric permittivities (ϵ_{r}'). To convert measured two-way-travel times (TWT) of the GPR signal to depth of the reflecting layer transition, the velocity of propagation of the wave has to be determined.

In cryospheric sciences, GPR is used to detect permafrost within rocky ground (e.g. Hinkel et al., 2001; Otto et al., 2012), to determine ice thickness (e.g. Span et al., 2005; Fischer et al., 2007), ice and firn layering (e.g. Spikes et al., 2004) as well as subglacial structures and liquid water content of snow and ice (e.g. Murray et al., 1997; Lundberg et al., 2000). The spatial variability in snow depth and the temporal evolution in snowpack stratigraphy have been analyzed utilizing GPR technology as well (e.g. Machguth et al., 2006; Heilig et al., 2009; Mitterer et al., 2011).

2.2.1 Field campaigns

Three GPR campaigns were conducted close to the respective date of ALS surveys. Dates and snow pit information are displayed in Table 1.

At the end of accumulation season 2010/2011, GPR data were recorded on Vernagtferner by the Commission of Glaciology, Bavarian Academy of Sciences and Humanities (Munich, Germany). A RIS One GPR instrument from IDS (Pisa, Italy) with shielded 600 MHz antennas was used. The

Table 1. Attributes of the field measurements in spring. Number of snow depth probings and snow pits shown in brackets indicate measurements not directly located at the GPR profiles. The GPR signal velocities derived from snow depths of the reflecting snow layer in the snow pits (v_p) and GPR signal velocities calculated from measured snow densities (v_k) using Eqs. (7) and (8) with the corresponding coefficient of variation (CV, Eq. 6).

ID	Date	GPR profile length (10^3 m)	Mean trace distance (m)	Snow depth probings	Snow pits	Signal velocity (m ns^{-1})			
						v_p	CV_p	v_k	CV_k
VF	28–29 Apr 2011	15.32	0.35	102	1(6)	–	–	0.2254	0.04
GF	28 Apr 2011	5.39	0.30	33	3(1)	0.2270	0.02	0.2263	0.01
KWF	28 Apr 2011	1.84	0.45	0(17)	2				
HEF	9 May 2012	7.66	0.25	49	5	0.2254	0.96	0.2254	0.77
KWF	8 May 2012	5.19	0.18	31	2				

GPR was pulled by a snowmobile. At the first and the last point of the sections snow depths were measured by snow depth probing and coordinates were recorded with a handheld GPS. Between these points the GPR data were continuously received while driving the snowmobile at uniform speed, which, however, differ between single sections due to topography. In total more than 15 km of GPR profiles were conducted, divided into four longitudinal profiles, three additional shorter profiles in the accumulation zone and one cross profile (Fig. 6). Snow densities were measured at six locations.

A GPR measurement campaign was conducted on Gepatschferner and Kesselwandferner in 2011. GPR data were recorded with a 3-D system from MALÅ (Malå, Sweden) with shielded 500 MHz antennas along a cross profile in the accumulation zone of Gepatschferner and a longitudinal profile on Kesselwandferner (Fig. 7). In total more than seven kilometres of profiles, divided into 33 sections, were investigated on skis. First and last points of the individual sections were georeferenced by DGPS (Fig. 7 – black dots). Three snow pits were dug on Gepatschferner and one on the tongue of Kesselwandferner. While snow depth probings were performed along the GPR sections on Gepatschferner, snow depth probings and GPR data do not spatially coincide on Kesselwandferner. Sample spacing is larger on Kesselwandferner than on Gepatschferner, due to greater velocities while skiing downhill the glacier (Table 1).

At the end of the accumulation season 2011/2012, a GPR campaign was conducted to measure snow depths along longitudinal sections and cross sections on Hintereisferner and Kesselwandferner (Figs. 8 and 9). Skis were used to pull another IDS RIS One system with shielded 400 MHz antennas. In total over 12 km of GPR data were recorded. Snow depth probings were conducted at every first and last point of the individual sections and seven snow pits were dug. All data points were georeferenced by DGPS.

For reflectors being located at a distance of half the wavelength in the respective medium, constructive and destructive interferences will occur (Daniels, 2004). The applied GPR systems with a centre frequency of 400–600 MHz, hence are capable of distinguishing various reflectors in a vertical distance of more than 0.19 m (600 MHz) to 0.28 m (400 MHz) (Trabant, 1984). All GPR campaigns were conducted in time mode. The trace spacing for each transect was chosen such that 2 consecutive traces are always separated less than 0.5 m. So quasi-continuous information could be tracked along the profiles based on known reflector depths from snow depth probing and from snow pits.

2.2.2 Processing

GPR raw data were processed applying the ReflexW software from Sandmeier Scientific Software (Karlsruhe, Germany). Coordinates for each trace were calculated at equal distances. The surface signal reflection was set to time zero. Low frequency parts and noise in the spectrum were filtered applying a dewow and bandpass filter. In a next step temporally consistent signals were eliminated utilising a background removal. To compensate for spherical divergence losses, we applied gain functions. No migration was used due to the spatial homogeneity and the low reflector depth relative to the high density of GPR measurements. The boundaries of the snow layers were picked and corrected to the zero phase change. The picks were exported with the attribute of the two-way-travel time. Finally, we merged individual sections to continuous longitudinal and cross profiles.

Snow depth probing and snow pits were used to identify the depth of the seasonal snow layer according to the guidelines for measuring glacier mass balance (Kaser et al., 2003). GPR signal velocities were calculated from measured reflector depths and snow densities at the snow pits only. Assuming that the snow surface and the ice surface are parallel, the slope of the surface has to be considered. The recorded GPR signal corresponds to the snow depths perpendicular to the surface, but snow depths derived from snow probings and snow pits (h_m) are measured vertically. We corrected vertically measured snow depths h_m by

$$h_{cor} = h_m \cdot \cos(\alpha) \tag{4}$$

in accordance to the respective slope angle α. Signal velocities (v) for GPR measurements on Gepatschferner and

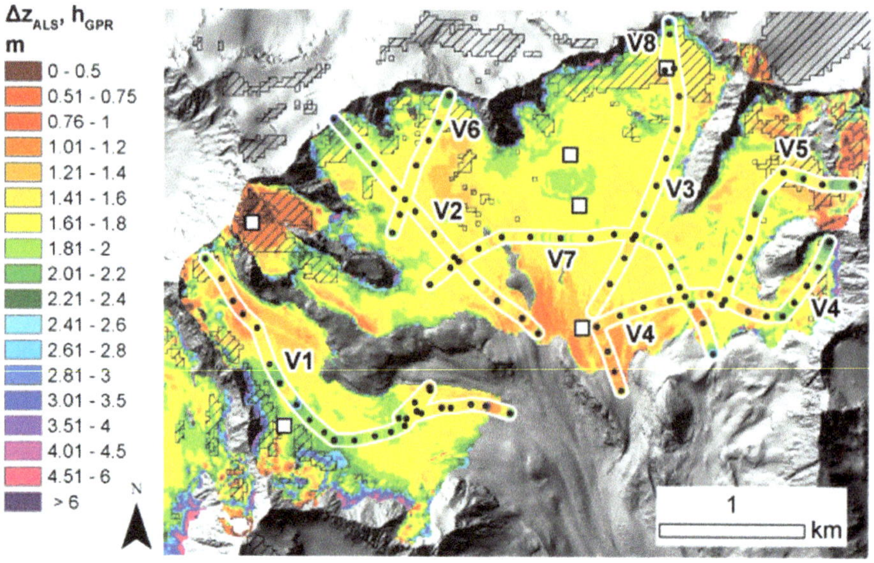

Fig. 6. Spatial distribution of ALS-derived surface elevation changes for for 10 October 2010 to 23 April 2011. Snow depths calculated from GPR measurements along the profiles at Vernagtferner plotted in the same colour-scale. Black dots indicate locations of snow depth probing. Locations of snow pits are shown as white squares. Snow covered areas derived from Landsat image of 31 August 2009 (LSC) are striped.

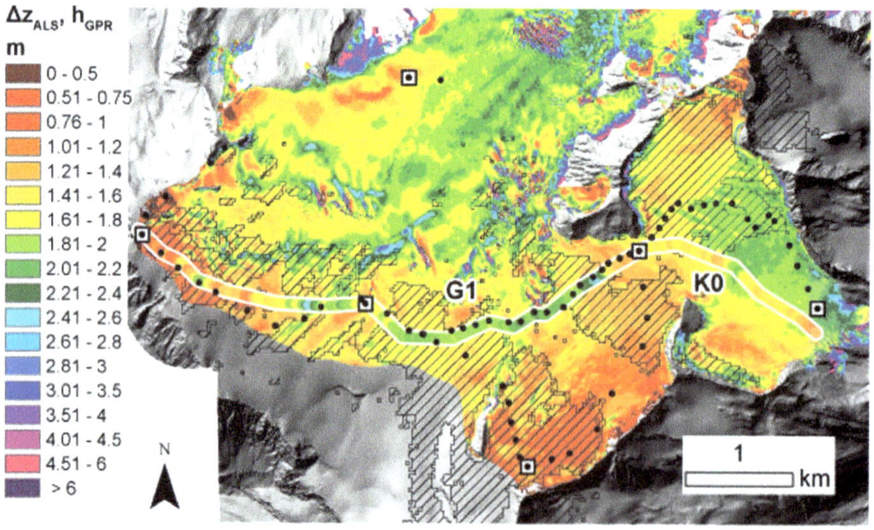

Fig. 7. Spatial distribution of ALS-derived surface elevation changes for 10 October 2010 to 23 April 2011. Snow depths calculated from GPR measurements along the profiles at Gepatschferner and Kesselwandferner shown in the same colour-scale. Black dots indicate locations of snow depth probing. Locations of snow pits are shown as white squares. Snow covered areas derived from Landsat image of 31 August 2009 (LSC) are striped.

Kesselwandferner in 2011 and GPR measurements on Hintereisferner and Kesselwandferner in 2012 were calculated following Daniels (2004):

$$v = \frac{2 \cdot h_{cor}}{\tau} \tag{5}$$

using h_{cor} and the TWT (τ) at the snow pit locations. A mean signal velocity ($\overline{v}$) was calculated for each measurement campaign individually (Table 1). The variation of the signal velocities (CV) is given by

$$\mathrm{CV} = \frac{v_{\max} - v_{\min}}{\overline{v}} \tag{6}$$

as the ratio of the range between the derived maximum velocity $v_{\max}$ and minimum velocity $v_{\min}$ to the mean velocity $\overline{v}$. According to Kovacs et al. (1995), we calculated ϵ_r' of the snow cover

$$\epsilon_r' = (1 + 0.845 \cdot \rho_s)^2, \tag{7}$$

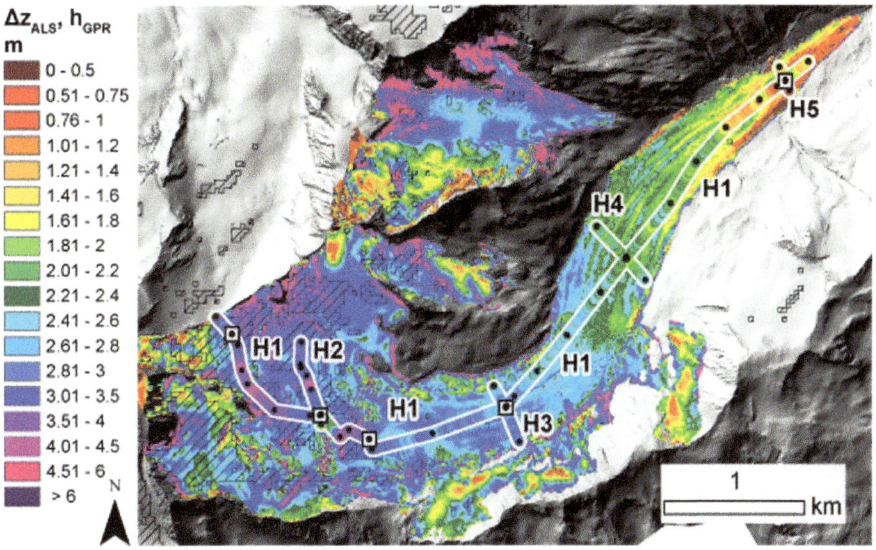

Fig. 8. Spatial distribution of ALS-derived surface elevation changes from 4 October 2011 to 11 May 2012. Snow depths calculated from GPR measurements along the profiles at Hintereisferner shown in the same colour-scale. Black dots indicate locations of snow depth probing. Locations of snow pits are shown as white squares. Snow covered areas derived from Landsat image of 31 August 2009 (LSC) are striped.

where ρ_s is the snow density in $\mathrm{g\,cm^{-1}}$. Based on ϵ'_r GPR signal velocities v were calculated

$$v = \frac{c}{\sqrt{\epsilon'_r}}, \tag{8}$$

where c is the velocity of propagation of the electromagnetic wave in a vacuum of approx. $0.3\,\mathrm{m\,ns^{-1}}$.

On Vernagtferner snow pits were not located along the GPR profiles. Therefore GPR signal velocities were calculated from measured snow densities only following Eqs. (7) and (8). Signal velocities derived from measured snow depths in the snow pits (Eq. 5) on Hintereisferner and Kesselwandferner were validated with signal velocities calculated from snow densities (Eq. 8) (Table 1). The CV of the signal velocities is larger for 2012 data due to higher variations in snow densities measured in the snow pits. Snow depth probings performed along the GPR profiles were used to validate the respective calculated wave speed. All exported TWT τ were converted into snow depths (h_v)

$$h_v = \frac{\tau \cdot \overline{v}}{2} \tag{9}$$

using the mean velocities $\overline{v}$ for each measurement campaign. Again to account for vertically measured Δz_{ALS}, GPR-determined h_v had to be corrected for the prevailing slope α resulting in the final snow depth h_{GPR}

$$h_{\mathrm{GPR}} = \frac{h_v}{\cos(\alpha)}. \tag{10}$$

Figure 5 shows the magnitude of corrections which have to be applied to h_v as a function of snow depth and slope. In the same figure the frequency distribution of snow depths measured with GPR and the frequency distribution of the occurring slopes are shown. The most frequent snow depths were between 1.55 m and 1.65 m. The most frequent slopes were between 5° and 7°. Corrections, which therefore had to be applied, were mainly of the order of 0.01 m.

2.3 Lidar

Lidar is an active remote sensing technology (Baltsavias, 1999; Wehr and Lohr, 1999; Kraus, 2004). Airborne laser scanning (ALS) can be applied in mountain areas on the catchment scale, because it is hardly affected by topographic shading and travel distance of the signal can be controlled by flight height (e.g. Deems et al., 2013). The DGPS referenced data provide 3-D information of the reflecting surface with a high point density and accuracy. Based on these 3-D point data (the so-called point cloud) digital elevation models (DEMs) can be produced (Lui, 2008), while high point densities allow simple interpolation schemes (Deems et al., 2013). Surface elevation changes can be detected from the difference of the DEMs of multi-temporal laser scan surveys.

Accuracy of the 3-D location of each point is affected by the accuracy of the recording of the position of the scanning platform and its orientation (e.g. Joerg et al., 2012; Deems et al., 2013). Accuracy of the resulting DEMs depends on point density, the adjustment of overlapping areas of the stripes and the interpolation algorithm used (e.g. Deems et al., 2013). For DEMs derived from ALS data of the Hintereisferner region, Bollmann et al. (2011) showed that for slopes below 35° a vertical accuracy of ±0.15 m can be assumed.

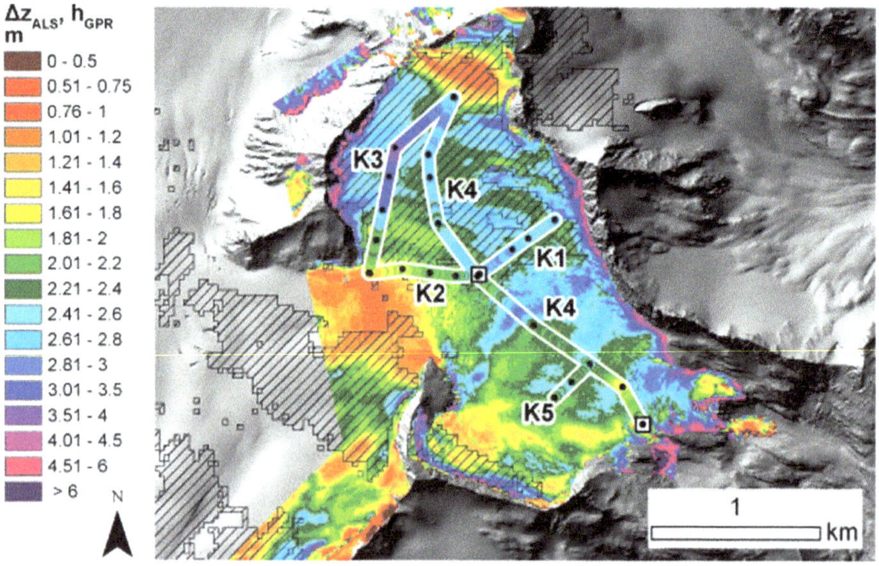

Fig. 9. Spatial distribution of ALS-derived surface elevation changes from 4 October 2011 to 11 May 2012. Snow depths calculated from GPR measurements along the profiles at Kesselwandferner shown in the same colour-scale. Black dots indicate locations of snow depth probing. Locations of snow pits are shown as white squares. Snow covered areas derived from Landsat image of 31 August 2009 (LSC) are striped.

2.3.1 Data and processing

In this study four laser scan surveys were used. They were conducted by TopScan using Optech devices mounted on Cesna aircrafts. Technical details and the accuracy measured at a reference surface are listed in Table 2. The last pulses of all ALS points within one pixel of the resulting DEMs were averaged to a grid size of 1 m. Data were analyzed on generally homogeneous glacier surfaces, which, except for very small areas, have a slope of less than 20° along the investigated profiles (Fig. 5). Surface elevation changes (Δz_{ALS}) were calculated by subtracting a DEM of an almost snow-free ALS survey in fall (t_1) from a DEM of an ALS survey at the end of accumulation season in spring (t_2), respectively, following Eq. (1), with $z_1 = z_{t1}$ and $z_2 = z_{t2}$ (Fig. 1).

2.4 Calculation of differences and uncertainties

Absolute differences (Δh_{abs}) were calculated between ALS surface elevation changes (Δz_{ALS}) and snow depths calculated from GPR (h_{GPR}) by

$$\Delta h_{\mathrm{abs}} = \Delta z_{\mathrm{ALS}} - h_{\mathrm{GPR}}. \tag{11}$$

A moving window approach was used to perform a statistical Mann–Whitney–Wilcoxon test with a number of 300 samples of Δh_{abs} around each single data point of GPR measurements. According to the mean trace distances shown in Table 1, the number of 300 samples corresponds to a distance of approx. 100 m, which is an adequate spatial scale for differences analyzed here. The null hypothesis was tested, if the median of the two samples (Δz_{ALS} and h_{GPR}) is equal with a significance level of $p = 0.05$. If the null hypothesis is rejected, the medians of the two data sets differ from each other. Additionally the mean value of Δh_{abs} of the 300 samples was compared to an assessment of absolute uncertainty U_{abs} calculated as follows

$$U_{\mathrm{abs}} = \sqrt{2 \cdot U_{\mathrm{ALS}}^2 + (U_{\mathrm{v}} \cdot h_{\mathrm{GPR}})^2}. \tag{12}$$

Uncertainties of gridded lidar data ($U_{\mathrm{ALS}} = \pm 0.15$ m) have to be considered twice, because two elevation models are used to investigate surface elevation changes. Uncertainties caused by the calculated GPR signal velocities (U_{v}) correspond to half of maximum CV (Eq. 6, Table 1), which is ± 0.05. U_{abs} are also a function of snow depths (h_{GPR}). The results of the statistical analysis show where Δz_{ALS} and h_{GPR} differ significantly. Furthermore it was calculated where Δh_{abs} exceed U_{abs} along the profiles. In these areas, Δz_{ALS} is increasingly influenced by processes more than only snow depth uncertainties (see Sect. 1).

2.5 Accumulation areas from Landsat

The densification of firn is supposed to be one source of differences of Δz_{ALS} from actual snow depths (Sect. 1). Also submergence ice flow is very likely in the central parts of typical accumulation areas. Hence, typical accumulation areas on glaciers were expected to show larger differences since both vertical ice flow and densification work vertically in the same direction. Snow covered areas of glaciers can be delineated by application of spaceborne optical data for

Table 2. Overview of the ALS flight campaigns investigated in this study. The vertical accuracy is obtained from differences between ALS surface elevations and elevations of a known reference surface. The vertical accuracy is shown in terms of a mean offset (mean) and the standard deviation (σ).

Timestamp	Date	Laser system	Mean Flight speed ($m\,s^{-1}$)	flight altitude above ground (m)	Maximum scan angle (°)	Laser repetition rate ($10^3\,s^{-1}$)	Scan frequency (s^{-1})	Mean point density (m^{-2})	Vertical accuracy (m) mean	σ
t_1	7–10 Oct 2010	ALTM Gemini	65	1000	25	70	36	3.6	0.071	0.047
t_2	20–23 Apr 2011	ALTM Gemini	65	1000	25	70	36	3.8	−0.007	0.041
t_1	4 Oct 2011	ALTM 3100	70	1200	20	70	40	2.9	0.001	0.042
t_2	11 May 2012	ALTM 3100	65	1200	20	70	40	2.8	0.005	0.057

whole catchments (Hall et al., 1987; Nolin, 2011). Landsat satellite images are available in a spatial resolution of 30 m (http://glovis.usgs.gov). A Landsat 5 Thematic Mapper (TM) image of the Ötztal Alps on the 31 August 2009 was retrieved from USGS. Snow covered areas on glacier surfaces were delineated by applying the Normalized Difference Snow Index (NDSI, Hall et al., 1987). The application of Landsat imagery is limited by the discrepancy of the date of surveys to the date of interest and the occurrence of clouds masking the surface information. The conditions of this scene fulfill all requirements in terms of no clouds, recent new-snow precipitation on the glaciers and an advanced state of snow ablation in late summer. The low accumulation area ratios (AAR, Cogley et al., 2011) derived using the NDSI for separation of snow and bare ice are comparable to those observed within the annual mass balance measurements on the glaciers in recent years. Hence the delineated snow covered areas were assumed to indicate recent accumulation areas on the investigated glaciers and are referred to as Landsat snow cover (LSC) hereafter.

3 Prevailing local conditions during lidar surveys

Surface elevation changes between field campaigns and the dates of ALS surveys (t_1, t_2) contribute to uncertainties in Δh_{abs}. Snow accumulation before the ALS surveys in the higher elevated glacier areas in fall has to be considered when identifying the snow layers in the GPR analysis. Photographs and measurements of snow depths and snow densities taken during annual glacier mass balance field campaigns on Hintereisferner, Kesselwandferner and Vernagtferner were included for data analysis. Information on the weather conditions before the ALS surveys, between the ALS surveys and the field campaigns as well as on the weather conditions during the accumulation seasons were obtained from two automatic weather stations located next to Hintereisferner (46°47′55″ N,10°45′37″ E; 3027 m a.s.l.) and next to Vernagtferner (46°51′23″ N, 10°49′43″ E; 2640 m a.s.l., Fig. 2). The course of daily mean temperatures and cumulative precipitation at the two automatic weather stations is presented in Figs. 3 and 4.

On 12 October 2010 snow depths and snow densities of a snow layer covering the 2010 firn layer on Gepatschferner and Kesselwandferner was measured by 43 snow depth probings and three snow pits. This snow layer was accumulated in September 2010. In a warm period before the ALS survey (Fig. 3), this layer was transformed into a snowpack with a high bulk density and a melt-freeze crust at the surface. This snow layer could be identified in the stratigraphies of spring snow pits and in the snow layering derived from GPR data in 2011. Snow depths of the same snow layer were recorded by 34 snow probings within the annual mass balance measurements at Vernagtferner. Between the ALS survey in spring 2011 and the corresponding GPR field campaigns a shallow snow cover of about 0.1 m with a low density was accumulated.

In September 2011 glacier tongues and steep southerly exposed parts on Hintereisferner and Kesselwandferner appeared to be snow free on photographs taken alongside the annual mass balance measurements on 1 October 2011. Only a shallow snow layer was covering the typical accumulation areas (A. Fischer, personal communication, 2012). Shortly after the ALS survey in October 2011 a snowfall event preserved the surface recorded by ALS from melt.

4 Results

The spatial distribution of Δz_{ALS} and snow depths calculated from GPR measurements (h_{GPR}) along the profiles are illustrated in Figs. 6–9. Locations of snow pits and snow depth probings of the spring field campaigns are included. Data corresponding to the individual profiles are listed in Table 3.

Figure 10 shows the total number of analyzed data points for LSC areas and bare ice areas (Sect. 2.5) from the profiles and the statistical distribution of all absolute differences (Δh_{abs}, Eq. 11) in terms of the median (red line), the scatter (25 % and 75 % percentiles, blue box), the range of extreme values (1.5 times the interquartile range added to the 25 % percentile and to the 75 % percentile, black whiskers) and the outliers (red crosses).

The profiles H1 (Fig. 11) and K4 (Fig. 12) along the central flow lines of the glaciers show distinct negative Δh_{abs} (Eq. 11) in the LSC areas. Δh_{abs} are constantly small along most of the glacier tongue of Hintereisferner. At the very front of the glacier tongue, Δh_{abs} increases towards positive values. Along the profile K4 on KWF distinctly less h_{GPR} match with the corresponding Δz_{ALS}. Towards the tongue of the glacier, Δh_{abs} is constantly positive for about 500 m in

Table 3. Properties of the GPR profiles in terms of ID, profile length, number of sections, the mean section length ($\overline{l_S}$), minimum elevation and maximum elevation. Mean value of the surface elevation changes derived from ALS data $\overline{\Delta z_{ALS}}$ (Eq. 1) and the mean absolute difference of this value from mean GPR snow depth $\overline{\Delta h_{abs}}$ (Eq. 11) are shown for LSC areas and ice areas of each profile.

Profile	Length	Sections	$\overline{l_S}$	Elevation (m a.s.l.)		Ice areas (m)		LSC areas (m)	
ID	(m)	n	(m)	max	min	$\overline{\Delta z_{ALS}}$	$\overline{\Delta h_{abs}}$	$\overline{\Delta z_{ALS}}$	$\overline{\Delta h_{abs}}$
H1	5788	25	232	3364	2532	2.32	−0.01	3.10	−0.50
H2	586	5	117	3159	3096	3.00	0.31	2.95	−0.38
H3	414	2	207	2905	2895	2.73	0.02	2.46	−0.47
H4	498	2	249	2754	2736	2.27	0.10	–	–
H5	176	2	88	2565	2557	1.33	0.07	–	–
K1	550	3	183	3175	3164	–	–	2.26	−0.32
K2	6576	4	164	3210	3174	1.72	−0.19	1.94	−0.15
K3	1207	5	241	3293	3210	–	–	2.23	−0.60
K4	2485	8	311	3293	3019	2.43	0.23	2.32	−0.26
K5	291	2	146	3087	3072	2.33	0.02	–	–
G1	5391	26	207	3484	3154	1.38	−0.28	1.23	−0.49
K0	1844	7	263	3212	3035	1.76	0.29	1.52	0.04
V1	3458	25	138	3370	2803	1.61	−0.01	1.25	−0.35
V2	2022	11	184	3303	2892	1.54	0.00	1.79	−0.23
V3	2093	9	233	3307	2961	1.56	0.02	1.55	−0.10
V4	2412	17	142	3327	2864	1.47	0.01	1.48	−0.22
V5	1488	9	165	3355	3091	1.67	−0.08	1.36	−0.28
V6	1007	6	168	3264	3131	1.54	−0.05	1.66	−0.03
V7	2526	14	180	3060	3020	1.62	0.01	1.50	−0.05
V8	314	1	314	3303	3255	–	–	1.57	−0.22

distance of the profile. For most parts of the profiles from Vernagtferner, Δh_{abs} do not exceed the uncertainty range (Sect. 2.4).

The profiles H2 on Hintereisferner, K1, K2, K3 on Kesselwandferner, V8 on Vernagtferner and G1 on Gepatschferner are mostly located in the accumulation area of the glaciers. Only very few parts of the respective profiles cover bare ice areas. For each of those profiles, the median of Δh_{abs} is distinctly negative. The cross profiles H3, H4, H5, K5 and V7, located in areas with bare ice at the surface, show up with a median of Δh_{abs}, which is close to 0. The interquartile ranges of these profiles are comparably small in contrast to most LSC areas (Fig. 10).

Figure 13 shows the comparison of Δh_{abs} to calculated uncertainties U_{abs} (Eq. 12) in combination with the results of the Mann–Whitney–Wilcoxon test (Sect. 2.4). Δh_{abs} calculated for LSC areas of several profiles exceed U_{abs}, which is ± 0.24 m for a mean snow depth of 2 m (Table 3). Along most parts of the profiles located in ice areas on the investigated glaciers, differences between Δz_{ALS} and h_{GPR} are significant ($p = 0.05$) but within the range of the calculated uncertainties (Eq. 12). Distinct positive Δh_{abs} were found on the very front of the glacier tongue of Hintereisferner and on the tongue of Kesselwandferner. Small areas of positive Δh_{abs} were found in crevassed zones located in the higher elevated parts of the glaciers.

Table 3 highlights the increased negative values of Δh_{abs} calculated for ALS measurements in LSC areas compared to Δh_{abs} for ice areas. Figure 10 shows that only a small number of Δh_{abs} values in ice areas are as negative as the majority of Δh_{abs} determined in LSC areas. In these LSC areas mean Δh_{abs} is −0.40 m with a standard deviation of 0.34 m, which is approximately 20 % of the mean value of h_{GPR} with a standard deviation of 17 %.

Very locally, absolute differences are in the magnitude of about 1 m, which corresponds to relative differences of up to −40 % or more. For the ice areas of the profiles, the mean absolute difference is 0.004 m with a standard deviation of 0.27 m, in relative values −0.002 % with a standard deviation of 14.4 %. Depending on the location of the profiles, negative Δh_{abs} were also found for ice areas (e.g. K2, G1; Table 3) or increased positive Δh_{abs} of up to 1 m (K0, K4; Fig. 12) were calculated.

5 Discussion

5.1 Differences in accumulation areas

The Landsat image of snow covered areas from 31 August 2009 cannot give direct proof for the general distinction between areas, where Δz_{ALS} is considerably influenced by snow densification and vertical ice flow, and areas with only small Δh_{abs}. In addition, only the areal extent of the snow

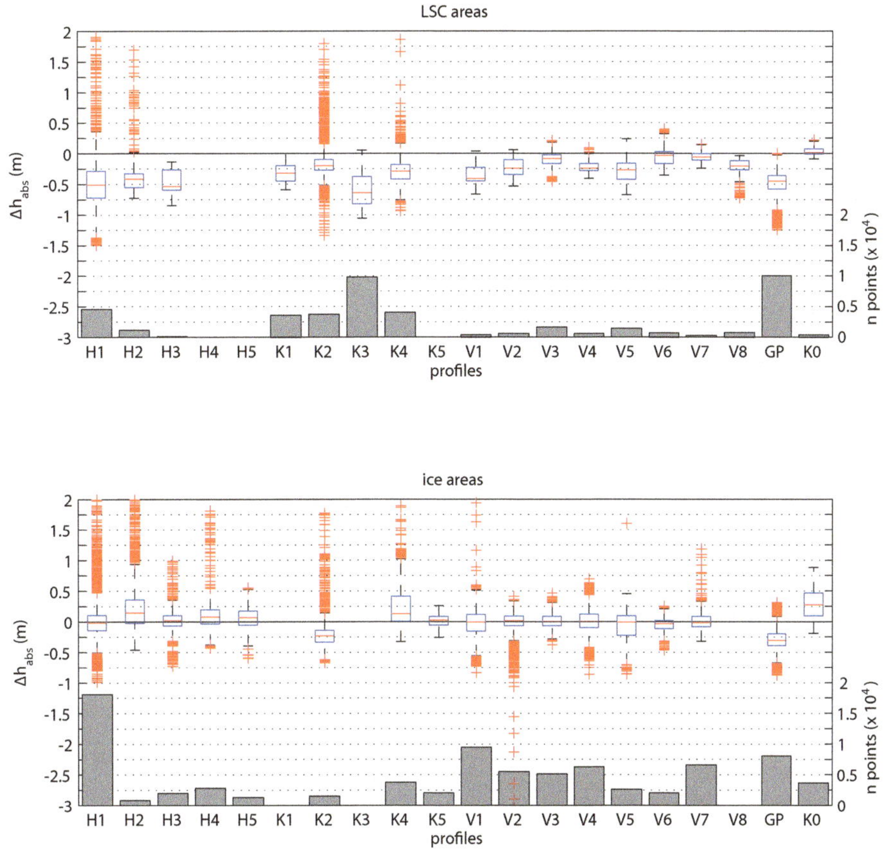

Fig. 10. Box plots of the absolute differences Δh_{abs} at the profiles for snow covered areas derived from Landsat (LSC, upper plot) and ice areas (lower plot) separately. Statistical distribution is shown in terms of the median (red line), the scatter (25 % and 75 % percentiles, blue box), the range of values within 1.5 times the interquartile range added to the 25 % percentile and to the 75 % percentile (black whiskers) and the outliers (red crosses). The number of points included in the statistics are plotted as grey bars (right scale).

cover (LSC) can be derived from Landsat imagery (30 m pixel resolution). No information on snow depths and, hence, spatial distribution of densification is provided by this data source. It does, however, give a broad picture of the distribution of typical firn areas and ice in the period of investigation. Areal snow cover extent on 31 August 2009 is generally in good agreement with the small firn areas observed for the last decade, which caused low accumulation area ratios (AAR) on these Alpine glaciers (e.g. http://www.glaziologie.de/) LSC area probably do not display the total zone of submergence ice flow, because flow dynamics are controlled by a variety of properties and usually adapt to mass changes within decades. Superimposed ice zones, which are part of the accumulation zone, are likely not visible to Landsat either. Furthermore, the applicability of Landsat scenes to distinguish typical accumulation areas on glaciers has to be proved carefully in terms of the state of seasonal ablation and the presence of new snow at the date of survey.

However, LSC areas at the end of ablation season used here show consistently higher Δh_{abs} compared to ice areas (Table 3,Fig. 13), since, as mentioned in Sect. 2.5, both vertical ice flow and densification of firn work in the same direction in typical accumulation areas of the glaciers. On Vernagtferner firn areas detected from Landsat data are very small. Even for these small firn areas the distinctly negative Δh_{abs} are evident. The total profile on Gepatschferner shows negative differences for subtracting h_{GPR} from Δz_{ALS}. This can be attributed to the location of the profile in the accumulation zone of the glacier. However, even in this accumulation zone different magnitudes of Δh_{abs} calculated for ice areas and Δh_{abs} calculated for LSC areas could be detected (Table 3).

Ambach and Eisner (1966) measured snow densities in a firn pit of 20 m depth on Kesselwandferner at an elevation of 3200 m a.s.l. Firn densities increased roughly linear from 630 kg m^{-3} in the uppermost layer to 836 kg m^{-3} in a layer at the age of 10 yr. The firn compression rate of 3 % per year in this 10 yr firn cover lead to a vertical compression of 16 m firn depth by approx. 0.5 m per year. Similar rates of densification are presented in a density profile of upper Seward Glacier (Yukon, Canada) in Cuffey and Paterson (2010) (page 17, Fig. 2.3). However, percolating meltwater

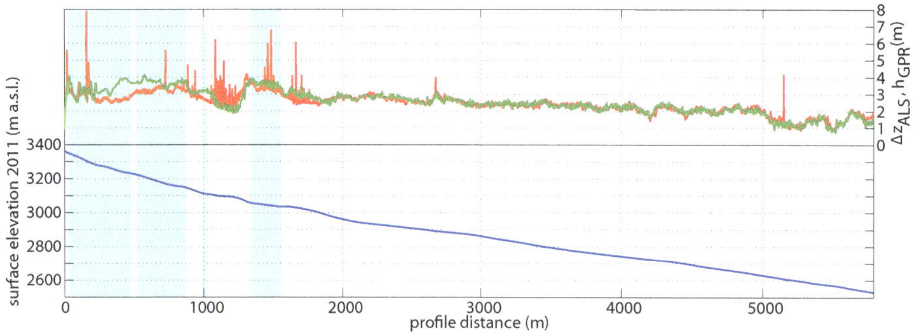

Fig. 11. Elevation of the surface in fall 2011 (blue curve), ALS surface elevation change Δz_{ALS} (red curve) and GPR snow depth h_{GPR} (green curve) along the longitudinal profile H1 at Hintereisferner. Snow covered areas derived from Landsat image of 31 August 2009 (LSC) are shaded in blue.

may contribute to annual densification of firn on Alpine glaciers. Because Δh_{abs} were investigated for the comparatively shorter accumulation season, the contribution of densification of firn to Δh_{abs} is assumed to be less than 0.5 m.

5.2 Comparison to measured ice flow velocities

Our data comparison does not allow to assess the quantitative proportion of the particular processes contributing to differences of Δz_{ALS} from actual snow depths (Fig. 1). However, the four investigated glaciers show different dynamical behaviour, which is also the case for annual ice flow velocities presented in Sect. 2.1.

In the uppermost parts of Hintereisferner negative Δh_{abs} (Eq. 11) occur in between the crevassed areas (Fig. 11). These Δh_{abs} exceed the uncertainties of the applied methods (Eq. 12). However, no measurements of vertical ice flow exist for this area. Positive Δh_{abs} could be detected only at the very front of the glacier tongue of Hintereisferner. These results coincide with the findings that mean annual vertical ice flow velocities are around 0.5 m s^{-1} and less apart from the front of the glacier tongue (Sect. 2.1) .

Distinct positive and negative Δh_{abs} are evident on Kesselwandferner (Fig. 12). Maximum Δh_{abs} in the LSC areas are within the magnitude of the measured annual vertical ice flow velocity (Sect. 2.1). Δh_{abs} of the profile K3 of 0.6 m (Table 3) is slightly more than half of the measured annual vertical ice flow velocities. Locally, derived Δh_{abs} are larger than half of the observed annual emergence of the ice surface which is about 1 m at the glacier tongue of Kesselwandferner (Sect. 2.1 and Fig. 9). For the accumulation zone of Gepatschferner no measurements of ice flow velocities exists. On this glacier Δh_{abs} are comparable to Δh_{abs} of the uppermost profiles on Hintereisferner (H1) and Kesselwandferner (K3). Differences between Δz_{ALS} and h_{GPR} are within the uncertainty range of the measurements for most parts of Vernagtferner. This coincides with the decreased ice flow of the glacier observed by Mayer et al. (2013).

5.3 Influence of prevailing conditions

In glacierized mountain catchments, it is generally difficult to coordinate the dates of maximum ablation in fall and maximum accumulation in spring with the ALS surveys. In this study, both the temporal offsets between GPR and ALS measurements and the dates of ALS surveys in relation to the weather conditions are in good accordance to the required accuracy based on the previous assumptions for this study (Sect. 3, Figs. 3 and 4).

Due to the short delay between ALS surveys and GPR campaigns, simple assumptions were made for processes causing surface elevation changes within this time period. The thin layer of new snow in spring 2011 was neglected, because this layer was moved aside when pulling the GPR antenna over the snow surface on Gepatschferner and Kesselwandferner. On Vernagtferner this new-snow layer was assumed to be depleted by the snowmobile. In 2012 the GPR campaign took place before the ALS survey. The snow depth reduction due to melt conditions is assumed to be in the magnitude of the immersion of the GPR antenna into the uppermost snow layer during the measurements.

To account for the densification of snow accumulated before the ALS survey in fall 2010, measurements of this snow layer can give a rough estimate. From snow depth and density measurements in fall 2010 and spring 2011, a compaction of approx. 0.07 m of a corresponding snow depth at ALS (t_1) of 1 m was calculated. However, measured snow depths at time of ALS(t_1) in 2010 was considerably less with mean values of 0.64 m on Gepatschferner and Kesselwandferner and 0.38 m on Vernagtferner.

5.4 Uncertainties

U_{abs} was calculated accounting for stochastic uncertainties of the combination of the applied measurement methods (Eq. 12). Further sources for systematic errors and uncertainties have to be mentioned: data were analyzed in the spatial resolution of 1 m for DEM data and for each GPR trace

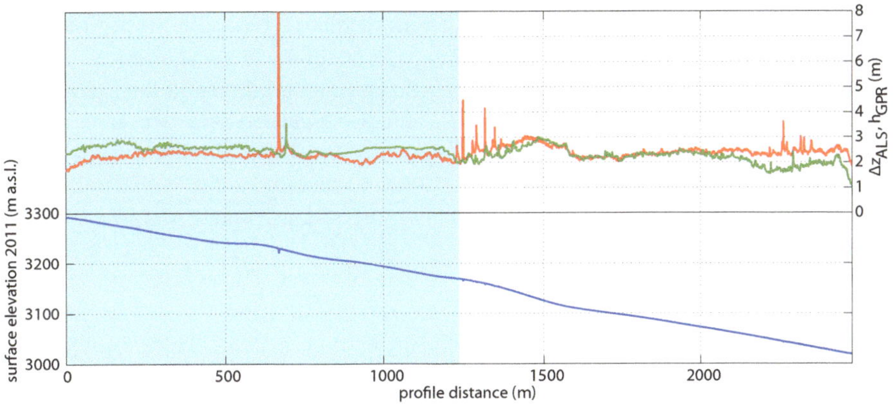

Fig. 12. Elevation of the surface in fall 2011 (blue curve), ALS surface elevation change Δz_{ALS} (red curve) and GPR snow depth h_{GPR} (green curve) along the longitudinal profile K4 at Kesselwandferner. Snow covered areas derived from Landsat image of 31 August 2009 (LSC) are shaded in blue.

separated by the trace distance (Table 1). Both, in ALS and in GPR data, surface heterogeneities (e.g. crevasses) in the scale of 1 m are mapped. Horizontal ice flow shifts surface heterogeneities along the flow direction within one accumulation season. Hence, location of crevasses in fall are included in the data of Δz_{ALS}, whereas GPR data show the location of the crevasse in spring (e.g. Fig. 12 at a profile distance of 700 m). In the analysis of Δh_{abs} the horizontal shift is not considered, because ALS data are validated with respect to all processes contributing to surface elevation change. Resulting maximum differences are displayed as outliers in the Fig. 10.

As shown by Joerg et al. (2012), stochastic uncertainties of ALS-derived surface elevations are strongly reduced on less inclined, homogeneous glacier surfaces. Mean systematic offset of each ALS survey was calculated based on a reference surface (Table 2). However, this systematic offset is not identical with a systematic offset of the ALS data at the investigated glaciers. Due to the fact that the systematic offsets are smaller than the assumed vertical uncertainty of the DEMs of ± 0.15 m, no corrections have been applied to ALS data.

The quality of the GPR data analysis depends strongly on the amount of additional information. While stochastic uncertainties are due to the vertical resolution of the used GPR frequency, systematic errors and stochastic uncertainties result from velocity assumptions and the non-automated analysis of the GPR data. In a snowpack, ϵ_r' is a function of the mixing ratio between ice, air and liquid water (e.g. Lundberg et al., 2006). However, the proportion of liquid water in a wet snowpack can alter ϵ_r' and hence the signal velocity in a significant way (Frolov and Macheret, 1999; Sundström et al., 2012). Snow depths calculated from GPR data (h_v, Eq. 9) and slope corrected snow depths (h_{cor}, Eq. 4) from snow probings were compared to validate the used GPR velocities. On Vernagtferner 95 snow probings show a mean offset of $+0.08$ m with a standard deviation of 0.13 m, accounting for lower snow depths calculated from GPR data. This can be related to the usage of a snowmobile for GPR measurements. While GPR data were recorded in the track of the snowmobile, snow depth probings were performed beside the snowmobile in the undisturbed snow. On Hintereisferner and Kesselwandferner in 2012, the comparison of snow depths calculated from GPR measurements and snow depth probings show a mean difference of 0.01 m with a standard deviation of 0.13 m.

5.5 Consequence for ALS snow cover studies on glaciers

For most of the ice areas of the four investigated glaciers differences between Δz_{ALS} and observed "ground truth" snow depths are less than the assumed uncertainties of the applied methods. The corresponding relative differences are less than 15 % for a mean snow depth of 1.5 m, decreasing with increasing snow depths. This is considerably less than the relative errors of measurements of solid precipitation, which may be up to 50 % when using rain gauges in high mountain catchments (Sevruk, 1985). In contrast, rain gauge measurements have the advantage of a high temporal resolution in comparison to ALS surveys. Hence the combination of both, ALS data for total volume and spatial distribution of the snow cover and rain gauge data for the timing of precipitation events, ensure a high potential for simulating the seasonal hydrology of mountain catchments more realistically.

Typical accumulation areas have to be considered when relating Δz_{ALS} to snow depth. For these areas, Δz_{ALS} has to be considered as a minimum snow depth value, as deviations to h_{GPR} appear to be systematic. However, at the end of ablation seasons, recent firn areas cover considerably smaller parts of the investigated glaciers than bare ice areas. Accumulation-area ratios (AAR) derived from LSC are 0.14 for Vernagtferner, 0.15 for Hintereisferner and 0.41 for Kesselwandferner. Mean Δh_{abs} of -0.4 m on average corresponds to a

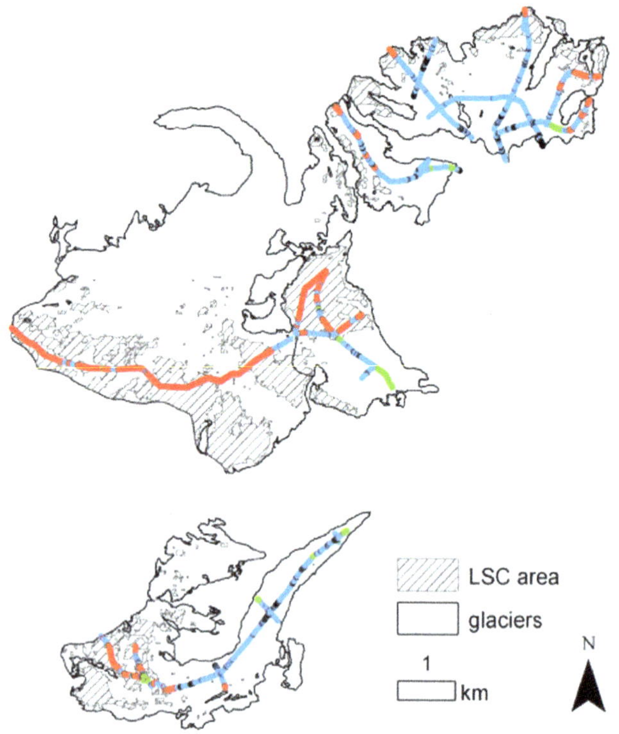

Fig. 13. Results of similarity between Δz_{ALS} and h_{GPR} applying a Mann–Whitney–Wilcoxon test (level of significance $p = 0.05$) and the comparison of Δh_{abs} (Eq. 11) to uncertainties of the two methods: ALS and GPR (Eq. 12) are given. Each midpoint of a window including $n = 300$ samples is coloured indicating not significant Δh_{abs} (black), significant Δh_{abs} (blue), significant and negative Δh_{abs} larger than uncertainty (red) and significant and positive Δh_{abs} larger than uncertainty (green). The outlines of the investigated glaciers (black) and the snow cover derived from Landsat image of 31 August 2009 (LSC, striped) are shown.

mean relative underestimation of approx. 20 % for the investigated LSC areas. Regarding the whole investigated glacierized areas, this systematic underestimation results in a mean relative error of 3 to 7 % for the respective glaciers. Hence, the information on snow depths from Δz_{ALS} is supposed to be a more reliable measurement not only of the spatial distribution of snow depths, but also for total volume of the snow cover in the investigated glacierized region at the time of the snow-on ALS survey (t_2).

In addition ablation in fall or already prevailing seasonal accumulation before the ALS survey at t_1 influence the accuracy of the derived ALS-derived snow depths on glacier surfaces. Furthermore, in contrast to non-glaciated areas, the ALS survey for t_1 has to be repeated annually, because glacier surface elevation varies annually. To calculate SWE from Δz_{ALS}, additional assumptions on the relation of snow density on snow depth (e.g. Jonas et al., 2009) or a spatially uniform snow density (e.g. Lehning et al., 2011) have to be made. This results in additional uncertainties of the derived accumulation based on the conversion used and the prevailing conditions of the snowpack.

The data presented in this study were analyzed for differences between Δz_{ALS} and actual snow depths, because recently an increased number of research projects using ALS data to investigate the spatial and temporal variability of the cryosphere are conducted in the mountain range of Ötztal Alps. The results may be transferable in the study region and to Alpine glaciers of similar size and with similar mass turnover. Nevertheless, various process parameters contributing to Δz_{ALS} (Sect. 1) have to be considered, for example, on larger or steeper glaciers, on glaciers with faster ice flow, on glaciers gaining more accumulation or on glaciers with a less distinctive seasonal cycle of ablation and accumulation period.

In difficult mountain terrain, the acquisition of actual snow depths by GPR along profiles delivers less information on spatial snow distribution compared to ALS surveys covering total mountain catchments. However, systematic differences between Δz_{ALS} and actual snow depths have to be analyzed separately for individual glaciers when using ALS for recording snow depths even in glacierized catchments. This study demonstrated that the usage of ALS data and GPR measurements appears to be a promising combination of methods to identify differences of Δz_{ALS} from actual snow depths and corresponding uncertainties not only for single points, but along profiles on glacier surfaces worldwide.

6 Conclusions

The study presented here was accomplished to evaluate surface elevation changes derived from airborne laser scanning (Δz_{ALS}) on glacier surfaces for snow cover studies in the Ötztal Alps. Therefore, differences of Δz_{ALS} from snow depths calculated from more than 35 km of GPR profiles (h_{GPR}) on the four glaciers Hintereisferner, Kesselwandferner, Vernagtferner and Gepatschferner were analyzed. It is shown that GPR can be applied to validate Δz_{ALS} regarding snow depths and be used to analyze systematic differences. The sum of all processes contributing to Δz_{ALS} in addition to seasonal snow depths on glacier surfaces could be detected by combining ALS and GPR data. A more extensive measurement set-up would be necessary for detailed analysis of the partitioning and magnitude of the individual processes causing these differences. The delineation of snow covered areas at the end of ablation season from Landsat imagery highlights areas of increased differences between Δz_{ALS} and "ground truth" snow depth derived from h_{GPR}, snow pits and snow probing. This study shows that on four different glaciers systematic differences between Δz_{ALS} and h_{GPR} greater than calculated uncertainties of the applied techniques were observable almost exclusively within the respective accumulation areas. Most of ice areas on the investigated glaciers do not show differences between Δz_{ALS} and actual snow depths larger than the uncertainty of the used combination of measurements. Distinct overestimation of actual snow depths by Δz_{ALS} could be detected at the very front of the glacier tongues. In summary, relative differences of Δz_{ALS} from actual snow depths are small for analysis of the mountain snow cover in the study area and comparable glacierized catchments. Hence, spatial distributed snow depths derived from Δz_{ALS} at glacier surface provide an information on minimum snow depths which must be simulated by hydro-meteorological models, particularly in the higher elevated parts of the glaciers in this region.

Acknowledgements. This study was carried out within the frame of the alpS project "H03 MUSICALS A – Multiscale Snow/Ice Melt Discharge Simulations for Alpine Reservoirs". We want to thank the Austrian Research Promotion Agency (FFG), and the TIWAG – Tiroler Wasserkraft AG, who promote this project. The surveys and processing of the ALS data were performed by TopScan GmbH (Rheine, Germany). ALS data management was supported by the lidar research group of the Institute of Geography, University of Innsbruck. We thankfully appreciate the provision of GPR profiles on Vernagtferner and data of the automatic weather station near this glacier by the Commission for Geodesy and Glaciology, Bavarian Academy of Sciences and Humanities (Munich, Germany). Also we appreciate the effort of the two anonymous referees, whose constructive comments and critical remarks helped to improve the paper. And many thanks to all of the field workers.

Edited by: A. Kääb

References

Abermann, J., Schneider, H., and Lambrecht, A.: Analysis of surface elevation changes on Kesselwand glacier – comparison of different methods, Z. Gletsch.kd. Glazialgeol., 41, 147–167, 2007.

Abermann, J., Lambrecht, A., Fischer, A., and Kuhn, M.: Quantifying changes and trends in glacier area and volume in the Austrian Ötztal Alps (1969–1997–2006), The Cryosphere, 3, 205–215, doi:10.5194/tc-3-205-2009, 2009.

Abermann, J., Fischer, A., Lambrecht, A., and Geist, T.: On the potential of very high-resolution repeat DEMs in glacial and periglacial environments, The Cryosphere, 4, 53–65, doi:10.5194/tc-4-53-2010, 2010.

Ambach, W. and Eisner, H.: Analysis of a 20 m firn pit on the Kesselwandferner (Ötztal Alps), J. Glaciol., 6, 223–231, 1966.

Aschwanden, H., Weingartner, R., and Leibundgut, C.: Zur regionalen Übertragung von Mittelwerten des Abflusses – Teil 1: Raumtypisierung der Abflussregime der Schweiz, Dtsch. gewässerkdl. Mitt. , 30, 52–61, 1986.

Baltsavias, E.: Airborne laser scanning: basic relations and formulas, ISPRS J. Photogramm. and Remote Sens., 54, 199–214, doi:10.1016/S0924-2716(99)00015-5, 1999.

Blöschl, G.: Scaling issues in snow hydrology, Hydrol. Process., 13, 2149–2175, doi:10.1002/(SICI)1099-1085(199910)13:14/15<2149::AID-HYP847>3.0.CO;2-8, 1999.

Bollmann, E., Sailer, R., Briese, C., Stötter, J., and Fritzmann, P.: Potential of airborne laser scanning for geomorphologic feature and process detection and quantifications in high alpine mountains, Z. Geomorphol. Supp., 55, 83–104, doi:10.1127/0372-8854/2011/0055S2-0047, 2011.

Cogley, J., Hock, R., Rasmussen, L., Arendt, A., Bauder, A., Braithwaite, R., Jansson, P., Kaser, G., Möller, M., Nicholson, L., and Zemp, M.: Glossary of Glacier Mass Balance and Related Terms, Tech. Rep., IHP-VII Technical Documents in Hydrology No. 86, IACS Contribution No. 2, UNESCO-IHP, Paris, 2011.

Cuffey, K. and Paterson, W.: The Physics of Glaciers, Academic Press, 4th Ed., 2010.

Dadic, R., Mott, R., Lehning, M., and Burlando, P.: Wind influence on snow depth distribution and accumulation over glaciers, J. Geophys. Res., 115, F01012, doi:10.1029/2009JF001261, 2010.

Daniels, D.: Ground Penetrating Radar, The Institution of Electrical Engineers, London, UK, 2nd Ed., 2004.

Deems, J., Paintner, T., and Finnegan, D.: Lidar measurement of snow depth: a review, J. Glaciol., 59, 467–479, doi:10.3189/2013JoG12J154, 2013.

Deems, J. S., Fassnacht, S. R., and Elder, K. J.: Interannual consistency in fractal snow depth patterns at two Colorado mountain sites, J. Hydrometeorol., 9, 977–988, doi:10.1175/2008JHM901.1, 2008.

Escher-Vetter, H., Kuhn, M., and Weber, M.: Four decades of winter mass balance of Vernagtferner and Hintereisferner, Austria: methodology and results, Ann. Glaciol., 50, 87–95, doi:10.3189/172756409787769672, 2009.

Fierz, C., Armstrong, R., Durand, Y., Etchevers, P., Greene, E., McClung, D., Nishimura, K., Satyawali, P., and Sokratov, S.: The International Classification for Seasonal Snow on the Ground, Tech. Rep., IHP-VII Technical Documents in Hydrology No. 83, IACS Contribution No. 1, UNESCO-IHP, Paris, 2009.

Fischer, A.: Glaciers and climate change: Interpretation of 50 years of direct mass balance of Hintereisferner, Global Planet. Change, 71, 13–26, doi:10.1016/j.gloplacha.2009.11.014, 2010.

Fischer, A.: Comparison of direct and geodetic mass balances on a multi-annual time scale, The Cryosphere, 5, 107–124, doi:10.5194/tc-5-107-2011, 2011.

Fischer, A., Span, N., Fischer, A., Kuhn, M., Massimo, M., and Butschek, M.: Radarmessungen der Eisdicke Österreichischer Gletscher. Band II: Messungen 1999 bis 2006, Österreichische Beiträge zu Meteorologie und Geophysik, 39, 142 pp., 2007.

Fischer, A., Markl, G., Kuhn, M., Schneider, H., and Abermann, J.: Glacier Mass Balance of Kesselwandferner, Ötztal Alps, Austria, from 1952/53–2009/11, World Data Center for Climate, doi:10.1594/WDCC/MB_KWF_1953-2011, 2011.

Fischer, A., Markl, G., and Kuhn, M.: Glacier Mass Balance of Hintereisferner, Ötztal Alps, Austria, from 1952/53–2010/11, doi:10.1594/WDCC/MB_HEF_1953-2011, 2012.

Frolov, A. D. and Macheret, Y. Y.: On dielectric properties of dry and wet snow, Hydrol. Process., 13, 1755–1760, doi:10.1002/(SICI)1099-1085(199909)13:12/13<1755::AID-HYP854>3.0.CO;2-T, 1999.

Geist, T. and Stötter, J.: Documentation of glacier surface elevation change with multi-temporal airborne laser scanner data – case study: Hintereisferner and Kesselwandferner, Tyrol, Austria, Z. Gletsch.kd. Glazialgeol., 41, 77 – 106, 2007.

Grünewald, T. and Lehning, M.: Altitudinal dependency of snow amounts in two small alpine catchments: can catchment-wide snow amounts be estimated via single snow or precipitation stations?, Ann. Glaciol., 52, 153–158, doi:10.3189/172756411797252248, 2011.
Haberkorn, A.: Assessing the contribution of basal melt processes to the local mass balance at the tongue of Hintereisferner, Ötztal, Austria, Master's thesis, Institute of Meteorology and Geophysics, University of Innsbruck, http://imgi.uibk.ac.at/sites/default/files/migration/thesisAnnaHaberkorn.pdf, 2011.
Hall, D. K., Ormsby, J. P., Bindschadler, R. A., and Siddalingaiah, H.: Characterization of snow and ice reflectance zones on glaciers using Landsat thematic mapper data, Ann. Glaciol., 9, 104–108, 1987.
Hall, D. K., Riggs, G. A., Salomonson, V. V., DiGirolamo, N. E., and Bayr, K. J.: MODIS snow-cover products, Remote Sens. Environ., 83, 181–194, doi:10.1016/S0034-4257(02)00095-0, 2002.
Heilig, A., Schneebeli, M., and Eisen, O.: Upward-looking ground-penetrating radar for monitoring snow-pack stratigraphy, Cold Reg. Sci. Technol., 59, 152–162, doi:10.1016/j.coldregions.2009.07.008, 2009.
Helfricht, K., Schöber, J., Seiser, B., Fischer, A., Stötter, J., and Kuhn, M.: Snow accumulation of a high alpine catchment derived from LiDAR measurements, Adv. Geosci., 32, 31–39, doi:10.5194/adgeo-32-31-2012, 2012.
Hinkel, K. M., Doolittle, J. A., Bockheim, J. G., Nelson, F. E., Paetzold, R., Kimble, J. M., and Travis, R.: Detection of subsurface permafrost features with ground-penetrating radar, Barrow, Alaska, Permafrost Periglac., 12, 179–190, doi:10.1002/ppp.369, 2001.
Hoinkes, H.: Methoden und Möglichkeiten von Massenhaushaltsstudien auf Gletschern, Z. Gletsch.kd. Glazialgeol., 6, 37–90, 1970.
Huss, M., Farinotti, D., Bauder, A., and Funk, M.: Modelling runoff from highly glacierized alpine drainage basins in a changing climate, Hydrol. Process., 22, 3888–3902, doi:10.1002/hyp.7055, 2008.
Huss, M., Bauder, A., and Funk, M.: Homogenization of long-term mass-balance time series, Ann. Glaciol., 50, 198–206, doi:10.3189/172756409787769627, 2009.
Joerg, P. C., Morsdorf, F., and Zemp, M.: Uncertainty assessment of multi-temporal airborne laser scanning data: A case study on an Alpine glacier, Remote Sens. Environ., 127, 118–129, doi:10.1016/j.rse.2012.08.012, 2012.
Jonas, T., Marty, C., and Magnusson, J.: Estimating the snow water equivalent from snow depth measurements in the Swiss Alps, J. Hydrol., 378, 161–167, doi:10.1016/j.jhydrol.2009.09.021, 2009.
Kaser, G., Fountain, A., and Jansson, P.: A manual for monitoring the mass balance of mountain glaciers, International Hydrological Programme (IHP-VI. Technical Documents in Hydrology 59), UNESCO, Paris, 2003.
Kirnbauer, R., Achleitner, S., Schöber, J., Asztalos, J., and Schönlaub, H.: Hochwasservorhersage Inn: Modellierung der Gletscherabflüsse, Mitteilungsblatt des Hydrographischen Dienstes in Österreich, Nr. 86, 109–130, 2009.
Koblet, T., Gärtner-Roer, I., Zemp, M., Jansson, P., Thee, P., Haeberli, W., and Holmlund, P.: Reanalysis of multi-temporal aerial images of Storglaciären, Sweden (1959–99) – Part 1: Determination of length, area, and volume changes, The Cryosphere, 4, 333–343, doi:10.5194/tc-4-333-2010, 2010.
Kovacs, A., Gow, A. J., and Morey, R. M.: The in-situ dielectric constant of polar firn revisited, Cold Reg. Sci. Technol., 23, 245–256, doi:10.1016/0165-232X(94)00016-Q, 1995.
Kraus, K.: Geometrische Informationen aus Photographien und Laserscanaufnahmen, Band 1, Walter de Gruyter, Berlin – New York, 2004.
Kuhn, M.: Verification of a hydrometeorological model of glacierized basins, Ann. Glaciol., 31, 15–18, doi:10.3189/172756400781820228, 2000.
Kuhn, M.: Redistribution of snow and glacier mass balance from a hydrometeorological model, J. Hydrol., 282, 95–103, doi:10.1016/S0022-1694(03)00256-7, 2003.
Kuhn, M. and Batlogg, N.: Glacier runoff in Alpine headwaters in a changing climate, Hydrology, Water Resources and Ecology in Headwaters (Proceedings of the HeadWater'98 Conference held at Meran/Merano, Italy, April 1998), IAHS Publication, 248, 79–88, 1998.
Kuhn, M., Dreiseitl, E., Hofinger, S., Markl, G., Span, N., and Kaser, G.: Measurements and models of the mass balance of Hintereisferner, Geogr. Ann. A, 81, 659–670, doi:10.1111/1468-0459.00094, 1999.
Lehning, M., Grünewald, T., and Schirmer, M.: Mountain snow distribution governed by an altitudinal gradient and terrain roughness, Geophys. Res. Lett., 38, L19504, doi:10.1029/2011GL048927, 2011.
Lui, X.: Airborne LiDAR for DEM generation: some critical issues, Prog. Phys. Geog., 32, 31–49, doi:10.1177/0309133308089496, 2008.
Lundberg, A., Thunehed, H., and Bergström, J.: Impulse Radar Snow Surveys – Influence of Snow Density, Nord. Hydrol., 3, 1–14, 2000.
Lundberg, A., Richardson-Näslund, C., and Andersson, C.: Snow density variations: consequences for ground-penetrating radar, Hydrol. Process., 20, 1483–1495, doi:10.1002/hyp.5944, 2006.
Lundberg, A., Granlund, N., and Gustafsson, D.: Towards automated "Ground truth" snow measurements: a review of operational and new measurement methods for Sweden, Norway, and Finland, Hydrol. Process., 24, 1955–1970, doi:10.1002/hyp.7658, 2010.
Machguth, H., Eisen, O., Paul, F., and Hoelzle, M.: Strong spatial variability of accumulation observed with helicopter-borne GPR on two adjacent Alpine glaciers, Geophys. Res. Lett., 33, L13503, doi:10.1029/2006GL026576, 2006.
Mayer, C., Lambrecht, A., Blumthaler, U., and Eisen, O.: Vermessung und Eisdynamik des Vernagtferners, Ötztaler Alpen, Z. Gletsch.kd. Glazialgeol., 45–46, 259–280, 2013.
Melvold, K. and Skaugen, T.: Multiscale spatial variability of lidar-derived and modeled snow depth on Hardangervidda, Norway, Ann. Glaciol., 54, 273–281, doi:10.3189/2013AoG62A161, 2013.
Mitterer, C., Heilig, A., Schweizer, J., and Eisen, O.: Upward-looking ground-penetrating radar for measuring wet-snow properties, Cold Reg. Sci. Technol., 69, 129–138, doi:10.1016/j.coldregions.2011.06.003, 2011.

Murray, T., Gooch, D., and Stuart, G.: Structures within the surge front at Bakaninbreen, Svalbard, using ground-penetrating radar, Ann. Glaciol. 24, 122–129, 1997.

Nolin, A. W.: Recent advances in remote sensing of seasonal snow, J. Glaciol., 56, 1141–1150, doi:10.3189/002214311796406077, 2011.

Otto, J.-C., Keuschnig, M., Götz, J., Marbach, M., and Schrott, L.: Detection of mountain permafrost by combining high resolution surface and subsurface information: an example from the Glatzbach Catchment, Austrian Alps, Geogr. Ann. A, 94, 43–57, doi:10.1111/j.1468-0459.2012.00455.x, 2012.

Rott, H. and Markl, G.: Improved snow and glacier monitoring by Landsat Thematic Mapper, in: Proceedings of Workshop on Earthnet Pilot Projects on Landsat Thematic Mapper Applications, 3–12, ESA SP-1102, 1989.

Schöber, J., Achleitner, S., Kirnbauer, R., Schöberl, F., and Schönlaub, H.: Hydrological modelling of glacierized catchments focussing on the validation of simulated snow patterns – applications within the flood forecasting system of the Tyrolean river Inn, Adv. Geosci., 27, 99–109, doi:10.5194/adgeo-27-99-2010, 2010.

Sevruk, B.: Correction of precipitation measurements: Swiss experience, WMO/TD, 104, 187–196, 1985.

Sold, L., Huss, M., Hoelzle, M., Andereggen, H., Joerg, P. C., and Zemp, M.: Methodological approaches to infer end-of-winter snow distribution on alpine glaciers, J. Glaciol., 59, 1047–1059, doi:10.3189/2013JoG13J015, 2013.

Span, N. and Kuhn, M.: Simulating annual glacier flow with a linear reservoir model, J. Geophys. Res.-Atmos., 108, 4313, doi:10.1029/2002JD002828, 2003.

Span, N., Kuhn, M., and Schneider, H.: 100 years of ice dynamics of Hintereisferner, Central Alps, Austria, 1894–1994, Ann. Glaciol., 24, 297–302, 1997.

Span, N., Fischer, A., Kuhn, M., Massimo, M., and Butschek, M.: Radarmessungen der Eisdicke Österreichischer Gletscher. Band I: Messungen 1995 bis 1998, Österreichische Beiträge zu Meteorologie und Geophysik, 33, 145 pp., 2005.

Spikes, V. B., Hamilton, G. S., Arcone, S. A., Kaspari, S., and Mayewski, P. A.: Variability in accumulation rates from GPR profiling on the West Antarctic plateau, Ann. Glaciol., 39, 238–244, doi:10.3189/172756404781814393, 2004.

Sundström, N., Kruglyak, A., and Friborg, J.: Modeling and simulation of GPR wave propagation through wet snowpacks: Testing the sensitivity of a method for snow water equivalent estimation, Cold Reg. Sci. Technol., 74–5, 11–20, doi:10.1016/j.coldregions.2012.01.006, 2012.

Trabant, P.: Applied high resolution geophysical methods: Offshore geoengineering hazards, IHRDC Press, Boston, MA, IHRDC Press, 137 Newbury Street, Boston, MA 02116, 1984.

Wehr, A. and Lohr, U.: Airborne laser scanning – an introduction and overview, ISPRS J. Photogramm. Remote Sens., 54, 68–82, doi:10.1016/S0924-2716(99)00011-8, 1999.

Weingartner, R., Barben, M., and Spreafico, M.: Floods in mountain areas – an overview based on examples from Switzerland, J. Hydrol., 282, 10–24, doi:10.1016/S0022-1694(03)00249-X, 2003.

Spatial patterns in glacier characteristics and area changes from 1962 to 2006 in the Kanchenjunga–Sikkim area, eastern Himalaya

A. E. Racoviteanu[1], Y. Arnaud[1,2], M. W. Williams[3], and W. F. Manley[4]

[1]Laboratoire de Glaciologie et Géophysique de l'Environnement, 54 rue Molière, Domaine Universitaire, BP 96, 38402 Saint Martin d'Hères CEDEX, France
[2]Laboratoire d'Étude des Transferts en Hydrologie et Environnement, BP 53, 38401 Saint Martin d'Hères CEDEX, France
[3]Department of Geography and Institute of Arctic and Alpine Research, University of Colorado, Boulder, CO 80309, USA
[4]Institute of Arctic and Alpine Research, University of Colorado, Boulder, CO 80309, USA

Correspondence to: A. Racoviteanu (racovite@gmail.com)

Abstract. This study investigates spatial patterns in glacier characteristics and area changes at decadal scales in the eastern Himalaya – Nepal (Arun and Tamor basins), India (Teesta basin in Sikkim) and parts of China and Bhutan – based on various satellite imagery: Corona KH4 imagery, Landsat 7 Enhanced Thematic Mapper Plus (ETM+) and Advanced Spaceborne Thermal Emission Radiometer (ASTER), QuickBird (QB) and WorldView-2 (WV2). We compare and contrast glacier surface area changes over the period of 1962–2000/2006 and their dependency on glacier topography (elevation, slope, aspect, percent debris cover) and climate (solar radiation, precipitation) on the eastern side of the topographic barrier (Sikkim) versus the western side (Nepal).

Glacier mapping from 2000 Landsat ASTER yielded $1463 \pm 88\,km^2$ total glacierized area, of which $569 \pm 34\,km^2$ was located in Sikkim and $488 \pm 29\,km^2$ in eastern Nepal. Supraglacial debris covered 11 % of the total glacierized area, and supraglacial lakes covered about 5.8 % of the debris-covered glacier area alone. Glacier area loss (1962 to 2000) was $0.50 \pm 0.2\,\%\,yr^{-1}$, with little difference between Nepal ($0.53 \pm 0.2\,\%\,yr^{-1}$) and Sikkim ($0.44 \pm 0.2\,\%\,yr^{-1}$). Glacier area change was controlled mostly by glacier area, elevation, altitudinal range and, to a smaller extent, slope and aspect. In the Kanchenjunga–Sikkim area, we estimated a glacier area loss of $0.23 \pm 0.08\,\%\,yr^{-1}$ from 1962 to 2006 based on high-resolution imagery. On a glacier-by-glacier basis, clean glaciers exhibit more area loss on average from 1962 to 2006 (34 %) compared to debris-covered glaciers (22 %). Glaciers in this region of the Himalaya are shrinking at similar rates to those reported for the last decades in other parts of the Himalaya, but individual glacier rates of change vary across the study area with respect to local topography, percent debris cover or glacier elevations.

1 Introduction

Himalayan glaciers have generated a lot of concern in the last few years, particularly with respect to potential consequences of glacier changes on the regional water cycle (Kaser et al., 2010; Immerzeel et al., 2010, 2012; Racoviteanu et al., 2013a). In the last decades, the availability of low-cost data from optical remote sensing platforms with global coverage provided opportunities for glacier mapping at regional scales. Remote sensing techniques have considerably helped improve estimates of glacier area changes (Bhambri et al., 2010; Bolch, 2007; Bajracharya et al., 2007; Kamp et al., 2011; Bolch et al., 2008a), glacier lake changes (Bajracharya et al., 2007; Bolch et al., 2008b; Gardelle et al., 2011; Wessels et al., 2002) and region-wide glacier mass balance (Berthier et al., 2007; Bolch et al., 2011; Gardelle et al., 2013; Kääb et al., 2012), but significant gaps do remain. The new global Randolph Glacier Inventory (RGI) v.4 (Pfeffer et al., 2014) provides a global data set of glacier outlines intended for large-scale studies; however, in some regions the quality varies and the outlines may not be suitable for detailed regional analysis of glacier parameters. A new Landsat-based inventory

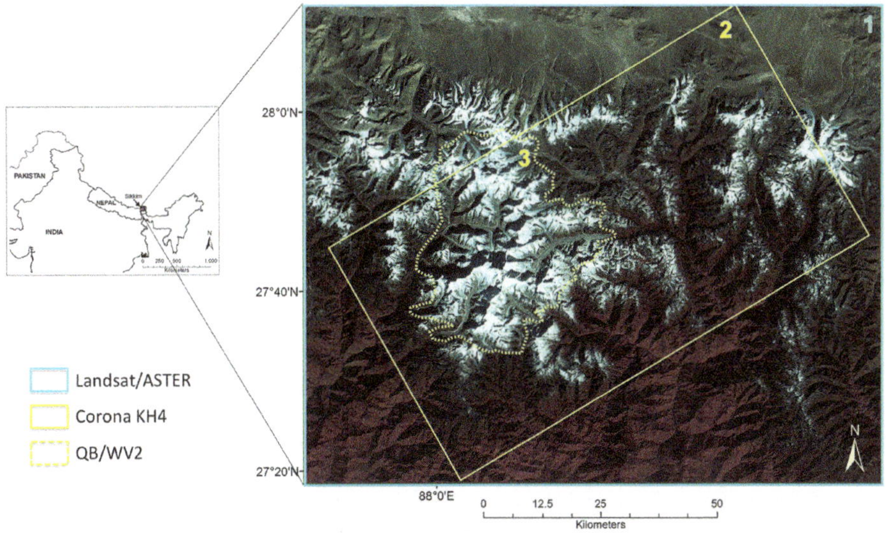

Figure 1. Location map of the study area, with spatial domains 1, 2 and 3 corresponding to the coverage of each satellite data. The background is a Landsat ETM+ color composite (432) overlaid on shaded relief from the SRTM DEM.

has been complied using imagery from 1999 to 2003 that, along with the current study, may help improve the accuracy in some areas of RGI (Nuimura et al., 2014). Some other regional glacier inventories have been constructed in the past, for example for the western part of the Himalaya (e.g., Kamp et al., 2011; Bhambri et al., 2011; Frey et al., 2012), but only a few are available for the eastern extremity of the Himalaya (e.g., Krishna, 2005; Bahuguna, 2014; Basnett et al., 2013; Bajracharya et al., 2014; Kulkarni and Narain, 1990). The use of remote sensing for glacier mapping in this area is limited by frequent cloud cover and sensor saturation due to unsuitable gain settings and the persistence of seasonal snow, which hampers quality satellite image acquisition. Furthermore, this area has very limited reliable baseline topographic data needed for glacier change detection, as discussed in detail in Bhambri and Bolch (2009). The earliest Indian glacier maps date from topographic surveys conducted by expeditions in the mid-nineteenth century (i.e. Mason, 1954), but these are limited to a few glaciers. The Geologic Survey of India (GSI) inventory based on 1970s Survey of India maps (Sangewar and Shukla, 2009; Shanker, 2001) is not in the public domain. For eastern Nepal, 1970s topographic maps from Survey of India on a 1 : 63 000 scale are available, but their accuracy is not known with certainty. Given these limitations, declassified Corona imagery from the 1960s and 1970s has increasingly been used to develop baseline glacier data sets, for example in the Tien Shan (Narama et al., 2010), Nepal Himalaya (Bolch et al., 2008a) and parts of Sikkim Himalaya (Raj et al., 2013).

With a wide variety of satellite data becoming available, the topographic and climatic controls on glacier surface area have received increasing attention in recent studies, particularly with respect to debris-covered glaciers (Basnett et al., 2013; Thakuri et al., 2014; Bolch et al., 2008a; Salerno et al., 2008). Some studies have characterized the small-scale glacier surface topography of debris-covered glaciers using field-based surveys (Iwata et al., 2000; Sakai and Fujita, 2010; Zhang et al., 2011), while other studies focused on understanding patterns at the mountain-range scale (Scherler et al., 2011; Bolch et al., 2012; Gardelle et al., 2013; Racoviteanu et al., 2014). Glacier shrinkage and mass loss has been documented in the Himalaya concomitantly with an increase in debris cover (Bolch et al., 2011; Nuimura et al., 2012). However, the influence of debris cover on glacier mass balance remains debatable (Scherler et al., 2011; Kääb et al., 2012), and modeling of melt under the debris cover is subject to uncertainties due to limited field-based measurements of debris thickness needed for model parameterization (Zhang et al., 2011; Foster et al., 2012; Mihalcea et al., 2008a, b).

While significant progress has been made in recent years in remote sensing glacier mapping in the Himalaya, some of the subregions still need updated glacier area and surface characteristics including debris cover. The objective of this study is twofold: (1) to present the current glacier distribution and characteristics in a data-scarce area of the eastern Himalaya based on an updated 2000 Landsat 7 Enhanced Thematic Mapper Plus (ETM+) and Advanced Spaceborne Thermal Emission Radiometer (ASTER) inventory, along with elevation data from the Shuttle Radar Topography Mission (SRTM); and (2) to investigate spatial patterns in glacier surface area changes from 1962 (Corona KH4) to 2000 (Landsat/ASTER) and 2006 QuickBird (QB)/2009 WorldView-2 (WV2) and their dependence on topographic and climatic factors, with a particular emphasis on debris-covered glacier tongues. These updated glacier data sets will help filling a gap in global glacier inventories such as the RGI

(Pfeffer et al., 2014), as well as for subsequent future mass balance applications at regional scales.

2 Study area

The study area encompasses glaciers in the eastern Himalaya (27°04′52″–28°08′26″ N latitude and 88°00′57″–88°55′50″ E longitude), located on either side of the border between Nepal and India in the Kanchenjunga–Sikkim area (Fig. 1). Based on SRTM data, relief in this area ranges from 300 m at the bottom of the valleys to 8598 m (Mt. Kanchenjunga). Valley glaciers cover about 68 % of the glacierized area, mountain glaciers cover 28 % and the remaining area is made up of cirque glaciers and aprons (Mool et al., 2002). The glacier ablation area is typically covered by heavy debris cover originating from rockfall on the steep slopes (Mool et al., 2002), reaching up to a thickness of several meters at the glacier termini (Kayastha et al., 2000). The eastern part of this area constitutes the Sikkim province of India, and the western part is located in eastern Nepal and encompasses the Tamor basin and parts of the Arun basin. Climatically, this area of the Himalaya is dominated by the South Asian summer monsoon circulation system (Bhatt and Nakamura, 2005) caused by the inflow of moist air from the Bay of Bengal to the Indian subcontinent during the summer (Benn and Owen, 1998; Yanai et al., 1992). The Himalaya and Tibetan plateau act as a barrier to the monsoon winds, bringing about 77 % of precipitation on the south slopes of the Himalaya during the summer months (May–September) (Fig. 2). This climatic particularity causes a "summer-accumulation" glacier regime type, with accumulation and ablation occurring simultaneously in the summer (Ageta and Higuchi, 1984). In Sikkim, rainfall amounts range from 500 to 5000 mm yr^{-1}, with annual averages of 3580 mm recorded at Gangtok station (1812 m; 1951–1980) (IMD, 1980), and 164 rainy days per year (Nandy et al., 2006). Mean minimum and maximum daily temperatures at Gangtok station were reported as 11.3 and 19.8 °C, with an average of 15.5 °C based on the same observation record (IMD, 1980).

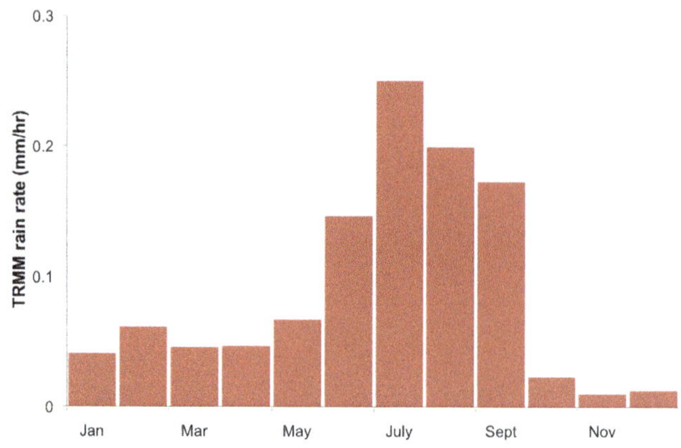

Figure 2. Precipitation regime over domain 1 expressed as rain rate, from the TRMM 2B31 data averaged for the period 1998–2010. The graph shows the monsoon period from June to September, with a peak precipitation in July, and the influence of the northeastern monsoon during the winter/early spring (January–March).

3 Methodology

3.1 Data sources

3.1.1 Satellite imagery

Remote sensing data sets used in this study are summarized in Table 1 and included (1) baseline remote sensing data from Corona declassified imagery (year 1962), (2) "reference" data sets for 2000s from Landsat ETM+ and ASTER and (3) high-resolution imagery from QB (2006) and WV2 (2009), all described below.

1. Corona KH4 scenes (1962) were obtained from the US Geological Survey EROS Data Center (USGS-EROS, 1962). The Corona KH4 system was equipped with two panoramic cameras (forward-looking and rear-looking with 30° separation angle) and acquired imagery from February 1962 to December 1963 (Dashora et al., 2007). We chose images from the end of the ablation season (October/November in this part of the Himalaya), suitable for glaciologic purposes. Six Corona stripes were scanned at 7 microns by USGS from the original film strips, with a reported nominal ground resolution of 7.62 m (Dashora et al., 2007). Corona images are known to contain significant geometric distortions due to cross-path panoramic scanning. The Frame Ephemeris Camera and Orbital Data camera/spacecraft parameters (roll, pitch, yaw, speed, altitude, azimuth, sun angle and film scanning rate) for Corona missions, needed to construct a camera model and to correct these distortions, are not easily available. To orthorectify the scenes, we defined a non-metric camera model in ERDAS Leica Photogrammetric Suite (LPS), with focal length, air photo scale and flight altitude extracted from the declassified documentation of the KH4 mission (Dashora et al., 2007). We used the bundle block adjustment procedure in LPS to simultaneously estimate the orientation of all the CORONA stripes on the basis of 117 ground control points (GCPs). Latitude and longitude (x, y) information (of the GCPs) were extracted from the panchromatic band of the 2000 Landsat ETM+ image (15 m spatial resolution) on non-glacierized terrain including moraines, river crossings and outwash areas, whereas elevation information (z) were extracted from the SRTM DEM v.4 (CGIAR-CSI 2004).

Table 1. Summary of satellite imagery used in this study.

Sensor	Scene ID	Date	Spatial resolution	Image type
Corona	DS009048070DA244	25 Oct 1962	7.5 m	Panchromatic
KH4	DS009048070DA243			
	DS009048070DA242			
Landsat ETM+	L7CPF20001001_20001231_07	26 Dec 2000	15 m	Pancromatic
			28.5 m	Visible, shortwave
			90 m	Thermal infrared
ASTER	AST_L1A#003_12012000051205_07292001131755	1 Dec 2000	15 m	Visible
	AST_L1A#003_12012000051214_07292001131813	1 Dec 2000	30 m	Shortwave
	AST_L1A_00311272001045729_02222004173619	27 Nov 2001	90 m	Thermal infrared
	AST_L1A#00301052002050207_01302002193030	5 Jan 2002	90 m	Surface kinetic
	AST_L1A#00301052002050216_01302002193046	5 Jan 2002		temperature
	AST_08_00310292002045428_20101212181710_16443	29 Oct 2002		
QuickBird	1010010004BD8700	1 Jan 2006	2.4 m	Visible,
	1010010004BB8F00	6 Jan 2006		shortwave
WorldView -2	102001000FBA1D00	2 Dec 2010	0.50 m	Panchromatic
	102001000586E700	1 Dec 2009		

Tie points were automatically extracted in LPS on overlapping Corona strips, and visually checked on the Landsat image. The Corona stripes were mosaicked in ERDAS LPS to produce the final orthorectified image, with a horizontal accuracy (RMSE x, y) of the bundle block adjustment of 10.5 m. The orthorectification process of the 1962 Corona yielded a RMSE x, y error of $\pm$10 m and the actual "ground" RMSE x, y of the Corona block of $\sim$ 60 m. A trend analysis on the horizontal shifts between Corona and the reference Landsat scene showed that the largest errors occurred towards the edges of the images, mostly outside the glaciers, and did not impact the area change analysis.

2. The orthorectified Landsat ETM+ scene from December 2000, obtained from the USGS Eros Data Center, was the main data set for the updated glacier inventory. In addition, six orthorectified ASTER products (2000–2002) were obtained at no cost through the Global Land Ice Monitoring from Space (GLIMS) project (Raup et al., 2007). Images were selected at the end of the ablation season for minimal snow and had little or no clouds. Five of these scenes were used for on-screen manual corrections of the Landsat-based glacier outlines in challenging areas where shadows or clouds obstructed the view of the glaciers. In addition, the surface kinetic temperature product (AST08) from the 27 November 2001 ASTER scene was used for clean ice delineation of debris cover along with topographic information using a decision-tree algorithm (Racoviteanu and Williams, 2012). The 29 October 2002 scene, covering the Kanchenjunga–Sikkim area east and west of the topographic divide, was used to investigate the spatial distribution of surface temperature over selected debris-covered tongues.

3. Two QB scenes from January 2006 were obtained from Digital Globe as ortho-ready standard imagery (radiometrically calibrated and corrected for sensor and platform distortions) (Digital Globe, 2007). These scenes, covering an area of 1107 km^2 were well contrasted and mostly snow-free outside the glaciers. We orthorectified these scenes in ERDAS Imagine Leica Photogrammetry Suite (LPS) using rational polynomial coefficients (RPCs) provided by Digital Globe and the SRTM DEM and mosaicked them in ERDAS Imagine. The scenes were resampled to 3 m pixel size during the orthorectification process using the cubic convolution method suitable for continuous raster data in order to reduce disk space and processing time. One WV2 panchromatic, ortho-ready scene at 50 cm spatial resolution from 2 December 2010 was also obtained to cover the terminus of Zemu glacier, which was missing from the QB extent. All data sets were registered to UTM projection zone 45N, with elevations referenced to the WGS84 datum.

3.1.2 Elevation data sets

Two elevation data sets were used in this study:

1. The hydrologically sound, void-filled CGIAR SRTM DEM (90 m spatial resolution) (CGIAR-CSI, 2004) was used to extract glacier parameters for 2000. The SRTM data set is known to have biases on steep slopes and at higher elevations (Fujita et al., 2008; Berthier et al., 2006; Nuth and Kääb, 2011), as well as due to radar penetration on snow (Gardelle et al., 2012b). For this area, the vertical accuracy of the SRTM DEM, calculated as root mean square (RMSE z) with respect to 25 field-based GCPs, was 31 m $\pm$ 10 m. The GCPs were obtained in the field on non-glacierized terrain including roads and bare land outside the glaciers using a Trimble Geoexplorer XE series.

Table 2. Spatial domains used for analysis and their characteristics.

Spatial domain	Number of glaciers	Area in 2000 (km^2)	Details
1. Landsat/ASTER	487	1463 ± 88	The entire study area extending from Sikkim to China, as well as into parts of western Bhutan and eastern Nepal
2. Corona KH4	232	777 ± 46	Glaciers of eastern Nepal (Tamor basin) and Sikkim (Zemu basin)
3. QB/WV2	50	551 ± 34	Selected glaciers from the Kanchenjunga–Sikkim area

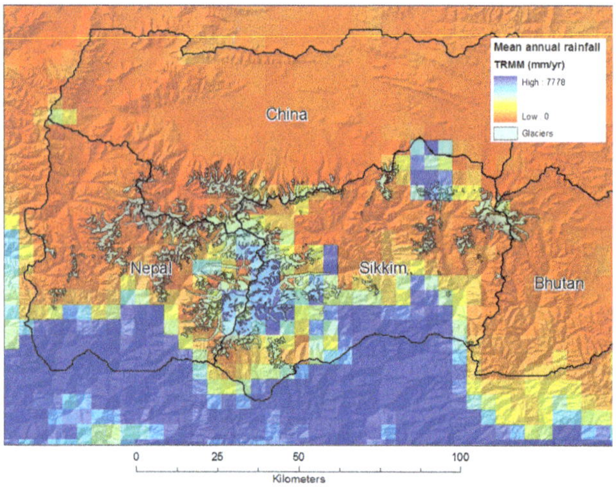

Figure 3. Spatial patterns in TRMM annual precipitation rate derived from the 3B43 data set for spatial domain 1. Also shown are the four main basins delineated based on topography and watershed functions. 2000 glacier outlines are also shown. We note several cells of high precipitation at high altitudes over the Kanchenjunga summits and parts of Tibet, most likely errors in TRMM data.

2. The Swiss topographic map (1 : 150 000 scale), compiled from Survey of India maps from the 1960s and published by the Swiss Foundation for Alpine Research, was used for manual corrections of the 1962 Corona glacier outlines to discard any seasonal snow and to correct shadow areas or bright water bodies that could be misclassified as ice. The exact month or year of each quadrant or of the original air photos is not known with certainty because the original large-scale Indian topographic maps at this scale are restricted to within 100 km of the Indian border and are therefore inaccessible (Srikantia, 2000; Survey of India, 2005); however this map was useful for manual corrections of the Corona outlines.

3.2 Analysis extents

We defined three analysis extents for our study area (Fig. 1 and Table 2):

1. The Landsat/ASTER domain includes the Sikkim province of India, parts of eastern Nepal (Tamor and Arun basins) and parts of Bhutan and China (Table 2). This domain was used to construct the updated 2000 glacier inventory.

Table 3. Topographic zones in spatial domain 1.

	N side (China)	W side (Nepal)	E side (Sikkim)	E side (Bhutan)
Mean basin elevation (m)	4931	4819	4658	4491
Mean rainfall TRMM ($mm\,yr^{-1}$)	146	805	977	383

The Landsat/ASTER domain was split into four subregions on the basis of east–west and north–south climate/topographic/political barriers, as shown in Fig. 3. Rainfall averages from the Tropical Rainfall Measuring Mission (TRMM) data 2B31 product (Bhatt and Nakamura, 2005; Bookhagen and Burbank, 2006) were used to characterize the subregions climatically. The data set contains rainfall estimates calibrated with ground-control stations derived from local and global gauge stations (Bookhagen and Burbank, 2006) with a spatial resolution of 0.4°, or $\sim$ 5 km. Given the well-known biases in the TRMM data (Bookhagen and Burbank, 2006; Palazzi et al., 2013; Andermann et al., 2011), here we are not concerned with the absolute values of gridded precipitation but only with characterizing the subregions in our study area using relative rainfall values. TRMM data integrated over 10 years (1998–2007) show differences in precipitation patterns among the four regions and justifies our choice of spatial domains (Table 3). The eastern side of the study area (Sikkim) receives higher precipitation amounts than the western side (Nepal) (977 versus 805 $mm\,yr^{-1}$). There is a pronounced north–south gradient in precipitation, with the lowest amount of precipitation noticeable on the Chinese side (146 $mm\,yr^{-1}$) (Table 3).

2. The Corona spatial domain is a subset of the Landsat/ASTER domain, which was covered by the 1962 Corona image. Glacier surface area changes and their dependence on climate and topography were computed for this extent between two time steps: the 1960s (rep-

resented by Corona imagery) and 2000s (represented by Landsat/ASTER). Glaciers from Bhutan in the east and China in the north were not covered by the Corona image, so the area change analysis only focused on glaciers of Sikkim and eastern Nepal, east and west of the topographic divide.

3. The Kanchenjunga–Sikkim domain is a smaller subset covered by all three data sets (Landsat/ASTER, Corona and QuickBird), allowing us to extend the glacier change analysis to 2006. It comprises of 50 glaciers from the Tamor basin (Nepal) and Zemu basin (Sikkim, India), located on the southern slopes of the Himalaya. The high-resolution 1962 and 2006 imagery was used here to illustrate glacier area changes at a smaller scale, to show surface characteristics of debris-covered glaciers and to evaluate mapping of debris-covered glaciers.

3.3 Glacier delineation and analysis

For the 1960s, clean glacier outlines were extracted from the panchromatic Corona imagery by thresholding the digital numbers ($DN > 200 = $ snow/ice), chosen based on visual interpretation. Debris-covered glacier tongues were delineated manually on the basis of lateral moraines and other visual clues such as supraglacial lakes. A 5×5 median filter was used to remove noise (isolated pixels from snowfields or internal rocks), as recommended in other studies (Racoviteanu et al., 2009; Andreassen et al., 2008). Ice polygons with area $< 0.02\,km^2$ were not considered valid glaciers and were excluded from the analysis. Manual corrections were applied subsequently on the basis of the topographic map using on-screen digitizing in areas of poor contrast or transient snow/clouds, which obstructed the view of glaciers.

For the 2000s, glaciers were delineated from the Landsat ETM+ scene using the normalized difference snow index (NDSI) (Hall et al., 1995), with a threshold of 0.7 ($NDSI > 0.7 = $ snow/ice). The NDSI algorithm relies on the high reflectivity of snow and ice in the visible to near infrared wavelengths (0.4–1.2 μm), compared to their low reflectivity in the shortwave infrared (1.4–2.5 μm) (Dozier, 1989; Rees, 2003). Compared to other band ratios (Landsat 3/4 and 3/5), the NDSI glacier map was cleaner and less noisy and was therefore preferred (Racoviteanu et al., 2008b). A 5×5 median filter was used here as well to remove remaining noise, and a few areas were adjusted manually on the basis of the ASTER images, notably frozen lakes misclassified as snow/ice and some glaciers underneath low clouds in the southern part of the image. Some transient snow persisting in the deep shadowed valleys was manually removed from the glacier outlines on the basis of the topographic map. Debris-covered glacier tongues were delineated using multispectral data (band ratios, surface kinetic temperature and texture) from the 27 November 2001 ASTER scene combined with topographic variables in a decision tree, as described in Racoviteanu and Williams (2012).

For the QB (2006) image, clean ice surfaces were delineated using band ratios 3/4, then ISODATA clustering with a threshold of 1.07 (snow/ice > 1.07) and a majority filter of 7×7 to remove noise. Debris-covered tongues for this data set were delineated manually on the basis of supraglacial features (lakes and ice walls), along with lateral and frontal moraines visible on the high-resolution images. We also mapped supraglacial lakes from this high-resolution data based on band ratios along with texture analysis.

For all inventories in the Landsat/ASTER domain, ice masses were separated into glaciers on the basis of the SRTM DEM, using hydrologic functions in an algorithm developed by Manley (2008), described in Racoviteanu et al. (2009). Glacier area, terminus elevation, maximum and median elevation, average slope angle and aspect were extracted on a glacier-by-glacier basis using zonal functions on the SRTM DEM. Average glacier thickness and length were calculated from mass turnover principles and ice flow mechanics by Huss and Farinotti (2012), based on the approach of Farinotti (2009). Their method used our glacier outlines and the SRTM DEM to derive thickness estimates iteratively based on Glen's flow law and a shape factor (Paterson, 1994). For simplicity and consistency for change analysis, we assumed no shift in the ice divides over the period of analysis and excluded all nunataks and snow-free steep rock walls from the glacier area calculations. Bodies of ice above the bergschrund were considered part of the glacier (Racoviteanu et al., 2009; Raup and Khalsa, 2007). Glacier area changes (1962–2000) and their dependency on topographic and climatic variables were calculated on a glacier-by-glacier basis for the 232 glaciers in spatial domain 2 using linear regression.

3.4 Uncertainty estimates

Glacier outlines derived from remote sensing data at various spatial and temporal resolution are subject to various degrees of uncertainty, as discussed in recent studies (Paul et al., 2013; Racoviteanu et al., 2009). This becomes an important issue in glacier change analysis, where errors from various data sources accumulate at each processing step. The main sources of uncertainty considered here are (1) image classification errors (positional errors and/or errors due to the semi-automated glacier mapping method) and (2) conceptual errors associated with the definition of a glacier, including mapping of ice divides, mixed pixels of snow and clouds and internal rock differences, which propagate to the glacier change analysis, all described in detail in Racoviteanu et al. (2009).

1. The errors in remote sensing glacier surface areas due to classification ($E_{classif}$) were estimated using Perkal's epsilon band around each glacier outline data

Table 4. Topographic parameters for glaciers in spatial domain 1 and subregions based on 2000s Landsat/ASTER analysis. All parameters are presented on a (a) region-by-region and (b) glacier-by-glacier basis from the SRTM DEM. Debris cover fraction is calculated as a percent of glacier area of debris-covered glaciers only.

Parameter	All	Nepal	Sikkim	Bhutan	China
(a) Region-wide averages					
Number of glaciers	487	162	186	30	109
Glacierized area (km^2)	1463 ± 88	488 ± 29	569 ± 34	106 ± 6	300 ± 18
Number of debris-covered tongues	68	30	27	7	4
Debris cover area (km^2)	161 ± 10	64 ± 4	78 ± 5	14 ± 1	6 ± 0.4
Debris cover (% total glacier area)	11	13	14	13	2
(b) Glacier averages					
Minimum elevation (m)	4908	4760	4702	4926	5425
Median elevation (m)	5702	5715	5569	5652	5950
Maximum elevation (m)	6793	6928	6908	6685	6530
Slope (°)	23	24	23	27	21
Aspect (°)	177	236	131	134	180
Mean glacier size (km^2)	3	3	3	4	3
Length (km)	2	2	2	3	2
Thickness (m)	24	23	23	31	27
Debris cover fraction (%)	23	21	23	32	17

set (Racoviteanu et al., 2009; Bolch et al., 2010), with a ~ 1-pixel variability (Congalton, 1991). Using ±30 m for Landsat/ASTER, ±6 m for Corona and ±3 m for QB outlines, the area uncertainty was ±3, ±6 and ±2 % of the glacierized area for Corona, Landsat/ASTER and QuickBird, respectively. The Perkal method is known to slightly overestimate the errors, as described in Burrough and McDonnel (1998). Recent glacier analysis comparison experiments reported a range of uncertainty of < 5 % for remote sensing glacier outlines compared to high-resolution imagery (Raup et al., 2007; Paul et al., 2013). For manually adjusted glacier outlines, particularly debris-covered tongues, we used screen digitizing in streaming mode with a high density of vertices to minimize area errors (B. Raup, National Snow and Ice Data Center, personal communication, 2014).

2. Uncertainties due to different digitization of internal rocks (E_{rocks}) were derived by comparing area changes computed with internal rocks specific to each data set versus "merged" internal rocks from all data sets. The differences in glacier data sets due to rock inconsistencies amounted to ~ 2 % of the glacier area. To minimize uncertainties in the glacier area change, we merged rock outcrops from each data set and removed them from all the area calculations. The "inactive" bodies of ice above the bergschrund were included as part of the glacier (Racoviteanu et al., 2009). For simplicity, we neglected the area change that might be due to exposure of new internal rock due to glacier ice thinning.

Total errors in glacier area estimate for each data set (E) were calculated as RMSE of the classification ($E_{classif}$) and the internal rocks (E_{rock}):

$$E = \sqrt{E_{classif}^2 + E_{rocks}^2}. \tag{1}$$

Errors in glacier surface area change (E_{change}) from 1962 to 2000 were computed as RMSE of the total error for each time step calculated in Eq. 1:

$$E_{1962-2000} = \sqrt{E_{1962}^2 + E_{2000}^2}. \tag{2}$$

4 Results

4.1 The 2000 Landsat/ASTER glacier characteristics

The 2000 glacier inventory based on Landsat and ASTER yielded 487 glaciers (of which 162 were situated in Nepal, 186 in Sikkim, 30 in Bhutan and 109 in China), covering a total surface area of $1463 \pm 88\,km^2$ (Table 4a). Of the 487 glaciers in this spatial domain, 68 glaciers (13 %) had debris cover on their ablation areas. Supraglacial debris covered $160 \pm 10\,km^2$ (11 % of the glacierized area in spatial domain 1), with some differences between north and south slopes of the study area (discussed in Sect. 5.1). In Sikkim, supraglacial debris covered an area of $78 \pm 5\,km^2$ in 2000 (14 % of the glacierized area).

In 2000, glacier size ranged from 0.02 to $105\,km^2$, with an average size of $3\,km^2$ and a median size of $0.9\,km^2$ (Table 4b). The histogram of glacier area (Fig. 4a) is skewed to

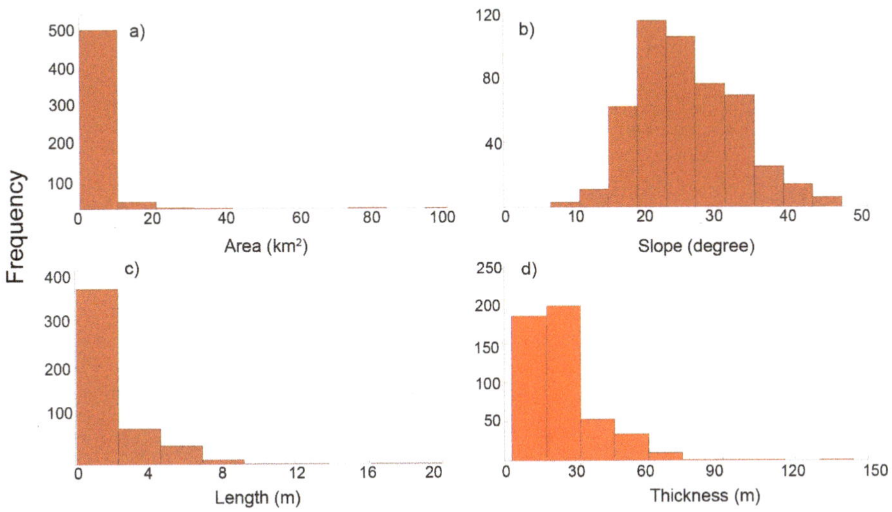

Figure 4. Frequency distribution of glacier parameters for the 487 glaciers in spatial domain 1 based on Landsat/ASTER analysis: **(a)** area, **(b)** slope, **(c)** length and **(d)** thickness. Glaciers smaller than 10 km^2 in area, < 2 km in length and < 30 m thickness are prevalent, with an average slope of 23°.

the right (skewness = 8.4), showing that glaciers with area < 10 km^2 are predominant in this region, and glacier size decreases non-linearly. The long right-tail extremes represent only a few glaciers, such as Zemu with an area > 100 km^2.

The average slope of all glaciers in the inventory was 23° with a positive skew (skewness = 0.38) (Table 4b and Fig. 4b) and no significant differences among the four regions ($p > 0.05$) (Table 4b). Glacier length ranged from 0.08 to 23 km (Zemu glacier), with an average of 2 km (Table 4b and Fig. 4c). Glacier thickness ranged from 3 to 144 m, with the highest frequency for thicknesses less than 30 m (Fig. 4d). The frequency distribution of both glacier length and thickness were positively skewed with long tails, indicating the prevalence of short, shallow valley-type glaciers. Glacier aspect shows two predominant orientations: west-northwest (W-NW) and east-northeast (E-NE), following the topographic divide (Fig. 5). On average, glaciers on the Nepal side had an average aspect of 237° (SW), whereas glaciers on the Sikkim side had an average aspect of 131° (SE), consistent with local topography (Table 4b).

Glacier terminus elevations in the Landsat/ASTER domain ranged from 3990 to 5777 m, with a mean of 4908 m; median glacier elevation ranged from 4515 to 6388 m, with a mean of 5702 m (Table 4b). Considering glacier median elevation as a coarse approximation of glacier equilibrium line altitude (ELA), our results are in agreement with Benn and Owen (2005), who documented higher ELAs on the northern slopes of the Himalaya (6000–6200 m) compared to ELAs on the southern slopes (4600–5600 m).

4.2 Glacier area changes 1962–2000/2006

Overall, glaciers in the Corona spatial domain 2 lost 182.5 ± 40 km^2 of their area (19 ± 7 %, or 0.5 ± 0.2 % yr^{-1}) from

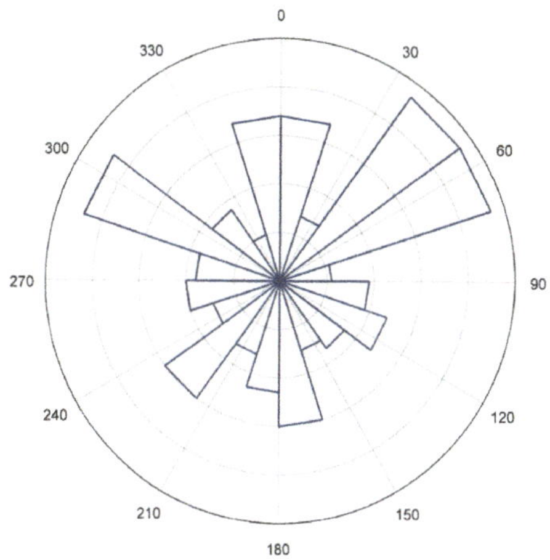

Figure 5. Aspect frequency distribution of the 487 glaciers in spatial domain 1 based on Landsat/ASTER analysis. On average, glaciers in this area are preferentially oriented towards NW (300°) and NE (60°).

1962 to 2000 (Table 5). Overall, the average glacier area changes were slightly smaller on the western side of the divide (Nepal, 16.9 ± 4 %, 1962–2000 or 0.44 ± 0.2 % yr^{-1}) compared to the eastern side (Sikkim, 20.1 ± 8 %, 1962–2000 or 0.52 ± 0.2 % yr^{-1}). When focusing on a smaller glacier subset in the Kanchenjunga–Sikkim subset area (50 glaciers), we obtained an area loss of 10 % ±3 % (0.23 ± 0.08 % yr^{-1}) based on high-resolution imagery (1962–2006) (Table 5). The rates of glacier area loss for this group are overall 50 % lower than the rates of loss in the larger spatial domain 1 perhaps due to higher percentage of debris (21 %) compared to the entire Landsat/ASTER spatial domain 2 (11 %).

Table 5. Overall glacier area changes (loss) (east versus west) for the 232 glaciers in spatial domain 2 from 1962 (Corona) to 2000 (Landsat/ASTER).

Subregion	Area (km^2)		Area loss 1962–2000		
	1962	2000	km^2	%	$\%\,yr^{-1}$
Nepal	323.9 ± 10	269.1 ± 16	54.8 ± 19	16.9 ± 6	0.44 ± 0.2
Sikkim	634.7 ± 19	507.0 ± 35	127.7 ± 42	20.1 ± 8	0.52 ± 0.2
All spatial domain	958.7 ± 31	776.1 ± 47	182.5 ± 40	19.0 ± 4	0.50 ± 0.2

Table 6. Glacier area loss for debris-covered versus clean glaciers in spatial domain 2 from 1962 to 2000. Area loss is shown as a percent of glacier area on a glacier-by-glacier basis.

	Sikkim		Nepal		All	
Glacier type/ Subregion	Number of glaciers	Area loss (%)	Number of glaciers	Area loss (%)	Number of glaciers	Area loss (%)
Clean glaciers	144	34.7	53	31.6	197	33.9
Debris-covered glaciers	20	20.8	15	23.8	35	22.1
Both types	164	33.0	68	29.9	232	32.1

On a glacier-by-glacier basis, glaciers in the Corona domain lost 2–95 % of their area, with a mean of 32 % from 1962 to 2000 (Fig. 6). The spatial distribution of these area changes, illustrated in Fig. 6, shows that the largest area changes (> 70 % area loss) occurred for only a few isolated glaciers in the northern and southern extremities of the study area (17 glaciers). A closer examination of these glaciers revealed that these were small clean glaciers (< 0.1 km^2) with steep slopes (mean of 26°), indicating a need to investigate the topographic controls on area change and clean versus debris-covered glaciers separately.

Clean glaciers lost more of their area from 1962 to 2000 (34 %) compared to debris-covered glaciers (22 %) across the region, with few differences from east to west (Nepal and Sikkim) (Table 6). The difference in mean rates of area change between clean and debris-covered glaciers was statistically significant based on a two-sample F test ($p < 0.05$).

Figure 7a and b show a larger spread and a higher percentage of surface area loss of clean glaciers compared to debris-covered glaciers. For both glacier types, however, there is a high variability in percent area change, perhaps due to other factors such as local topography.

Linear regression analysis showed that percent area change per glacier was negatively correlated to glacier area, altitudinal range, and glacier median and maximum elevation (significant correlations at 99 % confidence interval, $p < 0.01$) (Table 7). Glacier minimum elevation and slope were significant positive controls on glacier area change at 95 % confidence interval ($p < 0.05$). Solar radiation, precipitation and percent debris were not statistically significant controls on glacier area change ($p > 0.1$, confidence interval 90 %) (Table 7). These are discussed in Sect. 5.3.

Table 7. Linear regression of area change on topographic and climatic variables for the 232 glaciers in the spatial domain 2.

Regression	Coefficient	p value
Glacier area	−0.47	0.0003^b
Altitudinal range	−0.01	$<0.001^b$
Minimum elevation	0.008	0.02^a
Median elevation	−0.01	0.001^b
Maximum elevation	−0.01	$<0.001^b$
Percent debris	−0.004	0.83
Slope	0.47	0.01^a
Aspect	0.03	0.007^b
Solar radiation	0.01	0.74
Latitude	−19.8	0.04^a
Longitude	−4.79	0.93
Precipitation	−0.002	0.26

[a] Significant at the 0.05 level (two-tailed); [b] significant at the 0.01 level (two-tailed).

A further analysis of the clean and debris-covered glaciers showed significant differences in terms of glacier area, area change, minimum elevation, altitudinal range and length based on a two-sample F test for variances) ($p < 0.05$) (Table 8). Clean glaciers in this area are ~ 12 times smaller (1 km^2 on average) than debris-covered glaciers (15 km^2) and they have higher termini elevations (+391 m) and an overall altitudinal range about 3 times smaller than debris-covered glaciers (Table 7). On a glacier-by-glacier basis, clean glaciers lost more area (34 %) than debris-covered glaciers (22 %) from 1962 to 2000. Clean glaciers with smaller altitudinal range tend to display more area loss compared to debris-covered glaciers.

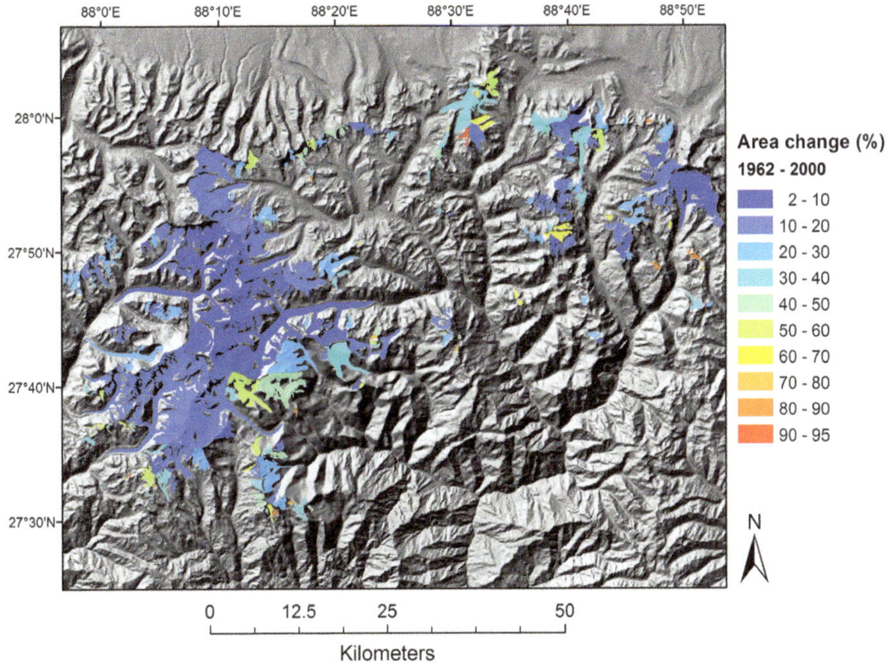

Figure 6. Spatial patterns in glacier area change derived from 1962 Corona and 2000 Landsat/ASTER data on a glacier-by-glacier basis.

Table 8. Comparison of glacier parameters for clean glaciers versus debris-covered glaciers in spatial domain 2.

Parameter	Clean glaciers	Debris-covered glaciers
Area (km^2)	1.2	15.0
Area loss (%)	33.9	22.1
Slope	25.8	24.5
Minimum elevation (m)	5105.6	4714.2
Median elevation (m)	5424.5	5538.9
Altitudinal range (m)	627.6	1928.6
Length (km)	1.3	6.7

5 Discussion

5.1 Spatial distribution of glacier characteristics across the study area

One of the important steps in utilizing our glacier inventory data is to understand spatial patterns in glacier characteristics across the region. Our study area displays region-wide consistency in glacier characteristics, notably glacier area, elevation and topography across the four subregions based on the 2000 glacier data (Table 4). For example, the prevalence of small glaciers noted in this area is consistent with worldwide patterns also observed for the Cordillera Blanca of Peru in a previous study (Racoviteanu et al., 2008a). There is however variability in glacier size within eastern Himalaya: for example, the mean glacier size reported in this study area (3 km^2) is double compared to the Khumbu region west of our study area (1.4 km^2) (Bajracharya and Shrestha, 2011). The glacier slope across our study area (23°) is consistent with average glacier slopes reported for the Khumbu region in Nepal (22°) (Bajracharya and Shrestha, 2011; Salerno et al., 2008), indicating a general tendency for steep glaciers across the region. There are only a few large, long glaciers in the area, such as Zemu glacier (103 km^2, 23 km in 2000). With respect to glacier aspect, we also note similar predominant orientations of glaciers southwards, in the direction of the prevailing monsoon circulation consistent with other studies such as the Khumbu region (average aspect 181°) (Mool et al., 2002; Salerno et al., 2008).

The comparison of glacier characteristics across subregions points to a pronounced gradient from north to south (Bhutan/China subregions compared to Sikkim/Nepal), particularly with respect to glacier elevations and debris cover. Glaciers on the northern side of the divide (China) have higher glacier termini and median elevations compared to the southern side (Nepal and Sikkim) (+700 and +400 m, respectively) (Table 4). These differences seem to be consistent with general air circulation patterns in the area. The Asian summer monsoon brings large amounts of precipitation on the southern slopes of the Himalaya, favoring glacier growth at lower elevations and a lower ELA. In contrast, in the upper reaches of the valleys and on the Tibetan plateau, the monsoon is blocked by the topographic barrier (Clift and Plumb, 2008), causing a drier climate and higher glacier ELAs. There is a much less pronounced east–west gradient in glacier elevations, with higher glacier minimum and me-

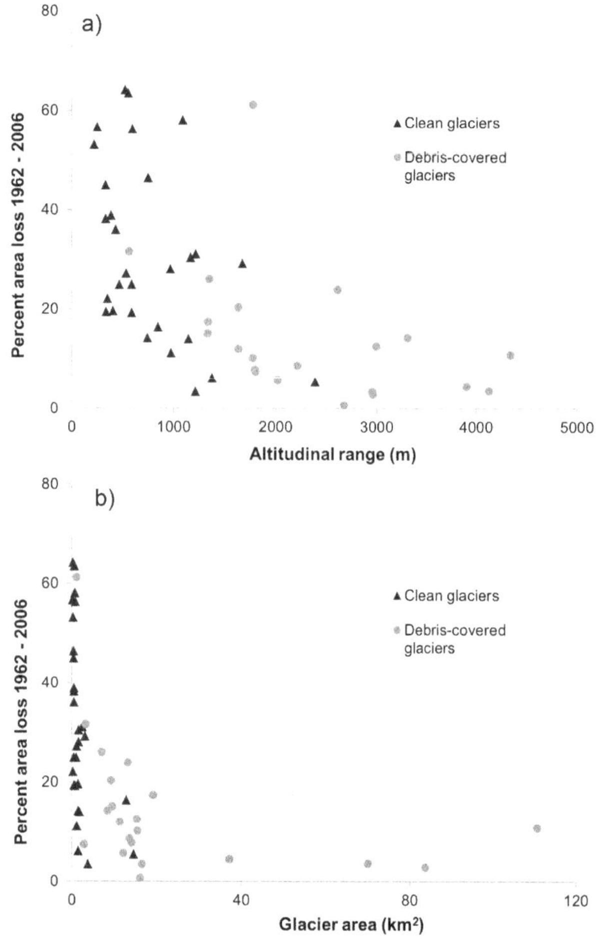

Figure 7. Dependency of glacier area change during 1962–2000 on (**a**) glacier altitudinal range (maximum–minimum elevation) and (**b**) glacier area. Debris-covered glaciers are shown as grey solid circles; clean glaciers are shown as black solid triangles.

dian elevations on the western side (Nepal) (+50 m) compared to the eastern side (Sikkim). This may be explained by the location of Nepalese glaciers on the western side of the topographic divide, away from the prevailing monsoon.

Debris coverage also shows a pronounced variability north to south of the topographic divide. Himalayan glaciers are often referred to as "heavily" debris-covered, but the percent glacierized area covered by supraglacial debris varies across the mountain range. In our study area, debris cover is more prevalent on the southern side of the divide (Sikkim, 14 % of glacierized area) compared to the northern one (China, 2 % of the glacierized area), perhaps due to different geologic and topographic patterns. The northern side of the divide, which is part of the Tibetan plateau, is situated in a monsoon shadow and is therefore dry; the gentler slopes induce lower rates of erosion. In contrast, the southern slopes of the Himalaya tend to be heavily covered with debris due to the abundance of rock material from the steep slopes. The steep slopes made of soft sedimentary rocks and Precambrian crystalline rocks (Mool et al., 2002) and are prone to high rates of erosion, particularly with large amounts of monsoon moisture. This north–south difference in debris cover amount was also noted in other studies (Scherler et al., 2011). In our study, we found a lower percent of debris coverage (21 %) than in the entire central/eastern Himalaya reported in Scherler et al. (2011) (36 % debris cover), or in the Khumbu region, west of our study area, reported by Fujii and Higuchi (1977), Nuimura et al. (2012) (34.8 %), Racoviteanu et al. (2013a) (27 %) and Thakuri et al. (2014) (32 %).

5.2 Regional glacier area changes

The overall rate of surface area loss of $0.5 \pm 0.2\ \%\ yr^{-1}$ from 1962 to 2000 for Sikkim and eastern Nepal obtained here is in agreement with other studies from the southern slopes of the Himalaya. Similar rates of area loss (0.1–$0.3\ \%\ yr^{-1}$) were reported from the Khumbu and Garhwal regions, west of our study area, for approximately the same time period (Thakuri et al., 2014; Basnett et al., 2013; Nuimura et al., 2012; Bolch et al., 2008a; Bhambri et al., 2011). Similarly, for glaciers of Bhutan, east of our study area, Karma et al. (2003) found an average surface glacier area loss of $0.3\ \%\ yr^{-1}$ from 1963 to 1993. It is worth mentioning that these rates of area change are lower than those previously reported for the drier monsoon-transition zone in the western Himalaya (Himachal Pradesh) ($0.7\ \%\ yr^{-1}$) by Kulkarni et al. (2007). In a more recent study, Bahuguna et al. (2014) found lower rates of glacier area loss ($0.4\ \%\ yr^{-1}$) for the same area, which is in agreement with rates of area loss we report here for eastern part. Updated glacier area changes from other recent studies (Racoviteanu et al., 2014; Bolch et al., 2012) also point at lower rates of area loss than previously reported, particularly for the Indian Himalaya. The similar overall rate of glacier area change in the eastern compared to the western part for both debris-covered glaciers and clean glacier types suggests consistent patterns across the region (Table 5).

The smaller glacier area loss for debris-covered glaciers noted in our study is in agreement with studies from Khumbu (Nuimura et al., 2012; Thakuri et al., 2014) or other studies in the central/eastern Himalaya (Bolch et al., 2008a; Thakuri et al., 2014; Bhambri et al., 2011). These studies also reported lower rates of glacier surface area loss and even stable or less retreating glacier termini for debris-covered tongues compared to clean glaciers (Scherler et al., 2011). Area changes for debris-covered glaciers need to be interpreted with caution due to the wide variability in debris cover characteristics such as thickness. Furthermore, these stagnating or less changing tongues may not reflect the true state of the glaciers, for example patterns of glacier thinning which may occur at similar rates to clean glaciers (Gardelle et al., 2012a; Kääb et al., 2012).

5.3 Topographic and climatic controls on area changes

While the consistent area change patterns across the subregions (east to west) are useful for comparison with larger areas, these patterns cannot be used to understand glacier-by-glacier variability in area changes, which may be controlled by local topography and climate. In this study, we found that topographic factors, notably glacier size, glacier altitude (maximum, median, altitudinal range) and aspect, were most important in determining rates of glacier area loss in spatial domain 1. Glacier size plays a significant role in determining area change, i.e., smaller glaciers experienced higher rates of area loss (Table 7). The tendency of larger glaciers to lose less area ($> 20\,km^2$) was observed in various studies (Racoviteanu et al., 2008a; Salerno et al., 2008; Loibl et al., 2014), though in the case of Salerno et al. (2008) the correlation was not statistically significant for the Khumbu. Higher glacier elevations and larger altitudinal ranges significantly reduce the rates of area loss, as was also noted in the Khumbu region in Nepal and elsewhere. The dependency of area change on glacier size and elevation is also consistent with observations from the Cordillera Blanca of Peru (Racoviteanu et al., 2008a) in the outer tropics, indicating consistent patterns in glacier area changes worldwide.

Glacier slope also plays a significant role in determining glacier area change, i.e., the steeper the glacier, the larger the area loss observed in our study. The same tendency was observed in the Khumbu area (Salerno et al., 2008), but the correlation is less significant than the glacier altitude ($p < 0.05$). The presence of gentle slopes covered with supraglacial debris in the ablation areas of glaciers, fairly common in this area, may have reduced the strength of the correlation. Glacier aspect was also found to be a significant control on area change, with more area loss for glaciers oriented southwards and southwest ($p < 0.01$). This is in agreement with findings from Salerno et al. (2008) for the Khumbu region but is in contrast with results from Loibl et al. (2014) for the Nyainqêntanglha Range in southeastern Tibet, about 600 km east of our study area, who found that south-facing glaciers experienced lower rates of terminus retreat. Percent debris cover was a negative control on area change, i.e., glaciers with more extensive debris cover on their areas tend to lose less area overall, but this was not statistically significant ($p > 0.05$). Debris-covered glaciers may benefit from the insulating effect of debris cover above a certain "critical" debris thickness (Mihalcea et al., 2008a; Zhang et al., 2011), which needs to be further investigated.

Geographic location (latitude and longitude) were negative controls on glacier area change, suggesting that glaciers located north and eastwards of the study area tend to lose more area, but only latitude was statistically significant ($p < 0.05$). Climate indices (precipitation and solar radiation) were not significant factors controlling glacier area loss. In contrast, Loibl et al. (2014) showed that glaciers located in a monsoon-influenced area were more sensitive to climate change. This is in agreement with larger-scale studies (Gardelle et al., 2013) that indicated a tendency for enhanced glacier wastage in the eastern, monsoon-influenced parts of the Himalaya. With respect to climatic factors in this area, Basnett et al. (2013) reported an increase in mean annual temperature, more significantly in the winter ($+2\,^{\circ}C\,yr^{-1}$ in the last four decades). Increasing temperatures on the south slopes of the Himalayas were also noted in other studies (Shrestha et al., 2000; Thakuri et al., 2014) based on instrumental data but were estimated to have less effect on glacier area than changes in precipitation because of the orientation of these glaciers towards the prevailing monsoon circulation. In our study, the climatic control on glacier area is not conclusive, and finer-resolution, more accurate temperature and precipitation data sets would be needed. Furthermore, similarly to areas further east (Loibl et al., 2014), average annual solar radiation and latitude were not found to be significant controls on glacier area change in our study.

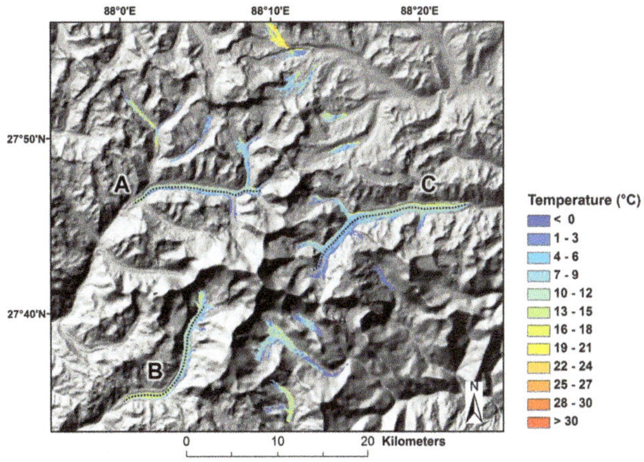

Figure 8. Daytime surface temperatures on several debris-covered tongues, extracted from ASTER data from the 29 October 2002 image. Labels point to: A – Kanchenjunga glacier, B – Yalung glacier and C – Zemu glacier.

5.4 Surface temperature distribution on debris-covered tongues

Lower rates of area change for the debris-covered glaciers may be further explained by surface characteristics of debris cover (thickness, thermal conductivity and resistance). However, these are not easily available in this area due to the lack of field-based measurements and the difficulty of conducting surveys on the debris-covered tongues. Therefore, in this study, we are only qualitatively showing the distribution of surface temperature on selected debris-covered tongues in spatial domain 3 based on the 2002 ASTER scene. Figure 8 shows a high variability in supraglacial surface temperature at 90 m spatial resolution, but there is no clear general temperature trend for the eastern slopes (Sikkim side) versus the western slopes (Nepal) side. The fluctuations in surface temperatures along transects are clearly visible in Fig. 9, with some sharp spikes of high and low temperatures, particularly for Kanchenjunga and Yalung glaciers (labeled "A" and "B" in Fig. 8). This strong variability in supraglacial debris temperatures may be due to the presence of surface features such as debris thickness, size of the debris particles and thermal resistance and conductivity of the debris. For the debris-covered tongues investigated here, the supraglacial temperatures range from 0 to 30 °C, suggesting that the supraglacial debris heats up considerably during the day. At the glacier scale, temperature drops over supraglacial features such as ice walls and supraglacial lakes, which tend to be colder than the surrounding debris, and this is visible even at the coarse spatial resolution of the temperature data (90 m). In Fig. 9 we note the slight upward trend of the supraglacial temperature towards the glacier termini, particularly for Zemu glacier ("C" in Fig. 8). For this glacier, the upper-middle part of the debris surface is colder (−3 to 5 °C) than the last 10 km towards the glacier terminus (5 to 14 °C) (Fig. 9). In a different paper (Racoviteanu and Williams, 2012), we found similar patterns of surface temperature increasing towards the glacier terminus for the same glacier but based on a different ASTER scene (November 2001), indicating consistent patterns for this glacier. The higher surface temperatures towards the glacier terminus may indicate a thicker debris cover that insulates the ice underneath, noted in other studies (Mihalcea et al., 2008a). The daytime debris temperature ranges and strong spatial variability noted here are similar to the ones we found for Khumbu, west of this study area (−3 to 17 °C) (Racoviteanu et al., 2013b). In Khumbu, we found that supraglacial debris had a distinct temperature signal compared to other surfaces such as non-ice moraine, clean ice and supraglacial/proglacial lakes, with more pronounced differences among these three during the daytime.

The suitability of ASTER-based surface for inferring debris characteristics, most notably thickness, has been demonstrated in other studies (Mihalcea et al., 2008a; Zhang et al., 2011). For this study area, there were no field measurements available to test the validity of ASTER temperatures for quantifying supraglacial debris characteristics. However, in a different study (Racoviteanu et al., 2013b), we validated ASTER-based surface temperatures extracted from nine night scenes from 2010 to 2011 for the Khumbu by inverting field-based long-wave radiation (L_{out}) using the Stefan–Boltzmann law ($L_{out} = \varepsilon T^4$). The measurements were from the automatic weather station installed on Changri Nup glacier (Wagnon et al., 2013). We found a good agreement between ASTER temperatures and field-based measurements ($R^2 = 0.92$) using a sensitivity analysis ($\varepsilon = 0.97 \pm 0.1$) to account for small-scale variability in emissivity. Given that the Kanchenjunga–Sikkim area has similar characteristics to Khumbu in terms of debris cover and geographic location and that the images were acquired around the same time of the year as the Khumbu (November–January), we consider

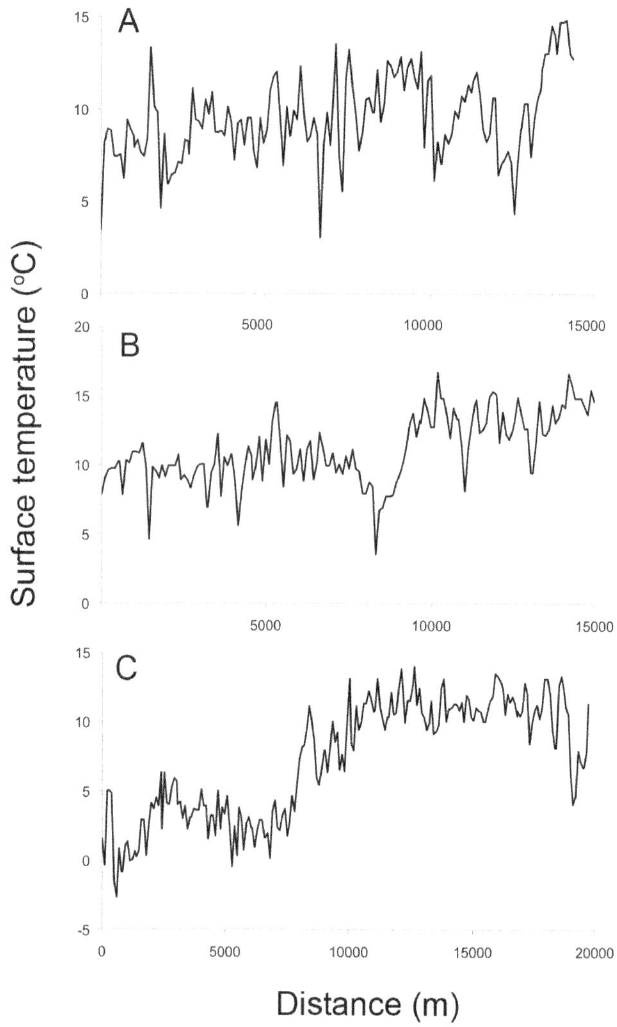

Figure 9. Surface temperature distribution along longitudinal transects from selected glaciers (shown in Fig. 8): A is Kanchenjunga glacier, B is Yalung glacier and C is Zemu glacier. Distance on the x axis is measured from the upper part of the debris-covered area down glacier to the terminus.

that this validation may be applicable to the present study area.

5.5 The role of glacier lakes

The role of supraglacial/proglacial lakes for glacier area change in this area of the Himalaya was addressed in detail in recent studies (Basnett et al., 2013; Bajracharya et al., 2014). Gardelle et al. (2011) also pointed out the increased formation of supraglacial lakes particularly in the eastern part of the Himalaya. A quantitative assessment of lake formation is beyond the scope of this paper; here we only illustrate qualitatively some of the changes occurring on glaciers with supraglacial or proglacial lakes using high-resolution Corona and QuickBird imagery. For the Teesta basin in Sikkim, Mool

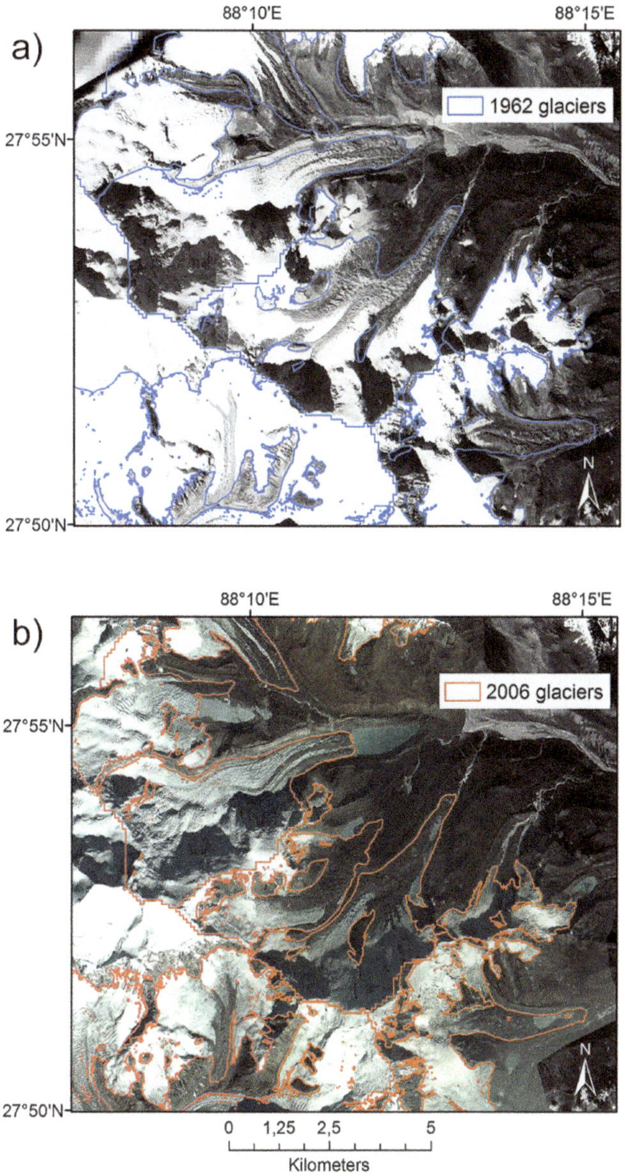

Figure 10. Area changes for some glaciers in the Zema Chhu basin Sikkim from 1962 to 2006: **(a)** 1962 Corona-based glacier outlines (in blue) and **(b)** 2006 QB glacier outlines (in orange).

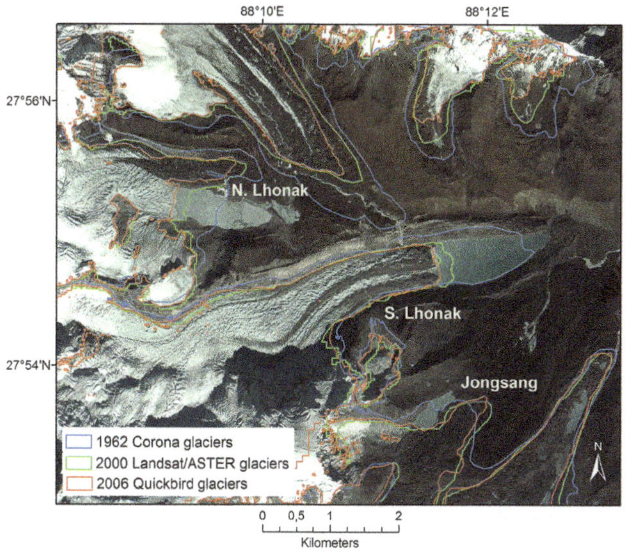

Figure 11. Close-up view of glacier area changes around the North and South Lhonak glaciers from 1962 to 2006, showing changes in the pro-glacial lakes.

and Bajracharya (2003) inventoried 266 glacier lakes covering a total area of 20 km^2 (3.5 % of the glacierized area) based on 2000 Landsat ETM+ imagery. For spatial domain 3, we estimated that glacier lakes covered 1.3 % of the total debris-covered glacier area, or 5.8 % of the area if we consider only the debris-cover (ablation) part, based on the QB/WV2 imagery. Salerno et al. (2012) reported similar percentage for the area of supraglacial lakes, i.e., 0.3–2 % of the overall glacierized area for the Khumbu region. While supraglacial lakes do not cover extensive areas of the glacierized surface, they were shown to increase surface ablation rates in this part of the Himalaya (Fujita and Sakai, 2014; Sakai et al., 2002). It was also shown that supraglacial lakes located at the glacier terminus tend to merge to create large, fast-growing proglacial lakes that accelerate glacier area loss (Basnett et al., 2013; Bajracharya et al., 2014).

Some of the proglacial lakes in our study area are visible in Figs. 10 and 11 for the northern part of spatial domain 3 (Changsang, east Langpo, Jongsang, middle Lhonak, south Lhonak). Most of these lakes are moraine-dammed lakes considered dangerous for potentially inducing glacier lake outburst flood events and were shown to accelerate the glacier area loss in recent decades (Bajracharya et al., 2014). Figure 10a–b show the evolution of the proglacial lake on North and South Lhonak glaciers in Sikkim, also noted in Basnett et al. (2013). A closer look at a subset area (Fig. 11) shows the visible growth of a proglacial lake for the adjacent North Lhonak and South Lhonak glaciers. We estimate that these two glaciers retreated $\sim$ 650 m and 1.3 km from 1962 to 2006, respectively. Another branch of the North Lhonak glacier has wasted significantly by $\sim$ 1.5 km from 1962 to 2006, and a glacier outlet is now clearly visible. The northern branch of Jongsang glacier was entirely covered by a supraglacial lake in 2006, while another part shows less significant rates of terminus retreat ($\sim$ 100 m in 44 years). A part of the Jongsang glacier shows a slight "false" glacier tongue advance, most likely due to uncertainties in the mapping of Corona imagery. While our purpose here is not to present glacier length changes, we note that these estimates are in agreement with trends of glacier thinning and increased glacier lake formation reported in this area of the Himalaya previously (Gardelle et al., 2011; Basnett et al., 2013; Kääb et al., 2012).

5.6 Uncertainty and limitations

Inconsistencies in glacier area change estimates have been pointed out in other studies, for the Himalaya and elsewhere (Racoviteanu et al., 2008a, b), and are also noted in the current study. Glacier area changes in the Himalaya are heterogeneous and depend on a variety of factors including local topography and climate, so some caution should be applied when comparing rates of area changes from one area to other areas, even in the same climatic zone. For example, for Sikkim we estimated a surface area loss of $88.9 \pm 5\,\text{km}^2$ (13.5 % from 1962 to 2006, or $0.36 \pm 0.17\,\%\,\text{yr}^{-1}$). Other studies in this area point to contrasting results. For the same geographic area, Basnett et al. (2013) reported an area loss of $0.16 \pm 0.10\,\%\,\text{yr}^{-1}$ from 1989/1990 to 2009/2010), which is about half of the area change in our findings. In contrast, a recent study (Bahuguna et al., 2014) reported the highest rates of area loss (about $0.8\,\%\,\text{yr}^{-1}$) for the last decade, even higher than rates reported previously for the western Himalaya by Kulkarni et al. (2007). We speculate that such large differences might be due to errors inherent in the baseline data sets coupled with misclassification due to snow cover or debris-covered areas.

Glacier area changes reported for Sikkim in different studies, using a variety of data including topographic maps (Table 9), illustrate this point. For example, for Sikkim, our study estimated $569 \pm 70\,\text{km}^2$ of glacierized area in 2000 based on Landsat/ASTER data. For the same time period, Mool et al. (2002) reported an area of $577\,\text{km}^2$ based on the same source imagery (Landsat ETM+) (Table 9). These two area estimates differ only by $8.2\,\text{km}^2$ (1.4 %) of our estimated area and only the number of glacier differs substantially (186 glaciers in our study compared to 285 glaciers in ICIMOD study) most likely due to the way in which ice masses were split and how glaciers were counted. Methodology differences and inconsistencies in glacier estimates are quite common in multi-temporal image analysis performed by different analysts and were previously noted in other areas of the world (Racoviteanu et al., 2009). Similarly, for the 1962 decade, our analysis of Corona 1962 imagery for Sikkim yielded 178 glaciers with an area of $658 \pm 20\,\text{km}^2$. In a recent publication (Racoviteanu et al., 2014), we reported 158 glaciers with an area of $742\,\text{km}^2$ for the 1960s based on the Swiss topographic map. The GSI (Sangewar and Shukla, 2009) reported 449 glaciers with an area of $706\,\text{km}^2$ for the 1970s based on topographic maps. Our 1962 Corona glacier inventory yields a smaller total glacier area than the one based on the topographic map ($84\,\text{km}^2$, or 11 %) (Racoviteanu et al., 2014) or the GSI inventory based on topographic maps ($48.3\,\text{km}^2$, or 7 % area) (Sangewar and Shukla, 2009). We consider that both of these latter mentioned studies overestimated the glacier area in the 1960s, perhaps due to the presence of persistent snow in the source aerial imagery. We consider the Corona 1962 data set to be more reliable than the inventories based on topographic maps, and hence we used this data set as baseline for comparison with the recent imagery.

Glacier inventories in Sikkim for the recent decades also point to contradictory patterns. For the 1980s, another study (Kulkarni, 1992) reported a glacierized area of $431\,\text{km}^2$ for 1987/1989 based on Indian IRS-1A and Landsat data. Considering our 1962 Corona inventory, this would imply an area loss of 42 % since 1962 ($2.1\,\%\,\text{yr}^{-1}$) followed by a strong increase in glacier area (+33.5 %, or $+3\,\%\,\text{yr}^{-1}$) from 1987/1988 to 2000 (based on our Landsat analysis), which is undocumented and unlikely in this area. We consider the 1987/1989 estimates to be highly unreliable, given that there are no glacier surges that might induce an apparent "glacier growth". In some areas, we noted omissions of some debris-covered tongues from the glacier maps, which might explain some of the differences.

6 Summary and outlook

In this study we combined remote sensing data from various sensors to construct a new glacier inventory for the Kanchenjunga–Sikkim region in the eastern Himalaya. Based on 1962 Corona and 2006 QuickBird imagery, we found an overall negative glacier surface area loss of $0.5 \pm 0.2\,\%\,\text{yr}^{-1}$ since 1962, in agreement with those noted in other studies in the Himalaya. The area loss rates reported here are lower than the average rate of $0.7\,\%\,\text{yr}^{-1}$ reported in other glacierized areas of the world such as the Alps (Kääb et al., 2002), the Tien Shan (Bolch, 2007) and the Peruvian Andes (Racoviteanu et al., 2008a). Glaciers exhibit heterogeneous patterns of area change depending on topographic and climatic factors, most notably glacier altitude (maximum, median, altitudinal range), glacier size, slope and aspect. Glacier area changes depend strongly on glacier size and elevation, which is consistent with other areas in the central/eastern Himalaya (Thakuri et al., 2014) or elsewhere, for example in the outer tropics (Racoviteanu et al., 2008a). The conclusions drawn with respect to spatial patterns in glacier characteristics, glacier area loss and their topographic and climatic dependency are as follows:

- strong north–south gradient in terms of glacier elevations and debris cover, with larger percent of area covered by debris and higher glacier elevations on the northern side for the divide but less east–west gradient in these characteristics;

- supraglacial debris cover is prevalent on the southern slopes of the Himalaya (14 % of the glacierized area) compared to northern slopes (2 %);

- supraglacial lakes constitute about 6 % of the debris-covered area, and some of these supraglacial lakes have merged to form proglacial lakes;

Table 9. Glacier area change in Sikkim based on previous studies. The percent area change is given with respect to the 1962 Corona glacier inventory from this study.

Study	Year	Data source	Number of glaciers	Area (km^2)	Area change since 1960s	
					%	% yr^{-1}
This study	1962	Corona KH4	178	658 ± 20	–	–
Geological Survey of India (1999)	~ 1960–1970s	Indian 1 : 63 000 topographic maps	449	706	+7.3	+0.9
Kulkarni and Narain (1990)	1987/1989	IRS-1C satellite images	n/a	426	−35.0	−1.4
ICIMOD Mool et al. (2002)	2000	Landsat TM, IRS-1C, topographic maps	285	577	−11.4	−0.3
This study	2000	Landsat TM, ASTER	185	569 ± 34	-13.5 ± 6.4	-0.3 ± 0.1

- glacier area loss of 0.5 % ±0.2 % yr^{-1} from 1962 to 2000, with some differences on the eastern side of the divide (Sikkim, 0.52 % ±0.2 % yr^{-1}) versus the western part (Nepal, 0.44 % ± 0.2 % yr^{-1});
- higher rates of area loss for clean glaciers (34 %, or 0.7 % yr^{-1}) compared to debris-covered glaciers (14.3 % or 0.3 % yr^{-1}) across the subregions on a glacier-by-glacier basis;
- the amount of glacier area loss is partly controlled by a glacier's headwater elevation, altitudinal range, glacier area, slope and aspect, with the largest area loss observed for small, steep glaciers with a smaller altitudinal range and less debris cover;
- while Himalayan glaciers are undoubtedly undergoing negative area change, the rates of area loss noted in this study (0.5 % yr^{-1}) as well as other recent studies in the area (0.2–0.4 % yr^{-1} since the 1960s) are lower than other glacierized areas worldwide (0.7 % yr^{-1}).

The glacier area change estimates reported here are subject to uncertainties most notably with respect to early topographic maps and declassified Corona imagery; therefore a considerable effort was given to minimizing errors by multiple re-iterations of the glacier outlines. The understanding of the spatial patterns of glacier changes in the current study is limited by (1) a lack of a baseline elevation data set for the 1960 to compute glacier elevation changes from 1960s to 2000 and (2) lack of field-based measurements to validate debris-cover mapping and surface temperature distribution. With respect to the latter, while surface temperature trends show a slight increase towards the terminus, suggesting a thicker debris cover, the supraglacial surface temperatures are highly heterogonous and require additional investigation. A further improvement in the current study will be to include the supraglacial and proglacial lakes and surface temperature as determinant factors for the glacier area change, perhaps in a more sophisticated multivariate regression model. The glacier data sets constructed in this study can be further utilized to understand the behavior of glaciers in this little-investigated area of the Himalaya, particularly with respect to spatial patterns of glacier melt, and the contribution of glaciers to water resources.

Acknowledgements. This research was funded by a NASA Earth System Sciences (ESS) fellowship (NNX06AF66H), a National Science Foundation doctoral dissertation improvement grant (NSF DDRI award BC 0728075), a CIRES research fellowship, a graduate fellowship from CU-Boulder and a post-doc fellowship from Centre national d'études spatiales (CNES), France. Participation of M. Williams was supported by the NSF-funded Niwot Ridge Long-Term Ecological Research (LTER) program and the USAID Cooperative Agreement AID-OAA-A-11-00045. ASTER imagery was obtained through the NASA-funded Global Land Ice Measurements from Space (GLIMS) project. We are grateful to Jonathan Taylor at University of California, Fullerton, for facilitating access to high-resolution imagery through the NASA Appropriations grant no. NNA07CN68G. We thank Damodar Lamsal and an anonymous reviewer for their thorough comments, which helped improve the quality of the manuscript.

Edited by: T. Bolch

References

Ageta, Y. and Higuchi, K.: Estimation of Mass Balance Components of a Summer-Accumulation Type Glacier in the Nepal Himalaya, Geogr. Ann. A, 66, 249–255, 1984.

Andermann, C., Bonnet, S., and Gloaguen, R.: Evaluation of precipitation data sets along the Himalayan front, Geochem. Geophy. Geosy., 12, Q07023, doi:10.1029/2011gc003513, 2011.

Andreassen, L. M., Paul, F., Kääb, A., and Hausberg, J. E.: Landsat-derived glacier inventory for Jotunheimen, Norway, and deduced glacier changes since the 1930s, The Cryosphere, 2, 131–145, doi:10.5194/tc-2-131-2008, 2008.

Bahuguna, I. M., Rathore, B. P., Brahmbhatt, R., Sharma, M., Dhar, S., Randhawa, S. S., Kumar, K., Romshoo, S., Shah, R. D., and Ganjoo, R. K.: Are the Himalayan glaciers retreating?, Curr. Sci., 106, 1008–1013, 2014.

Bajracharya, S. R. and Shrestha, B.: The status of glaciers in the Hindu-Kush Himalayan Region, International Centre for Integrated Mountain Development (ICIMOD), Kathmandu, Nepal, 127 pp., 2011.

Bajracharya, S. R., Mool, P. K., and Shresta, A.: Impact of climate change on Himalayan glaciers and glacial lakes, ICIMOD, Kathmandu, Nepal, 2007.

Bajracharya, S. R., Maharjan, S. B., and Shresta, F.: The status and decadal change of glaciers in Bhutan from 1980's to 2010 based on the satellite data, Ann. Glaciol., 55, 159–166, doi:10.3189/2014AoG66A125, 2014.

Basnett, S., Kulkarni, A., and Bolch, T.: The influence of debris cover and glacial lakes on the recession of glaciers in Sikkim Himalaya, India, J. Glaciol., 59, 1035–1046, 2013.

Benn, D. and Owen, L. A.: The role of the Indian summer monsoon and the mid-latitude westerlies in Himalayan glaciation: review and speculative discussion, J. Geol. Soc., 155, 353–363, 1998.

Benn, D. and Owen, L. A.: Equilibrium-line altitudes of the Last Glacial Maximum for the Himalaya and Tibet: an assessment and evaluation of results, Quat.Int., 138–139, 55–58, 2005.

Berthier, E., Arnaud, Y., Vincent, C., and Remy, F.: Biases of SRTM in high-mountain areas: Implications for the monitoring of glacier volume changes, Geophys. Res. Lett., 33, L08502, doi:10.01029/02006GL025862, 2006.

Berthier, E., Arnaud, Y., Kumar, R., Ahmad, S., Wagnon, P., and Chevallier, P.: Remote sensing estimates of glacier mass balances in the Himachal Pradesh (Western Himalaya, India), Remote Sens. Environ., 108, 327–338, 2007.

Bhambri, R. and Bolch, T.: Glacier mapping: a review with special reference to the Indian Himalayas, Programm. Phys. Geogr., 33, 672–702, 2009.

Bhambri, R., Bolch, T., Chaujar, R. K., and Kulshreshth, S. C.: Glacier changes in the Garhwal Himalayas, India during the last 40 years based on remote sensing data., J. Glaciol., 54, 543–556, 2010.

Bhambri, R., Bolch, T., Chaujar, R. K., and Kulshreshtha, S. C.: Glacier changes in the Garhwal Himalaya, India, from 1968 to 2006 based on remote sensing, J. Glaciol., 57, 543–556, 2011.

Bhatt, B. C. and Nakamura, K.: Charactieristics of Monsoon rainfall around the Himalayas revealed by TRMM precipitation radar, Mon. Weather Rev., 133, 149–165, doi:10.1175/MWR-2846.1, 2005.

Bolch, T.: Climate change and glacier retreat in northern Tien Shan (Kazakhstan/Kyrgyzstan) using remote sensing data, Global Planet. Change, 56, 1–12, 2007.

Bolch, T., Buchroithner, M. F., Pieczonka, T., and Kunert, A.: Planimetric and Volumetric Glacier Changes in the Khumbu Himalaya since 1962 Using Corona, Landsat TM and ASTER Data, J. Glaciol., 54, 592–600, 2008a.

Bolch, T., Buchroithner, M. F., Peters, J., Baessler, M., and Bajracharya, S.: Identification of glacier motion and potentially dangerous glacial lakes in the Mt. Everest region/Nepal using spaceborne imagery, Nat. Hazards Earth Syst. Sci., 8, 1329–1340, doi:10.5194/nhess-8-1329-2008, 2008b.

Bolch, T., Menounos, B., and Wheate, R.: Landsat-based inventory of glaciers in western Canada, 1985–2005, Remote Sens. Environ., 114, 127–137, 2010.

Bolch, T., Pieczonka, T., and Benn , D. I.: Multi-decadal mass loss of glaciers in the Everest area (Nepal Himalaya) derived from stereo imagery, The Cryosphere, 5, 349–358, doi:10.5194/tc-5-349-2011, 2011.

Bolch, T., Kulkarni, A., Kääb, A., Huggel, C., Paul, F., Cogley, J. G., Frey, H., Kargel, J. S., Fujita, K., Scheel, M., Bajracharya, S., and Stoffel, M.: The State and Fate of Himalayan Glaciers, Science, 336, 310–314, doi:10.1126/science.1215828, 2012.

Bookhagen, B. and Burbank, D. W.: Topography, relief and TRMM-derived rainfall variations along the Himalaya, Geophys. Res. Lett., 33, L08405, doi:10.1029/2006gl026037, 2006.

Burrough, P. A. and McDonnel, R. A.: Principles of Geographic Information Systems, Oxford University Press, Okford, UK, 1998.

CGIAR-CSI: Void-filled seamless SRTM data V1, available at: http://srtm.csi.cgiar.org;http://www.ambiotek.com/topoview (last access: 1 April 2014), 2004.

Clift, P. D. and Plumb, R. A.: The Asian Monsoon: Causes, History and Effects, Cambridge University Press, New York, USA, 88 pp., 2008.

Congalton, R. G.: A review of assessing the accuracy of classifications of remotely sensed data, Remote Sens. Environ., 37, 35–46, 1991.

Dashora, A., Lohani, B., and Malik, J. N.: A repository of Earth resource information-CORONA satellite programme, Current Science, 92, 926–932, 2007.

Dozier, J.: Spectral Signature of Alpine Snow Cover from the Landsat Thematic Mapper, Remote Sens. Environ., 28, 9–22, 1989.

Farinotti, D., Huss, M., Bauder, A., and Funk, M.: An estimate of the glacier ice volume in the Swiss Alps, Glob. Planet. Change, 68, 225–231, 2009.

Foster, L. A., Brock, B. W., Cutler, M. E. J., and Diotri, F.: A physically based method for estimating supraglacial debris thickness from thermal band remote-sensing data, J. Glaciol., 58, 677–690, 2012.

Frey, H., Paul, F., and Strozzi, T.: Compilation of a glacier inventory for the western Himalayas from satellite data: methods, challenges, and results, Remote Sens. Environ., 124, 832–843, doi:10.1016/j.rse.2012.06.020, 2012.

Fujii, Y. and Higuchi, K.: Statistical analyses of the forms of the glaciers in Khumbu Himal, Japanese Society of Snow and Ice, 39, 7–14, 1977.

Fujita, K. and Sakai, A.: Modelling runoff from a Himalayan debris-covered glacier, Hydrol. Earth Syst. Sci., 18, 2679–2694, doi:10.5194/hess-18-2679-2014, 2014.

Fujita, K., Suzuki, R., Nuimura, T., and Sakai, A.: Performance of ASTER and SRTM DEMs, and their potential for assessing glacial lakes in the Lunana region, Bhutan Himalaya, J. Glaciol., 54, 220–228, 2008.

Gardelle, J., Arnaud, Y., and Berthier, E.: Contrasted evolution of glacial lakes along the Hindu Kush Himalaya mountain range between 1990 and 2009, Glob. Planet. Change, 75, 47–55, doi:10.1016/j.gloplacha.2010.10.003, 2011.

Gardelle, J., Berthier, E., and Arnaud, Y.: Slight mass gain of Karakoram glaciers in the early twenty-first century, Nat. Geosci., 5, 322–325, 2012a.

Gardelle, J., Berthier, E., and Arnaud, Y.: Impact of resolution and radar penetration on glacier elevation changed computed from DEM differencing, J. Glaciol., 58, 419–422, 2012b.

Gardelle, J., Berthier, E., Arnaud, Y., and Kääb, A.: Region-wide glacier mass balances over the Pamir-Karakoram-Himalaya during 1999–2011, The Cryosphere, 7, 1263–1286, doi:10.5194/tc-7-1263-2013, 2013.

Hall, D. K., G. Riggs, G. A., and Salomonson, V. V.: Development of methods for mapping global snow cover using moderate resolution imaging spectroradiometer data, Remote Sens. Environ., 54, 127–140, 1995.

Huss, M. and Farinotti, D.: Distributed ice thickness and volume of all glaciers around the globe, J. Geophys. Res., 117, F04010, doi:10.1029/2012JF002523, 2012.

IMD: Climatological tables 1951–1980, Government of India, New Delhi, India, 1980.

Immerzeel, W., van Beek, L. P. H., and Bierkens, M. F. P.: Climate Change Will Affect the Asian Water Towers, Science, 328, 1382–1385, doi:10.1126/science.1183188, 2010.

Immerzeel, W. W., Beek, L. P. H. v., Konz, M., Shrestha, A. B., and Bierkens, M. F. P.: Hydrological response to climate change in a glacierized catchment in the Himalayas, Clim. Change, 110, 721–736, 2012.

Iwata, S., Aoki, T., Kadota, T., Seko, K., and Yamaguchi, S.: Morphological evolution of the debris cover on Khumbu Glacier, Nepal, between 1978 and 1995, in: Debris-covered glaciers, edited by: Nakawo, M., Raymond, C. F., and Fountain, A., IAHS, 264, 3–11, 2000.

Kääb, A., Paul, F., Maisch, M., Hoelzle, M., and Haeberli, W.: The new remote-sensing-derived Swiss glacier inventory: II. First results, Ann. Glaciol., 34, 362–366, 2002.

Kääb, A., Berthier, E., Nuth, C., Gardelle, J., and Arnaud, Y.: Contrasting patterns of early twenty-first-century glacier mass change in the Himalayas, Nature, 488, 495–498, 2012.

Kamp, U., Byrne, M., and Bolch, T.: Glacier fluctuations between 1975 and 2008 in the Greater Himalaya Range of Zanskar, southern Ladakh, J. Mt. Sci., 8, 374–389, 2011.

Karma, Ageta, Y., Naito, N., Iwata, S., and Yabuki, H.: Glacier distribution in the Himalayas and glacier shrinkage from 1963 to 1993 in the Bhutan Himalayas, B. Glaciol. Res., 20, 29–40, 2003.

Kaser, G., Grosshauser, M., and Marzeion, B.: Contribution potential of glaciers to water availability in different climate regimes, Proc. Natl. Acad. Sci. USA, 107, 20223–20227, doi:10.1073/Pnas.1008162107, 2010.

Kayastha, R. B., Takeuchi, Y., Nakawo, M., and Ageta, Y.: Practical prediction of ice melting beneath various thickness of debris cover on Khumbu Glacier, Nepal, using a positive degree-day factor, in: Debris-Covered Glaciers, edited by: Raymond, C. F., Nakawo, M., and Fountain, A., IAHS, Wallingford, UK, 71–81, 2000.

Krishna, A. P.: Snow and glacier cover assessment in the high mountains of Sikkim Himalaya, Hydrol. Process., 19, 2375–2383, 2005.

Kulkarni, A. V.: Glacier inventory in the Himalaya, Natural resources management – a new perspective, NNRMS, Bangalore, 474–478, 1992.

Kulkarni, A. and Narain, A.: Remote sensing based glacier inventory in the Sikkim state, Scientific note no. SAC/RSA/RSAG-MWRD/ SN/01/90, 11, 1990.

Kulkarni, A. V., Bahuguna, I. M., Rathore, B. P., Singh, S. K., Randhawa, S. S., Sood, R. K., and Dhar, S.: Glacial retreat in Himalaya using Indian Remote Sensing satellite data, Current Science, 92, 69–74, 2007.

Loibl, D. M., Lehmkuhl, F., and Grießinger, J.: Reconstructing glacier retreat since the Little Ice Age in SE Tibet by glacier mapping and equilibrium line altitude calculation, Geomorphology, 214, 22–39, doi:10.1016/j.geomorph.2014.03.018, 2014.

Manley, W. F.: Geospatial inventory and analysis of glaciers: A case study for the eastern Alaska Range, in Glaciers of Alaska, in: Satellite Image Atlas of Glaciers of the World: USGS Professional Paper 1386-K, edited by: Williams, R. S. Jr. and Ferrigno, J. G., US Geological Survey, Books Express Publishing, K424 K439,2008.

Mason, L. E.: Abode of Snow: A History of Himalayan Exploration and Mountaineering, Dutton, New York City, USA, 372 pp., 1954.

Mihalcea, C., Mayer, C., Diolaiuti, G., D'Agata, C., Smiraglia, C., Lambrecht, A., Vuillermoz, E., and Tartari, G.: Spatial distribution of debris thickness and melting from remote-sensing and meteorological data, at debris-covered Baltoro glacier, Karakoram, Pakistan, Ann. Glaciol., 48, 49–57, 2008a.

Mihalcea, C. E., Brock, B. W., Diolaiuti, G. A., D'Agata, C., Citterio, M., Kirkbride, M. P., Cutler, M., and Smiraglia, C.: Using ASTER satellite and ground-based surface temperature measurements to derive supraglacial debris cover and thickness patterns on Miage Glacier (Mont Blanc Massif, Italy), Cold Reg. Sci. Technol., 52, 341–354, doi:10.1016/j.coldregions.2007.03.004, 2008b.

Mool, P. K. and Bajracharya, S. R.: Tista Basin, Sikkim Himalaya: Inventory of Glaciers and Glacial Lakes and the Identification of Potential Glacial Lake Outburst Floods (GLOFs) Affected by Global Warming in the Mountains of Himalayan Region, International Center for Integrated Mountain Development, Kathmandu, Nepal, 145 pp., 2003.

Mool, P. K., Bajracharya, S. R., Joshi, S. P., Sakya, K., and Baidya, A.: Inventory of Glaciers, Glacial Lakes and Glacial Lake Outburst Floods Monitoring and Early Warning Systems in the Hindu-Kush Himalayan region, Nepal, International Center for Integrated Mountain Development, Nepal, 2002.

Narama, C., Kääb, A., Duishonakunov, M., and Abdrakhmatov, K.: Spatial variability of recent glacier area and volume changes in Central Asia using Corona ($\sim$ 1970), Landsat ($\sim$ 2000), and ALOS ($\sim$ 2007) optical satellite data, Glob. Planet. Change, 71, 42–54, doi:10.1016/j.gloplacha.2009.08.002, 2010.

Nuimura, T., Fujita, K., Yamaguchi, S., and Sharma, R. R.: Elevation changes of glaciers revealed by multitemporal digital elevation models calibrated by GPS survey in the Khumbu region, Nepal Himalaya, J. Glaciol., 58, 648–656, 2012.

Nuimura, T., Sakai, A., Taniguchi, K., Nagai, H., Lamsal, D., Tsutaki, S., Kozawa, A., Hoshina, Y., Takenaka, S., Omiya, S., Tsunematsu, K., Tshering, P., and Fujita, K.: The GAMDAM Glacier Inventory: a quality controlled inventory of Asian glaciers, The Cryosphere Discuss., 8, 2799–2829, doi:10.5194/tcd-8-2799-2014, 2014.

Nuth, C. and Kääb, A.: Co-registration and bias corrections of satellite elevation data sets for quantifying glacier thickness change, The Cryosphere, 5, 271–290, doi:10.5194/tc-5-271-2011, 2011.

Palazzi, E., von Hardenberg, J., and Provenzale, A.: Precipitation in the Hindu-Kush Karakoram Himalaya: Observations and future scenarios, J. Geophys. Res.-Atmos., 118, 85–100, doi:10.1029/2012JD018697, 2013.

Paterson, W. S. B.: The Physics of Glaciers, 3rd ed., Pergamon, Oxford, UK, 480 pp., 1994.

Paul, F., Barrand, N., Berthier, E., Bolch, T., Casey, K., Frey, H., Joshi, S. P., Konovalov, V., Bris, R. L., Mölg, N., Nosenko, G., Nuth, C., Pope, A., Racoviteanu, A., Rastner, P., Raup, B., Scharrer, K., Steffen, S., and Winsvold, S.: On the accuracy of glacier outlines derived from remote sensing data, Ann. Glaciol., 54, 171–182, doi:10.3189/2013AoG63A296, 2013.

Pfeffer, W. T., Arendt, A. A., Bliss, A., Bolch, T., Cogley, J. G., Gardner, A. S., Hagen, J. O., Hock, R., Kaser, G., Kienholz, C., Miles, E. S., Moholdt, G., Mölg, N., Paul, F., Radić, V., Rastner, P., Raup, B. H., Rich, J., and Sharp, M. J.: The Randolph Glacier Inventory: a globally complete inventory of glaciers, J. Glaciol., 60, 537–552, doi:10.3189/2014JoG13J176, 2014.

Racoviteanu, A. and Williams, M. W.: Decision tree and texture analysis for mapping debris-covered glaciers: a case study from Kangchenjunga, eastern Himalaya, Remote Sensing Special issue, 4, 3078-3109, doi:3010.3390/rs4103078, 2012.

Racoviteanu, A., Arnaud, Y., and Williams, M.: Decadal changes in glacier parameters in Cordillera Blanca, Peru derived from remote sensing, J. Glaciol., 54, 499–510, 2008a.

Racoviteanu, A., Williams, M. W., and Barry, R.: Optical remote sensing of glacier mass balance: a review with focus on the Himalaya, Remote Sens. Environ., 8, 3355–3383, doi:10.3390/s8053355, 2008b.

Racoviteanu, A., Paul, F., Raup, B., Khalsa, S. J. S., and Armstrong, R.: Challenges and recommendations in mapping of glacier prameters from space: results of the 2008 Global Land Ice Measurements frpom Space (GLIMS) workshop, Boulder, Colorado, USA, Ann. Glaciol., 50, 53–69, 2009.

Racoviteanu, A., Armstrong, R., and Williams, M.: Evaluation of an ice ablation model to estimate the contribution of melting glacier ice to annual discharge in the Nepal Himalaya, Water Resour. Res., 49, 5117–5133, doi:10.1002/wrcr.20370, 2013a.

Racoviteanu, A., Nicholson, L., and Arnaud, Y.: Surface characteristics of debris-covered glacier tongues in the Khumbu Himalaya derived from remote sensing texture analysis, European Geophysical Union (EGU) General Assembly, Vienna, Austria, 7–12 April, 2013b.

Racoviteanu, A., Arnaud, Y., Bahuguna, I. M., Bajracharya, S., Berthier, E., Bhambri, R., Bolch, T., Byrne, M., Chaujar, R. K., Kääb, A., Kamp, U., Kargel, J., Kulkarni, A. V., Leonard, G., Mool, P., Frauenfelder, R., and Sossna, I.: Himalayan glaciers, in: Global Land and Ice Monitoring from Space: Satellite Multispectral Imaging of Glaciers, edited by: Kargel, J. S., Bishop, M. P., Kääb, A., and Raup, B. H., Springer-Praxis, Heidelberg, Germany, 549–582, 2014.

Raj, K. B. G., Remya, S. N., and Kumar, K. V.: Remote sensing-based hazard assessment of glacial lakes in Sikkim Himalaya, Current Science, 104, 359–364, 2013.

Raup, B. and Khaksa, S. J. S.: GLIMS Analysis tutorial: http://www.glims.org/MapsAndDocs/assets/GLIMS_Analysis_Tutorial_a4.pdf (last access: 4 January 2014), 2007.

Raup, B. H., Kääb, A., Kargel, J. S., Bishop, M. P., Hamilton, G., Lee, E., Paul, F., Rau, F., Soltesz, D., Khalsa, S. J. S., Beedle, M., and Helm, C.: Remote sensing and GIS technology in the Global Land Ice Measurements from Space (GLIMS) Project, Comput. Geosci., 33, 104–125, 2007.

Rees, W. G.: Remote sensing of snow and ice, Taylor & Francis Group, Boca Raton, FL, USA, 2003.

Sakai, A. and Fujita, K.: Correspondence: Formation conditions of supraglacial lakes on debris covered glaciers in the Himalaya, J. Glaciol., 56, 177–181, 2010.

Sakai, A., Nakawo, M., and Fujita, K.: Distribution Characteristics and Energy Balance of Ice Cliffs on Debris-Covered Glaciers, Nepal Himalaya, Arct. Antarct. Alp. Res., 34, 12–19, 2002.

Salerno, F., Buraschi, E., Bruccoleri, G., Tartari, G., and Smiraglia, C.: Glacier surface-area changes in Sagarmatha pational park, Nepal, in the second half of the 20th century, by comparison of historical maps, J. Glaciol., 54, 738–752, 2008.

Salerno, F., Thakuri, S., D'Agata, C., Smiraglia, C., Manfredi, E. C., Viviano, G., and Tartari, G.: Glacial lake distribution in the Mount Everest region: Uncertainty of measurement and conditions of formation, Glob. Planet. Change, 92–93, 30–39, 2012.

Scherler, D., Bookhagen, B., and Strecker, M. R.: Spatially variable response of Himalayan glaciers to climate change affected by debris cover, Nat. Geosci., 4, 156–159, 2011.

Shanker, R.: Glaciological studies in India, Contribution from Glaciological Survey of India, in: Proceesings of Symposium on Snow, Ice and Glacier, March 1999, Geological Survey of India Special Publication 53, 11–15, 2001.

Shrestha, A., Wake, C. P., Dibb, J. E., and Mayevski, P. A.: Precipitation fluctuations in the Nepal Himalaya and its vicinity and relationship with some large-scale climatologic patterns, Int. J. Climatol., 20, 317–327, 2000.

Srikantia, S. V.: Restriction on maps: A denial of valid geographic information, Current Science, 79, 484–488, 2000.

Thakuri, S., Salerno, F., Smiraglia, C., Bolch, T., D'Agata, C., Viviano, G., and Tartari, G.: Tracing glacier changes since the 1960s on the south slope of Mt. Everest (central Southern Himalaya) using optical satellite imagery, The Cryosphere, 8, 1297–1315, doi:10.5194/tc-8-1297-2014, 2014.

USGS-EROS: Declassified Imagery-1, available at: https://lta.cr.usgs.gov/declass_1 (last access: 21 February 2015), 1962.

Wagnon, P., Vincent, C., Arnaud, Y., Berthier, E., Vuillermoz, E., Gruber, S., Ménégoz, M., Gilbert, A., Dumont, M., Shea, J. M., Stumm, D., and Pokhrel, B. K.: Seasonal and annual mass balances of Mera and Pokalde glaciers (Nepal Himalaya) since 2007, The Cryosphere, 7, 1769–1786, doi:10.5194/tc-7-1769-2013, 2013.

Wessels, R. L., Kargel, J. S., and Kieffer, H. H.: ASTER measurement of supraglacial lakes in the Mount Everest region of the Himalaya, Ann. Glaciol., 34, 399–408, 2002.

Yanai, M., Li, C., and Song, Z.: Seasonal heating of the Tibetan plateau and its effect on the evolution of the Asian summer monsoon, J. Meteorol. Soc. Jpn., 70, 319–351, 1992.

Zhang, Y., Fujita, K., Liu, S., Liu, Q., and Nuimura, T.: Distribution of debris thickness and its effect on ice melt at Hailuogou glacier, southeastern Tibetan Plateau, using in situ surveys and ASTER imagery, J. Glaciol., 57, 1147–1157, 2011.

Brief Communication: Contending estimates of 2003–2008 glacier mass balance over the Pamir–Karakoram–Himalaya

A. Kääb[1], D. Treichler[1], C. Nuth[1], and E. Berthier[2]

[1]Department of Geosciences, University of Oslo, P.O. Box 1047, Oslo, Norway
[2]CNRS, Université de Toulouse, LEGOS, 14 avenue Ed. Belin, 31400 Toulouse, France

Correspondence to: A. Kääb (kaeaeb@geo.uio.no)

Abstract. We present glacier thickness changes over the entire Pamir–Karakoram–Himalaya arc based on ICESat satellite altimetry data for 2003–2008. We highlight the importance of C-band penetration for studies based on the SRTM elevation model. This penetration seems to be of potentially larger magnitude and variability than previously assumed. The most negative rate of region-wide glacier elevation change ($< -1\,\mathrm{m\,yr^{-1}}$) is observed for the eastern Nyainqêntanglha Shan. Conversely, glaciers of the western Kunlun Shan are slightly gaining volume, and Pamir and Karakoram seem to be on the western edge of this mass-gain anomaly rather than its centre. For the Ganges, Indus and Brahmaputra basins, the glacier mass change reaches $-24 \pm 2\,\mathrm{Gt\,yr^{-1}}$, about 10 % of the current glacier contribution to sea-level rise. For selected catchments, we estimate glacier imbalance contributions to river run-off from a few percent to greater than 10 %.

1 Introduction and methods

Region-wide measurements of glacier volume or mass change are limited for the Pamir–Karakoram–Himalaya region, leaving room for speculation about the glacier response to climate change and its hydrological significance. Glacier mass change in high mountain Asia (or some part of it) have been obtained by (i) extrapolating the few existing in situ mass balance series (Cogley, 2011; Bolch et al., 2012; Yao et al., 2012), (ii) space gravimetry (Jacob et al., 2012; Gardner et al., 2013), (iii) laser altimetry (Kääb et al., 2012; Gardner et al., 2013; Neckel et al., 2014) and (iv) the differencing of digital elevation models (Gardelle et al., 2013). Between these studies that narrow down the range of uncertainties for core parts of this remote mountain region, significant inconsistencies remain.

The aims of this study are (i) to provide a new consistent regional-scale data set from the ICESat autumn laser campaigns (2003–2008) by extending Kääb et al. (2012) to completely cover the study region by Gardelle et al. (2013) and several major river basins, (ii) to compare the results to other previous estimates of the Pamir–Karakoram–Himalaya glacier volume change and (iii) to roughly evaluate the contribution of glacier mass change to river run-off.

We follow the methods explained in Kääb et al. (2012) with a considerable extension towards the eastern Nyainqêntanglha Shan, the Pamir and part of the Tibetan Plateau (Fig. 1). In short, ICESat footprints are intersected with the February 2000 SRTM DEM and overlaid on the most snow-free multispectral Landsat images over $\sim$2000–2013 to manually classify footprints into three classes: glaciers, non-glaciers and water. Glacier elevation difference trends are then estimated regionally and at a $1^\circ \times 1^\circ$ geographic grid by fitting a robust linear temporal trend to the time series of elevation differences between the SRTM DEM and individual ICESat footprint elevations. Trends are derived from autumn ICESat campaigns only (2009 ICESat winter campaigns excluded), because combined autumn and winter trends are sensitive to temporal variations in accumulation amount and timing, potentially introducing bias (see Supplement of Kääb et al., 2012). We confirm that our trends are not due to sampling bias of ICESat elevations by comparing ICESat elevation histograms with glacier hypsometry. The resulting elevation difference trends for all our zones are given in Table 1.

Table 1. Glacier elevation difference trends over the Pamir–Karakoram–Himalaya from this and other studies. Note that Gardelle et al. (2013) cover the period 2000 to ~2010, while the other studies cover 2003 to 2008/2009. Note also that the zones of this study and Gardelle et al. (2013) coincide whereas the zones of the other do so only roughly, which can potentially explain parts of the disagreements. See text in Sects. 3 and 4 for an explanation of how the glacier areas were estimated.

Zone	Glacier area (km^2)	This study ($m\,yr^{-1}$, ± at 1σ-level)	Gardner et al. (2013; $m\,yr^{-1}$, ± at 2σ-level)	Neckel et al. (2014; $m\,yr^{-1}$, ± at 1σ-level)	Gardelle et al. (2013; $m\,yr^{-1}$, ± at 1σ-level)
Eastern Nyainqêntanglha[a]	6000	−1.34 ± 0.29	−0.30 ± 0.13 −0.40 ± 0.41[b]	−0.81 ± 0.32	−0.39 ± 0.16
Bhutan	3500	−0.89 ± 0.16	−0.89 ± 0.18	−0.78 ± 0.27	−0.26 ± 0.15
Everest	8500	−0.37 ± 0.10	−0.44 ± 0.20		−0.30 ± 0.16
West Nepal	7500	−0.43 ± 0.09		−0.44 ± 0.26	−0.38 ± 0.16
Spiti–Lahaul	9500	−0.49 ± 0.12	−0.53 ± 0.13		−0.53 ± 0.16
Karakoram	21 000	−0.10 ± 0.06	−0.12 ± 0.15		+0.12 ± 0.19
Hindu Kush	5500	−0.49 ± 0.10			−0.14 ± 0.19
Pamir	6500	−0.48 ± 0.14	−0.13 ± 0.22		+0.16 ± 0.15
Western Kunlun Shan–Tarim	12 500	+0.05 ± 0.07	+0.17 ± 0.15	+0.04 ± 0.29	
Area-weighted mean	80 500	−0.37 ± 0.10			

[a] Named Hengduan Shan in Gardelle et al. (2013); [b] two zones of Gardner et al. (2013) overlap with our zone and both their values are given.

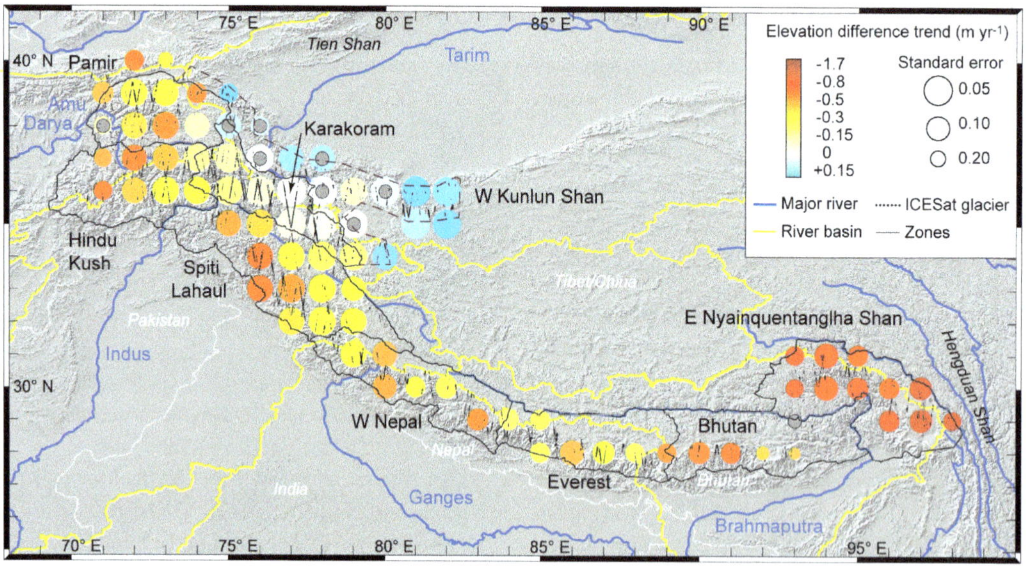

Figure 1. Study region and trends of elevation differences during 2003–2008. Data are shown on a 1° grid with overlapping rectangular geographic averaging cells of 2° × 2°. Trends are based on autumn ICESat acquisitions. Only ICESat footprints over glaciers are indicated. The zones indicated by black outlines are equivalent to the ones of Gardelle et al. (2013) with the western Kunlun Shan–Tarim zone (dashed outline) being the only additional one. Trends for all cells (coloured data circles) are statistically significant except for the cells that are marked with grey centres. The uncertainty of the temporal trends per cell is indicated through circle sizes indirectly proportional to the standard error of trends at 68 % level.

2 Glacier thickness changes

2.1 Thickening in the Karakoram and western Kunlun Shan

A first striking feature in the regional map of elevation difference trends (Fig. 1) is glacier thickness gain in the western Kunlun Shan (~ +0.1 $m\,yr^{-1}$), agreeing with in situ mass balance and length change measurements (Yao et al., 2012). There is a southwest-to-northeast gradient from considerably negative glacier mass balances in Hindu Kush and Spiti–Lahaul to positive values in the Pamir–Karakoram–western-Kunlun-Shan region (Fig. 1). This suggests the so-called Karakoram glacier mass-balance anomaly (Hewitt,

2011; Gardelle et al., 2012a), or Pamir–Karakoram anomaly (Gardelle et al., 2013), is rather the edge or southwest limit of an anomaly centred more to the northeast over the western Kunlun Shan, or Tarim Basin. The anomaly seems thus indeed the result of a larger-scale meteorological or climatic feature, and peculiarities of the Karakoram topography or glaciers (e.g. surge type, hypsometry, avalanche contribution; Hewitt, 2011) do not necessarily play a decisive role. Combined, the results by Gardner et al. (2013), Neckel et al. (2014) and the glacier elevation change pattern of Fig. 1 suggest the centre of the anomaly could be located over the Tibetan Plateau.

Direct precipitation measurements in this region are scarce and thus trends are uncertain. Satellite-retrieved precipitation and gauge data (Global Precipitation Climatology Project) suggest an increase of precipitation over the study region north of Karakoram and east of Pamir (Yao et al., 2012). Chinese measurements show increased precipitation over the Tibetan Plateau (C.-Y. Xu, personal communication, 2014), and Tao et al. (2014) suggest wetter conditions over the Tarim Basin since the mid 1980s. A number of abnormally wet years occurred during the early 21st century over the Tarim Basin and the Tibetan Plateau (Becker et al., 2013), in particular for the hydrological years 2003/2004 and 2005/2006. A recent climate modelling study proposes that stable or increasing snowfalls characterise the Karakoram anomaly on a background of increasing air temperatures (Kapnick et al., 2014). Despite the available studies and data, further research seems necessary to consolidate the precipitation and temperature trends and the reason behind the slight glacier volume gains.

2.2 Massive thinning in eastern Nyainqêntanglha Shan and Spiti–Lahaul

The other striking feature in Fig. 1 is the massive glacier thickness loss in the eastern Nyainqêntanglha Shan (between -1 and $-1.7\,\mathrm{m\,yr^{-1}}$), also consistent with the large negative mass balances and frontal retreats in this zone (Yao et al., 2012). The glaciers of eastern Nyainqêntanglha Shan have the smallest total elevational range in our study region, indicating a large sensitivity to fluctuations in the equilibrium line altitude (Pelto, 2010; Loibl et al., 2014). The few available in situ mass balance measurements in the area suggest that the equilibrium line was over the vertical limits of the monitored glaciers in the late 2000s, and precipitation in this zone shows the strongest long-term decrease over the entire Pamir–Karakoram–Himalaya region (Yao et al., 2012; Becker et al., 2013). A similar pattern of glacier shrinkage, though less distinct, is found at the western end of the Great Himalaya Range within our Spiti–Lahaul zone and forms the cluster of second-largest thickness loss rates in this study (-0.5 to $-0.7\,\mathrm{m\,yr^{-1}}$). Also here, Landsat data indicate that firn lines have risen in several years towards high glacier elevations, resulting in very small accumulation areas or even their complete loss.

The 2003–2008 glacier thickness changes in the other study zones are all similar, on the order of ~ -0.4 to $-0.5\,\mathrm{m\,yr^{-1}}$ (Table 1), with more negative values in the Bhutan zone at the transition between the eastern Nyaiqêntanglha and Everest zones. We note that glaciers dominated by the summer monsoon (i.e. east of the Spiti–Lahaul) all show thickness losses (summer-accumulation type glaciers; Fujita, 2008; Kapnick et al., 2014; Maussion et al., 2014). Eastern Nyaiqêntanglha Shan, the zone with strongest glacier thickness loss, receives most accumulation during March–May (spring-accumulation type; Maussion et al., 2014). The glaciers with considerable winter accumulation under influence of the westerlies show a more mixed picture, with stable or growing thicknesses in the Karakoram and western Kunlun Shan but thickness losses for instance in the Hindu Kush.

2.3 Comparison to previous thickness change studies

The following comparison to other studies uses average glacier thickness changes rather than total mass changes in order to minimise effects from different delineations of study zones, glacier cover areas and density assumptions. From Hindu Kush and Karakoram in the west to Nepal in the east, results of all studies agree within their errors (Table 1). Results are most sensitive to zone delineation in the Hindu Kush, reflecting the strong spatial variability of glacier thickness change rates in this area (Fig. 1) and presumably also locally heterogeneous glacier behaviour (Sarikaya et al., 2012; see also below for Pamir).

Significant differences among the results of all studies are found over eastern Nyainqêntanglha Shan. Our results and those from Neckel et al. (2014) agree within the errors but not with Gardner et al. (2013), although all three studies are based on ICESat. While our study and Neckel et al. (2014) use ICESat footprint classifications from contemporary satellite images, Gardner et al. (2013) use Randolph Glacier Inventory outlines (RGI version 2.0; Pfeffer et al., 2014), which contain considerable errors of commission and omission in this zone (see Table 1 in Gardelle et al., 2013). Repeating our analysis with footprint classifications based on the Randolph Glacier Inventory results in less negative elevation difference trends on glaciers ($\sim 20\,\%$ less negative) due to inclusion of non-glacier footprints. Contrarily, the elevation difference trends on land ($-0.10 \pm 0.06\,\mathrm{m\,yr^{-1}}$ when using our own footprint classification) become more negative if ICESat footprints are classified using RGI due to inclusion of glacier footprints ($-0.16 \pm 0.07\,\mathrm{m\,yr^{-1}}$). The remaining discrepancy is presumably due to the fact that the ICESat-based results of Gardner et al. (2013) are averaged from three different methods. Their results based on autumn footprints only (method B, Gardner et al., 2013) suggest a thickness change rate of $-0.86\,\mathrm{m\,yr^{-1}}$, which is in closer agreement with our results.

At a first glance, eastern Nyainqêntanglha Shan results from Gardelle et al. (2013; zone called Hengduan Shan) and Gardner et al. (2013) seem to agree, but we believe this might be a coincidence. Previously, we argued why the Gardner et al. (2013) results might be less negative. Additionally, the results in Gardelle et al. (2013) rely crucially on an estimate of SRTM C-band penetration. Over any glacier globally, the SRTM radar waves will typically have penetrated into the snow and ice, with potential largest penetration depths through snow and firn and smallest through ice (Kääb et al., 2012; Dall et al., 2001; Rignot et al., 2001). As a consequence, SRTM glacier elevations do not, in general, reflect real mid-February 2000 glacier surface elevations but some lower horizon, the elevation of which depends on, among others, the dielectric properties and structure of the penetrated glacier volume during the SRTM campaign. For elevation difference studies where one of the elevation data sets is the SRTM, its penetration depth needs to be estimated for correction, and biases in this estimate translate directly into offsets in thickness change. Gardelle et al. (2013) used an average C-band penetration of 1.7 m for eastern Nyainqêntanglha Shan estimated from the difference of SRTM C-band and X-band DEMs (Gardelle et al., 2012b). Here, we extrapolate our ICESat elevation trends over 2003–2008 and their uncertainty back in time to the SRTM acquisition period in February 2000. Under the coarse assumption that the 2000–2003 trends equal to those of 2003–2008, the extrapolation of February 2000 should result in a zero elevation difference to ICESat since the SRTM DEM was used as elevation reference. Offsets in this elevation difference for February 2000 are mainly attributed to SRTM radar penetration into ice and snow (for method and discussion see Kääb et al., 2012). For eastern Nyainqêntanglha Shan this analysis indicates an average penetration of 8–10 m (7–9 m if based on the winter trends that might alternatively be assumed to reflect February conditions), much more than the 1.7 m assumed in Gardelle et al. (2013), while the corresponding off-glacier penetration is not discernible from zero. Clearly, our penetration depth lies at the high end but remains within the range of possible C-band phase-centre penetrations (Kääb et al., 2012; Dall et al., 2001; Rignot et al., 2001). Sakai et al. (2014) suggest the highest accumulation rates of the entire study region occur in eastern Nyainqêntanglha Shan, together with Hindu Kush. Correction of the Gardelle et al. (2013) results by our present C-band penetration estimate completely reconciles their results with ours. Note, however, that extrapolation of our 2003–2008 elevation difference trend back to 2000 is based on the risky assumption that the 2000–2003 trend equals the 2003–2008 trend.

For the Bhutan zone, Gardelle et al. (2013) estimated a C-band penetration for February 2000 of 2.4 m, whereas our extrapolation of ICESat trends suggests around 6 m, which again reconciles the results of both studies for this zone.

In the Pamir, our results are more negative than Gardner et al. (2013) and in particular Gardelle et al. (2013). As above, we suggest that our manual classification of ICESat footprints versus the Randolph Glacier Inventory contributed to the difference between this study and Gardner et al. (2013) (remark: Gardelle et al. (2013) used their own inventory). Also, the difference between our study and Gardner et al. (2013) is reduced when only the results from their Method B (similar to ours) are considered. Gardelle et al. (2013) find glacier thickness changes of $+0.16 \pm 0.15\,\mathrm{m\,yr^{-1}}$ over the Pamir, whereas the present study suggests $-0.48 \pm 0.08\,\mathrm{m\,yr^{-1}}$. Again, we find larger SRTM C-band penetration of 5–6 m compared to 1.8 m (Gardelle et al., 2013). Applying the average C-band penetration from the present study again reconciles the results of both studies. However, comparison of both studies in Pamir is complicated by a number of glacier surges (Gardelle et al., 2013) in connection with particularly sparse ICESat glacier coverage. Superimposing ICESat tracks over Landsat images and the elevation change map of Gardelle et al. (2013) reveals that they cross areas of either strongly positive or negative elevation change zones from surge waves. The ICESat trends thus become biased depending on where they sample surges, and the total ICESat sample size over Pamir is not large enough to compensate for these effects. The different observation periods for both studies (2000–2011 versus 2003–2008) may also have considerable impact due to surge activities and climate interannual variability (Yi and Sun, 2013).

3 Glacier mass changes and water resources

We assume an average density of $850\,\mathrm{kg\,m^{-3}}$ for all 2003–2008 volume changes to convert the thickness changes to water equivalent quantities (Huss (2013); see Kääb et al. (2012), for different density scenarios). The total glacier area is estimated using a simple cross product: we multiply the number of ICESat glacier footprints in each zone with the ratio between the total zone area and total number of ICESat footprints. Our method to estimate the total glacier areas is certainly open to discussion, but we prefer the above procedure over using areas from the Randolph Glacier Inventory because of the large deviations to our estimates, mainly for eastern Nyainqêntanglha and Pamir, from obviously outdated glacier outlines and voids in the Randolph Inventory (Nuimura et al., 2014). The uncertainty of water equivalent quantities includes the standard error of the elevation difference trend fit, the off-glacier trends, an error due to temporal offset of the ICESat autumn campaigns from maximum cumulative ablation conditions, an uncertainty of $\pm 20\,\%$ for the glacier cover areas and an uncertainty of $\pm 60\,\mathrm{kg\,m^{-3}}$ for density (Kääb et al., 2012; Huss, 2013). The effects of these individual sources of uncertainty, all converted to error in mass change, are combined through the root sum of squares to arrive at the total uncertainty. Note that water equivalent results from this study are not identical to Kääb et al. (2012),

Table 2. Glacier thickness and mass changes over the major river basins of the study area. The discharge equivalent is a unit conversion from mass change and neglects any losses such as by evaporation or to groundwater.

Major river basin	Glacier area (km^2)	Elevation difference trend ($m\,yr^{-1}$)	Mass change ($Gt\,yr^{-1}$)	Discharge equivalent DE ($m^3\,s^{-1}$)
Tarim[a]	15 000	$+0.06 \pm 0.08$	$+0.7 \pm 1.0$	$+24 \pm 33$
Amu Darya[b]	11 000	-0.43 ± 0.08	-4.0 ± 0.8	-128 ± 25
Indus	25 000	-0.33 ± 0.04	-7.0 ± 0.8	-220 ± 26
Ganges	11 000	-0.44 ± 0.07	-4.1 ± 0.6	-130 ± 20
Brahmaputra	14 000	-1.06 ± 0.15	-12.6 ± 1.9	-400 ± 60

[a] The Tarim Basin is endorheic. Only parts of the glacier area (~ 40 %) within the Tarim Basin are covered in this study. [b] Endorheic basin.

even when elevation difference trends agree, due to the simplified density assumption and the different glacier area estimates used.

3.1 Comparison to gravimetric mass loss

For the Pamir, Kunlun Shan and Karakoram (zone 8b of Jacob et al. (2012); note that the Karakoram is part of their zone 8b, not 8c as suggested by their zone names) we estimate a glacier mass change of $-6 \pm 2\,Gt\,yr^{-1}$ for 2003–2008 that agrees well within the error with Jacob et al. (2012) results from satellite gravimetry of $-5 \pm 10\,Gt\,yr^{-1}$ (January 2003–December 2007) and $-8 \pm 9\,Gt\,yr^{-1}$ (January 2004–December 2008). For the Himalaya and eastern Nyainqêntanglha Shan (zone 8c of Jacob et al., 2012) we estimate a 2003–2008 glacier mass change of $-19 \pm 3\,Gt\,yr^{-1}$ that compares to $-3 \pm 12\,Gt\,yr^{-1}$ (January 2003–December 2007) and $-2 \pm 10\,Gt\,yr^{-1}$ (January 2004–December 2008) from satellite gravimetry. Given their fundamentally different approaches, it is challenging to discuss potential sources of disagreement between the two studies in the Himalaya and eastern Nyainqêntanglha Shan. Groundwater depletion (Rodell et al., 2009), glacier imbalance run-off into endorheic basins (Zhang et al., 2013) and errors and biases in the ICESat-derived trends as discussed above and in Kääb et al. (2012) are all likely explanations. Note that Gardner et al. (2013) offer a second, more negative gravimetric estimate for the entire combined high mountain Asia that is, however, not spatially resolved enough to compare to our results. The uncertainties of our results in this entire paragraph are given at 2σ confidence level to better agree with the uncertainty level in Jacob et al. (2012), whereas elsewhere in this contribution uncertainty is provided at 1σ confidence level.

3.2 River run-off

The glaciers of the Tarim Basin (only 40 % of its total glacier area is covered here, with notably Tien Shan missing) and the Amu Darya basin (all glacier areas covered) drain into endorheic basins and thus their mass changes do not contribute to sea-level changes (Table 2). The glacier mass changes in the Indus, Ganges and Brahmaputra basins from the present study contributed together $\sim 0.06 \pm 0.01\,mm\,yr^{-1}$ to eustatic sea-level rise, that is ~ 10 % of the current sea-level contribution of $0.71 \pm 0.08\,mm\,yr^{-1}$ from glaciers outside the ice sheets (Gardner et al., 2013).

The discharge equivalent of these mass changes, that is the annual average glacier imbalance contribution to river run-off, is given in Table 2 for the major river basins covered. Note that computation of our discharge equivalents is a pure unit conversion from $Gt\,yr^{-1}$ to $m^3\,s^{-1}$, neglecting any hydrological processes and with the sole aim to roughly evaluate the relative importance of glacier mass changes for river flow in the catchments.

The Tarim Basin glaciers most likely stored water over 2003–2008 ($+24 \pm 33\,m^3\,s^{-1}$ discharge equivalent, DE). The glacier imbalance contribution to run-off is largest for Brahmaputra ($-400 \pm 60\,m^3\,s^{-1}$ DE), followed by the Indus ($-220\,m^3\,s^{-1}$ DE), Ganges and Amu Darya (each $-130\,m^3\,s^{-1}$ DE). Comparison of the discharge equivalent of glacier imbalance to measured river run-off is increasingly biased the further downstream the gauging stations are situated from the glaciers due to cumulative natural and man-made losses. It is also important to note that the available run-off data from literature and databases refer to various time periods, in parts considerably older than the ICESat period. Figure 2 illustrates thus only roughly the hydrological significance of the 2003–2008 glacier mass change in selected gauged catchments. For details on the gauging stations used and the uncertainty of the contributions see Supplement. As an example, the 2003–2008 glacier imbalance within the upper Indus basin at Besham Qila contributes ~ 6 % to annual average river discharge (Fig. 2; Supplement), and we roughly estimate a very similar number for the Amu Darya (Supplement). For the upper Indus basin, the hydrological balance is under ongoing discussion (Reggiani and Rientjes, 2014), and we hope that our glacier mass change estimates can contribute towards balance closure and better understanding of spatial–temporal patterns of run-off or high-elevation precipitation amounts in the region (e.g. Immerzeel et al., 2012).

The modelling results for "non-renewable glacier run-off" of Savoskul and Smakhtin (2013) agree well with ours for Amu Darya, less for Indus (they obtain $-0.55\,m\,w.e.\,yr^{-1}$ specific mass loss rate over 2001–2010; we have $-0.28\,m\,w.e.\,yr^{-1}$) and Ganges (they obtain $-0.77\,m\,w.e.\,yr^{-1}$; we have $-0.37\,m\,w.e.\,yr^{-1}$) and not very well for Brahmaputra (they obtain $-0.36\,m\,w.e.\,yr^{-1}$; we have $-0.90\,m\,w.e.\,yr^{-1}$).

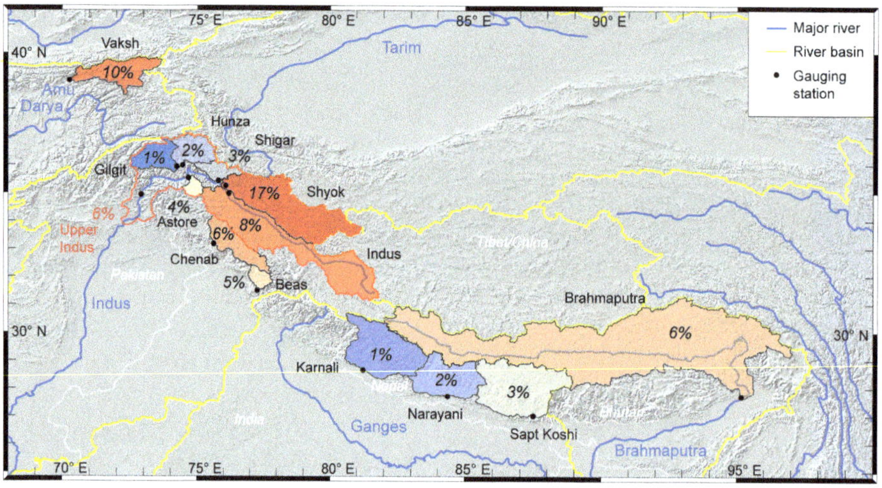

Figure 2. The percentage of discharge equivalent from annual glacier imbalance to measured average river run-off for selected catchments. Note that the actual numbers will be somewhat lower due to unaccounted water losses such as from evaporation or to groundwater. For details on the gauging stations used and the uncertainty of the contributions see the Supplement.

4 Conclusions

From 2003 to 2008 ICESat-derived elevation difference trends over Pamir–Karakoram–Himalaya and from comparisons to geographically overlapping studies, we draw the following conclusions:

- Glacier thickness loss over the study region is most pronounced for the eastern Nyainqêntanglha Shan, followed by the western end of the Great Himalaya Range. Glaciers in and around the western Kunlun Shan are in balance or even gaining volume, and Pamir and Karakoram seem to be on the western limit of this mass balance anomaly rather than its centre. This suggests it is a meteorological or climatic anomaly (rise in precipitation). However, the cause and duration of this regional glacier anomaly is not fully understood yet.

- Our glacier volume changes seem especially uncertain in Pamir and, to a lesser extent, Hindu Kush. The heterogeneous behaviour of individual glaciers in these two zones, for instance from glacier surges, may lead to biases when extrapolating elevation difference trends from particularly sparse ICESat tracks, or areas covered by differential DEMs, to the entire zones.

- Extrapolation of ICESat trends back in time to the SRTM acquisition date suggests a much larger potential magnitude and variability of SRTM C-band phase-centre penetration than often assumed. Given the crucial importance of radar penetration for glacier thickness change studies based on radar DEMs, such as the SRTM or the upcoming TanDEM-X, we recommend being critical of penetration assumptions used in previous studies and to investigate the issue more extensively and systematically (Langley et al., 2007; chapter 7 in Müller, 2011). The problem is complicated by the fact that radar penetration has to be known specifically for certain dates from the past.

- The glacier mass changes in the Tarim and Amu Darya basins of $+0.7 \pm 1.0$ and $-4.0 \pm 0.8\,\mathrm{Gt\,yr^{-1}}$ do not contribute to sea-level rise. The combined Ganges, Indus and Brahmaputra basin glacier mass change is $-23.7 \pm 2.1\,\mathrm{Gt\,yr^{-1}}$, almost 10 % of the glacier contribution to sea-level rise.

- Neglecting water losses downstream of the glaciers, the 2003–2008 glacier imbalances amount to $\sim 6\,\%$ of the annual discharge of Amu Darya and upper Indus where they leave the mountains. This is a considerable amount given the significance of the rivers for the Aral Sea (Amu Darya) and massive irrigation schemes and household use in these dry climate regions. Maximum glacier imbalance contributions to annual average river run-off of up to $\sim 17\,\%$ are found for the Shyok (Indus) and $\sim 10\,\%$ for Vakhsh (Amu Darya), while minimum contributions are only ~ 1–$3\,\%$ for the monsoon-type catchments in Nepal.

- Our results on glacier mass loss agree with those from satellite gravimetry (Jacob et al., 2012) over Pamir, western Kunlun Shan and Karakoram but significantly diverge over the Himalaya and eastern Nyainqêntanglha Shan.

It is important to note that our results only cover 5 years, 2003–2008, and it remains open to what extent those years are representative for longer periods, such as the 10 years covered by Gardelle et al. (2013). For short mass balance series, single anomalous years may have large impacts on trends. Our water equivalent results are also sensitive to density and glacier area assumptions. We find that glacier outlines and areas in the study region are still quite uncertain and invite the reader to use improved glacier area estimates for upscaling our results and own assumptions for the conversion of volume changes to mass changes.

Author contributions. A. Kääb designed the study, performed the data analysis and wrote the paper. D. Treichler, C. Nuth and E. Berthier contributed to data analysis, performed supporting analyses and edited the paper.

Acknowledgements. Sincere thanks are due to Duncan Quincey, another anonymous referee, colleagues that commented on the study, and the editor Andreas Vieli for their valuable feedback that certainly improved our work. The study was funded by the European Research Council under the European Union's Seventh Framework Programme (FP/2007-2013)/ERC grant agreement no. 320816, the ESA project Glaciers_cci (4000109873/14/I-NB) and the Department of Geosciences, University of Oslo. E. Berthier acknowledges support from TOSCA (CNES). We are very grateful to NASA and NSIDC for free provision for the ICESat data and the USGS for the SRTM DEM and Landsat imagery.

Edited by: A. Vieli

References

Becker, A., Finger, P., Meyer-Christoffer, A., Rudolf, B., Schamm, K., Schneider, U., and Ziese, M.: A description of the global land-surface precipitation data products of the Global Precipitation Climatology Centre with sample applications including centennial (trend) analysis from 1901–present, Earth Syst. Sci. Data, 5, 71–99, doi:10.5194/essd-5-71-2013, 2013.

Bolch, T., Kulkarni, A., Kääb, A., Huggel, C., Paul, F., Cogley, J. G., Frey, H., Kargel, J. S., Fujita, K., Scheel, M., Bajracharya, S., and Stoffel, M.: The state and fate of Himalayan glaciers, Science, 336, 310–314, 2012.

Cogley, J. G.: Present and future states of Himalaya and Karakoram glaciers, Ann. Glaciol., 52, 69–73, 2011.

Dall, J., Madsen, S. N., Keller, K., and Forsberg, R.: Topography and penetration of the Greenland Ice Sheet measured with airborne SAR interferometry, Geophys. Res. Lett., 28, 1703–1706, 2001.

Fujita, K.: Effect of precipitation seasonality on climatic sensitivity of glacier mass balance, Earth Planet. Sc. Lett., 276, 14–19, doi:10.1016/j.epsl.2008.08.028, 2008.

Gardelle, J., Berthier, E., and Arnaud, Y.: Slight mass gain of Karakoram glaciers in the early 21st century, Nat. Geosci., 5, 322–325, 2012a.

Gardelle, J., Berthier, E., and Arnaud, Y.: Impact of resolution and radar penetration on glacier elevation changes computed from multi-temporal DEMs, J. Glaciol., 58, 419–422, 2012b.

Gardelle, J., Berthier, E., Arnaud, Y., and Kääb, A.: Region-wide glacier mass balances over the Pamir-Karakoram-Himalaya during 1999–2011, The Cryosphere, 7, 1263–1286, doi:10.5194/tc-7-1263-2013, 2013.

Gardner, A. S., Moholdt, G., Cogley, J. G., Wouters, B., Arendt, A. A., Wahr, J., Berthier, E., Hock, R., Pfeffer, W. T., Kaser, G., Ligtenberg, S. R. M., Bolch, T., Sharp, M. J., Hagen, J. O., van den Broeke, M. R., and Paul, F.: A reconciled estimate of glacier contributions to sea level rise: 2003 to 2009, Science, 340, 852–857, 2013.

Hewitt, K.: Glacier change, concentration, and elevation effects in the Karakoram Himalaya, Upper Indus Basin, Mount. Res. Develop., 31, 188–200, 2011.

Huss, M.: Density assumptions for converting geodetic glacier volume change to mass change, The Cryosphere, 7, 877–887, doi:10.5194/tc-7-877-2013, 2013.

Immerzeel, W. W., Pellicciotti, F., and Shrestha, A. B.: Glaciers as a proxy to quantify the spatial distribution of precipitation in the Hunza basin, Mount. Res. Develop., 32, 30–38, 2012.

Jacob, T., Wahr, J., Pfeffer, W. T., and Swenson, S.: Recent contributions of glaciers and ice caps to sea level rise, Nature, 482, 514–518, 2012.

Kääb, A., Berthier, E., Nuth, C., Gardelle, J., and Arnaud, Y.: Contrasting patterns of early twenty-first-century glacier mass change in the Himalayas, Nature, 488, 495–498, 2012.

Kapnick, S. B., Delworth, T. L., Ashfaq, M., Malyshev, S., and Milly, P. C. D.: Snowfall less sensitive to warming in Karakoram than in Himalayas due to a unique seasonal cycle, Nat. Geosci., 7, 834–840, doi:10.1038/ngeo2269, 2014

Langley, K., Hamran, S. E., Hogda, K. A., Storvold, R., Brandt, O., Hagen, J. O., and Kohler, J.: Use of C-band ground penetrating radar to determine backscatter sources within glaciers, IEEE T. Geosci. Remote, 45, 1236–1246, doi:10.1109/Tgrs.2007.892600, 2007.

Loibl, D., Lehmkuhl, F., and Griessinger, J.: Reconstructing glacier retreat since the Little Ice Age in SE Tibet by glacier mapping and equilibrium line altitude calculation, Geomorphology, 214, 22–39, doi:10.1016/j.geomorph.2014.03.018, 2014.

Maussion, F., Scherer, D., Mölg, T., Collier, E., Curio, J., and Finkelnburg, R.: Precipitation seasonality and variability over the Tibetan Plateau as resolved by the High Asia Reanalysis, J. Climate, 27, 1910–1927, doi:10.1175/Jcli-D-13-00282.1, 2014.

Müller, K.: Microwave penetration in polar snow and ice: implications for GPR and SAR, Dissertation, Faculty of Mathematics and Natural Sciences. University of Oslo, available from http://www.duo.uio.no/ (last access: 17 March 2015), 2011.

Neckel, N., Kropacek, J., Bolch, T., and Hochschild, V.: Glacier mass changes on the Tibetan Plateau 2003–2009 derived from ICESat laser altimetry measurements, Environ. Res. Lett., 9, 014009, doi:10.1088/1748-9326/9/1/014009, 2014.

Nuimura, T., Sakai, A., Taniguchi, K., Nagai, H., Lamsal, D., Tsutaki, S., Kozawa, A., Hoshina, Y., Takenaka, S., Omiya, S., Tsunematsu, K., Tshering, P., and Fujita, K.: The GAMDAM Glacier Inventory: a quality controlled inventory of Asian glaciers, The Cryosphere Discuss., 8, 2799–2829, doi:10.5194/tcd-8-2799-2014, 2014.

Pelto, M. S.: Forecasting temperate alpine glacier survival from accumulation zone observations, The Cryosphere, 4, 67–75, doi:10.5194/tc-4-67-2010, 2010.

Pfeffer, W. T., Arendt, A. A., Bliss, A., Bolch, T., Cogley, J. G., Gardner, A. S., Hagen, J. O., Hock, R., Kaser, G., Kienholz, C., Miles, E. S., Moholdt, G., Molg, N., Paul, F., Radic, V., Rastner, P., Raup, B. H., Rich, J., Sharp, M. J., Andeassen, L. M., Bajracharya, S., Barrand, N. E., Beedle, M. J., Berthier, E., Bhambri, R., Brown, I., Burgess, D. O., Burgess, E. W., Cawkwell, F., Chinn, T., Copland, L., Cullen, N. J., Davies, B., De Angelis, H., Fountain, A. G., Frey, H., Giffen, B. A., Glasser, N. F., Gurney, S. D., Hagg, W., Hall, D. K., Haritashya, U. K., Hartmann, G., Herreid, S., Howat, I., Jiskoot, H., Khromova, T. E., Klein, A., Kohler, J., Konig, M., Kriegel, D., Kutuzov, S., Lavrentiev, I., Le Bris, R., Li, X., Manley, W. F., Mayer, C., Menounos, B., Mercer, A., Mool, P., Negrete, A., Nosenko, G., Nuth, C., Osmonov, A., Pettersson, R., Racoviteanu, A., Ranzi, R., Sarikaya, M. A., Schneider, C., Sigurdsson, O., Sirguey, P., Stokes, C. R., Wheate, R., Wolken, G. J., Wu, L. Z., Wyatt, F. R., and Consortium, R.: The Randolph Glacier Inventory: a globally complete inventory of glaciers, J. Glaciol., 60, 537–552, doi:10.3189/2014jog13j176, 2014.

Reggiani, P. and Rientjes, T. H. M.: A reflection on the long-term water balance of the Upper Indus Basin, Hydrol. Res., in press, 2014.

Rignot, E., Echelmeyer, K., and Krabill, W.: Penetration depth of interferometric synthetic-aperture radar signals in snow and ice, Geophys. Res. Lett., 28, 3501–3504, 2001.

Rodell, M., Velicogna, I., and Famiglietti, J. S.: Satellite-based estimates of groundwater depletion in India, Nature, 460, 999–1002, 2009.

Sakai, A., Nuimura, T., Fujita, K., Takenaka, S., Nagai, H., and Lamsal, D.: Climate regime of Asian glaciers revealed by GAMDAM Glacier Inventory, The Cryosphere Discuss., 8, 3629–3663, doi:10.5194/tcd-8-3629-2014, 2014.

Sarikaya, M. A., Bishop, M. P., Shroder, J. F., and Olsenholler, J. A.: Space-based observations of Eastern Hindu Kush glaciers between 1976 and 2007, Afghanistan and Pakistan, Remote Sens. Lett., 3, 77–84, 2012.

Savoskul, O. S. and Smakhtin, V.: Glacier systems and seasonal snow cover in six major Asian river basins: hydrological role under changing climate, Research Report no. 150, International Water Management Institute (IWMI), Colombo, Sri Lanka, 45 pp., 2013.

Tao, H., Borth, H., Fraedrich, K., Su, B., and Zhu, X.: Drought and wetness variability in the Tarim River Basin and connection to large-scale atmospheric circulation, Int. J. Climatol., 34, 2678–2684, 2014.

Yao, T. D., Thompson, L., Yang, W., Yu, W. S., Gao, Y., Guo, X. J., Yang, X. X., Duan, K. Q., Zhao, H. B., Xu, B. Q., Pu, J. C., Lu, A. X., Xiang, Y., Kattel, D. B., and Joswiak, D.: Different glacier status with atmospheric circulations in Tibetan Plateau and surroundings, Nat. Clim. Change, 2, 663–667, 2012.

Yi, S. and Sun, W.: Evaluation of glacier changes in high-mountain Asia based on 10-year GRACE RL05 models, J. Geophys. Res.-Solid, 119, 2504–2517, doi:10.1002/2013JB010860, 2013.

Zhang, G. Q., Yao, T. D., Xie, H. J., Kang, S. C., and Lei, Y. B.: Increased mass over the Tibetan Plateau: From lakes or glaciers?, Geophys. Res. Lett., 40, 2125–2130, 2013.

Permissions

All chapters in this book were first published in The Cryosphere (TC), by Copernicus Publications; hereby published with permission under the Creative Commons Attribution License or equivalent. Every chapter published in this book has been scrutinized by our experts. Their significance has been extensively debated. The topics covered herein carry significant findings which will fuel the growth of the discipline. They may even be implemented as practical applications or may be referred to as a beginning point for another development.

The contributors of this book come from diverse backgrounds, making this book a truly international effort. This book will bring forth new frontiers with its revolutionizing research information and detailed analysis of the nascent developments around the world.

We would like to thank all the contributing authors for lending their expertise to make the book truly unique. They have played a crucial role in the development of this book. Without their invaluable contributions this book wouldn't have been possible. They have made vital efforts to compile up to date information on the varied aspects of this subject to make this book a valuable addition to the collection of many professionals and students.

This book was conceptualized with the vision of imparting up-to-date information and advanced data in this field. To ensure the same, a matchless editorial board was set up. Every individual on the board went through rigorous rounds of assessment to prove their worth. After which they invested a large part of their time researching and compiling the most relevant data for our readers.

The editorial board has been involved in producing this book since its inception. They have spent rigorous hours researching and exploring the diverse topics which have resulted in the successful publishing of this book. They have passed on their knowledge of decades through this book. To expedite this challenging task, the publisher supported the team at every step. A small team of assistant editors was also appointed to further simplify the editing procedure and attain best results for the readers.

Apart from the editorial board, the designing team has also invested a significant amount of their time in understanding the subject and creating the most relevant covers. They scrutinized every image to scout for the most suitable representation of the subject and create an appropriate cover for the book.

The publishing team has been an ardent support to the editorial, designing and production team. Their endless efforts to recruit the best for this project, has resulted in the accomplishment of this book. They are a veteran in the field of academics and their pool of knowledge is as vast as their experience in printing. Their expertise and guidance has proved useful at every step. Their uncompromising quality standards have made this book an exceptional effort. Their encouragement from time to time has been an inspiration for everyone.

The publisher and the editorial board hope that this book will prove to be a valuable piece of knowledge for researchers, students, practitioners and scholars across the globe.

List of Contributors

D. V. Divine
Norwegian Polar Institute, Fram Centre, 9296 Tromsø, Norway
Department of Mathematics and Statistics, Faculty of Science and Technology, University of Tromsø, 9037 Tromsø, Norway

M. A. Granskog
Norwegian Polar Institute, Fram Centre, 9296 Tromsø, Norway

S. R. Hudson
Norwegian Polar Institute, Fram Centre, 9296 Tromsø, Norway

C. A. Pedersen
Norwegian Polar Institute, Fram Centre, 9296 Tromsø, Norway

T. I. Karlsen
Norwegian Polar Institute, Fram Centre, 9296 Tromsø, Norway

S. A. Divina
Norwegian Polar Institute, Fram Centre, 9296 Tromsø, Norway

A.H. H. Renner
Norwegian Polar Institute, Fram Centre, 9296 Tromsø, Norway
Institute of Marine Research, Sykehusveien 23, 9019 Tromsø, Norway

S. Gerland
Norwegian Polar Institute, Fram Centre, 9296 Tromsø, Norway

A.Kriegsmann
Meteorological Institute, University of Hamburg, Hamburg, Germany
Climate Service Center, Hamburg, Germany

B. Brümmer
Meteorological Institute, University of Hamburg, Hamburg, Germany
Climate Service Center, Hamburg, Germany

I.Joughin
Polar Science Center, Applied Physics Lab, University of Washington, 1013 NE 40th St., Seattle, WA 98105-6698, USA

B. E. Smith
Polar Science Center, Applied Physics Lab, University of Washington, 1013 NE 40th St., Seattle, WA 98105-6698, USA

D. E. Shean
Polar Science Center, Applied Physics Lab, University of Washington, 1013 NE 40th St., Seattle, WA 98105-6698, USA
Earth & Space Sciences, University of Washington, Seattle, WA 98195, USA

D. Floricioiu
German Aerospace Center (DLR), Remote Sensing Technology Institute, SAR Signal Processing, Muenchenerstr. 20, 82230 Wessling, Germany

A.Belleflamme
Laboratory of Climatology, Department of Geography, University of Liège, Allée du 6 Août, 2, 4000 Liège, Belgium

X. Fettweis
Laboratory of Climatology, Department of Geography, University of Liège, Allée du 6 Août, 2, 4000 Liège, Belgium

M. Erpicum
Laboratory of Climatology, Department of Geography, University of Liège, Allée du 6 Août, 2, 4000 Liège, Belgium

M. Fischer
Department of Geosciences, University of Fribourg, 1700 Fribourg, Switzerland

M. Huss
Department of Geosciences, University of Fribourg, 1700 Fribourg, Switzerland
Laboratory of Hydraulics, Hydrology and Glaciology (VAW), ETH Zurich, 8093 Zurich, Switzerland

M. Hoelzle
Department of Geosciences, University of Fribourg, 1700 Fribourg, Switzerland

J. M. vanWessem
Institute for Marine and Atmospheric Research Utrecht, Utrecht University, Utrecht, the Netherlands

C. H. Reijmer
Institute for Marine and Atmospheric Research Utrecht, Utrecht University, Utrecht, the Netherlands

J. T. M. Lenaerts
Institute for Marine and Atmospheric Research Utrecht, Utrecht University, Utrecht, the Netherlands

W. J. van de Berg
Institute for Marine and Atmospheric Research Utrecht, Utrecht University, Utrecht, the Netherlands

M. R. van den Broeke
Institute for Marine and Atmospheric Research Utrecht, Utrecht University, Utrecht, the Netherlands

E. van Meijgaard
Royal Netherlands Meteorological Institute, De Bilt, the Netherlands

F. Hüsler
Institute of Geography and Oeschger Center for Climate Change Research, University of Bern, Bern, Switzerland

T. Jonas
WSL Institute for Snow and Avalanche Research SLF, Davos Dorf, Switzerland

M. Riffler
Institute of Geography and Oeschger Center for Climate Change Research, University of Bern, Bern, Switzerland

J. P. Musial
Institute of Geography and Oeschger Center for Climate Change Research, University of Bern, Bern, Switzerland

S. Wunderle
Institute of Geography and Oeschger Center for Climate Change Research, University of Bern, Bern, Switzerland

C. Lang
Département de Géographie, Université de Liège, Liège, Belgium

X. Fettweis
Département de Géographie, Université de Liège, Liège, Belgium

M. Erpicum
Département de Géographie, Université de Liège, Liège, Belgium

D. Notz
Max Planck Institute for Meteorology, Hamburg, Germany

B. Bereiter
Climate and Environmental Physics, Physics Institute, University of Bern, Sidlerstrasse 5, 3012 Bern, Switzerland
Oeschger Centre for Climate Change Research, University of Bern, 3012 Bern, Switzerland

H. Fischer
Climate and Environmental Physics, Physics Institute, University of Bern, Sidlerstrasse 5, 3012 Bern, Switzerland
Oeschger Centre for Climate Change Research, University of Bern, 3012 Bern, Switzerland

J. Schwander
Climate and Environmental Physics, Physics Institute, University of Bern, Sidlerstrasse 5, 3012 Bern, Switzerland
Oeschger Centre for Climate Change Research, University of Bern, 3012 Bern, Switzerland

T. F. Stocker
Climate and Environmental Physics, Physics Institute, University of Bern, Sidlerstrasse 5, 3012 Bern, Switzerland
Oeschger Centre for Climate Change Research, University of Bern, 3012 Bern, Switzerland

J. Nilsson
Department of Geodynamics, DTU Space, Technical University of Denmark, Elektrovej 327, 2800 Lyngby, Denmark

L. Sandberg Sørensen
Department of Geodynamics, DTU Space, Technical University of Denmark, Elektrovej 327, 2800 Lyngby, Denmark

V. R. Barletta
Department of Geodynamics, DTU Space, Technical University of Denmark, Elektrovej 327, 2800 Lyngby, Denmark
Ohio State University, 275 Mendenhall Lab, 125 S. Oval Mall, Columbus, Ohio, 43214, USA

R. Forsberg
Department of Geodynamics, DTU Space, Technical University of Denmark, Elektrovej 327, 2800 Lyngby, Denmark

H. Pang
Key Laboratory of Coast and Island development of Ministry of Education, School of Geographic and Oceanographic Sciences, Nanjing University, Nanjing 210093, China

S. Hou
Key Laboratory of Coast and Island development of Ministry of Education, School of Geographic and Oceanographic Sciences, Nanjing University, Nanjing 210093, China

S. Kaspari
Department of Geological Sciences, Central Washington University, Ellensburg, WA 98926, USA

P. A. Mayewski
Climate Change Institute and Department of Earth Sciences, University of Maine, Orono, ME 04469, USA

D. M. Chandler
Bristol Glaciology Centre, School of Geographical Sciences, Bristol University, Bristol, UK

J. D. Alcock
Bristol Glaciology Centre, School of Geographical Sciences, Bristol University, Bristol, UK

J. L. Wadham
Bristol Glaciology Centre, School of Geographical Sciences, Bristol University, Bristol, UK

S. L. Mackie
School of Earth Sciences, Bristol University, Bristol, UK

J. Telling
Bristol Glaciology Centre, School of Geographical Sciences, Bristol University, Bristol, UK

H. A. Dugan
Department of Earth and Environmental Sciences, University of Illinois at Chicago, Chicago, IL, USA
Center for Limnology, University of Wisconsin-Madison, Madison, WI, USA

P. T. Doran
Department of Earth and Environmental Sciences, University of Illinois at Chicago, Chicago, IL, USA

B. Wagner
Institute of Geology and Mineralogy, University of Cologne, Cologne, Germany

F. Kenig
Department of Earth and Environmental Sciences, University of Illinois at Chicago, Chicago, IL, USA

C. H. Fritsen
Division of Earth and Ecosystem Sciences, Desert Research Institute, Reno, NV, USA

S. A. Arcone
US Army Cold Regions Research and Engineering Laboratory, Hanover, NH, USA

E. Kuhn
Division of Earth and Ecosystem Sciences, Desert Research Institute, Reno, NV, USA

N. E. Ostrom
Department of Zoology, Michigan State University, East Lansing, MI, USA

J. P. Warnock
Department of Geology and Environmental Geosciences, Northern Illinois University, DeKalb, IL, USA

A. E. Murray
Division of Earth and Ecosystem Sciences, Desert Research Institute, Reno, NV, USA

A. Gafurov
Department of Hydrology, GFZ German Research Centre for Geosciences, Potsdam, Germany

S. Vorogushyn
Department of Hydrology, GFZ German Research Centre for Geosciences, Potsdam, Germany

D. Farinotti
Department of Hydrology, GFZ German Research Centre for Geosciences, Potsdam, Germany

D. Duethmann
Department of Hydrology, GFZ German Research Centre for Geosciences, Potsdam, Germany

A. Merkushkin
Uzbek Hydrometeorological Service (Uzhydromet), Tashkent, Uzbekistan

B. Merz
Department of Hydrology, GFZ German Research Centre for Geosciences, Potsdam, Germany

K. Helfricht
alpS - Centre for Climate Change Adaptation, Grabenweg 68, 6020 Innsbruck, Austria
Institute of Meteorology and Geophysics, University of Innsbruck, Innrain 52, 6020 Innsbruck, Austria

M. Kuhn
Institute of Meteorology and Geophysics, University of Innsbruck, Innrain 52, 6020 Innsbruck, Austria

M. Keuschnig
alpS - Centre for Climate Change Adaptation, Grabenweg 68, 6020 Innsbruck, Austria
Department of Geography and Geology, University of Salzburg, Hellbrunnerstrasse 34, 5020 Salzburg, Austria

A. Heilig
Commission for Geodesy and Glaciology, Bavarian Academy of Sciences and Humanities, Alfons-Goppel-Str. 11, 80539 Munich, Germany
Institute of Environmental Physics, University of Heidelberg, Im Neuenheimer Feld 229, 69120 Heidelberg, Germany

A. E. Racoviteanu
Laboratoire de Glaciologie et Géophysique de l'Environnement, 54 rue Molière, Domaine Universitaire, BP 96, 38402 Saint Martin d'Hères CEDEX, France

Y. Arnaud
Laboratoire de Glaciologie et Géophysique de l'Environnement, 54 rue Molière, Domaine Universitaire, BP 96, 38402 Saint Martin d'Hères CEDEX, France
Laboratoire d'Étude des Transferts en Hydrologie et Environnement, BP 53, 38401 Saint Martin d'Hères CEDEX, France

M. W. Williams
Department of Geography and Institute of Arctic and Alpine Research, University of Colorado, Boulder, CO 80309, USA

W. F. Manley
Institute of Arctic and Alpine Research, University of Colorado, Boulder, CO 80309, USA

A. Kääb
Department of Geosciences, University of Oslo, P.O. Box 1047, Oslo, Norway

D. Treichler
Department of Geosciences, University of Oslo, P.O. Box 1047, Oslo, Norway

C. Nuth
Department of Geosciences, University of Oslo, P.O. Box 1047, Oslo, Norway

E. Berthier
CNRS, Université de Toulouse, LEGOS, 14 avenue Ed. Belin, 31400 Toulouse, France